WITHDRAWN

Thermal and Moisture Protection Manual

Thermal and Moisture Protection Manual

For Architects, Engineers, and Contractors

Christine Beall, NCARB, CCS

McGraw-Hill
New York San Francisco Washington, D.C. Auckland Bogotá
Caracas Lisbon London Madrid Mexico City Milan
Montreal New Delhi San Juan Singapore
Sydney Tokyo Toronto

Library of Congress Cataloging-in-Publication Data

Beall, Christine.
    Thermal and moisture protection manual : for architects,
      engineers, and contractors / Christine Beall.
        p.   cm.
    Includes bibliographical references.
    ISBN 0-07-005155-0
    1. Waterproofing. 2. Insulation (Heat). I. Title.
  TH9031.B294 1998
  693.8'92—dc21
                                                          98-16859
                                                                  CIP

McGraw-Hill

A Division of The McGraw·Hill Companies

Copyright © 1999 by Christine Beall. All rights reserved. Printed in the United States of America. Except as permitted under the United States Copyright Act of 1976, no part of this publication may be reproduced or distributed in any form or by any means, or stored in a data base or retrieval system, without the prior written permission of the publisher.

1 2 3 4 5 6 7 8 9 0 DOC/DOC 9 0 3 2 1 0 9 8

ISBN 0-07-005155-0

The sponsoring editor for this book was Larry Hager, the editing supervisor was Peggy Lamb, and the production supervisor was Sherri Souffrance. It was set in New Century Schoolbook by McGraw-Hill's Professional Book Group composition unit, Hightstown, New Jersey.

Printed and bound by R.R. Donnelley & Sons Company.

 This book is printed on recycled, acid-free paper containing a minimum of 50% recycled de-inked fiber.

McGraw-Hill books are available at special quantity discounts to use as premiums and sales promotions, or for use in corporate training programs. For more information, please write to the Director of Special Sales, McGraw-Hill, 11 West 19th Street, New York, NY 10011. Or contact your local bookstore.

Information contained in this work has been obtained by The McGraw-Hill Companies, Inc. ("McGraw-Hill") from sources believed to be reliable. However, neither McGraw-Hill nor its authors guarantees the accuracy or completeness of any information published herein and neither McGraw-Hill nor its authors shall be responsible for any errors, omissions, or damages arising out of use of this information. This work is published with the understanding that McGraw-Hill and its authors are supplying information, but are not attempting to render engineering or other professional services. If such services are required, the assistance of an appropriate professional should be sought.

To Star

Contents

Preface　xiii

Chapter 1　Buildings and Moisture　1

1.1　Building Science　1
1.2　Thermal and Moisture Protection　2
1.3　Durability of Materials　5
　　1.3.1　Expansion and contraction　6
　　1.3.2　Mold and mildew　9
　　1.3.3　Corrosion　9
　　1.3.4　Decay　11
　　1.3.5　Freeze-thaw damage　14
　　1.3.6　Delamination and adhesion loss　14
　　1.3.7　Loss of thermal resistance　15
　　1.3.8　Efflorescence　16
　　1.3.9　Insect infestations　19
1.4　Increasing Durability against Moisture Damage　19

Chapter 2　Climate, Weather, and Microclimate　21

2.1　Weather and Climate　21
2.2　Wind and Air Movement　24
2.3　Precipitation　35
　　2.3.1　Rain　40
　　2.3.2　Dew　41
　　2.3.3　Snow　41
2.4　Microclimate　46
　　2.4.1　Orientation　47
　　2.4.2　Slope　47
　　2.4.3　Wind　49
2.5　Evaluating Risk　51

Chapter 3　Heat Flow and Insulation　59

3.1　Heat and Cold　59
　　3.1.1　Thermal conduction and convection　59
　　3.1.2　Evaporation　60
　　3.1.3　Thermal radiation　61
3.2　Thermal Resistance　71
　　3.2.1　Effect of air spaces　72
　　3.2.2　Effect of mass　81
　　3.2.3　Effect of insulation　84
　　3.2.4　Effect of thermal bridges　85

3.3 Types of Insulation — 86
    3.3.1 Loose-fill insulation — 87
    3.3.2 Flexible and semirigid insulation — 90
    3.3.3 Rigid insulation — 90
    3.3.4 Formed-in-place insulation — 92
    3.3.5 Insulation requirements — 93
    3.3.6 Loss of thermal resistance — 93
3.4 Calculating Thermal Gradients — 95
    3.4.1 Calculating surface temperatures — 99
    3.4.2 Calculating the effect of thermal bridges — 101
    3.4.3 Effect of insulation location — 103
    3.4.4 Slabs and below-grade walls — 112

Chapter 4 Water Penetration — 121

4.1 Rain and Groundwater — 121
4.2 Water + Opening + Force — 121
    4.2.1 Gravity — 122
    4.2.2 Air currents — 122
    4.2.3 Capillary suction — 123
    4.2.4 Surface tension — 124
    4.2.5 Kinetic energy (momentum) — 124
    4.2.6 Air pressure — 124
    4.2.7 Hydrostatic pressure — 126
4.3 Preventing Water Problems — 126
    4.3.1 Limit water penetration — 127
    4.3.2 Prevent water accumulation — 128
    4.3.3 Neutralize forces — 128
4.4 Barriers, Drains, and Rain Screens — 138
    4.4.1 Barrier systems — 138
    4.4.2 Drainage systems — 139
    4.4.3 Rain screen systems — 140
    4.4.4 Combining protective strategies — 141

Chapter 5 Vapor Condensation — 143

5.1 Psychrometrics — 143
5.2 Condensation in Buildings — 146
    5.2.1 Humidity, comfort, and health — 146
    5.2.2 Sources of moisture — 149
    5.2.3 Surface condensation — 150
    5.2.4 Concealed condensation — 153
5.3 Vapor Movement — 154
    5.3.1 Diffusion — 154
    5.3.2 Air flow — 156
    5.3.3 Radiant solar heating — 164
5.4 Vapor Resistance — 166
    5.4.1 Calculating the thermal gradient — 174
    5.4.2 Calculating the vapor pressure gradient — 176
5.5 Analyzing the Risk — 176
    5.5.1 Moisture accumulation — 178
    5.5.2 Wetting and drying cycles — 179
    5.5.3 Cold climates — 180
    5.5.4 Hot, humid climates — 180
5.6 Control Strategies — 180
    5.6.1 Limiting moisture sources — 185

		5.6.2	Ventilation and dehumidification	185
		5.6.3	Vapor retarders and air barriers	186
	5.7	Installing Air Barriers		192
		5.7.1	Membrane air barriers	196
		5.7.2	Gypsum drywall air barriers	199
		5.7.3	Exterior sheathing air barriers	205
		5.7.4	Curtain wall air barriers	205
		5.7.5	Face seal air barriers	208
		5.7.6	Roof-wall connections	208
		5.7.7	Weatherstripping	208

Chapter 6 Roofing   213

	6.1	Service Conditions		213
		6.1.1	Drainage	213
		6.1.2	Loads	214
		6.1.3	Wind and air movement	215
		6.1.4	Ultraviolet radiation	216
		6.1.5	Temperature cycles	218
		6.1.6	Water vapor	218
		6.1.7	Foot traffic	218
	6.2	Roof Components		218
		6.2.1	Structural deck	219
		6.2.2	Vapor retarders, air barriers, and underlayments	222
		6.2.3	Insulation	224
		6.2.4	Roof coverings and membranes	225
		6.2.5	Flashing and counterflashing	226
		6.2.6	Attachments	230
	6.3	Steep Roofing		230
		6.3.1	Asphalt shingles	236
		6.3.2	Metal roofing	239
	6.4	Low-Slope Roofing		256
		6.4.1	Conventional and protected membrane roofing systems	256
		6.4.2	Condensation in conventional low-slope roofs	262
		6.4.3	Built-up membranes	267
		6.4.4	Single-ply membranes	280

Chapter 7 Waterproofing   283

	7.1	Moisture Movement in Soils		284
		7.1.1	Water movement	284
		7.1.2	Vapor movement	288
	7.2	Basic Waterproofing Principles		290
		7.2.1	Surface drainage	291
		7.2.2	Subsurface drainage	293
		7.2.3	Waterproofing membranes and dampproof coatings	297
		7.2.4	Vapor retarders	301
	7.3	Materials and Properties		305
		7.3.1	Built-up membranes	305
		7.3.2	Fluid-applied membranes	308
		7.3.3	Sheet membranes	310
		7.3.4	Bentonite clay waterproofing	312
		7.3.5	Cementitious, metallic oxide, and crystalline coatings	314
		7.3.6	Waterstops and compression seals	315
		7.3.7	Flashing	315
	7.4	Slabs on Grade and Crawl Spaces		317

7.5	Below-Grade Walls and Floors	320
7.6	Plaza Decks	323
	7.6.1 Substrate	330
	7.6.2 Membrane	331
	7.6.3 Protection course	334
	7.6.4 Percolation layer	334
	7.6.5 Insulation	335
	7.6.6 Protection or working slab	342
	7.6.7 Wearing course/traffic surface	343
	7.6.8 Earth-covered plazas and planters	346

Chapter 8 Cladding — 351

8.1	Basic Design Principles	351
	8.1.1 Rain barrier systems	352
	8.1.2 Drainage wall systems	352
	8.1.3 Rain screen systems	353
	8.1.4 Second lines of defense	353
8.2	Cladding System Details	355
	8.2.1 Precast concrete cladding	355
	8.2.2 Masonry cladding	357
	8.2.3 Curtain walls and windows	368
	8.2.4 Exterior insulation and finish systems (EIFS)	377

Chapter 9 Joints and Sealants — 391

9.1	Building Movements	391
9.2	Joint Types	391
9.3	Joint Sizing	392
	9.3.1 Calculating thermal movement	393
	9.3.2 Calculating moisture movement	397
	9.3.3 Construction tolerances	399
	9.3.4 Sizing vertical joints for walls	400
	9.3.5 Sizing horizontal joints for walls	401
	9.3.6 Sizing joints for slabs and decks	403
9.4	Sealant Stresses	403
	9.4.1 Tension and compression	403
	9.4.2 Sealant shear stresses	404
9.5	Other Factors Affecting Sealant Performance	406
	9.5.1 Joint geometry	407
	9.5.2 Installation temperatures	409
	9.5.3 Movement during cure	409
	9.5.4 Three-sided adhesion	410
9.6	Sealant Properties	410
	9.6.1 Movement capability	410
	9.6.2 Recovery and stress relaxation	411
	9.6.3 Modulus of elasticity	412
	9.6.4 Hardness	412
	9.6.5 Ultraviolet and ozone resistance	412
	9.6.6 Chalking and dirt pickup	413
	9.6.7 Substrate compatibility	413
	9.6.8 Paintability	413
	9.6.9 VOC compliance	414
9.7	Sealant Materials	414
	9.7.1 Forms	414

	9.7.2 Oil-based compounds	415
	9.7.3 Butyl sealants	419
	9.7.4 Acrylic latex sealants	419
	9.7.5 Solvent-release acrylics	419
	9.7.6 Polysulfides	420
	9.7.7 Urethanes	421
	9.7.8 Silicones	422
	9.7.9 Tapes and gaskets	422
	9.7.10 Accessories	425
9.8	Sealant Installation	426
	9.8.1 Climatic conditions	426
	9.8.2 Substrate contaminants	427
	9.8.3 Joint cleaning	427
	9.8.4 Priming	428
	9.8.5 Sealant backing	428
	9.8.6 Sealant application	429
	9.8.7 Repair and remedial caulking	430
9.9	Quality Assurance and Quality Control	431
	9.9.1 Design considerations	432
	9.9.2 Testing	434
	9.9.3 Applicator qualifications	436
	9.9.4 Submittals	436
9.10	Avoiding Common Problems	436

Chapter 10 Coatings   439

10.1	Paints and Primers	439
	10.1.1 Paint and varnish	439
	10.1.2 Primers	441
	10.1.3 Cementitious coatings	441
	10.1.4 Elastomeric coatings	443
	10.1.5 Coating selection	444
10.2	Clear Water-Repellent Coatings	444
	10.2.1 Film formers	447
	10.2.2 Penetrants	448
	10.2.3 Water repellent selection and field tests	450
	10.2.4 Application procedures	451
10.3	Surface Preparation	452
	10.3.1 Concrete	452
	10.3.2 Stucco	454
	10.3.3 Concrete block	454
	10.3.4 Brick	454
	10.3.5 Exterior wood	454
	10.3.6 Exterior hardboard	454
	10.3.7 Steel	454
	10.3.8 Galvanized steel	456
	10.3.9 Aluminum	456
10.4	Common Problems	456
	10.4.1 Alligatoring	456
	10.4.2 Blistering	456
	10.4.3 Checking, cracking, and flaking	457
	10.4.4 Delamination or peeling	457
	10.4.5 Flatting	457
	10.4.6 Oxidation, chalking, and fading	458
	10.4.7 Pinholing	458
	10.4.8 Premature rusting	458
	10.4.9 Wrinkling	458

10.5	Maintenance Painting	458
	10.5.1 Concrete, stucco, and masonry surfaces	458
	10.5.2 Painted wood and metal surfaces	459
	10.5.3 Stained wood surfaces	459
Appendix A	Glossary	461
Appendix B	ASTM Standards	493
Appendix C	Bibliography	497
Index		501

Preface

Publications from many different industry associations and government agencies have been used to compile the information and recommendations in this book. Included in the long list of sources are the National Roofing Contractor's Association (NRCA), the American Society of Heating, Refrigeration and Air Conditioning Engineers (ASHRAE), the American Architectural Manufacturers Association (AAMA), the Sealant, Waterproofing and Restoration Institute (SWRI), the American Society for Testing and Materials (ASTM), the Building Thermal Envelope Coordinating Council (BTECC), the National Oceanic and Atmospheric Administration (NOAA), the Oak Ridge National Laboratory (ORNL), and the National Research Council of Canada (NRCC). A complete bibliography of sources and reference material is included at the back of the book.

Christine Beall, NCARB, CCS
Columbus, Texas

Chapter 1

Buildings and Moisture

Climate once played as much a part in determining the shape and form of buildings as culture. Weather protection relied solely on the physical form of the envelope itself. Steeply sloped roofs shed the winter snow in cold climates, while minimal roof overhangs increased winter sun exposure and prevented ice damming at cold eaves. In hot, arid climates, massive building walls absorbed and stored the day's heat, and flat roofs with parapets collected and held the rain for its evaporative cooling effect.

The intervention of modern heating and air conditioning systems has, to some extent, separated form from function and dissociated climate from design. Thermal insulation, high-tech glazing, mechanical equipment, and electrical systems can provide relatively efficient methods of controlling heat and air movement in contemporary buildings regardless of their physical form. Moisture, though, remains a persistent and insidious enemy. In fact, the more aggressively we have tried to control heat flow and air movement, the more we have aggravated some moisture problems. The environment in which a building must function, however, is still a primary determinant of the measures which must be taken to protect against moisture damage. Climate, weather, and microclimatic influences on building performance are discussed in Chap. 2.

1.1 Building Science

Building science is the study of the physical phenomena that affect building performance. Much of this science is focused on thermal and moisture protection. The success or failure of a building in terms of thermal and moisture control is predictable from the known behavior of physical forces and the characteristics of building systems and materials. The successful design of buildings can therefore be based on an analysis of occupancy requirements, environmental exposure, individual building materials and envelope components, and the integration of those components into a complete system of thermal and moisture protection.

The durability and integrity of the building envelope are primary considerations of good design, including protection from degradation caused by moisture, temperature, air movement, radiation, and biological attack. Rain, snow, ice, groundwater, and condensation each can cause physical damage and deterioration of building materials and contents, and may also contribute to increased energy consumption and poor indoor air quality. A balance must be achieved among the sometimes conflicting performance requirements for thermal comfort, resource management, indoor air quality, and building integrity (*Table 1.1*)

1.2 Thermal and Moisture Protection

Although mechanical systems play the leading role in controlling the interior environment of modern buildings, comprehensive thermal and moisture protection strategies must include careful analysis and design of the envelope which encloses that environment. A building envelope essentially separates two masses of air, moisture, and energy. Functional and durable envelopes should provide an efficient and effective separation between the interior and exterior environments by regulating the flow of heat, air, and moisture in and out of buildings.

The infiltration and exfiltration of heat, air, and moisture must be controlled in both quantity and location so that they do not adversely affect the interior conditions or damage the envelope as they pass through it. The degree of control that is necessary depends on the nature of the occupancy, the climate in which the building is located, the season, and the time of day. Depending on the effectiveness of the envelope, interior conditions may vary from day to night and from season to season as thermal and moisture conditions change. Most occupancies require that a building envelope provide protection from rain penetration, groundwater intrusion, vapor condensation, and excessive heat loss or gain.

Thermal and moisture conditions are closely related in building science. Cold temperatures can cause condensation to form on or within walls. Hot, humid conditions can support the growth of mold and mildew, promote wood decay, and adversely affect indoor air quality. Freezing temperatures can cause physical damage to saturated materials as the moisture they contain expands and contracts with winter freeze-thaw cycles. Heat flow and insulation are discussed in detail in Chap. 3.

In order to properly design and integrate the details of a building's moisture protection system, we must understand how moisture behaves. There are four common moisture movement mechanisms which may act independently or in combination with one another:

- Liquid flow
- Capillary suction
- Air movement
- Vapor diffusion

TABLE 1.1 Building Performance Requirements

Building integrity
   Moisture (rain, snow, ice, and vapor)
      Penetration
      Migration
      Condensation
      Moisture expansion & contraction
   Temperature
      Thermal resistance
      Thermal bridging
      Cyclic freezing and thawing
      Thermal expansion and contraction
   Air movement
      Exfiltration
      Infiltration
   Radiation and light
      Solar radiation
      Environmental radiation
      Visible light spectrum
   Biological attack
      Mold and mildew
      Insect infestation

Thermal comfort
   Air temperature
   Radiant temperature
   Humidity
   Air speed
   Occupancy factors

Resource management
   Energy
   Finance

Air quality
   Ventilation rate (fresh air supply, circulation)
   Indoor pollution (gases, vapors, microorganisms, fumes, smoke, dust)
   Occupancy factors

Liquid flow is primarily responsible for moving moisture into the building and the building envelope from the outside. Capillary suction moves rainwater and groundwater into the building and the envelope from outside, and redistributes moisture within porous materials. Air movement and vapor diffusion move moisture in its gaseous form from the outside in and from the interior space into and through the building envelope. A 1970 article in the World Meteorological Organization's *Building Climatology,* Technical Note 109, succinctly describes the mechanisms and effects of moisture penetration:

> Water may be brought into porous materials as liquid or as vapor. The driving force may originate from differences in hydrostatic pressure, osmotic pressure, vapor pressure, as well as from free surface energy. Considering rain penetration into building structures, this would be caused by 1) capillary action due to free surface energy; 2) kinetic energy of rain drops or water streams; 3) hydrostatic pressure due to gravity or to wind pressure on water films; or 4) air flow through the structures. The ability of a building structure to dry out and keep dry generally depends on the successful design of every detail of the building and its various structural components and draining systems above or under the ground as well as on certain physical properties of the applied materials and workmanship, all in relation to the ambient climate. Some of these principles are related to the design and construction of certain building components and are, in exposed areas, frequently reflected in a characteristic of local design and building practice.

Water penetration is discussed in detail in Chap. 4, and vapor condensation in Chap. 5.

The design of a building and its component details should include a realistic evaluation of the risk of moisture damage. The walls, roof, and floor of a building must form a true envelope of protection in which all components and accessories are integrated into a fully functional moisture protection system. Successful building designs must consider every possible source of water in every possible form—liquid, solid, and vapor. *Effective moisture protection cannot be based on assumptions of ideal weather, optimum fit, perfect installation, and impeccable maintenance.* More realistic scenarios include bad weather, poor fit, shoddy installation, and no maintenance. If you apply Murphy's law to the problem of moisture and buildings, it might be said that "Whatever possibly *can* leak usually will." On the other hand, in mild climates where environmental exposure is moderate, Mother Nature can sometimes be quite forgiving of design and construction errors.

Moisture-related failures occur most commonly because of design and construction errors which neglect the proper integration of components. Building envelopes should be designed as a complete *system* of protection. Every detail should be studied for possible moisture entry locations, and the path of any penetrated moisture traced to determine requirements for secondary seals, flashing, and drainage. Without safety factors, margins for error, and redundant protection, even small moisture problems have potentially disastrous effects. Redundancy in moisture protection systems provides a much needed factor of safety, and avoids reliance on a single line of defense. *What if* water penetrates a sealant joint—is there a backup for collecting and removing the moisture? *What if* a roof/parapet flashing fails—is there a secondary means of preventing leakage? *What if* moisture penetrates a protective coating—where will the water go? Protective measures must include

- Limiting or controlling moisture entry
- Preventing moisture accumulation
- Neutralizing the physical forces that transport moisture

A balanced approach is required, however, because strategies that are designed to prevent moisture entry may also prevent moisture removal and drying. By the same token, strategies that are effective in removing moisture may also allow moisture entry. Chapters 6 through 10 provide practical applications of thermal and moisture protection strategies for roofing, waterproofing, cladding, joint sealants, and coatings.

1.3 Durability of Materials

Durability is not an inherent property of any material. Durability results from choosing a material with the right physical characteristics for its environment. A publication by the American Society of Civil Engineers entitled *Failure Mechanisms in Building Construction* (Nicastro 1997) defines durability as "the quality of maintaining satisfactory aesthetic, economic, and functional performance for the useful life of a material or system." A material that failed on one building may perform quite well on another. The material that failed did so either because of the manner in which it was used or the environment to which it was exposed. That environment is a function of the combined effects of atmospheric conditions and the design of the building and its details. If we understand the nature of materials, we can learn to use them in ways that do not subject them to intolerable conditions. If we understand the mechanisms of deterioration, we can make rational assessments as to whether the conditions surrounding a material will be detrimental to it.

Nicastro defines weathering as "Degradation due to exposure to the weather," and lists as weathering factors ultraviolet radiation, temperature, moisture, wind, ozone, carbon dioxide, pollution, and freeze-thaw. Of the several factors, acting singly or in combination, which affect the durability of building materials, the two primary ones considered here are moisture and temperature. Of these, moisture is the most damaging, and moisture damage is the most pervasive problem we face in building design and construction.

Building envelope materials may periodically become wet during service, but this is acceptable so long as there is ample opportunity for drying or draining the moisture in a timely fashion. Cyclic wetting and drying are acceptable as long as the materials do not stay wet for long periods and begin to deteriorate. In order to select appropriate materials, we must be able to determine the thermal and moisture conditions at any given point in the envelope. Alternatively, if we want to use a particular material for economic or aesthetic reasons, it may sometimes be necessary to control or modify thermal and moisture conditions so that they can be tolerated by that material. Building science helps us determine what conditions can be tolerated by which materials, and what measures can be taken to alter conditions where necessary. Building materials may suffer from a variety of moisture-related problems:

- Expansion and contraction
- Mold and mildew growth
- Corrosion
- Decay
- Freeze-thaw damage
- Delamination and adhesion loss
- Reduced thermal resistance
- Efflorescence
- Insect infestation

1.3.1 Expansion and contraction

As a material warms up it expands, and as it cools off it contracts again. The rate of expansion or contraction for each degree of temperature change is called the *coefficient of thermal expansion*. When unrestrained by adjacent construction, building materials exhibit thermal movements that are linearly proportional to temperature changes. Metals generally have the highest rates of thermal expansion and contraction, while concrete and masonry materials have the lowest. The relative expansion and contraction between adjacent materials affects the ways in which the materials can be connected to one another, the size of the joints which must be left between the two, and the performance of sealants used to fill the joints. If two parts of a piece of material are at different temperatures, a stress will be induced in the material. Similarly, if two materials with different coefficients of expansion are bonded together and then subjected to the same range of temperatures, stresses are induced, and there is a potential for damage or distortion. When solar radiation strikes the surface of a building material, thermal expansion and contraction are increased significantly over what would be expected from variations in air temperature alone.

Temperature changes also affect the viscosity of some organic materials such as bitumens and sealants. As the material is heated it stretches more easily. As it cools, it thickens and, at a sufficiently low temperature, can become quite brittle. Polymers have an intermediate rubbery state. A much longer elongation before break occurs with rubbery polymers, and the force needed for deformation is much smaller (i.e., the modulus of elasticity is lower). See Chap. 9 for detailed methods of calculating thermal expansion and contraction.

Moisture expansion and contraction is a phenomenon separate and distinct from thermal expansion and contraction, although absorbed moisture can affect the rate and degree of thermal movement. Some building materials, such as glass, plastics, and metals, are nonporous. The outside surface can be wetted with water, but moisture cannot get inside the material. With other materials, there are internal spaces or pores which connect with one another and with the outside surface. Materials such as brick, concrete, wood, and stone fall into this category. Adding water to a porous material causes expansion, and removing it causes shrinkage. When different parts of such a material have different moisture contents, the material will expand and contract differentially and stresses are induced. It is such differential moisture conditions that can cause cracking, splitting, and warping in some materials.

Many materials have an irregular pore structure which gives them more or less isotropic characteristics because the arrangement of the pores is approximately the same in any direction. Other materials, such as wood, have a microstructure of regular shape and uniform arrangement. Wood cells are tubular and bonded together along their length. This gives wood non-isotropic characteristics, with high strength and low moisture shrinkage in the direction of the fibers and lower strength and high moisture shrinkage across the fibers.

The moisture content of a material at most times during its service life is determined by the relative humidity of the air to which it is exposed. If the relative humidity is allowed to fluctuate over a wide range, continual shrinking and swelling can damage articles made from moisture-sensitive materials, particularly if different components of the article shrink and swell to different degrees. It is this phenomenon that can cause wooden furniture to loosen at the joints and wood veneers to crack and peel off.

Warm, dry air passing over a moist material will dry it out. Moisture within the material evaporates at the surface until the material's moisture content reaches equilibrium with the surrounding air. Solar radiation also affects wetting and drying of building materials, roofs, and wall sections. Heating and evaporation of moisture at the surface causes moisture within the section to move toward the zone of higher pressure at the outside surface, where it can evaporate. Very often, such drying action is beneficial because it removes excessive moisture absorbed from rain or snow, but it can also be harmful in prematurely drying poured concrete, freshly laid masonry, or fresh plaster.

Cement-based products such as concrete, concrete masonry, and stucco experience initial moisture shrinkage as the cement hydrates and excess construction water evaporates. This initial *permanent shrinkage* is in addition to subsequent reversible thermal and moisture shrinkage and expansion. Clay masonry products such as brick and terra cotta experience initial moisture expansion as the units re-absorb atmospheric moisture after firing. This initial *permanent expansion* is in addition to subsequent reversible thermal and moisture expansion and contraction. Permanent moisture shrinkage and expansion must be taken into account in the detail of anchorages and the sizing and spacing of expansion and control joints.

Table 1.2 shows temperature and moisture deformations for some common building materials. The two temperature ranges of 80°F and 230°F are values for daily and seasonal variations for materials on the outside of a building,

TABLE 1.2
Temperature and Moisture Deformations for Some Common Building Materials

Material	Coefficient of thermal expansion, per °F	Deformation due to temperature change				Moisture deformation on wetting from dry to saturated (or vice versa)		Modulus of elasticity E	Failing stress, psi (compression or tension)	Deformation required to cause failure	
		Of 80°F		Of 230°F							
		%	in./10 ft	%	in./10 ft	%	in./10 ft			%	in./10 ft
Normal dense concrete	6×10^{-6}	0.05	0.06	0.14	0.17	0.03	0.04	2.5×10^6	2500 C 250 T	0.10 0.01	0.12 0.01
Brick	3×10^{-6}	0.024	0.03	0.07	0.08	0.007	0.008	3×10^6	6000 C 500 T	0.20 0.016	0.24 0.02
Marble and dense limestone	3×10^{-6}	0.024	0.03	0.07	0.08	<0.001		10×10^6	25,000 C 600 T	0.25 0.006	0.30 0.007
Sandstone	7×10^{-6}	0.056	0.07	0.16	0.19	0.07	0.08	5×10^6	12,000 C 400 T	0.24 0.008	0.29 0.01
Steel	7×10^{-6}	0.056	0.07	0.16	0.19	None		30×10^6	40,000 T (yield point)	0.13	0.15
Copper	10×10^{-6}	0.08	0.10	0.23	0.28	None		17×10^6	50,000 T	0.29	0.35
Aluminum	14×10^{-6}	0.11	0.13	0.32	0.38	None		10.3×10^6	40,000 T	0.39	0.47

SOURCE: Latta, *Walls, Windows and Roofs for the Canadian Climate*.

including the effects of solar radiation. Formulas and coefficients for calculating precise thermal and moisture expansion are discussed in Chap. 9.

1.3.2 Mold and mildew

One of the most commonly experienced surface moisture problems is the growth of mold or mildew. Mold is a wooly or powdery fungal growth that forms on the surface of materials in damp, stagnant atmospheres causing patchy discolorations that may be green, gray, or black, or occasionally pink, red, or yellow. Mildew is a type of mold that occurs on wood, concrete, or other absorbent materials. Mold can discolor surfaces, cause odor problems, deteriorate materials, and adversely affect indoor air quality. The following conditions are necessary and sufficient for mold growth to occur on surfaces:

- Mold spores
- A food source
- Temperatures between 40 and 100°F
- Relative humidity above 70% near the surface

Mold spores are almost always present in both outdoor and indoor air, almost all construction materials contain nutrients which can support mold growth, and occupant comfort requirements prohibit temperatures outside the range of 40 to 100°F, so it is most practical to control relative humidity near the surface. When the relative humidity near surfaces is kept below 70% by ventilation, heating, or air movement, mold and mildew growth can be eliminated (see Chap. 5 for more information).

1.3.3 Corrosion

Corrosion is "the deterioration of a metal by an electro-chemical reaction with its environment or with another material with which it is in contact" (Nicastro, 1997). Corrosion can be caused by chemical attack, galvanic action, exterior weathering, prolonged exposure to condensed moisture within a wall or roof, or humidity in excess of 75% within the hollow cores and cavities of a wall or roof.

Chemical attack of metals can occur in many different circumstances. Wet, uncured mortar, concrete, and stucco are alkaline and can be highly corrosive to aluminum and lead. Redwood and red cedar contain acid and are highly corrosive to aluminum and zinc alloys, and moderately corrosive to galvanized steel. Some set-accelerating concrete, mortar, and masonry grout admixtures contain calcium chloride and can cause rapid corrosion of reinforcing steel and metal accessories. Deep carbonation of mortar and concrete caused by carbon dioxide intrusion through cracks or voids may also accelerate corrosion of metal anchors, ties, or reinforcement. Water containing dissolved carbon dioxide can corrode copper. Pollutants in the air such as sulfur dioxide can combine with water to form sulfuric acid, which greatly accelerates corrosion. Salt spray in coastal areas is also very corrosive and can accelerate corrosion of exposed metals as well as reinforcing bars in concrete and masonry.

Galvanic corrosion occurs when metals are "exposed to a conductive solution that allows an electric (galvanic) current to flow between an anodic and a cathodic region" (Nicastro, 1997). Galvanic corrosion can occur in areas such as crevices and lap joints where metal-ion or oxygen cells are concentrated. It can also be caused by a thermal gradient in a metal where polarization forms an anode and a cathode. But galvanic action most commonly causes corrosion between dissimilar metals in the presence of an electrolyte such as water (bimetallic corrosion). The degree of galvanic corrosion which can occur between dissimilar metals depends on the intimacy of contact, the type of electrolyte, and the voltage developed between the two metals. An electric current is conducted through the electrolyte, corroding one metal (the anode) and plating the other (the cathode). The greater the potential difference between the two metals, the more severe the corrosion (*Table 1.3*). Less noble metals are subject to corrosion by more noble metals. The potential for galvanic corrosion between common metals and building materials is shown in *Table 1.4*.

Also important is the density of a galvanic corrosion current, or the size of the current relative to the surface area of the anode (less noble, or corroding, metal). A fastener's surface area is small compared to the metal which it fastens (such as a screw fastening an anchor to a metal stud, or fastening two pieces of sheet metal). The current density is concentrated at the fastener and the fastener is therefore subject to rapid corrosion. Corrosion of the anode (the fastener) can be 100 to 1000 times more severe than if its surface area were approximately the same as the cathode (the stud on sheet metal). Therefore, as a general rule, a fastener in a given environment should be more noble than the larger material to which it fastens. *Table 1.5* shows recommended fastener materials for use with various type of sheet metal.

Although zinc is susceptible to corrosive attack, it is used in the galvanizing process to provide a barrier coating to isolate steel from corrosive elements and to provide a sacrificial coating that is consumed to protect the steel at uncoated areas such as scratches and cut ends. Corroded metal occupies twice the volume of the original metal and exerts tremendous expansive forces. In masonry and concrete construction, corroded steel reinforcing and accessories can cause extensive cracking. The protective film of zinc used to galvanize steel is so thin that the pressure it exerts when it is sacrificially corroded is not sufficient to cause the same kind of cracking.

Mill galvanizing and electro-galvanizing of steel sections before the materials are fabricated into end use products do not provide protection at cut edges, wire ends, shop welds, and penetrations in the same way that hot-dip galvanizing the products after fabrication does. The life expectancy of the corrosion protection afforded by galvanizing is directly proportional to its thickness (*Fig. 1.1*). Stainless steel is less susceptible to corrosion than galvanized steel and provides greater long term durability in construction.

TABLE 1.3 Galvanic Series of Metals

Metal or alloy	Position
Magnesium	Anode (+) least noble
Zinc	
Aluminum, 5052 alloy	
Aluminum, 6061 alloy	
Cadmium	
Aluminum, 2024-T4 alloy	
Iron or carbon steel	
4–6% chromium steel	
Ferritic stainless steel, 400 series (active)	
Austenitic stainless steel, 18-8 series (active)	
Lead	
Tin	
Nickel (active)	
Brass	
Copper	
Bronze	
Monel	
Silver solder	
Nickel (passive)	
Ferritic stainless steel (passive)	
Austinitic stainless steel (passive)	
Silver	
Titanium	
Graphite	
Gold	
Platinum	Cathode (−) most noble

NOTE: The further apart two metals are in the galvanic series, the greater the corrosion of the less noble materials.

To protect against galvanic corrosion when dissimilar metals are used, isolation can be provided by an electrical insulator such as neoprene rubber or asphalt-impregnated felt.

1.3.4 Decay

Wood rots when attacked by a fungus under suitable temperature and moisture conditions. Dry rot (also called brown rot) leaves the wood lightweight, friable, and dull brown in color. The decayed wood cracks into cubical pieces

TABLE 1.4 Compatibility of Common Building Metals

	Copper	Aluminum	Stainless steel	Galvanized steel	Zinc alloy	Lead
Aluminum	1					
Stainless steel	1	3				
Galvanized steel	2	3	2			
Zinc	1	3	1	3		
Lead	2	2	2	3	3	
Brass	2	1	1	2	1	2
Bronze	2	1	1	2	1	2
Monel	2	3	1	2	1	2
Iron/steel	1	2	2	2	1	3

1: Galvanic action will occur.
2: Galvanic action may occur under certain conditions or over a period of time.
3: Galvanic action is insignificant under normal conditions.

TABLE 1.5
Recommended Fastener Materials to Prevent Galvanic Corrosion

Type of sheet metal	Nails	Screws	Rivets	Bolts and nuts
Aluminum	Aluminum	Aluminum	Aluminum	Aluminum
Copper	Copper	Bronze	Copper	Bronze
Copper-clad stainless steel	Copper or stainless steel	Bronze or stainless steel	Bronze, copper or stainless steel	Bronze or stainless steel
Galvanized steel	Galvanized steel	Cadmium plated	Cadmium plated	Cadmium plated
Lead	Galvanized steel	Cadmium plated	Cadmium plated	Cadmium plated
Lead-coated copper	Copper	Bronze	Bronze	Bronze
Stainless steel	Stainless steel or galvanized steel	Stainless steel	Stainless steel	Stainless steel
Terne	Galvanized steel	Cadmium plated	Cadmium plated	Cadmium plated
Weathering steel	Weathering steel or galvanized steel	Weathering steel or galvanized steel	Weathering steel or galvanized steel	Weathering steel or galvanized steel
Zinc alloy	Galvanized steel or aluminum	Cadmium plated	Cadmium plated	Cadmium plated

as the fungus destroys the cellulose in the cell walls. The brown lignin is largely unaffected, but the strength of the wood is almost entirely lost. Wet rot (also called white rot) destroys both the cellulose and the lignin producing a soft, spongy texture. Five conditions must occur for rotting to take place:

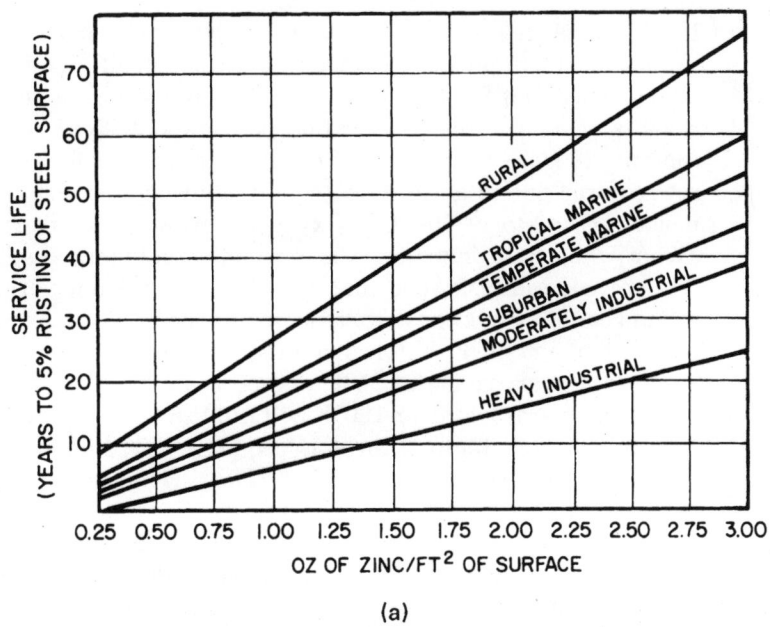

(a)

Probability of occurrence (%)	Corrosion rate (10^{-4} oz. zinc/sq ft/yr)	Life expectancy (yr)			
		ASTM A153, class B2		ASTM A153, class B1	
		Minimum	Average	Minimum	Average
5	2415	5.2	6.2	7.5	8.3
10	1791	7.0	8.4	10.1	11.2
20	1075	11.6	14.0	16.7	18.6
25	875	14.3	17.1	20.6	22.9
33	656	19.1	22.9	27.4	30.5
50	393	31.8	38.2	45.8	50.9

*Data taken in climatic areas with a driving rain index of 2.5 to 5.0 (see Chap. 2)

(b)

Figure 1.1 Life expectancy of galvanized coatings. (*From American Galvanizers Association.*)

- Suitable substrate (food)
- Moisture
- Oxygen
- Temperatures between 40 and 100°F
- A source of fungal infection

If any one of these conditions is absent, wood will not rot.

Food for the fungus is provided by the wood itself. There is always some moisture in wood, but fungal spores cannot germinate if the moisture content of the wood is below the fiber saturation point (about 25 to 30%). A lower moisture content, 20% or less, is usually required to prevent fungal attack entirely. Wet rot requires persistently damp conditions to occur, and thrives at an optimal moisture content of 50–60%. Oxygen is always present in sufficient quantity except when the construction is completely submerged such as the lower portion of wood pilings. Wood that is under water or below the water table in the ground cannot rot because there is insufficient oxygen. Fungus will become dormant at about 40°F, but is not killed by low temperatures. Fungal growth becomes less active above 95°F, and ceases entirely at temperatures just over 100°F. Temperatures over 120°F will kill the fungus, although prolonged exposure to the high temperatures is required. The higher the temperature, the more quickly the fungus is killed. Decay is most prolific at temperatures between 65 and 95°F.

Some wood species are less susceptible to fungal attack than others. Chemical preservatives are effective for a time, but complete chemical penetration is impossible, so subsequent checking opens paths for the infection to get to the untreated wood.

1.3.5 Freeze-thaw damage

As water freezes, it expands about 9%. If the freezing water has room to expand, no harm is done, but if expansion is restricted, the force is significant and can cause physical damage to the material or vessel which contains the water. The water in an uncapped bottle expands harmlessly when it freezes, but water which freezes in a sealed bottle will break the glass if there is not at least 9% free area in which to expand. Similarly, when a porous material is 100% saturated, there is no free pore space to accommodate expansion if the water freezes. Because of variations in the pore structure of different materials and the influence of the rate of freezing, a maximum moisture content of 75 to 80% of saturation is generally required to prevent freezing damage in porous materials.

1.3.6 Delamination and adhesion loss

Moisture can cause delamination and adhesion loss in many different building materials and systems. Condensation within a wall or roof section can be particularly damaging. As moisture vapor moves through the section, it may reach its dew point and condense, causing peeling of exterior paint, loosening of plaster or stucco, and blistering of roof membranes. Moisture in the substrate or backing material can also cause failure of some joint sealant materials. Water-soluble adhesives in plywood and other materials can also delaminate in the presence of moisture, and moisture-related corrosion or freeze-thaw damage can cause delamination or splitting into layers of concrete, stone, and other materials.

1.3.7 Loss of thermal resistance

One very important consideration of an insulating material is the stability of its thermal conductivity with time and under changing moisture conditions. As the moisture content of a material increases, thermal conductivity also increases and thermal resistance decreases proportionately. *Table 1.6* shows the increase in conductivity for increases in moisture content for some common building materials.

TABLE 1.6
Effect of Moisture on Thermal Conductivity

Material	Moisture content, % by volume	Increase in thermal conductivity, %, for 1% of water by volume	Increase in thermal conductivity, %, for 1% of water by weight
Lime mortar	5	13.5	
	10	9.7	
	15	6.4	
Brick wall	5	8.4	
	10	6.2	
	15	4.3	
Gravel concrete	1	27.0	
	5	13.4	
	10	9.6	
Lightweight concrete	1	22.0	
	5	10.0	
	10	8.4	
Pumice concrete	5	14.0	
	10	11.2	
	20	8.5	
Fiberboard insulation	3		1.44
Glass wool insulation			1.87

SOURCE: "Effect of Moisture on Heat Transmission Through Building Insulating Materials," Technical Translation TT-317, National Research Council of Canada.

Some types of building insulation are more susceptible to moisture accumulation and loss of thermal resistance than others. Porous materials like fiberglass and mineral fiberboard can absorb and hold large quantities of water, with a significant decrease in thermal resistance. In closed cell foam insulations, the gases used to form the insulation are in time displaced by air and water vapor, with a resulting loss in thermal resistance. To account for this loss of thermal resistance in closed cell foam insulations, manufacturers commonly list the "aged" R-value of urethane and isocyanurate insulations. If the air that displaces the gas in the insulation contains significant amounts of water vapor, then the insulation also becomes wet and loses its thermal resistance much faster. If there is no vapor drive across the insulation and no condensation, then the moisture content of the insulation simply reaches equilibrium with the air around it. If there is a vapor drive and condensation can accumulate within the insulation, the moisture content can be much higher than the surrounding atmosphere. Closed cell foam insulations, in fact, gain much more moisture when subjected to thermally induced vapor pressure gradients than when soaked in water under isothermal conditions.

1.3.8 Efflorescence

Efflorescence and calcium carbonate stains are two of the most common forms of surface stains on concrete and masonry. Both are white and both are activated by excessive moisture in the wall, but beyond that, there are few similarities. Efflorescence is a powdery salt residue, while calcium carbonate stains are hard, crusty, and much more difficult to remove.

Efflorescence occurs when soluble salts are taken into solution by water. As the moisture begins to dry out, the salt solution migrates toward the surface through capillary pores. When the water evaporates, the salts are deposited on the surface (*Fig. 1.2*). Hot summer months are not as conducive to efflorescence because wetting and drying is generally more rapid. Efflorescence is more likely to appear in late fall, winter, and early spring, particularly after rainy periods when evaporation is slower and temperatures cooler.

Three simultaneous conditions must exist in order for efflorescence to occur: (1) soluble salts must be present, (2) there must be a source of water in contact with the salts long enough to form a solution, and (3) the pore structure must be such that paths exist for the migration of the salt solution to a surface where evaporation can take place (*Fig. 1.3*). In conventional concrete and masonry construction exposed to weather, it is impossible to ensure that no salts are present, no water penetrates the construction, and no paths exist for migration. The most practical approach to the prevention and control of efflorescence is to avoid *trapping* moisture in the construction for extended periods of time.

The source of moisture necessary to produce efflorescence may be either rainwater or the condensation of water vapor within the assembly. Excessive water may also be present in masonry walls which were not properly protected from rain and snow during construction. "New building bloom" (efflorescence

Figure 1.2 Efflorescence. (*From Beall, Masonry Design and Detailing, 4th ed.*)

which occurs within the first year of the building's completion) is often traced to slow evaporation of excess moisture in walls that were unprotected during construction.

Efflorescence will often disappear with normal weathering if the source of moisture is located and stopped. Efflorescence can also be dry-brushed, washed away by a thorough flushing with clean water, or scrubbed away with a brush.

Calcium carbonate stains occur when calcium hydroxide is leached to the surface, where it reacts with atmospheric carbon dioxide to form calcium carbonate. The calcium hydroxide (lime) is a natural by-product of the cement hydration process. As cement cures, it produces 12 to 20% of its weight in calcium hydroxide. Extended saturation through construction defects prolongs the curing process and maximizes the amount of lime produced. As the moisture evaporates, it deposits the calcium hydroxide on the surface where it reacts with carbon dioxide in the air to form calcium carbonate. The stains usually occur as hard, encrusted streaks, and are sometimes referred to as "lime deposits" or "lime run" (*Fig. 1.4*).

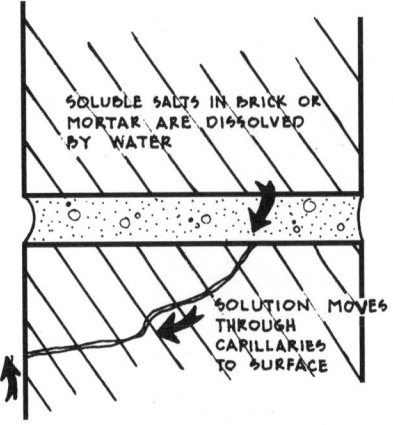

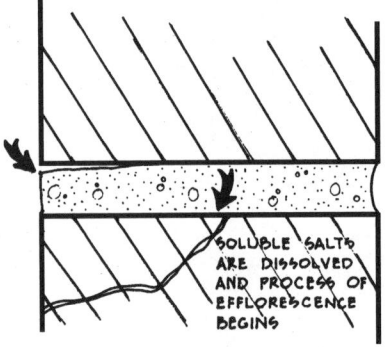

Figure 1.3 Efflorescence in masonry. (*From Acme Brick Co., Ft. Worth, Texas.*)

Figure 1.4 Calcium carbonate stain. (*From Beall, Masonry Design and Detailing, 4th ed.*)

1.3.9 Insect infestations

Termites, carpenter ants, cockroaches, and other insects frequently invade structures, creating millions of dollars in damage each year. One element common to a hospitable environment for insect infestation is moisture. Proper management of surface and subsurface drainage, control of interstitial condensation, drainage of penetrated moisture, and ventilation of attic and crawl spaces provide several means of limiting the moisture available in building microclimates and therefore controlling the invasion of insects. *Figure 1.5* shows the areas of the United States most susceptible to termite infestations. Approximately 95% of all termite damage in the United States is caused by subterranean termites where moist soil and subfloor conditions provide a thriving environment.

1.4 Increasing Durability against Moisture Damage

There are some basic design and construction factors which significantly affect the durability of buildings against damage from moisture in all its forms. The American Society for Testing and Materials (ASTM) develops and publishes many Standard Guides and Standard Practices which identify such factors and can assist the designer and contractor in achieving continued serviceability of a structure throughout its expected life. ASTM E241 *Standard Practices for Increasing Durability of Building Constructions Against Water-Induced*

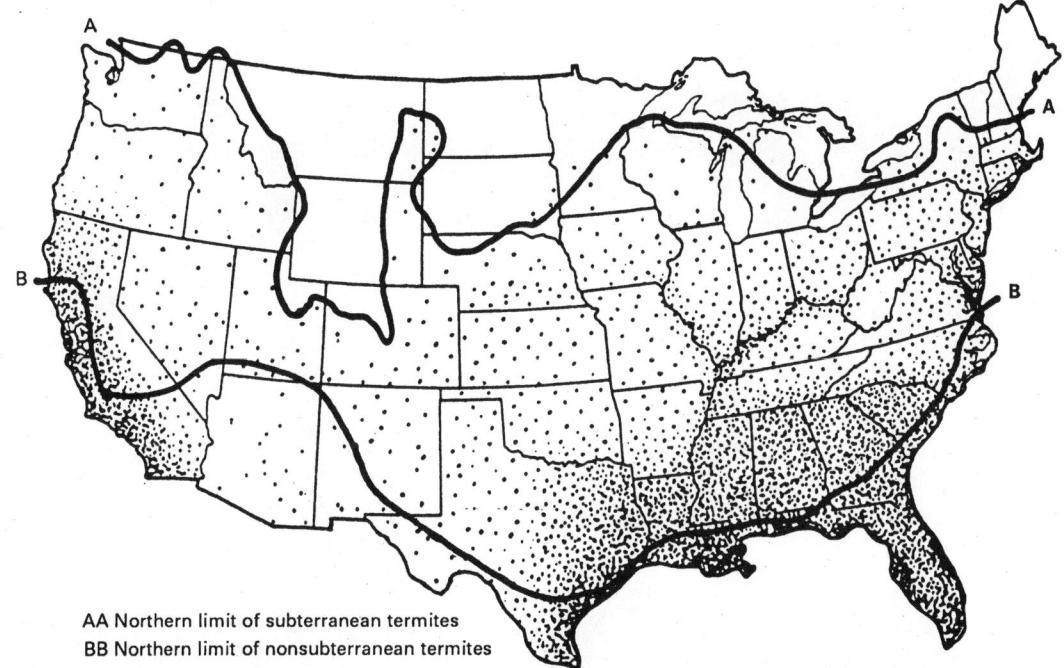

Figure 1.5 Areas of the United States most susceptible to termite infestations. (*From Olin, Construction Principles, Materials and Methods, 5th ed.*)

Damage cautions that time, moisture, and temperature should be considered in evaluating performance of materials and constructions and that both leakage and vapor condensation should be guarded against. Design elements that are understood to enhance durability are vapor diffusion retarders, air barriers, ground slope, flashings, drainage of accumulated moisture, coatings, thermal breaks, and controlled ventilation. Although some guidelines are universal in their application, each building must be analyzed individually in terms of climate, microclimate, occupancy conditions, interior environment, and building envelope components to produce designs and details that are both energy-efficient and effective in resisting moisture damage. The chapters which follow explain the physics of building science and the steps which must be taken to accommodate or control natural forces. Chapters 2 through 5 cover the theories of heat flow and moisture movement, the influence of climate on design, and the performance of various building materials. Chapters 6 through 10 cover the practical applications of those theories to roofing, waterproofing, cladding, joint sealants, and coating systems.

Chapter

2

Climate, Weather, and Microclimate

Not every building has the same weather exposure. Regional climatic conditions, macro environment, and micro environment all affect the risk of moisture damage, and must be taken into account in design.

2.1 Weather and Climate

Weather is defined as the general atmospheric conditions at a given place and a given time with respect to temperature, humidity, precipitation, wind, radiation, and other meteorological events. Weather is a dynamic process, changing from moment to moment, from day to night, and from season to season. *Climate* is defined as the characteristic weather conditions of an area averaged over an extended period of time.

Solar radiation, air temperature, wind, and moisture make up the parameters of weather and climate throughout the world. Weather is fueled primarily by solar heating and radiational cooling, which create temperature differentials between oceans and land masses and between the equator and the poles. Temperature differentials in turn create pressure differentials which generate air movement and winds. Moisture evaporated from the ocean and land surfaces rises into the sky, condenses, and rains on the land and oceans again.

Climate is primarily determined by latitude, altitude, distance from the ocean, and conditions of the ocean. Latitude determines the angle at which solar radiation penetrates the atmosphere and affects the amount of solar energy received on the surface. Altitude affects the density of the atmosphere. The air is heavier and more dense at sea level, becoming lighter and less dense as altitude increases. Since the thinner air at higher altitudes conducts and stores less heat, temperatures drop as elevation increases. The oceans serve as moderating elements in temperature cycles by storing and releasing heat. Since large bodies of water are usually warmer in winter than the adjacent land surface, coastal areas often have very mild winters. In summer the water is cooler than the land surface, so ocean breezes prevent excessive heat. Climate is also a function of the flow of air and the character of the earth's surface. Wind, clouds, snow, and rain are all influenced by the shape, direction, and formation of mountain ranges. The windward sides of mountains have high precipitation because the air is deflected upward and cooled as it flows over the top. The leeward sides of mountains are dry and relatively sunny.

Climate, like weather, is measured in terms of air temperature, wind, humidity, rainfall, snow, ice, and solar radiation (including sunlight, skylight, reflected light, and radiant heat). Climate interacts with buildings through the flow of energy in the same way that it does with people and organisms (*Fig. 2.1*). The climatic factors of significance in this energy flow are radiation (including sunlight), air temperature, humidity, and wind. Thermal energy can flow from one object to another by radiation, conduction, convection, and evaporation. When wind forces cold air across a surface, energy is lost by convection. If the air is hot, energy is gained by the same process.

There are more than 10 different types of climate in North America (*Fig. 2.2*). For building design purposes, the general climatic regions of the continental United States may be identified in several different ways. United States Weather Bureau maps refer to cool, temperate, hot-arid, and hot-humid climates (*Fig. 2.3*). Researchers at the Oak Ridge National Laboratory define a heating climate zone, a cooling climate zone, and a mixed climate zone (*Fig. 2.4*).

Detailed climatic data for a given location may be obtained from the National Oceanic and Atmospheric Administration (NOAA). Both daily reports and monthly weather summaries are available for most weather stations (*Fig. 2.5*). These reports are helpful in documenting weather conditions during construction. Also available from NOAA is a publication called *Climatic Averages and Extremes for U.S. Cities* (*Fig. 2.6*) which gives temperature, humidity, precipitation, and wind speed information for most U.S. cities (*Fig. 2.7*). This data can be used in the design and analysis of building envelope sections and particularly for the calculation of dew points and condensation.

Much of the climatic data that is available to the architectural and engineering professions is used in the design of active and passive heating and cooling systems. For specific project locations within a climatic region, a range of temperatures must be selected within which the building will be expected to perform satisfactorily. Should the temperature vary outside this range, some discomfort to the occupants and possibly some distress to the building envelope may occur. If the design temperatures have been well chosen, the distress should be minor and of short duration. The American Society of Heating, Refrigeration, and Air Conditioning Engineers (ASHRAE) *Handbook of Fundamentals* gives summer and winter design temperatures for most U.S. and Canadian cities. A 1 or 2½% winter design temperature can

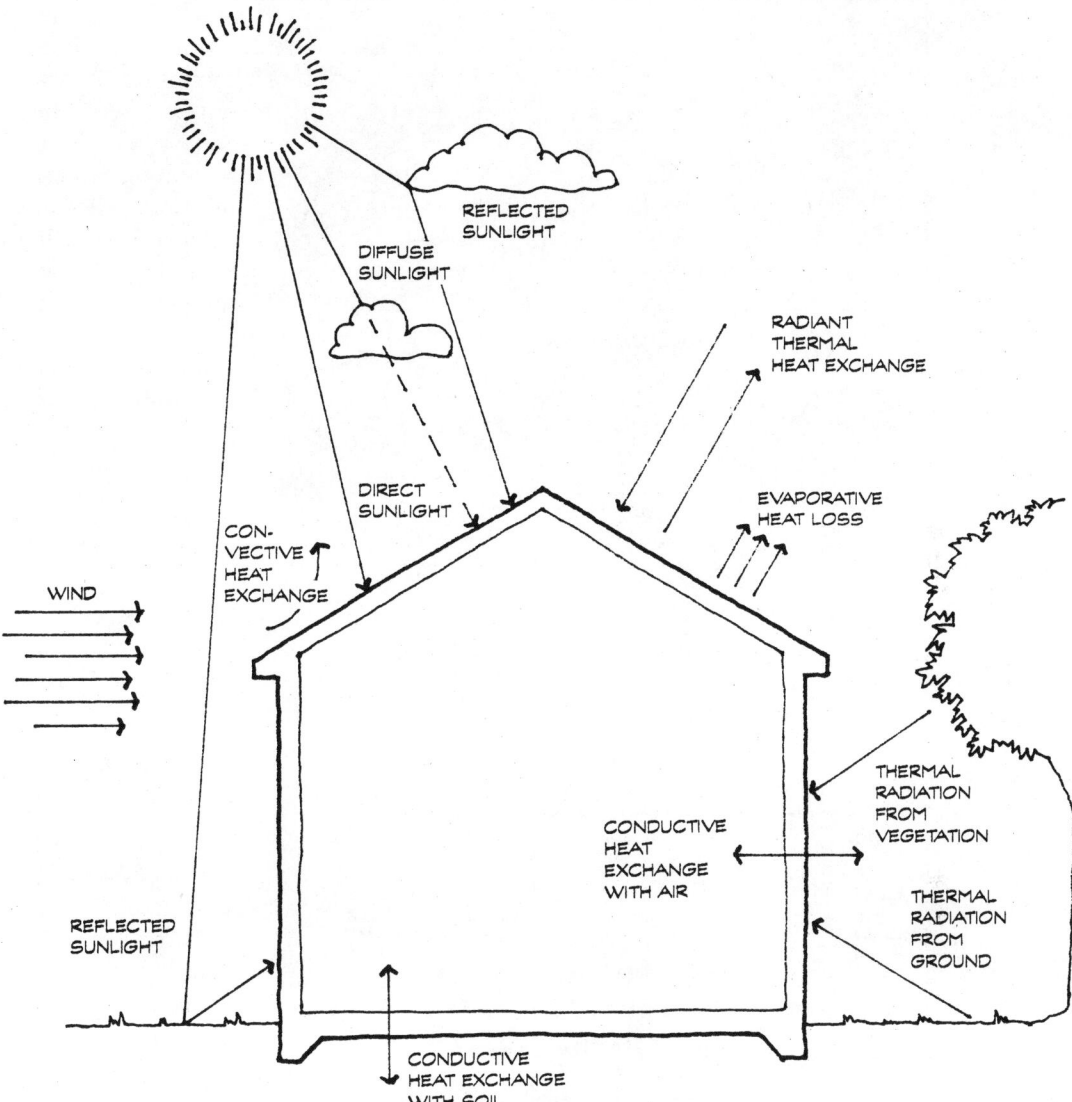

Figure 2.1 Climatic interaction between a building and its environment.

be selected for the design of heating systems. This means that low temperatures will be outside the range only 1 or 2½% of the time, respectively. For summer conditions, 1, 2½, or 5% design temperatures can be selected for the design of cooling systems. These ASHRAE design temperatures can also be used for calculating temperature differentials for thermal expansion and contraction in exterior walls, for locating movement joints, and sizing sealant joints.

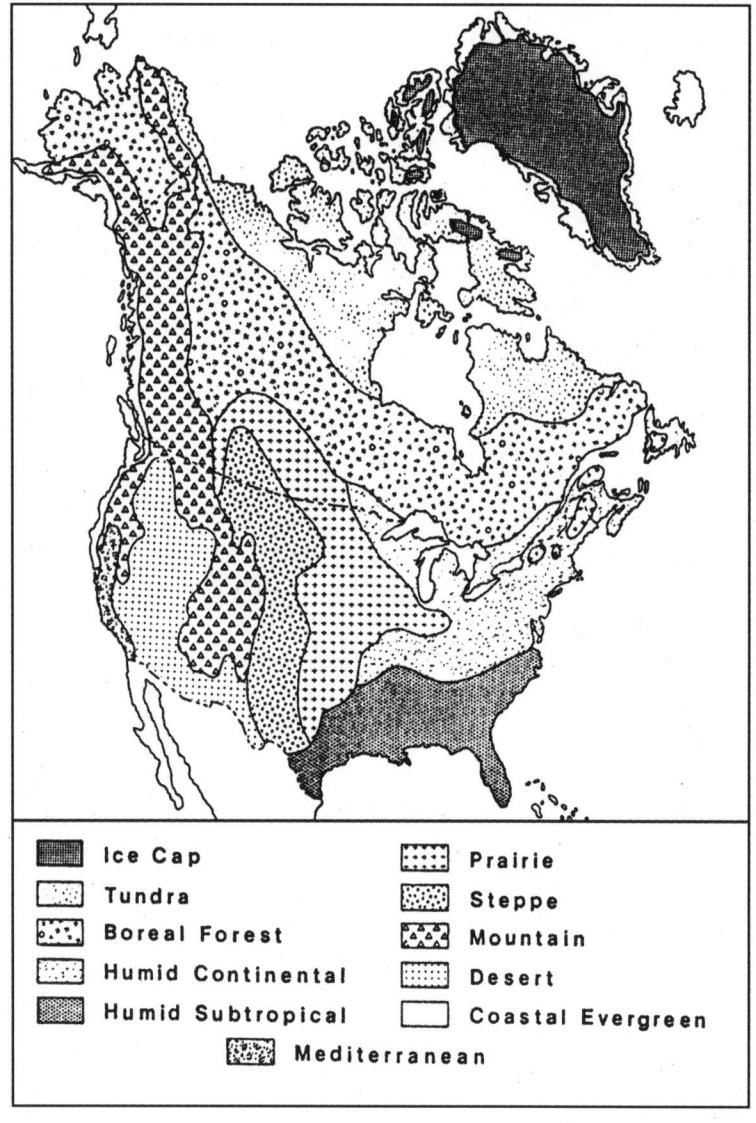

Figure 2.2 Climates of North America.

2.2 Wind and Air Movement

We refer to the motion of the air as wind. Global winds are driven by solar radiation. The differential absorption of sunlight between land and bodies of water, the existence of large ocean currents, and the rising of air masses over mountain barriers create complex circulation patterns. Temperature and pressure differences between air masses also create air motion and thus affect the transport of airborne moisture vapor. The same is true in the microenvironment of buildings. As its temperature changes, an air mass expands or contracts, and the air pressure rises or falls accordingly. The regions of high pressure rush toward the regions of low pressure, creating air movement and with it, moisture vapor movement.

Calm air is considered to have movement less than 1 mph. Light air movement is 1 to 3 mph, a light breeze 4 to 7 mph, a gentle breeze 19 to 24 mph, a strong breeze 25 to 31 mph, gale winds 32 to 63 mph, and storm winds 64 to 73 mph. Winds greater than 74 mph are considered to be of hurricane intensity. Coastal areas are windier than most forested inland regions, grasslands, or deserts. Well-watered regions tend to be less turbulent, while mountain tops are persistently windy, even when covered with trees. ANSI Standard A58.1 shows basic fastest-mile wind speeds at 33 ft above ground for the continental United States which range from 70 to 110 mph and ASCE maps show 50-year peak gusts from 85 to 150 mph. (*Fig. 2.8*). Mean wind speeds are always punctuated by gusts and lulls (*Fig. 2.9*).

Wind speed can be translated to equivalent static pressure using the formula

$$q = 0.00256 \, V^2 \tag{2.1}$$

where q = static pressure, psf, and V = wind velocity, mph.

Pressure equivalents for various wind speeds and velocity equivalents for various wind pressures are given in *Fig. 2.10*. Requirements for many different building envelope components are based on "design" wind speed or pressure as dictated by the governing building code.

Wind speed changes with height above the ground. A 10-mph wind at 6 ft above ground may be only 1 mph at a few inches, and near zero at the ground surface. The speed is reduced by the frictional effect of the ground surface on the

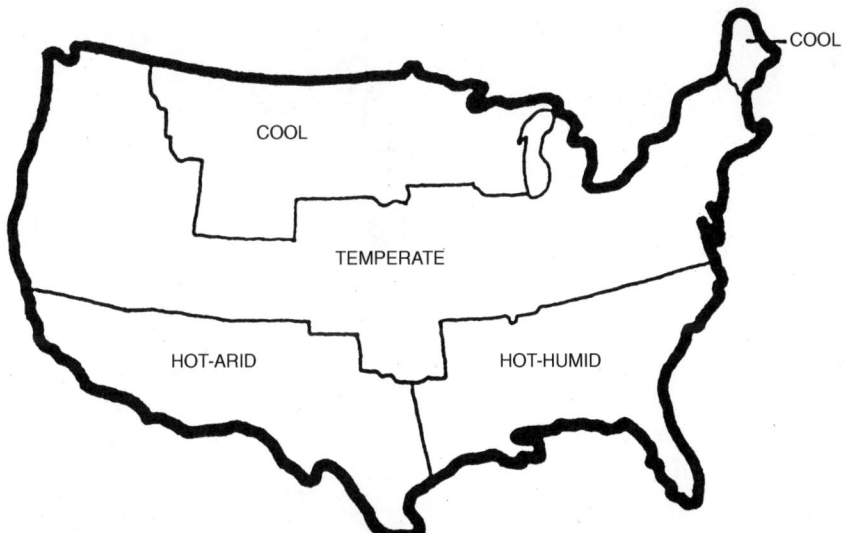

Figure 2.3 United States climatic regions. (*From U.S. Weather Bureau.*)

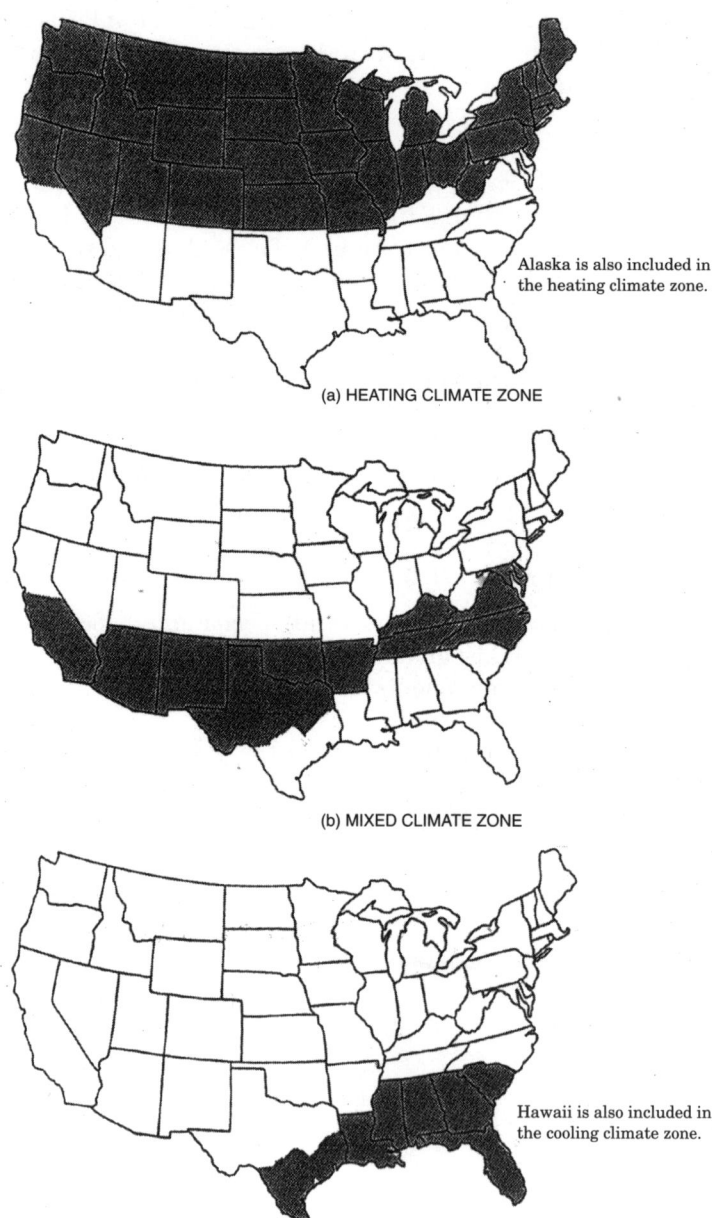

Figure 2.4 United States heating and cooling climate zones. (*From Lstiburek and Carmody, Moisture Control Handbook.*)

Climate, Weather, and Microclimate   27

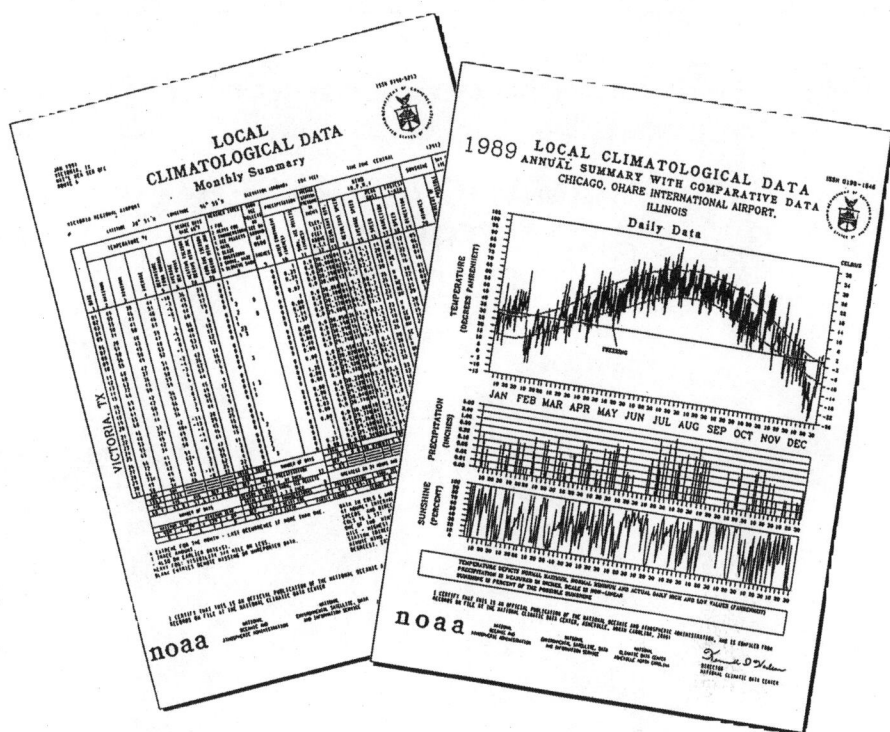

Figure 2.5 Local climatological data—monthly summary/daily data available from National Oceanic and Atmospheric Administration.

air flow (*Fig. 2.11*). Air very close to the surface is strongly retarded by this drag, and the retardation is transmitted to the air above with diminishing effect. Since there is usually very little wind in the boundary layer close to the ground surface, temperatures are hottest here during the day with the least convectional cooling, and coolest during the night with the least convectional warming.

Near the ground, wind is slowed and stirred into complex patterns because of surface drag and turbulence. The roughness of the earth's surface causes drag on the wind and converts some of its energy into mechanical turbulence. Turbulence is generated at ground level by surface irregularities, trees, large topographical features, and buildings. Since air is a viscous fluid, the turbulence is transmitted into the air above, and may persist downwind for a distance more than 100 times the height of the surface obstacles. In a large city, the air is forced to flow upward and over the buildings, thus creating greater wind speeds at higher levels. Valleys and city streets with tall buildings on each side can also create a strong funneling effect that increases wind speed along the axis of the valley or street. Wind conditions can change significantly within a few yards in such an environment. High-rise buildings create their own wind patterns because of temperature differences between the sunny elevations and the surfaces in the shade. The resulting air currents are often increased by the funnel effect created between two buildings.

Figure 2.6 Climatic averages and extremes data available from National Oceanic and Atmospheric Administration.

NORMALS, MEANS, AND EXTREMES
BILLINGS, MONTANA

LATITUDE: 45°48'N LONGITUDE: 108°32'W ELEVATION: FT. GRND 3567 BARO 3583 TIME ZONE: MOUNTAIN WBAN: 24033

	(a)	JAN	FEB	MAR	APR	MAY	JUNE	JULY	AUG	SEP	OCT	NOV	DEC	YEAR
TEMPERATURE °F:														
Normals														
-Daily Maximum		29.9	37.9	44.0	55.9	66.4	76.3	86.6	84.3	72.3	61.0	44.4	36.0	57.9
-Daily Minimum		11.8	18.8	23.6	33.2	43.3	51.6	58.0	56.2	46.5	37.5	25.5	18.2	35.4
-Monthly		20.9	28.4	33.8	44.6	54.9	64.0	72.3	70.3	59.4	49.3	35.0	27.1	46.7
Extremes														
-Record Highest	56	68	72	79	92	96	105	106	105	103	90	77	69	106
-Year		1953	1961	1986	1939	1936	1984	1937	1961	1983	1963	1983	1980	JUL 1937
-Record Lowest	56	-30	-38	-19	-5	14	32	41	40	22	2	-22	-32	-38
-Year		1937	1936	1989	1936	1954	1969	1972	1939	1984	1984	1959	1983	FEB 1936
NORMAL DEGREE DAYS:														
Heating (base 65°F)		1367	1025	967	612	318	111	9	27	214	487	900	1175	7212
Cooling (base 65°F)		0	0	0	0	0	81	235	191	46	0	0	0	553
% OF POSSIBLE SUNSHINE	51	47	53	61	61	61	65	76	76	68	61	46	45	60
MEAN SKY COVER (tenths)														
Sunrise - Sunset	51	7.1	7.2	7.2	7.1	6.6	5.9	4.3	4.3	5.2	5.8	6.9	6.8	6.2
MEAN NUMBER OF DAYS:														
Sunrise to Sunset														
-Clear	51	5.5	4.0	4.3	4.5	5.5	7.1	13.6	13.7	10.4	9.2	5.7	6.0	89.4
-Partly Cloudy	51	7.7	8.1	8.6	8.7	6.5	11.9	11.8	10.9	9.4	9.2	7.7	8.2	112.8
-Cloudy	51	17.8	16.1	18.1	16.8	15.0	11.0	5.6	6.4	10.3	12.6	16.6	16.7	163.1
Precipitation														
.01 inches or more	56	7.8	7.5	9.1	9.4	11.1	11.0	7.2	6.4	7.1	6.2	6.1	6.9	96.1
Snow, Ice pellets														
1.0 inches or more	48	3.3	2.6	3.5	2.2	0.5	0.*	0.0	0.0	0.4	1.2	2.2	2.8	18.6
Thunderstorms	51	0.*	0.*	0.1	1.2	4.2	7.2	7.5	5.6	1.8	0.2	0.0	0.*	27.9
Heavy Fog Visibility														
1/4 mile or less	43	1.5	2.2	2.0	2.5	1.2	0.7	0.3	0.3	1.2	2.0	2.1	1.6	17.6
Temperature °F														
-Maximum														
90° and above	31	0.0	0.0	0.0	0.*	0.4	4.3	12.6	10.6	1.9	0.*	0.0	0.0	29.9
32° and below	31	13.8	8.1	5.0	0.7	0.0	0.0	0.0	0.0	0.*	0.6	5.2	11.7	45.2
-Minimum														
32° and below	31	27.5	23.8	23.5	12.8	1.7	0.1	0.0	0.0	1.5	8.1	21.6	27.6	148.2
0° and below	31	8.1	3.5	1.3	0.0	0.0	0.0	0.0	0.0	0.0	0.0	1.2	4.7	18.9
AVG. STATION PRESS. (mb)	18	890.8	890.7	888.3	889.4	888.9	890.0	891.8	891.8	892.5	892.3	890.5	890.8	890.7
RELATIVE HUMIDITY (%)														
Hour 05	31	64	66	68	67	69	70	63	61	65	63	65	64	65
Hour 11	31	60	58	54	48	48	45	39	39	47	49	57	60	50
Hour 17 (Local Time)	31	56	52	46	40	42	39	30	30	37	41	53	56	44
Hour 23	31	63	62	62	58	59	58	49	47	54	56	61	62	58
PRECIPITATION (inches):														
Water Equivalent														
-Normal		0.97	0.71	1.05	1.93	2.39	2.07	0.85	1.05	1.26	1.16	0.85	0.80	15.09
-Maximum Monthly	56	2.35	1.77	2.70	4.42	7.71	7.64	3.12	3.50	4.99	3.80	2.34	2.00	7.71
-Year		1972	1978	1954	1955	1981	1944	1958	1965	1941	1971	1978	1973	MAY 1981
-Minimum Monthly	56	0.04	0.05	0.13	0.06	0.53	0.24	0.04	0.05	0.06	0.01	T	0.05	T
-Year		1941	1977	1936	1962	1937	1961	1988	1955	1964	1987	1954	1957	NOV 1954
-Maximum in 24 hrs	56	1.41	0.65	1.01	3.19	2.83	2.78	1.87	2.47	2.19	1.98	1.37	0.96	3.19
-Year		1972	1986	1973	1978	1952	1937	1958	1965	1966	1974	1959	1978	APR 1978
Snow, Ice pellets														
-Maximum Monthly	56	27.7	22.4	27.6	42.3	15.6	2.0	T	T	9.3	23.1	25.2	28.8	42.3
-Year		1963	1978	1935	1955	1981	1950	1990	1990	1984	1949	1978	1955	APR 1955
-Maximum in 24 hrs	52	16.6	9.0	10.5	23.7	15.3	2.0	T	T	7.5	11.2	15.3	13.7	23.7
-Year		1972	1944	1964	1955	1981	1950	1990	1990	1983	1980	1959	1978	APR 1955
WIND:														
Mean Speed (mph)	51	13.1	12.3	11.5	11.5	10.8	10.2	9.6	9.5	10.2	11.0	12.1	13.1	11.2
Prevailing Direction through 1963		SW	SW	SW	SW	NE	SW	SW	SW	SW	SW	SW	WSW	SW
Fastest Mile														
-Direction (!!)	47	W	W	NW	NW	NN	NW	N	NW	NW	NW	NW	NW	NW
-Speed (MPH)	47	66	72	61	72	68	79	73	69	61	68	63	66	79
-Year		1953	1963	1956	1947	1939	1968	1947	1983	1949	1949	1948	1953	JUN 1968
Peak Gust														
-Direction (!!)	7	NW	W	NW	NW	NW	W	NE	NW	NW	NW	SW	W	NW
-Speed (mph)	7	59	62	52	59	60	54	58	58	61	61	58	64	69
-Date		1986	1988	1990	1987	1988	1987	1985	1986	1989	1985	1990	1988	AUG 1986

Figure 2.7 Page excerpt from NOAA *Climatic Averages and Extremes for U.S. Cities.*

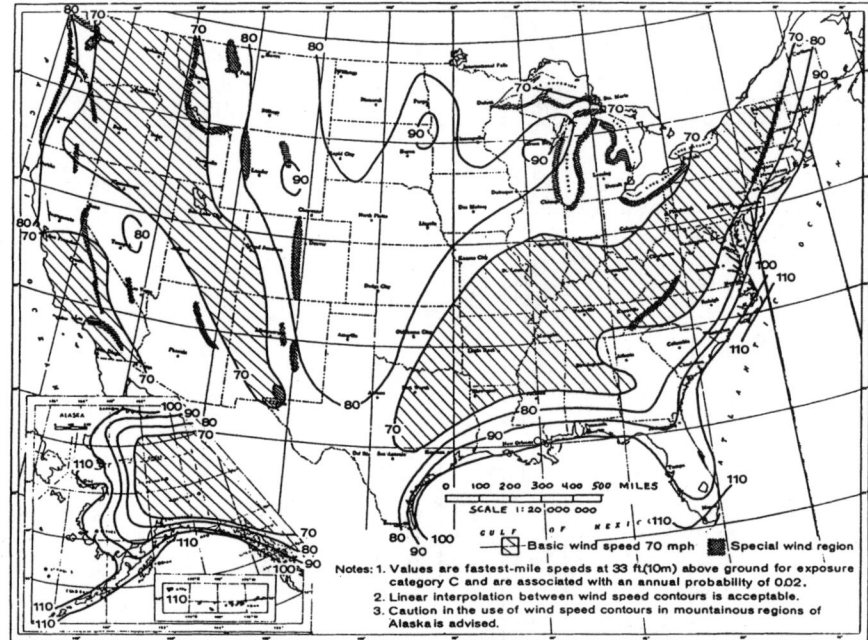

(a) ANSI A58.1 BASIC WIND SPEED MAP, MPH

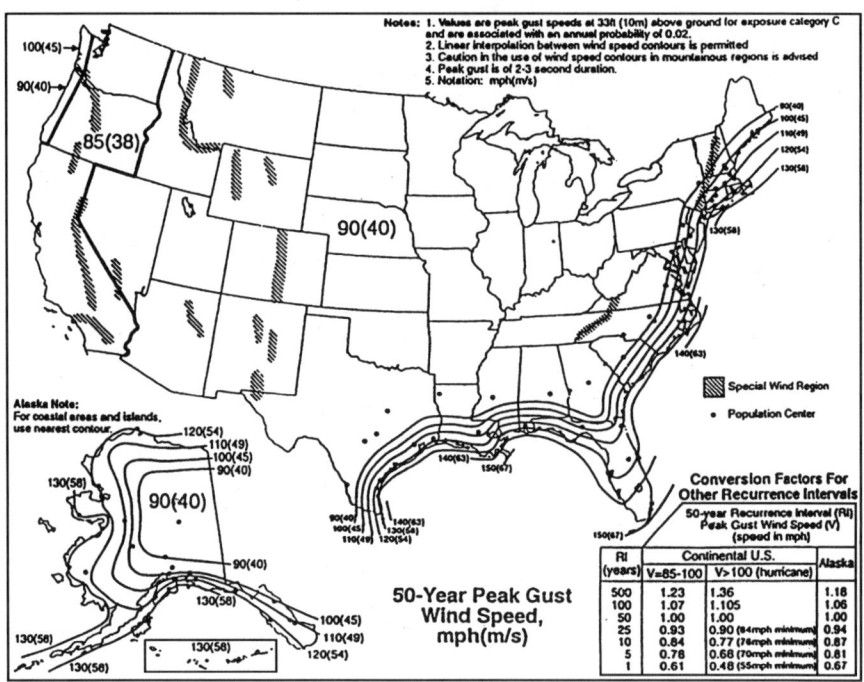

(b) ASCE PEAK GUST WIND SPEED MAP, MPH

Figure 2.8 Basic wind speed map from ANSI A58.1 and 50-year peak gust wind speed map from ASCE.

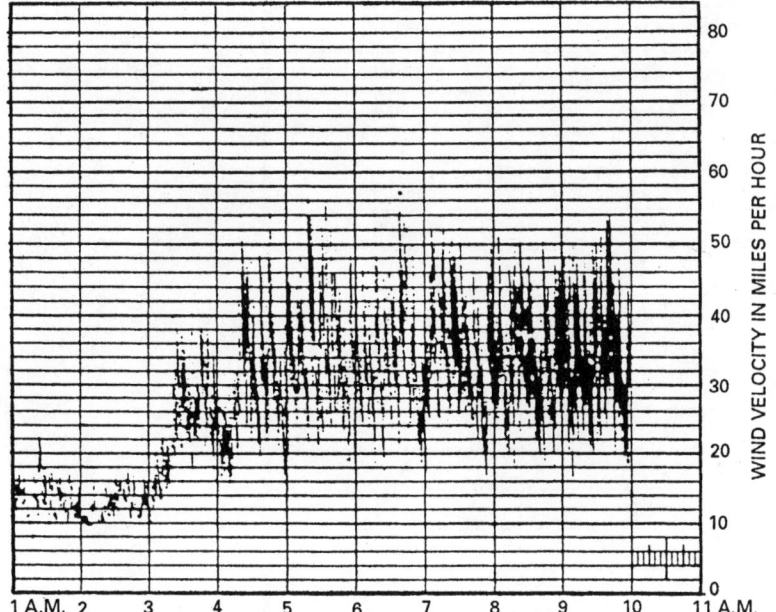

Figure 2.9 Typical wind speed variation. (*From Latta, Walls, Windows and Roofs for the Canadian Climate.*)

Turbulence includes vertical as well as horizontal air movement, because the effect of surface frictional drag is propagated upward. The mechanical turbulence and the effect of frictional drag gradually decrease with height, and at the "gradient" level around 1000 to 2000 ft, the frictional effect is negligible. *Figure 2.12* shows four hypothetical velocity profiles where the effect of variable surface roughness on the mean wind speeds is shown for an arbitrarily selected gradient wind of 100 mph.

Design wind pressure varies with location and height of building. The data in *Table 2.1* give stagnation pressures for basic wind speeds from 70 to 130 mph, and coefficients for combined height, exposure, and gust factors. The stagnation pressure is that which would occur near the center of a vertical wall when the wind strikes it head on. To obtain the design pressure, multiply the pressure coefficient times the stagnation pressure. Some building codes also add coefficients for different building elements and critical occupancies like hospitals and emergency facilities.

Wind is usually much stronger over the brow of a hill or ridge because the flow lines must converge and the speed increases in order to pass the same quantity of air. The same is true at building corners and edges. The distribution of pressure over a building depends on how it disturbs the air flow. Streamlined air flow can be represented by a series of parallel lines. In order to get over and around a building, these lines will be bent and crowded together or spread out at various locations (*Fig. 2.13*). Air flow increases over the brow of a hill or the edge of a building because the wind velocity increases as

Wind Pressure, psf	Wind Speed, mph
0.256	10.0
1.024	20.0
2.3	30.0
4.1	40.0
5.0	44.2
6.4	50.0
9.2	60.0
10.0	62.5
12.5	70.0
15.0	76.5
16.4	80.0
20.0	88.4
20.7	90.0
25.0	98.8
25.6	100.0
30.0	108.3
36.8	120.0
40.0	125.0
50.0	139.8
50.2	140.0
55.0	146.6
57.6	150.0
60.0	153.1
65.5	160.0
70.0	165.4
80.0	176.8
82.9	180.0
85.0	182.2
90.0	187.5
100.0	197.6
102.4	200.0

psf = 0.00256 (mph)^2

(a)

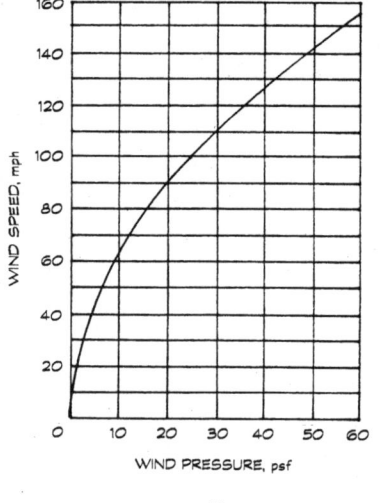

(b)

Figure 2.10 Wind speed and pressure equivalents.

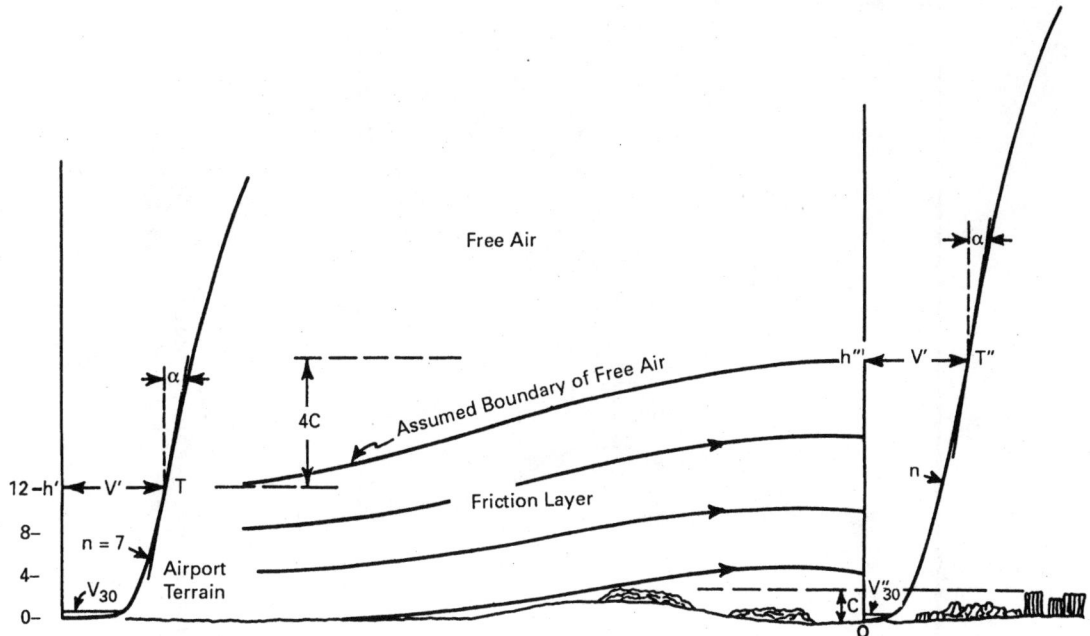

Figure 2.11 Frictional effect of ground surface on air flow. (*From National Institute of Standards and Technology, Wind Loads on Buildings and Structures.*)

the flow lines bunch together to pass the same quantity of air through a narrower space. When the building is on the convex side of the streamlines, the pressure exerted on the surface increases above the ambient barometric pressure in the undisturbed air flow. When the building is on the concave side of the streamlines, the pressure is reduced below the ambient, and there is a suction force on the surface. In either case, the sharper the bend in the streamlines, the greater will be the pressure or suction. The important factors in determining the shape of the wind flow around a building are the ratios of height to width, and width to length. The highest wind loads are at the corners and along the roof lines where the flow lines must converge.

Wind interacts with buildings in several different ways. For example, the movement of air across a building facade creates pressure differentials between the outside and inside atmospheres (*Fig. 2.14*). These pressure differentials affect moisture penetration through the building envelope and moisture vapor migration through the building. The orientation of the building with respect to wind direction affects the magnitude and direction of the pressure differentials (*Fig. 2.15*). In roofing design, wind uplift on roofing membranes is a significant factor in the design of parapet walls and in the selection of fastener spacing (*Fig. 2.16*). Wind has a drying effect on wet building surfaces as well as an evaporative cooling effect, but it can also drive moisture uphill in defiance of gravity. A 50-mph wind will raise a column of water 1.23 in. (*Fig. 2.17*). A 100-mph wind will raise a column of water nearly 5 in. The risk of high wind exposure therefore affects such things as the height of flashings and the design of joints.

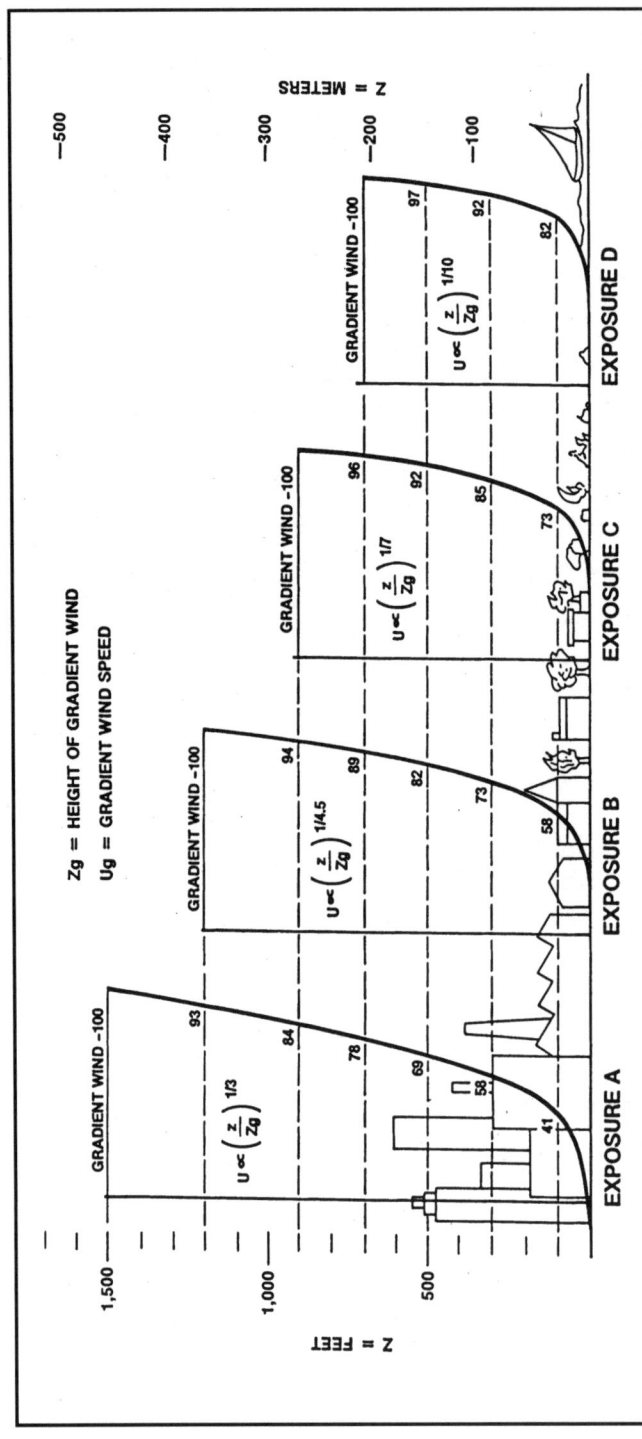

Figure 2.12 Profiles of mean wind velocity over level terrain of differing roughness. *(From ANSI A58.1.)*

EXPOSURE A: large city centers with at least 50% of the buildings having a height in excess of 70 feet.
EXPOSURE B: urban and suburban areas, wooded areas, or other terrain with numerous, closely spaced obstructions having the size of single family dwellings or larger.
EXPOSURE C: open terrain with scattered obstructions having heights generally less than 30 feet.
EXPOSURE D: flat, unobstructed coastal areas directly exposed to wind flowing over large bodies of water.

TABLE 2.1A Combined Height, Exposure and Gust Factor Coefficient for Design of Building Elements

Height above average level of adjoining ground, ft	Exposure B*	Exposure C†	Exposure D‡
0–15	0.62	1.06	1.39
20	0.67	1.13	1.45
25	0.72	1.19	1.50
30	0.76	1.23	1.54
40	0.84	1.31	1.62
60	0.95	1.43	1.73
80	1.04	1.53	1.81
100	1.13	1.61	1.88
120	1.2	1.67	1.93
160	1.31	1.79	2.02
200	1.42	1.87	2.10
300	1.63	2.05	2.23
400	1.80	2.19	2.34

*Exposure B has terrain with buildings, forest or surface irregularities 20 ft or more in height covering at least 20% of the area extending 1 mile or more from the site (i.e., centers of large cities and very rough, hilly terrain).

†Exposure C has terrain which is flat and generally open, extending one-half mile or more from the site in any full quadrant (i.e., suburban areas, towns, city outskirts, and rolling terrain).

‡Exposure D represents the most severe exposure in areas with basic wind speeds of 80 mph or greater, and has terrain which is flat and unobstructed facing large bodies of water over 1 mile or more in width relative to any quadrant of the building site. Extends inland from the shoreline $\frac{1}{4}$ mile or 10 times the building height, whichever is greater (i.e., flat, open country; open, flat coastal belts; and grassland).

TABLE 2.1B To Obtain Design Pressure, Multiply Pressure Coefficient from Table 2-1A by Wind Stagnation Pressure

Basic wind speed, mph*	70	80	90	100	110	120	130
Stagnation pressure, psf, at standard height of 33 ft	12.6	16.4	20.8	25.6	31.0	36.9	43.3

*NOTE: Obtain basic wind speed from map in Figure 2.8A.
SOURCE: Merritt, *Standard Handbook for Civil Engineers*.

2.3 Precipitation

Water is in the atmosphere as vapor, clouds, and precipitation, including rain, mist, dew, snow, sleet, and hail. Water is also underground, moving among rock strata or bound within the soil. Ponds, lakes, seas, oceans, rivers, and streams are global reservoirs for the water which washes the earth's surface.

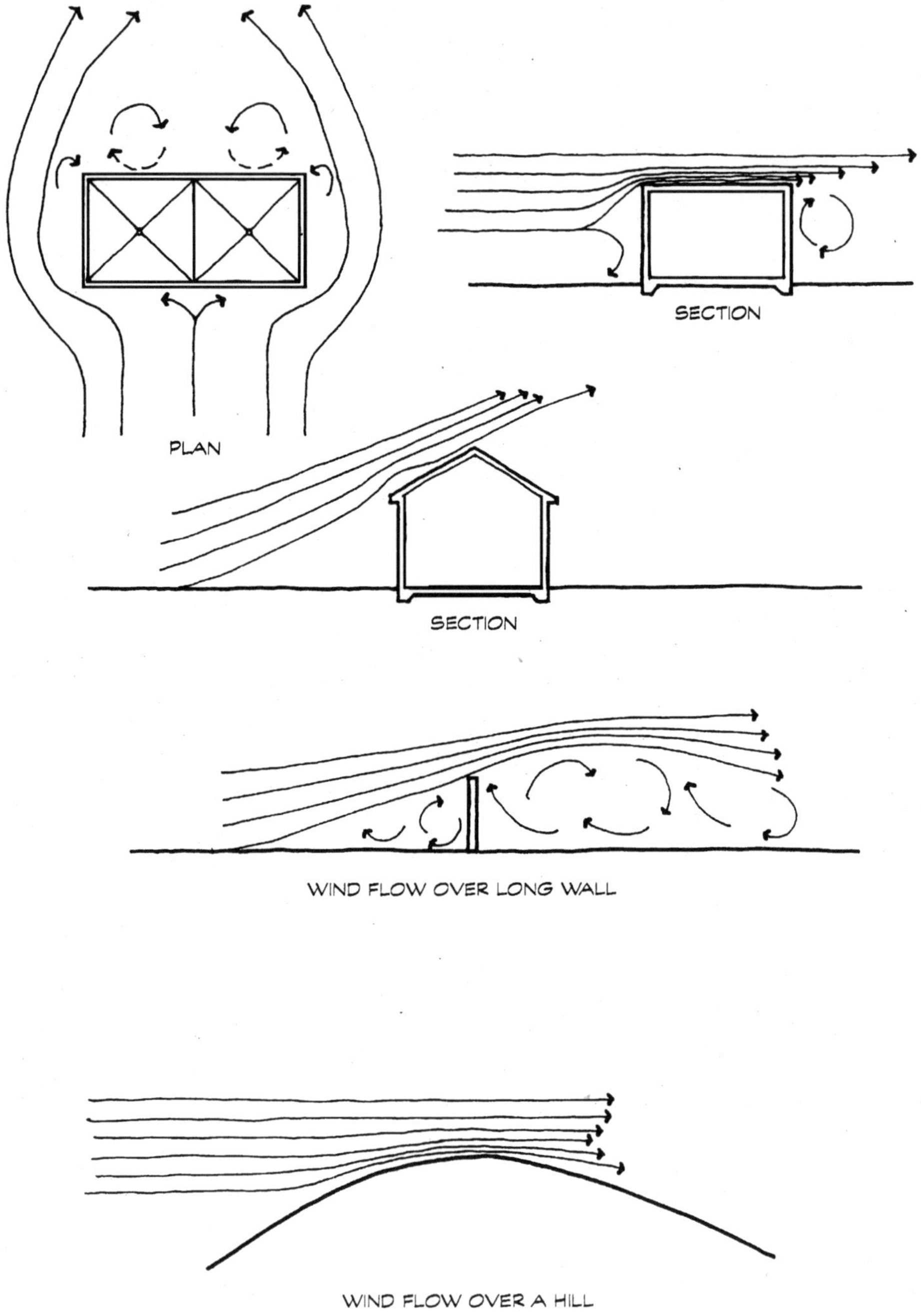

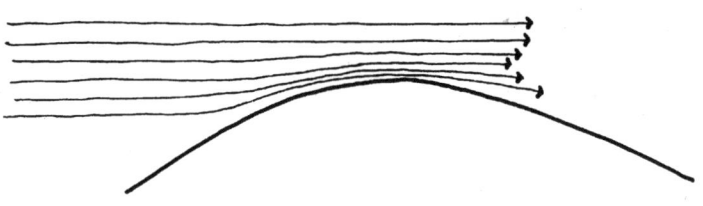

Figure 2.13 Air flow over and around buildings and objects.

Climate, Weather, and Microclimate   37

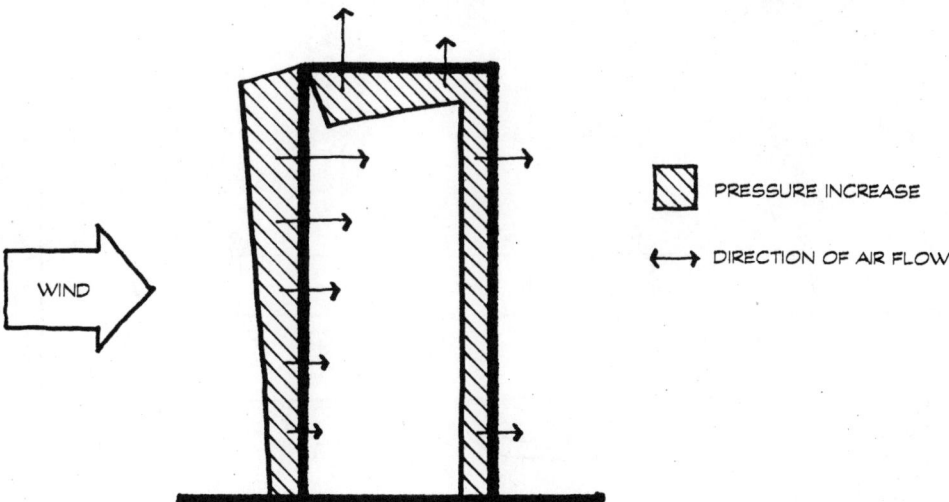

Figure 2.14 Pressure differentials caused by wind. (*From Latta, Walls, Windows and Roofs for the Canadian Climate.*)

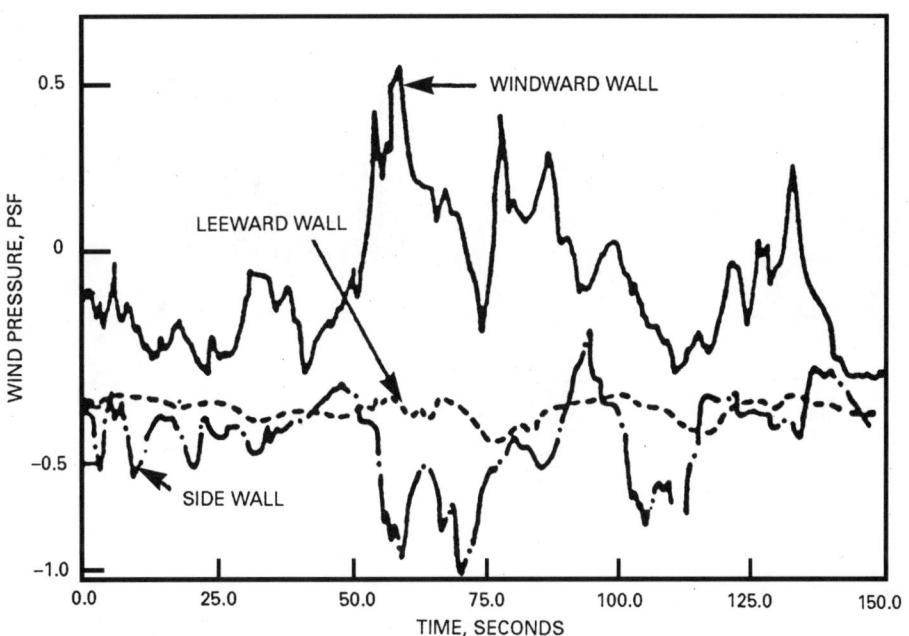

(A) WIND PRESSURE MEASURED ON THE EXTERIOR WALLS OF A BUILDING IN MONTREAL

Figure 2.15 Building orientation affects pressure differentials caused by wind. (*From Latta, Walls, Windows and Roofs for the Canadian Climate.*)

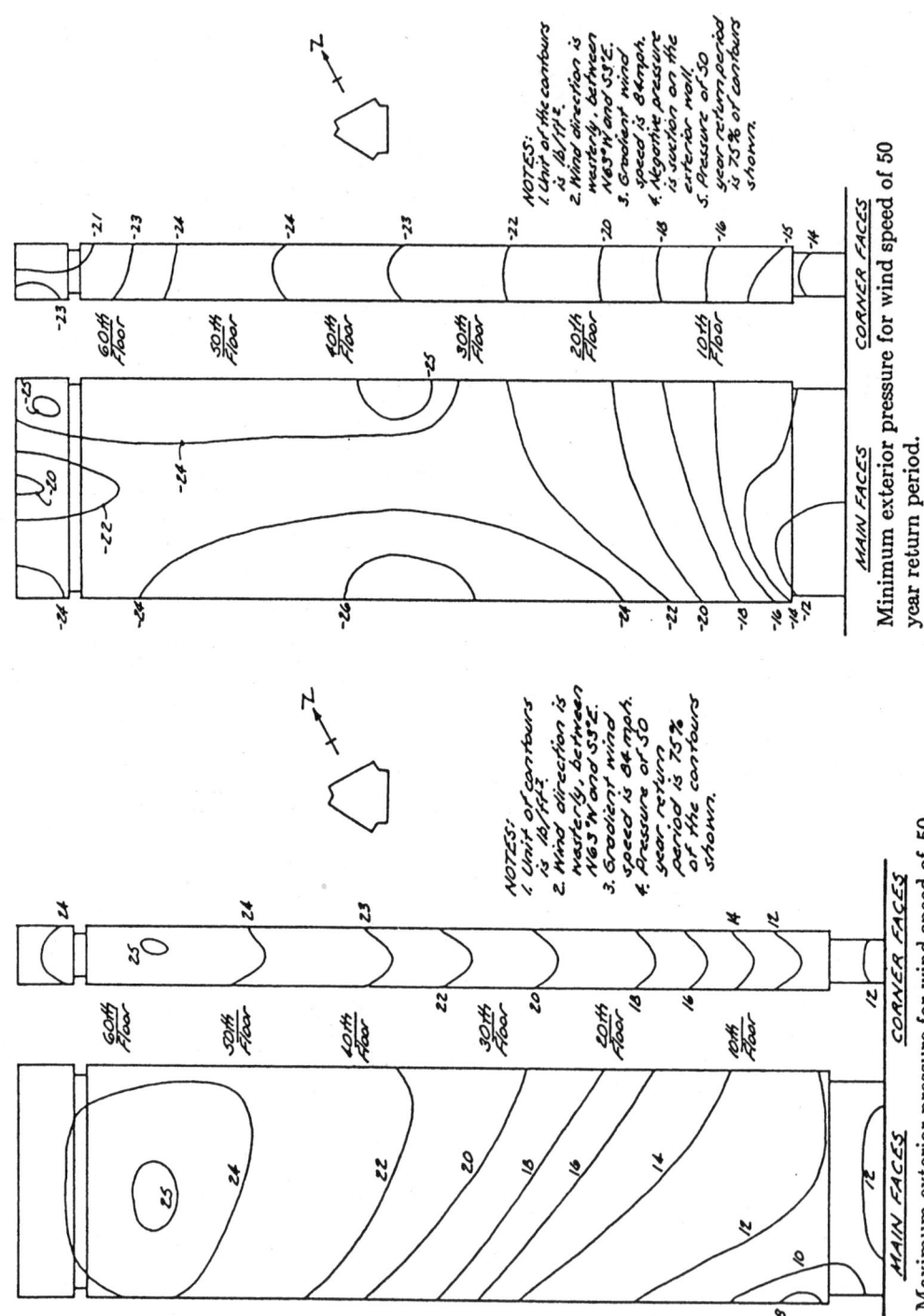

Figure 2.15 *(Continued)*

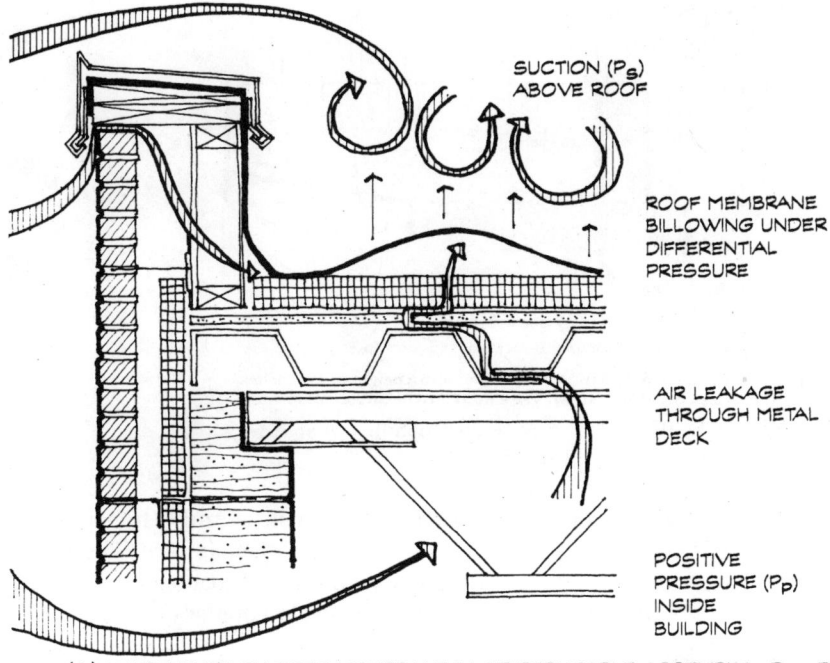

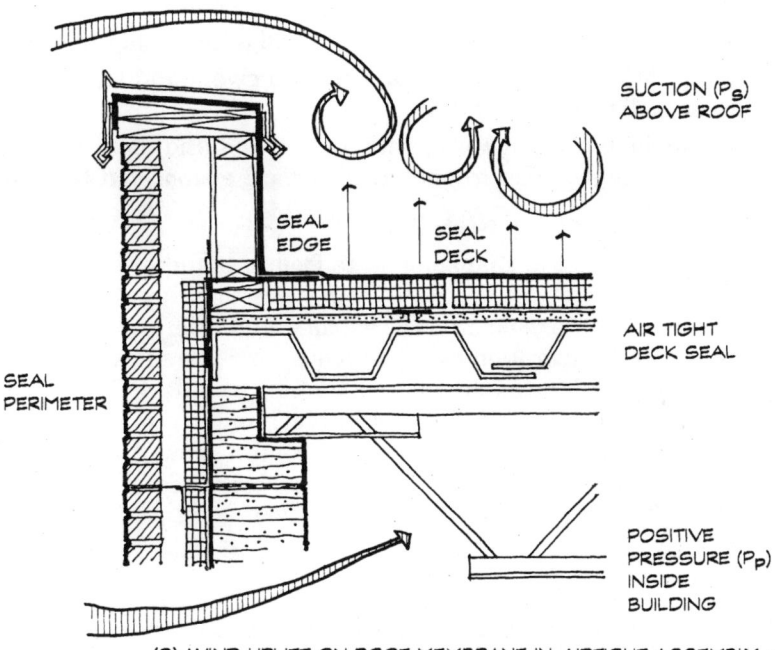

Figure 2.16 Wind uplift on roof membranes in air-permeable and airtight assemblies. (*From National Research Council of Canada, Roofs That Work.*)

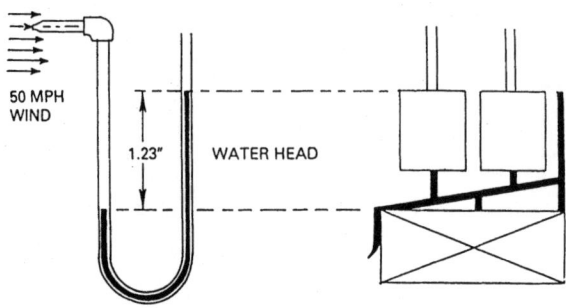

Water Head

h_w (inches) = 0.192 x static pressure (psf)

A 100 mph wind (25.6 psf) will raise water to a height of 4.92 inches
A 50 mph wind (6.4 psf) will raise water to a height of 1.23 inches

Figure 2.17 Wind can drive water uphill. (*From AAMA Window Selection Guide.*)

About 97% of earth's water is in the oceans and other bodies of water. This water is continuously evaporated into the atmosphere. The natural movement of water and vapor through the earth and its atmosphere is called the *hydrologic cycle* (*Fig. 2.18*). The hydrologic cycle consists of several processes.

- Water evaporates from oceans and other bodies of open water.
- Evaporated moisture condenses to form clouds.
- Moisture precipitates as rain, snow, sleet, and hail.
- Precipitation falls directly on oceans and other bodies of open water.
- Precipitation falls on land, runs off into rivers and streams, and flows into larger bodies of open water.
- Rain and melting snow seep into the soil and find their way to bodies of open water through underground strata or are evaporated through the respiration of plants.

Evaporation is the change of state from a liquid or solid to a gas, and condensation is the change of state from a gas to a liquid or solid. Water evaporates when it is warmed, and the warmer and drier the air, the more rapid the evaporation. Evaporation can take place from the surface of a liquid and from land surfaces so long as the relative humidity of the air is less than 100%. This vapor condenses as fog if the relative humidity of the air is near 100%. Evaporation can also take place from ice and snow. As moisture is lost to the air, the ice mass becomes smaller, just as ice cubes in a freezer will shrink if not used. This process is called *sublimation*.

2.3.1 Rain

When warm moist air rises into the atmosphere, it mixes with cooler air at higher altitudes, and the moisture condenses out as some form of precipitation. Rainfall amounts vary from the driest global regions where only 1 or 2 mm or less are recorded each year, to the wettest regions with an average annual rainfall of over 100 in. The average annual precipitation in the United States varies from 10 in or less to as much as 60, 80, or 100 in. (*Fig. 2.19*).

2.3.2 Dew

Dew is a form of precipitation that usually occurs when other precipitation is least likely. Dew is the condensation of moisture onto the ground and other surfaces when surface temperatures are cooler than the moisture laden air. Several millimeters of dew may condense during the night, and in many locations, this is an important source of moisture for vegetation.

2.3.3 Snow

Snow is another important form of precipitation. While rain is generally measured as inches of water on a horizontal surface, the amount of water contained in snowfall is measured in terms of the liquid equivalent if the snow were melted. The water content of snow varies considerably, but it usually requires 8 to 10 in. of snow to produce 1 in. of water. Wet snow may stick to some surfaces and build up on others, creating drifts against a building wall or heavy snow loads on a roof. Fine, powdery snow can be blown into attics or wall cavities through even very small openings, causing damage because of the accumulated moisture and slow drying. Freshly fallen snow is loose and light, with a specific gravity of about 0.05 to 0.1 (one-twentieth to one-tenth that of water). As soon as they hit the ground, though, snowflakes begin to change in shape and the snow settles and compacts. Even at temperatures below freezing, the specific gravity will have increased within a few days to about 0.2, and after about a month to 0.3. Periods of warmer weather and rain

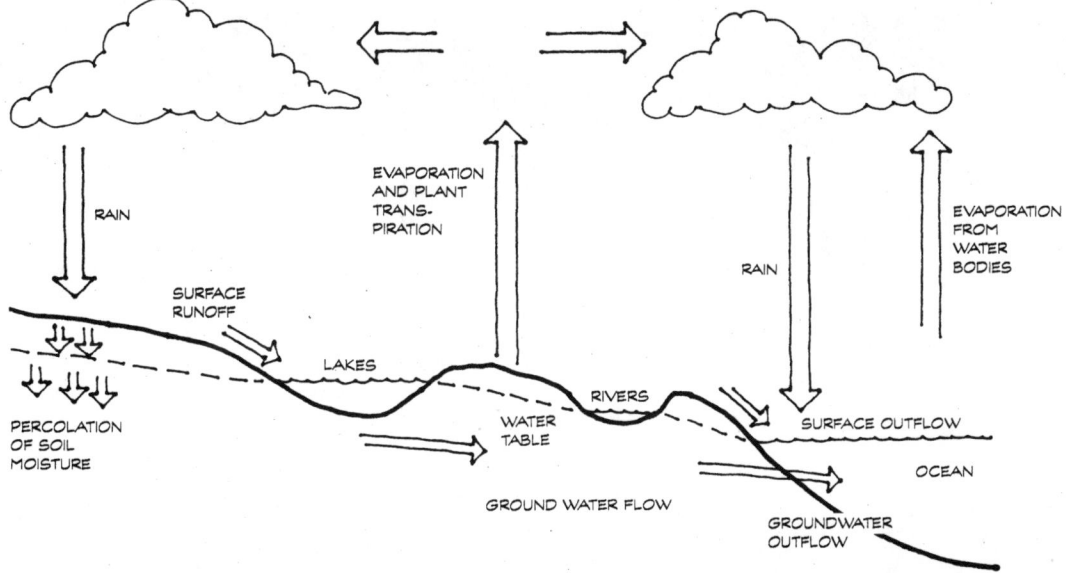

Figure 2.18 The hydrologic cycle.

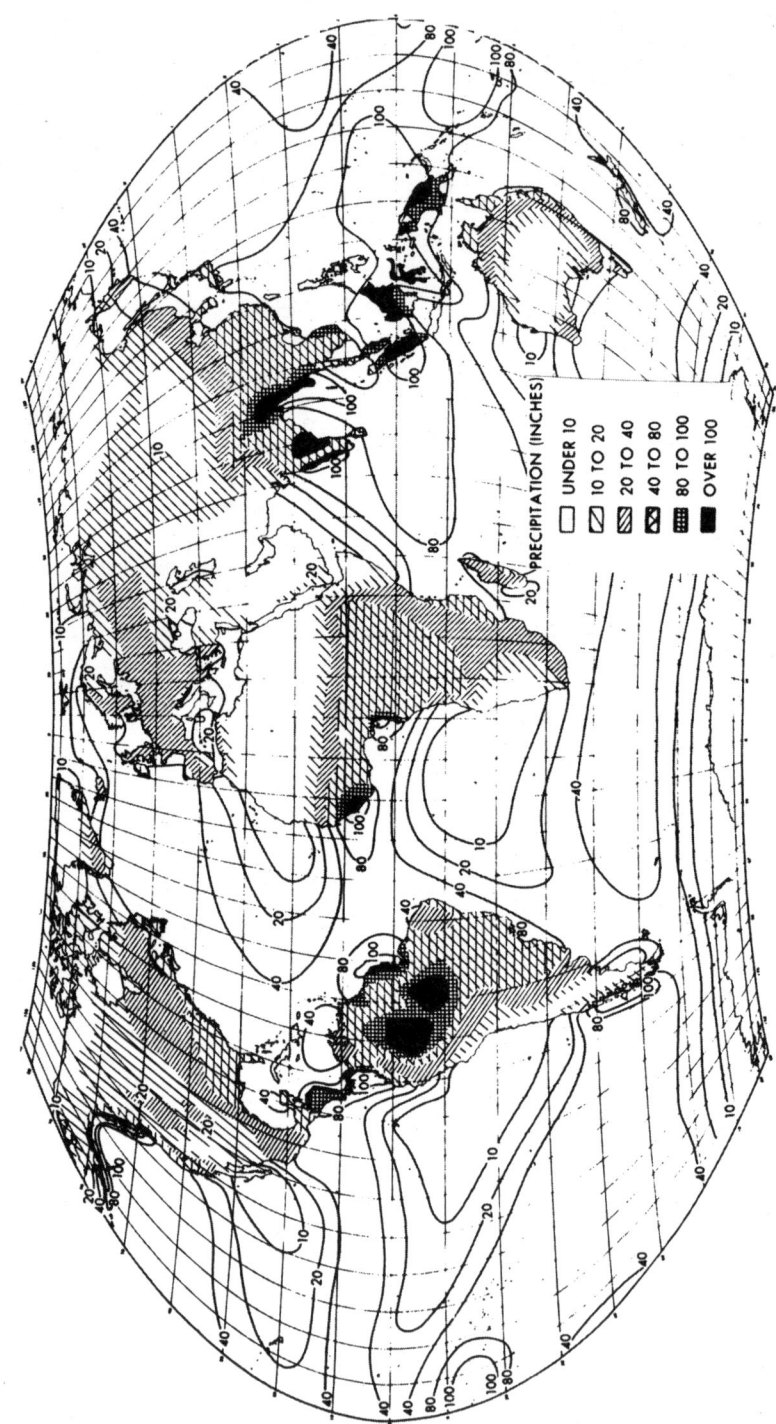

Figure 2.19 Rainfall maps.

falling on the snow will increase its density even further. An inch of snow cover on a roof or other horizontal building surface weighs about 1 to 1½ psf, depending on how much compaction it has undergone. When snow melts, it can release large quantities of water very rapidly, or it can cause slow seepage of moisture into a roof or the top of a parapet wall.

Average annual snowfall in some areas of the United States is more than 6 ft, and maximum annual amounts sometimes exceed 10 ft in the Rocky Mountains, Cascades, and Sierras (*Fig. 2.20*). Large lakes such as the Great

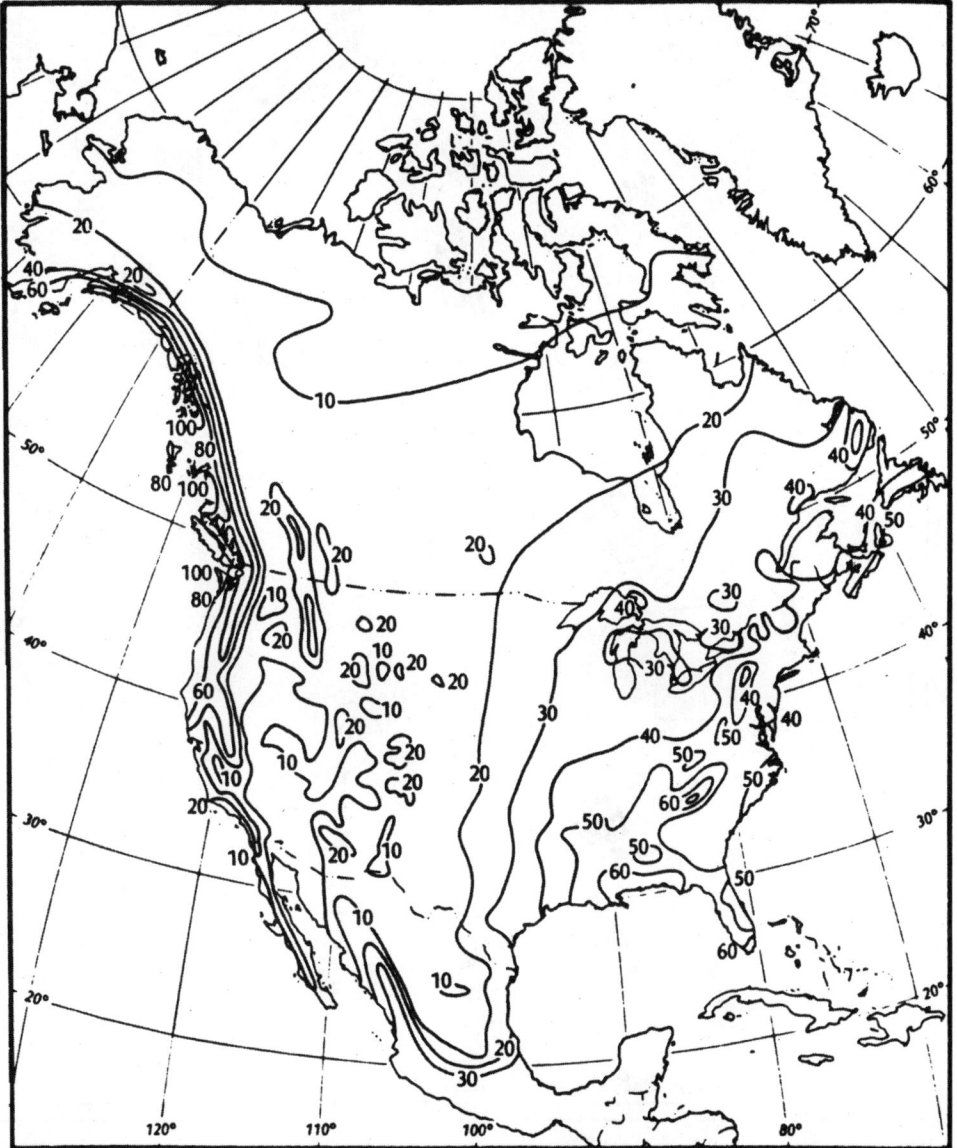

Figure 2.19 (*Continued*)

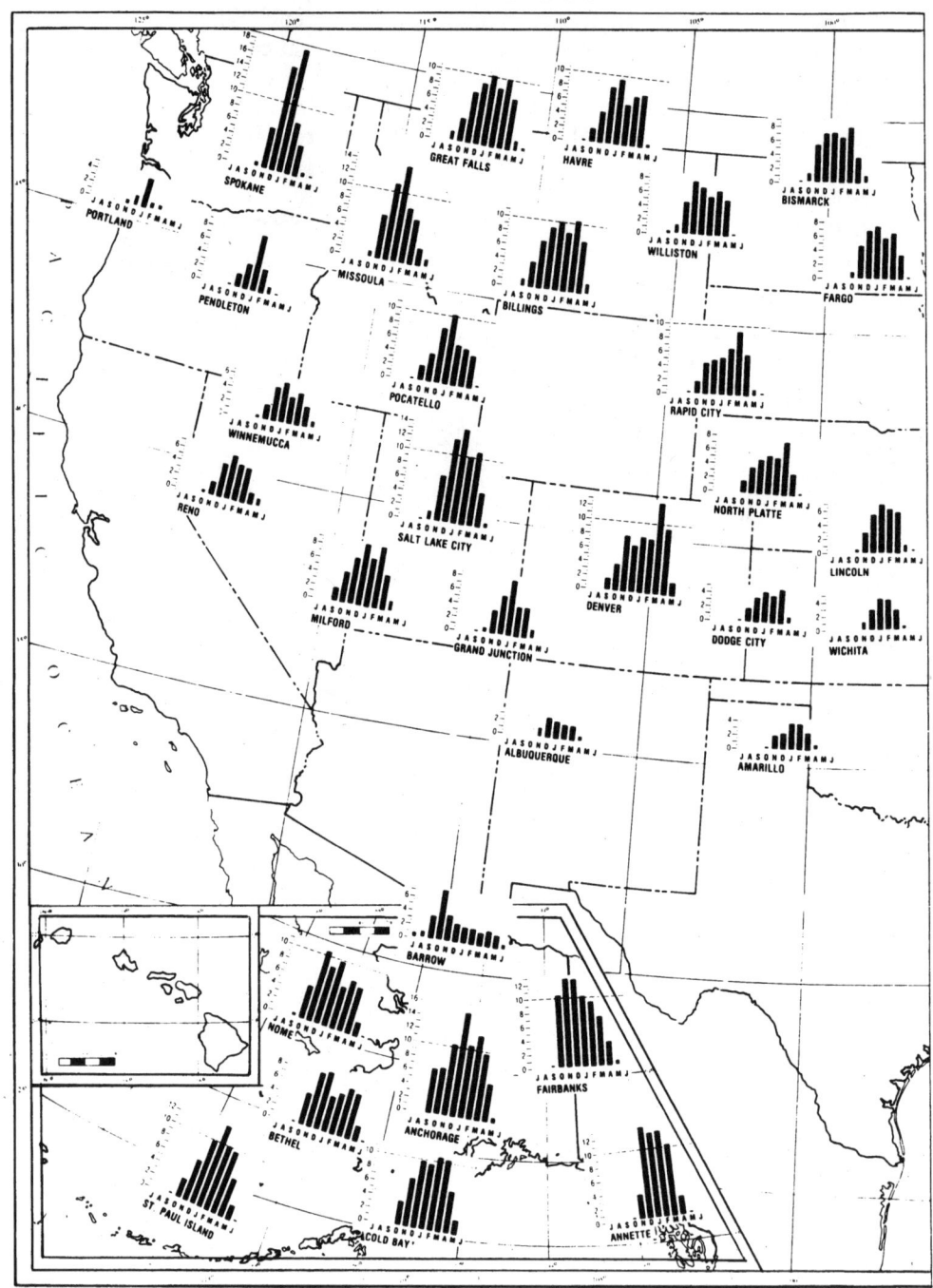

Figure 2.20 Mean monthly snowfall in the United States (inches).

Climate, Weather, and Microclimate 45

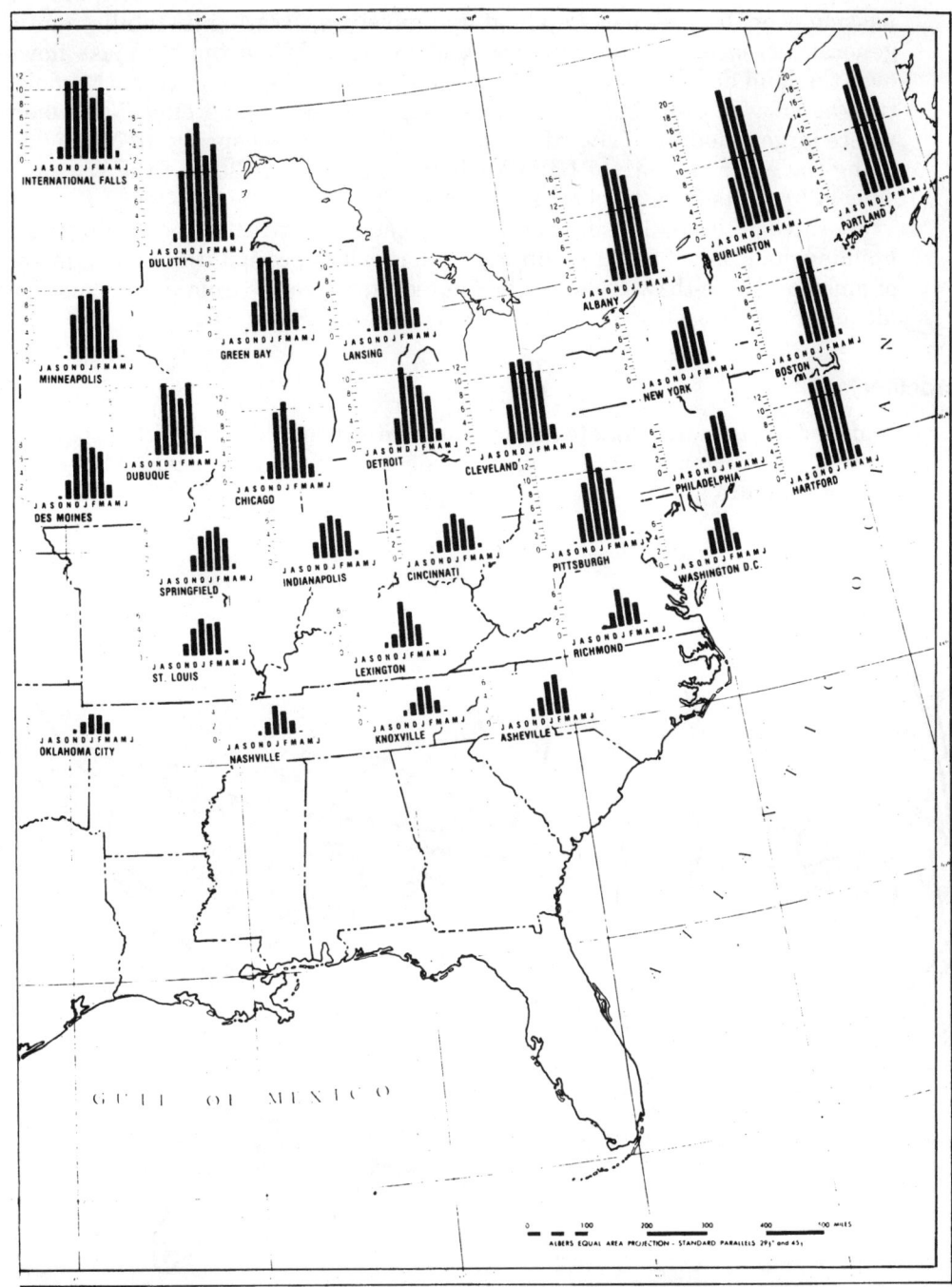

Figure 2.20 (*Continued*)

Lakes affect the amount of snowfall over land downwind from the lake. Throughout the autumn and winter, the lake is a source of heat. Cold air passing over the lake picks up heat and moisture, creating instability and a general increase in cloudiness and precipitation. When the air mass flows over the land downwind from the lake, it cools and dumps much of the moisture as snow. Downwind from Lake Erie, there is a "lake effect" snowbelt where large amounts of snow fall during the winter and spring. In Michigan, along the eastern shore of Lake Michigan, as much as 30% of the seasonal snowfall results from the lake's influence.

Repeated cycles of freezing and thawing can be destructive to buildings and building materials if they remain wet from previous precipitation. The number of ambient freeze-thaw cycles in the United States varies from fewer than five along the Gulf Coast to more than 130 in the west (*Fig. 2.21*).

2.4 Microclimate

A microclimate is the climate in the immediate vicinity of an object, organism, or building. Microclimates can vary considerably from the general climate of an area and are affected by

- Elevation
- Topography

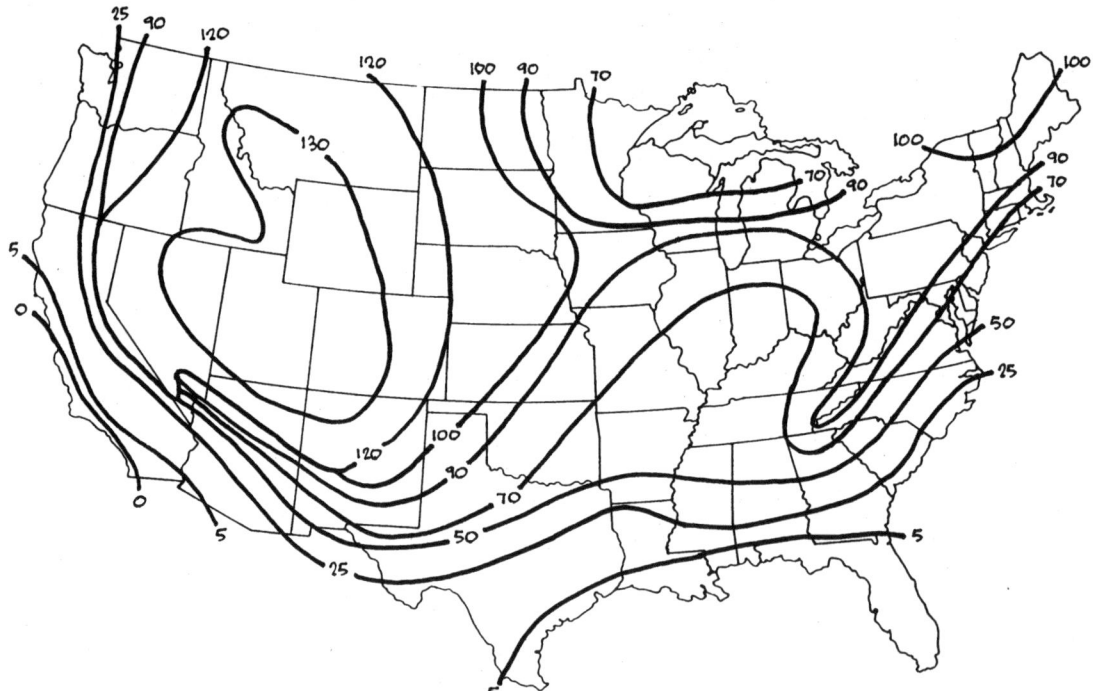

Figure 2.21 Annual number of freeze-thaw cycles for various parts of the United States.

- Orientation
- Slope
- Adjacent conditions
- Proximity to large bodies of water

Urban microclimates have temperature, humidity, wind speed, and precipitation characteristics that are significantly different from the rural areas surrounding them (*Table 2.2*). The warmer temperatures associated with urban areas are attributable to reductions in the natural tree canopy and extensive use of dark, heat-absorbing materials in buildings, parking lots, and roads. This phenomenon, called the *urban heat island,* can be mitigated through preservation of the natural landscape, development of parks, implementation of aggressive planting programs, and the use of light, reflective colors in buildings (refer to Chap. 3).

2.4.1 Orientation

In temperate climates such as that of the continental United States, orientation significantly affects microclimate, so that each side of a building has its own unique set of conditions. For example, south-facing building elevations are much warmer than those facing toward the north. This means that the microclimate on the south side of a building might be warm and dry in winter, while the north side is damp and shady and the regional climate is generally cold and wet. Southern exposures, even in mild climates, can develop summer surface temperatures well in excess of 100°F, creating adverse effects on materials such as coatings and sealants. Geographical orientation also affects the number of microclimatic freeze-thaw cycles experienced. In the example shown in *Fig. 2.22*, the measured temperature of the brick on the south wall rose above and fell below 32°F, causing a freezing and thawing cycle in the brick even though the air temperature remained below 32°F, as did the temperature of the brick on the north wall. During the three winter months, the north wall of one test building was subjected to only 81 freeze-thaw cycles while the south wall experienced 108 cycles (*Table 2.3*). Winters with below-normal temperatures would show the greatest difference in freeze-thaw cycling between north and south exposures.

2.4.2 Slope

In addition to geographic orientation, the slope of a surface affects the amount of radiant heat absorbed (*Fig. 2.23*). Solar radiation on a horizontal surface can elevate the temperature of the surface 85 to 90°F above the air temperature. At night, especially when the sky is clear, horizontal (and nearly horizontal) surfaces lose heat by radiational cooling. The surface temperature may drop 18°F below the air temperature by the early morning hours. As the air temperature drops, the relative humidity near the surface rises and dew may form on the cool surfaces. The combination of daytime solar heating and nighttime

TABLE 2.2 Climatic Changes in Urban Microclimates

Element	Compared to rural area
Contaminants	
Condensation nuclei	10 times more
Particulates	10 times more
Gaseous admixtures	5–25 times more
Radiation	
Total on horizontal surface	0–20% less
Ultraviolet, winter	30% less
Ultraviolet, summer	5% less
Sunshine duration	5–15% less
Cloudiness	
Clouds	5–10% more
Fog, winter	100% more
Fog, summer	30% more
Precipitation	
Amounts	5–15% more
Days with less than 5 mm	10% more
Snowfall, inner city	5–10% less
Snowfall, lee of city	10% more
Thunderstorms	10–15% more
Temperature	
Annual mean	1–5°F more
Winter minimums (average)	2–4°F more
Summer maximums	2–5°F more
Heating degree days	10% less
Relative humidity	
Annual mean	6% less
Winter	2% less
Summer	8% less
Wind speed	
Annual mean	20–30% less
Extreme gusts	10–20% less
Calm	5–20% more

SOURCE: Adapted from Landsberg, *Weather, Climate and Human Settlements,* Special Environmental Report 7, Geneva, World Meteorological Organization.

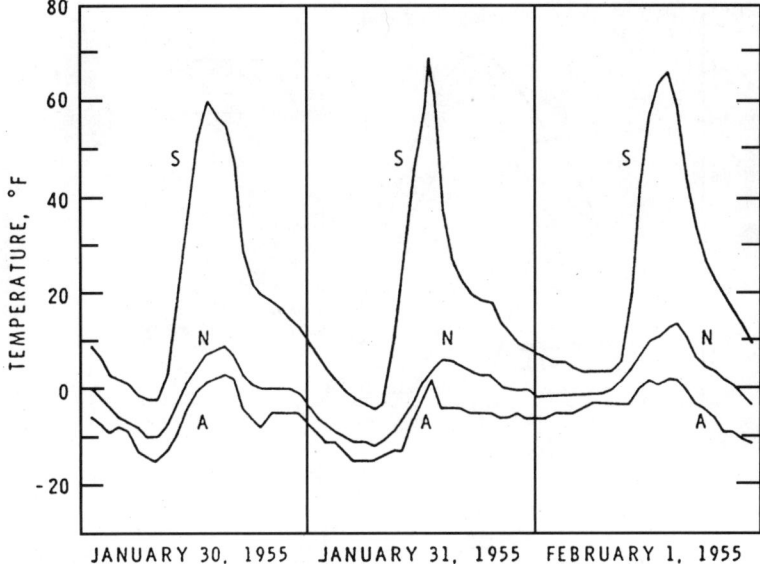

Figure 2.22 Effect of solar radiational heating on the surface temperature of a south-facing brick wall (S) compared with a north-facing brick wall (N) and the ambient air temperature (A). (*From Latta, Walls, Windows and Roofs for the Canadian Climate.*)

TABLE 2.3 Geographical and Directional Effects on Freeze-Thaw Cycles of Brick

Brick facing	Number of freeze-thaw cycles in one winter	
	Ottawa	Halifax
North	65	81
East	70	83
South	98	108
West	79	88

radiant heat loss significantly increases the maximum and minimum temperature range of horizontal surfaces such as roofs and plaza decks compared to the adjacent vertical surfaces of a building wall. Horizontal, south sloping, and south facing surfaces can vary 150°F or more in temperature from summer to winter, with an accompanying air temperature variation of only 50°F. Such wide surface temperature fluctuations affect thermal expansion and contraction characteristics and the subsequent size and spacing of movement joints.

2.4.3 Wind

Elevation generally causes a decrease in temperature and an increase in wind speed. Wind turbulence is lowest at the bottom of a hill and highest at the top.

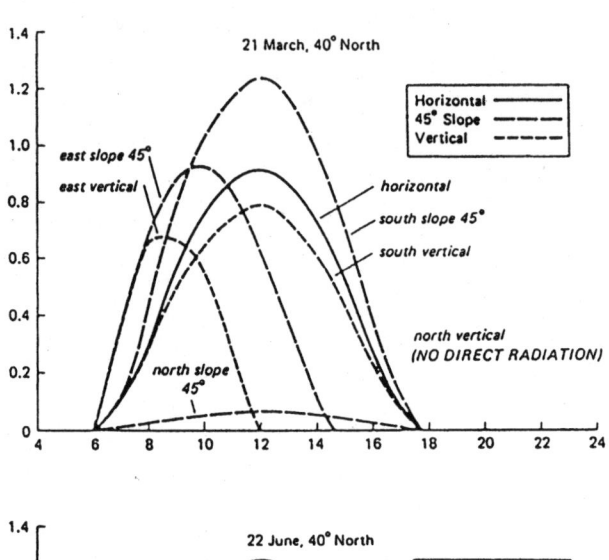
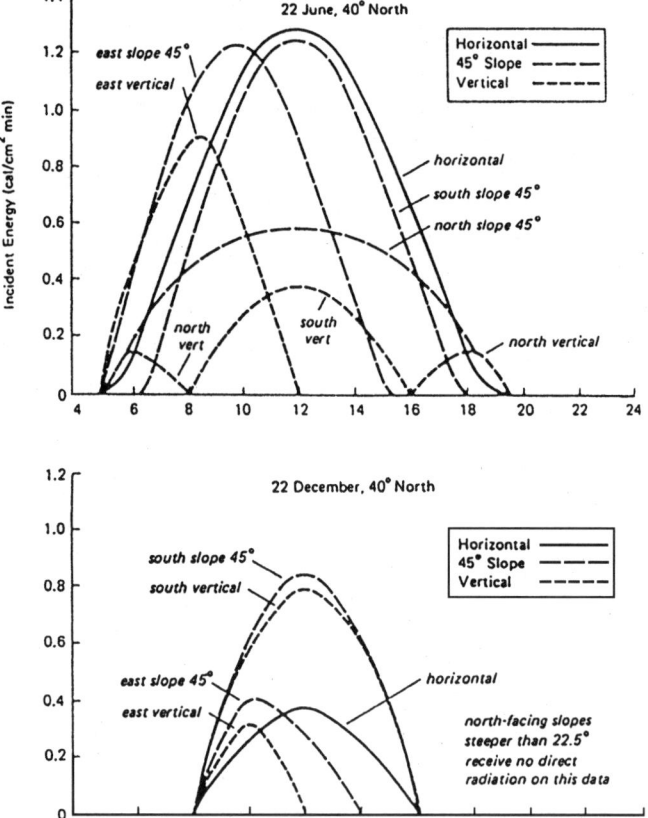

Figure 2.23 Incident radiant energy on vertical, horizontal and sloping surfaces. (*From Latta, Walls, Windows and Roofs for the Canadian Climate.*)

Wooded areas reduce wind speeds, and high-rise buildings create their own air flow patterns because of shape and temperature differences between the sunny and shady sides. The resulting air currents are often increased by the funnel effect created between two buildings. This means that the microclimate at the top of a high-rise building is different from that at the base. Increased wind speed means increased risk of rain penetration and roof damage, so buildings that have high microclimatic wind exposure on the side of a hill may require different design strategies from those on low tree-sheltered sites.

2.5 Evaluating Risk

General weather information can sometimes be translated into specific product or performance requirements. For example, winter temperatures and precipitation amounts were used to develop weathering indexes for the selection of weathering grades for clay face brick (*Fig. 2.24*). The timber industry used temperature and humidity data to develop an index of the potential for decay in wood structures (*Fig. 2.25*). And annual mean relative humidity is related to the allowable moisture content and shrinkage potential of concrete masonry units and the required spacing of control joints (*Fig. 2.26* and *Table 2.4*). But the use of climatic and weather data in the design of buildings is not always so precise. More often than not, the information merely provides general guidance as to the severity of climatic exposure. Wind-driven rain usually presents the highest risk of moisture penetration.

Rain driven by the force of wind is a formidable enemy. Under moderate wind conditions, rain that strikes a vertical building surface cascades down the face of the wall to the ground. Small cracks and openings that would otherwise be of little concern become vulnerable to moisture penetration when winds push that cascading sheet of water inward and sometimes upward. When wind creates pressure differentials between the inside and outside of the wall, this additional force acts to pull moisture through even the smallest of openings.

Wind exposure is directly related to the risk of moisture penetration. A building perched on the side of a hill has greater exposure to wind, and therefore to wind-driven rain, than one nestled in a depression or sheltered by adjacent buildings or terrain, and a high-rise building generally has greater exposure than a low-rise building (*Fig. 2.27*). Most buildings though, at one time or another, are subjected to heavy rain driven by strong winds. Buildings at low risk in dry climates should provide basic protection from damage caused by occasional moisture penetration and condensation as effectively as those at high risk in wet climates. Higher-risk exposures, though, present a greater potential for serious moisture problems, and building assemblies will generally be less forgiving. Exterior envelope designs for high-risk exposures should include redundant levels or layers of moisture protection, in effect adding a factor of safety as is common in structural design.

Weather exposures in terms of wind pressure and annual rainfall vary in the United States from severe along the Atlantic and Gulf coasts to moderate

TABLE (a) PHYSICAL REQUIREMENTS

Designation	Minimum compressive strength psi, (MPa) gross area		Maximum water absorption by 5-hr boiling (%)		C/B Maximum saturation coefficient*	
	Average of 5 brick	Individual	Average of 5 brick	Individual	Average of 5 brick	Individual
Grade SW	3000	2500	17.0	20.0	0.78	0.80
Grade MW	2500	2200	22.0	25.0	0.88	0.90

*The saturation coefficient is the ratio of absorption by 24-hr submersion in cold water to that after 5-hr submersion in boiling water.

TABLE (b) GRADE REQUIREMENTS FOR FACE EXPOSURES

Exposure	Weathering Index (Explanatory Note 1)	
	Less than 50	50 and greater
In vertical surfaces		
In contact with earth	MW	
Not in contact with earth	MW	
In other than vertical surfaces		
In contact with earth	SW	
Not in contact with earth	MW	

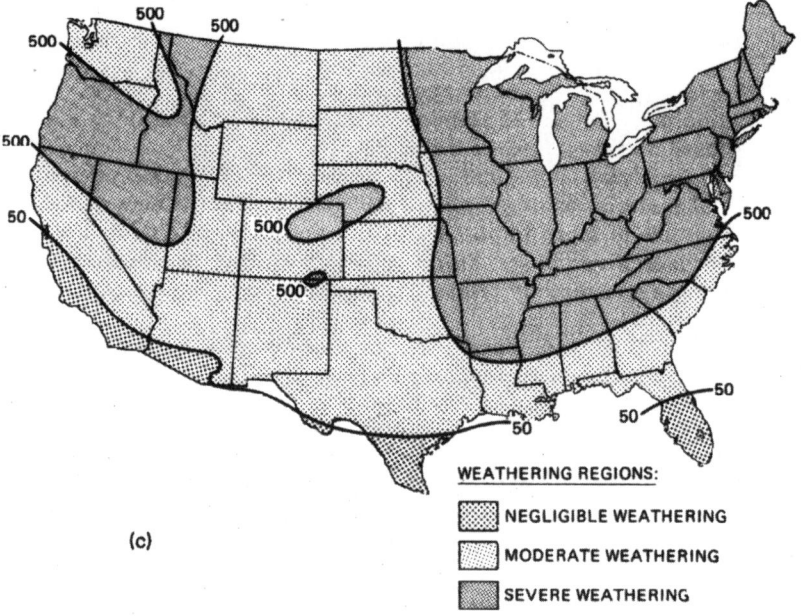

(c)

Figure 2.24 Weathering index for grades of face brick. [*From ASTM C216, Standard Specification for Facing Brick (Solid Masonry Units Made from Clay or Shale), ASTM, 100 Barr Harbor Drive, West Conshohocken, PA. Reprinted with permission*]

Climate, Weather, and Microclimate   53

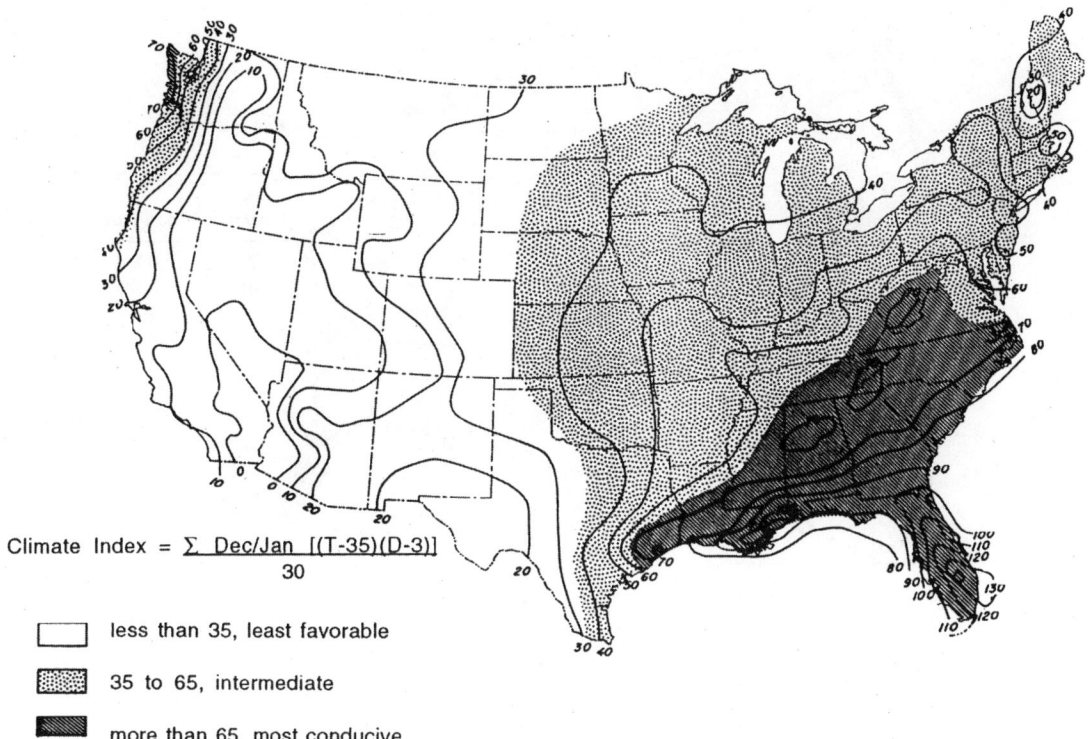

Climate Index = $\dfrac{\sum \text{Dec/Jan }[(T-35)(D-3)]}{30}$

▢ less than 35, least favorable

▨ 35 to 65, intermediate

▮ more than 65, most conducive

Figure 2.25 Climatic influence on biologic activity—Scheffer's index of potential for decay of "off the ground" wood structures. (*From Olin, Construction Principles, Materials, and Methods, 5th ed.*)

in the west and southwest (*Fig. 2.28*). Some measure of the combined effect of wind and rain is obtained by combining average annual rainfall amounts with average wind speed data into a *driving rain index,* in which exposures are classified as sheltered, moderate, or severe (*Fig. 2.29*) for buildings located more than 5 miles from a large body of water. All unprotected building corners, and all buildings within 5 miles of an ocean, sea, large lake or estuary are considered to have severe exposure. "Protected" walls are described as those where permanent buildings or terrain effectively block wind-driven rain. In areas with severe driving rain exposure, moisture protection systems should provide a factor of safety that may not be needed in climates where rains are less frequent and of shorter duration, and where there are longer drying periods between rains. As the risk of moisture penetration increases with exposure, or the opportunity for normal drying decreases, the design should add redundant protection. For example, window systems in high-risk exposures should include internal drainage provisions *in addition to* the so-called barrier seals at the perimeter of the glazing.

It is not possible to calculate resistance to moisture penetration or wind-driven rain in the same precise way that we can calculate resistance to temperature changes. The risks involved in climatic exposure can be generally

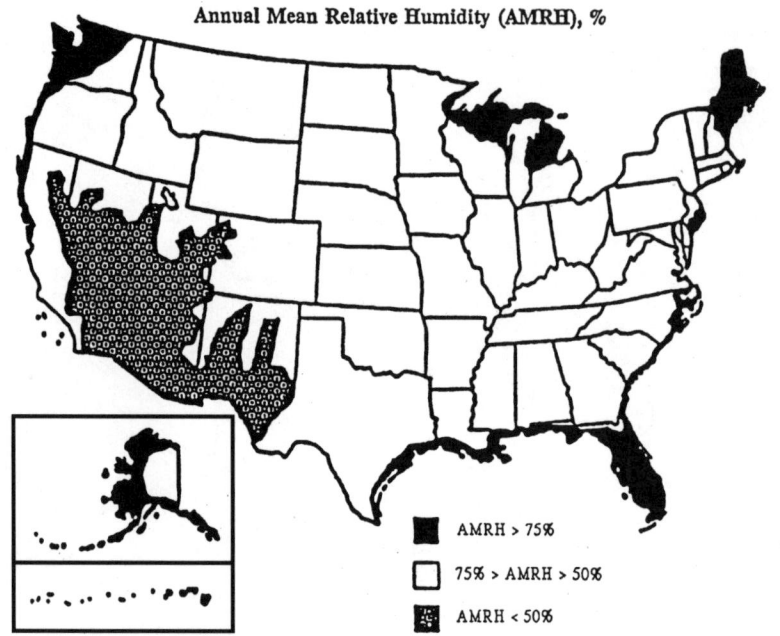

Figure 2.26 Relationship between annual mean relative humidity and the allowable moisture content and shrinkage potential of concrete masonry. (*From ASTM C90, Standard Specification for Loadbearing Concrete Masonry Units, ASTM, 100 Barr Harbor Drive, West Conshohocken, PA Reprinted with permission, and ACI/TMS/CMR Masonry Designers' Guide.*)

determined from data such as the driving rain index and an understanding of microclimatic influences. Evaluating those risks and providing the appropriate degree of weather protection, however, depends on experience supplemented by good judgment. An awareness of the variation in risk and a knowledge of recommended practices for different building systems is the foundation of good design.

Maximum Horizontal Spacing of Vertical Control Joints in Concrete Masonry Walls, Ft

Average annual relative humidity	Wall location	Vertical spacing of bed joint reinforcement, inches	Type of CMU, ASTM C90	
			I Moisture controlled	II Nonmoisture controlled
Less than 50%	Exerior	None	12	6
		16	18	10
		8	24	14
	Interior	None	16.5	9
		16	24	14
		8	31.6	19
Between 50 and 75%	Exterior	None	18	12
		16	24	16
		8	30	20
	Interior	None	22.5	15
		16	30	20
		8	37.6	25
Greater than 75%	Exterior	None	24	18
		16	30	22
		8	36	26
	Interior	None	28.5	21
		16	36	26
		8	43.6	31

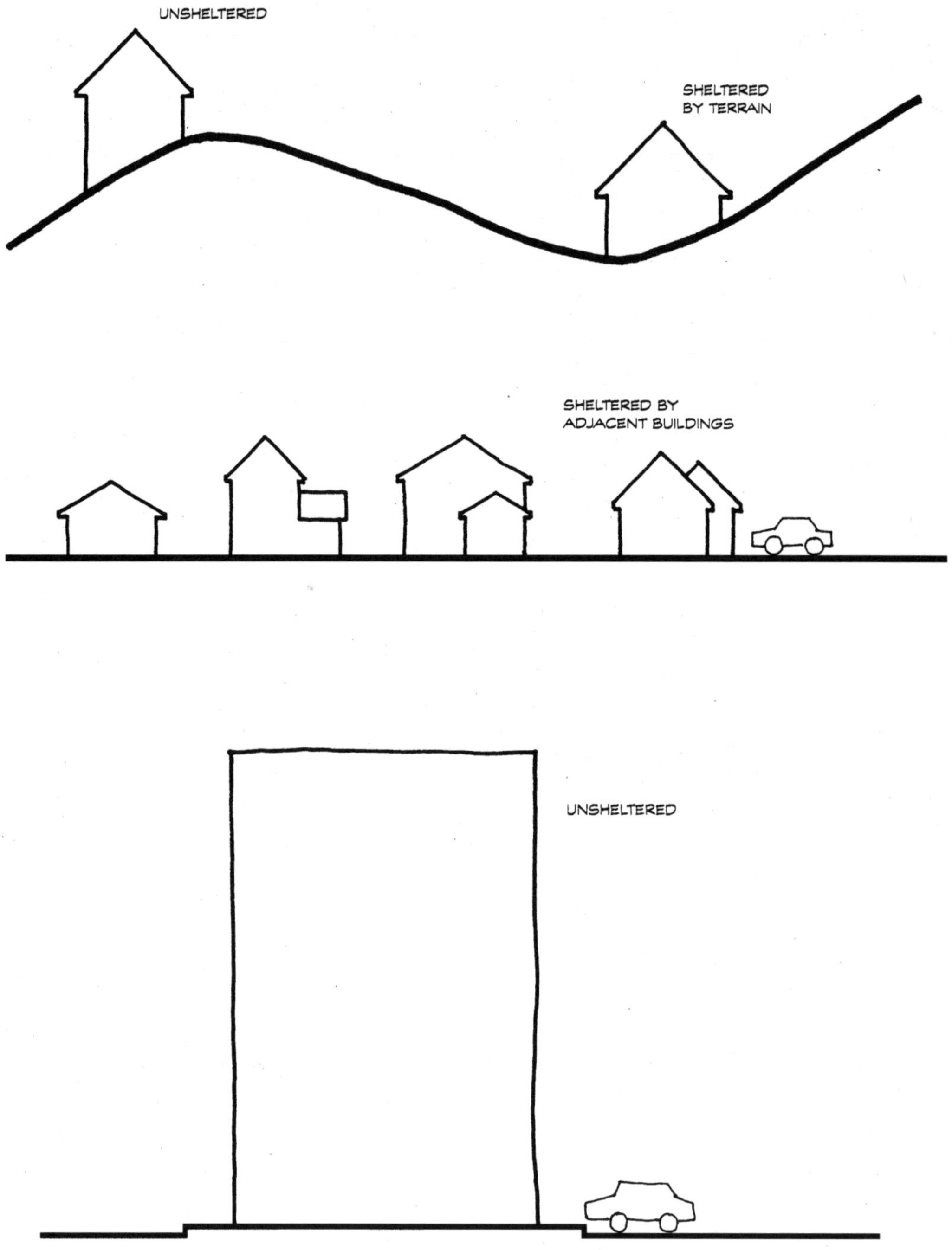

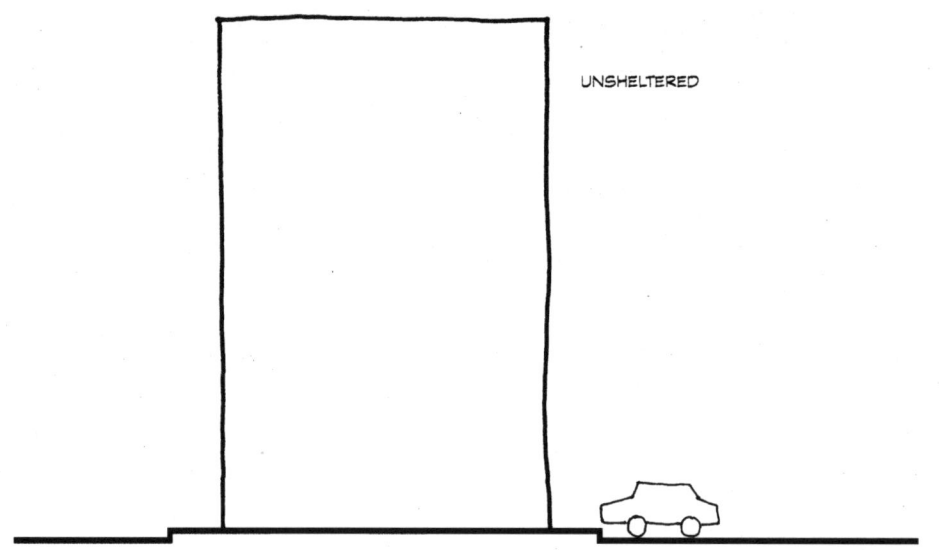

Figure 2.27 Exposure to wind-driven rain is affected by surroundings.

Climate, Weather, and Microclimate 57

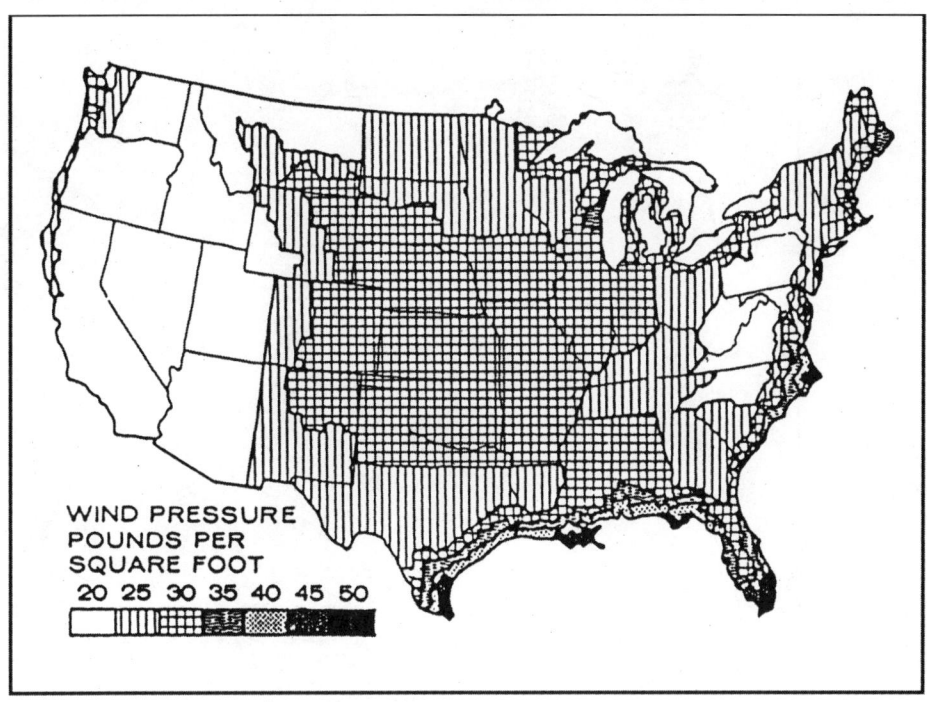

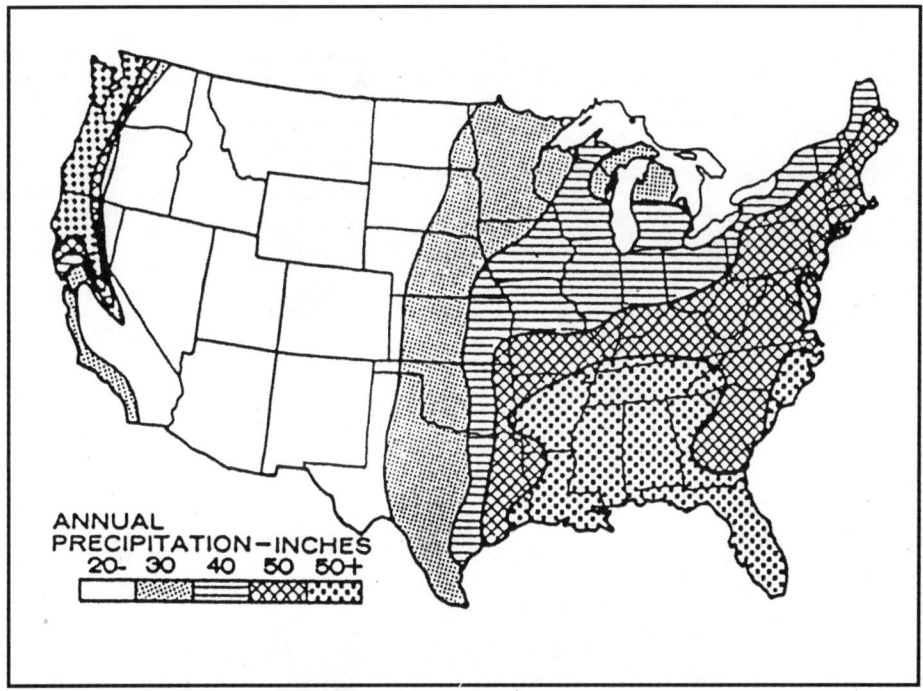

Figure 2.28 Wind pressure and rainfall maps of the United States. (*From BIA Tech Note 7, Brick Institute of America, 11490 Commerce Park Drive, Reston, VA.*)

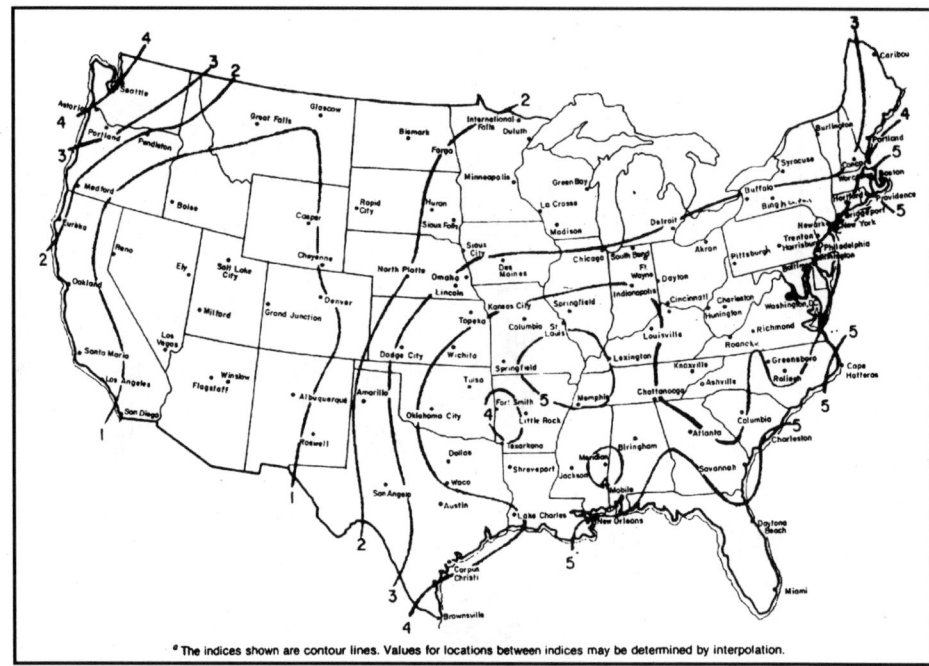

^a The indices shown are contour lines. Values for locations between indices may be determined by interpolation.

Driving Rain Index

Driving rain index†		Wall standing above surroundings			
		Yes (unprotected)‡		No (protected)‡	
		Wall near facade edge§		Wall near facade edge§	
Greater than	Less than	Yes	No	Yes	No
0	1.5	Severe	Moderate	Sheltered	Sheltered
1.5	3.0	Severe	Moderate	Moderate	Sheltered
3.0	5.0	Severe	Severe	Severe	Moderate
5.0	—	Severe	Severe	Severe	Severe

WALL EXPOSURE TO WIND-DRIVEN RAIN*

*Exposures are for walls on buildings located more than 5 miles (8 km) from a sea, large lake, or estuary. All walls on buildings located 5 miles (8 km) or less from a sea, large lake, or estuary have a severe exposure.

†See map.

‡A wall might be considered protected where permanent buildings or terrain face the wall in all directions and have a height above the top of the wall of more than 1.2 times their individual distances from the wall or where there is a permanent solid wall overhang at the top of the wall having a width of at least 85% of the wall height.

§Near facade edge is within one-tenth of the facade width from a corner or one-tenth of the facade height from the top.

Figure 2.29 Driving rain index for the United States. (*Grimm, "A Driving Rain Index for Masonry Walls," in Masonry: Materials, Properties, and Performance, ASTM STP 778, J. G. Borchelt, Ed., American Society for Testing and Materials, 1982. Reprinted with permission.*)

Chapter 3

Heat Flow and Insulation

Building envelopes must control the flow of heat not only to maintain comfortable conditions and reduce operating costs, but also because heat flow is related to water vapor movement, condensation, and the expansion and contraction of building materials and sealant joints.

3.1 Heat and Cold

Heat moves from one location to another. Whenever there is a difference in temperature, heat flows from the warmer area to the cooler area. Thermal energy can flow from one object to another by radiation, conduction, convection, and evaporation. The climatic factors of significance in the thermal energy flow into and out of buildings are solar radiation, air temperature, wind, and humidity.

Wind forces the movement of air around buildings. If the air is cold, the walls and roof lose thermal energy by conduction and convection, and if the air is hot, they absorb thermal energy by the same process. Solar radiation is absorbed by building surfaces, and radiant heat emitted from the surfaces is absorbed by carbon dioxide and water vapor molecules in the air. Wind, stack effect, and temperature differentials move heat and moisture around and within a building.

3.1.1 Thermal conduction and convection

Conduction is the flow of energy through a liquid, solid, or gas, or the direct transfer of heat from one material to another. Heat energy is transferred from particle to particle by the movement of molecules, and temperature is a measure of the intensity of that movement. Warm molecules move faster than colder ones. As they collide with one another, the faster molecules are slowed down and the slower ones are accelerated until all are moving at the same speed and the thermal energy is equalized at the same temperature throughout the material. Solids are good conductors. Liquids can be good conductors, but when the liquid is free to move, it will transfer heat more effectively by convection. Gases are very poor conductors but will transfer heat by convection.

Whenever there is a temperature difference between the opposite faces of a wall or roof section, heat flows by conduction from the warm side to the cool side. Heat can flow from one material in an assembly to another by thermal conduction if the materials are in intimate contact—the studs and sheathing in a wall, for example. Heat can also flow from a solid material such as siding or wallboard to the boundary layer of air adjacent to it.

Thermal *convection* is the transfer of heat by the movement of a fluid such as air or water. Since a mass of the fluid must move from one place to another in order to transfer the heat energy, there must be a force to cause this movement. If the force is provided mechanically by a pump or fan, the movement is called *forced convection*. If the movement of the fluid or air mass is generated by differences in temperature and density, it is called *natural convection*. Convective heat transfer cannot take place without air or fluid movement, but air or fluid movement can occur without heat transfer.

When air is heated, it expands, and this added buoyancy causes it to rise. Conversely, when air is cooled it becomes more dense and falls, displacing the warmer air below it. It is these thermal convective forces which create the chimney or *stack effect* in buildings. Warm air rises to the top of a building through stairwells, elevator shafts, plumbing chases, and other passages and is replaced by cooler air at the lower floors. Without proper control of air movement through the building envelope, upper-level exfiltration and lower-level infiltration can lead to serious condensation problems (see Chap. 5 for more information).

The movement of air over a surface involves both conductive and convective heat exchange. The actual heat exchange occurs between the solid surface and the air by conduction, but once the air is heated or cooled, it is convected away. Increasing air speed increases the rate of heat transfer, but the direction of heat flow depends upon the relative temperatures. If the air is warmer than the surface, heat will flow from the air to the surface by conduction. If the air is cooler than the surface, heat will flow from the surface to the air by conduction and then be convected away. Air movement is therefore related to thermal energy flow. The rate of conductive heating of the air depends on how long the air remains in contact with the surface—the shorter the time, the faster the rate of transfer. Air rapidly moving over a surface gives off or takes on much more heat than a layer of still air that clings to the surface. This phenomenon is the basis of the "wind chill factor" reported so often in weather reports.

3.1.2 Evaporation

A large amount of heat is absorbed by water in the phase change from liquid to vapor. Known as *latent heat,* this thermal energy is held in the vapor and released only when the phase change reverses through condensation.

Although water must be heated to 212°F before it will boil and create steam, evaporation can actually take place at any temperature. The evaporation of ice or snow at temperatures below freezing is called *sublimation*. Evaporation rates are governed by both air speed and vapor pressure. Increasing air speed always increases evaporative cooling, although at high vapor pressures, the effect may be small. Evaporation rates are higher when the vapor pressure (and therefore the relative humidity) is low. High vapor pressures and high relative humidities inhibit evaporation.

Heat and moisture move through building envelopes in vapor form, and although the cooling and drying of building surfaces by evaporation may often be desirable, condensation of vapor on or within the enclosure usually is not.

3.1.3 Thermal radiation

Radiation is the flow of energy in the form of electromagnetic waves of different length known as light, ultraviolet radiation, and infrared heat. Radiation that is propagated by temperature differences is called *thermal radiation*. All objects gain and lose energy continuously by radiation. Radiant heat can be transferred through space without warming the air in between. The outside surface of a building receives radiant solar energy independent of any temperature difference between the surface and the air. *Figure 3.1* shows the radiant energy flow between a building and its environment.

Unlike conduction and convection, radiant heat exchange is impeded by any medium interposed between the source and the object. Much of the sun's radiation is absorbed or dispersed by the atmosphere before it reaches the earth's surface. The ultraviolet radiation is absorbed by oxygen, ozone, and nitrogen, the infrared radiation by carbon dioxide and water vapor. The total incident solar radiation on a building's surface consists of direct, diffuse, and reflected radiation. Direct radiation penetrates the atmosphere in parallel rays. Diffuse radiation results from the scattering of sunlight among dust particles, water molecules, and other particles in the air. A building surface may also receive appreciable amounts of solar radiation reflected from adjacent roofs and walls, and other surrounding buildings or objects. *Table 3.1* shows the amounts of direct and diffuse radiation incident on a horizontal surface under various atmospheric conditions. *Insolation* (*in*cident *sol*ar radi*ation*) is a measure of total solar irradiation intensity, expressed in terms of $Btu/ft^2 \cdot h$. *Solar flux* is the total of direct and diffuse radiation received on a given surface.

Solar radiation is at a maximum when the sun is directly overhead, and decreases as the elevation of the sun above the horizon decreases, so latitude and time of year affect the potential for radiant heat energy on a building (Fig. 3.2). The slope of a surface and its orientation also affect the exchange of radiant heat energy. A north-facing wall receives a different amount of incident sunlight than do south-, east-, and west-facing walls. Roofs or walls of differing slopes also receive different amounts of incident sunlight. Figure 3.3 shows the amount of direct solar radiation that is received on various slopes as a function of the time of day at 40° north latitude for three dates during the year—the winter solstice, spring equinox, and summer solstice. The autumn equinox is the same as the spring. Such microclimatic radiational differences affect the way a building performs. Snow will accumulate and lay longer on

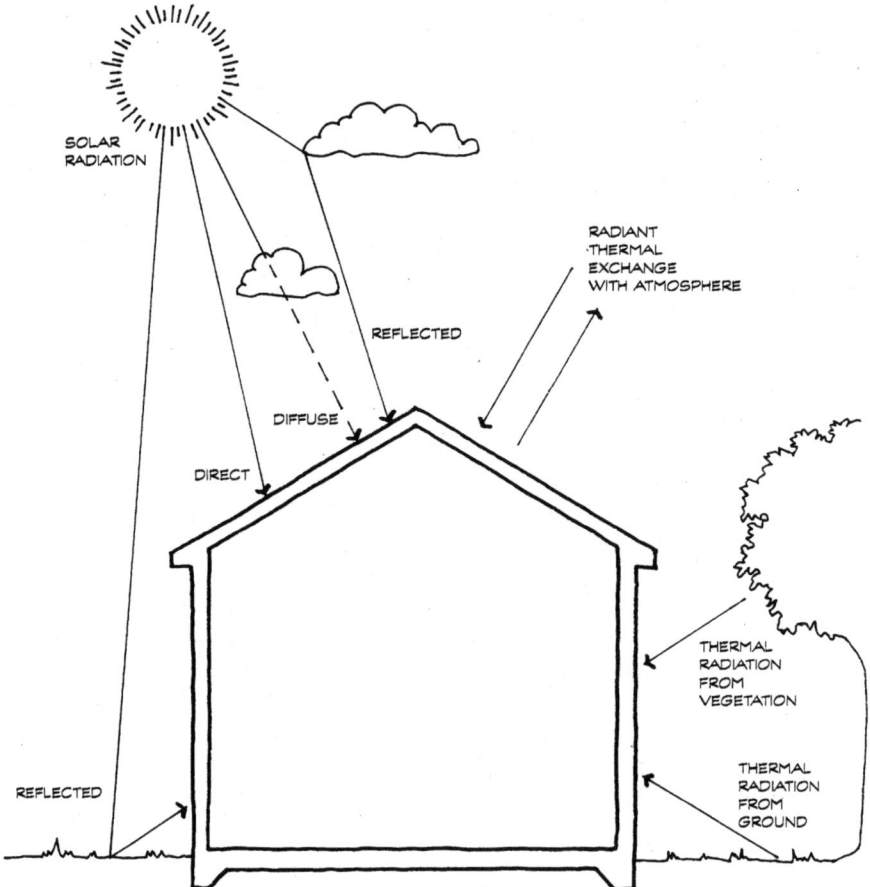

Figure 3.1 Radiant energy flow between a building and its environment.

TABLE 3.1 Diffuse Solar Radiation Incident on a Horizontal Surface

Sky condition	Ratio of actual direct to maximum direct radiation	Ratio of diffuse to maximum direct radiation
Clear	1.00	0.12
Clear, slightly hazy	0.80	0.25
Hazy	0.60	0.35
Overcast	0.40	0.55

SOURCE: Watson and Labs (1983).

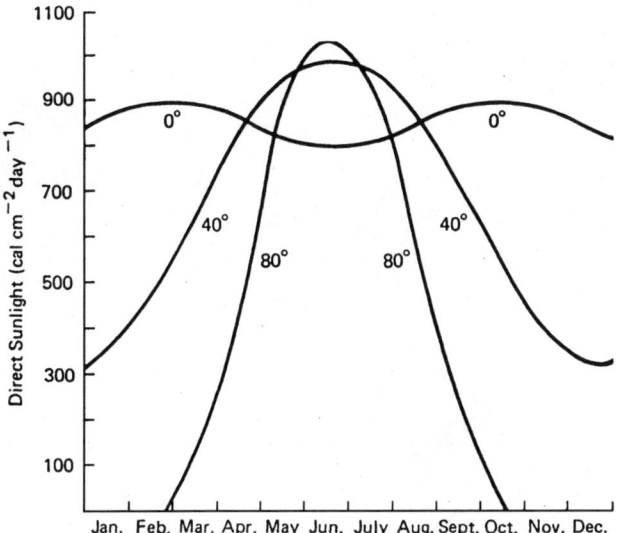

Figure 3.2 Amount of direct sunlight incident on a horizontal surface at the outer extremity of the earth's atmosphere as a function of the time of year for latitudes 0°, 40°, and 80°. (*From Gates, Man and His Environment: Climate.*)

north-facing roof slopes than on south-facing ones, subjecting the north-facing roofs to greater risk of moisture penetration. Soil temperatures near the surface will be 18°F different between north- and south-facing earth slopes, affecting heat loss from basement walls. A change in slope toward the north by only 5° will reduce soil temperatures as much as a change of 300 miles latitude to the north. Ivy can grow on a north-facing vertical wall, but not usually on a south-facing wall, holding moisture on the north wall's surface and retarding its evaporation.

The frequency and wavelength of radiant energy is determined by the temperature of the emitting body. The sun is hot so it emits short-wave radiation. Buildings and the objects in and around them are relatively cool by comparison, so they emit long-wave radiation. This difference in wavelengths is partly responsible for the "greenhouse effect." Short-wave solar radiation is transmitted through glass and some of it is absorbed by objects and surfaces inside. Being at a lower temperature, these objects and surfaces re-radiate long-wave energy which cannot pass through the glass and so is trapped inside. The temperature inside builds up until the heat gained by radiation is balanced by the heat lost by conduction and convection, and a state of equilibrium is achieved.

- *Solar absorptance* is the fraction of incident solar energy absorbed by a surface.

- *Solar reflectance* (sometimes called *albedo*) is the fraction of incident solar energy reflected by a surface.

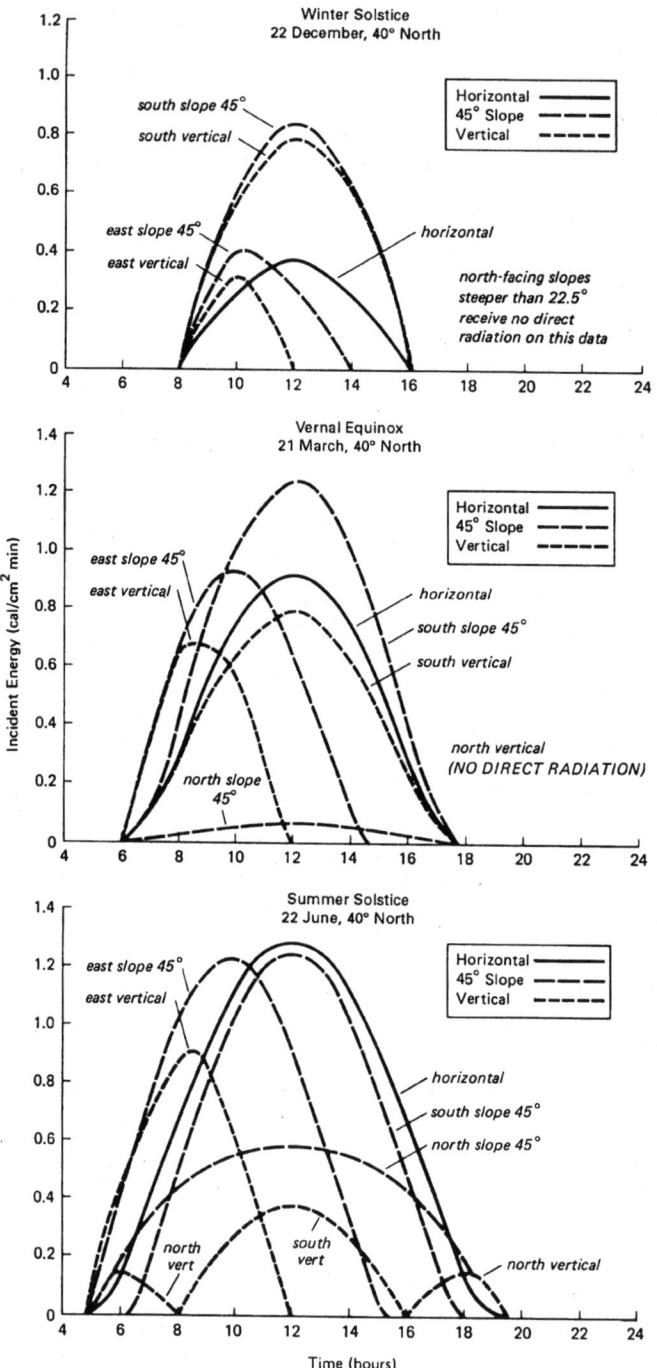

Figure 3.3 Amount of solar radiation incident upon sloping surfaces as a function of time of day at latitude 40° N for the winter and summer solstice and the vernal and autumnal equinox. (*From Gates, Man and His Environment: Climate.*)

- *Emittance, emissivity* is the fraction of the amount of heat radiated (emitted) by a surface at a specified wavelength and temperature to the emittance of an ideal blackbody at the same wavelength and temperature.
- *Solar transmittance* is the fraction of incident solar energy transmitted by a surface.

Solar radiation that strikes an opaque building surface is partly absorbed and partly reflected. Transparent and translucent materials also transmit part of the radiation. Absorptance, reflectance, and transmittance can be expressed as decimal fractions, in which case their sum must equal 1. If they are expressed as percentages, their sum must equal 100%. For opaque surfaces where transmittance is zero, the sum of solar reflectance and solar absorptance must always equal 1. Reflected radiation has no further effect on the reflecting surface, but it becomes part of the incident radiation on, and will raise the temperature of, any other surface onto which it is reflected. Solar radiation elevates temperatures based on the intensity of the irradiation, the mass of the object, the reflectance of its surface, and the balance between the rates of radiant heat gain and conductive and convective heat loss.

The amount of radiant energy absorbed and then re-radiated or emitted by a surface is related to the nature of the surface. An *ideal blackbody* is one that absorbs all the radiant energy which falls on its surface—none is reflected or transmitted. Building materials are not ideal blackbodies, not even those that look black. They absorb only a portion of the radiant energy that strikes them, and either reflect or transmit the remainder. Although different surfaces may have similar absorptance, they do not necessarily dissipate the heat they absorb in the same manner or to the same degree of effectiveness. Within a narrow band of radiation wavelengths, the emittance of a surface will approximate its absorptance and this is the basis of the concept that "a good absorber is a good emitter." But this rule does not apply to radiation received at one frequency (solar) and emitted at another (thermal). The result is that some surfaces have a "heat trapping" effect relative to other surfaces of similar absorbance and reflectance. *Figure 3.4* illustrates this effect. The polished aluminum surface will absorb a lower percentage of the solar radiation it receives, but having a low emittance, it will become hotter than the more absorptive painted surface with a higher thermal emittance. Although the painted surface absorbs more radiation, it is capable of emitting more, so its net heat gain is less. Values of reflectance, absorptance, and emittance for some typical building materials and surfaces are given in *Table 3.2*.

The temperature of a *clear night sky* is an average of 15 to 20°F below that measured near the ground. Horizontal and sloping surfaces like roofs emit radiant heat into this clear night sky until they reach equilibrium with it at temperatures 15 to 20°F below the surrounding ambient air temperature. The magnitude of the difference varies from one climatic region to another in relation to the moisture content of the air. The larger differences between air temperature and horizontal or sloping surface temperatures are characteristic of arid climates, and the smaller differences of humid climates. Clouds negate much of this radiant heat loss effect, but for precise calculations

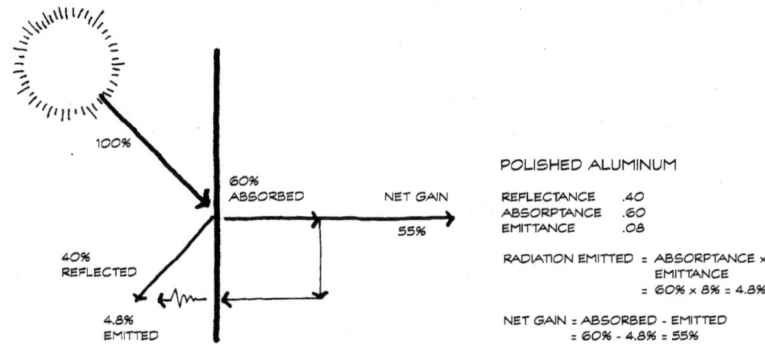

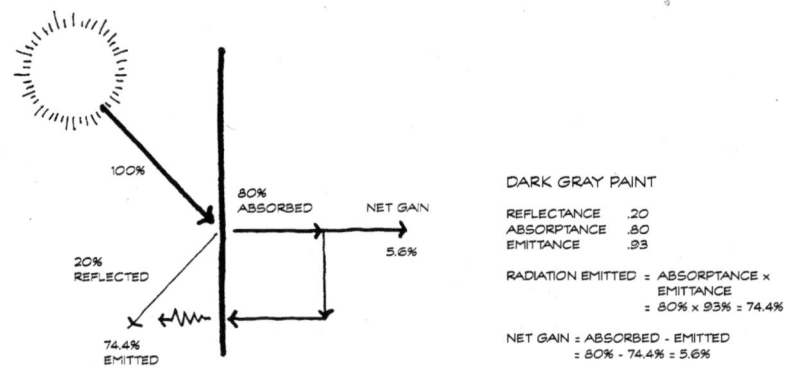

Figure 3.4 Heat trapping effect. (*From Watson and Labs, Climatic Building Design.*)

correction factors can be used to adjust the radiant cooling rate for low-level, midlevel, and high-level clouds. Radiant heat loss can cause condensation to occur on horizontal and sloping surfaces which ambient temperatures and vapor drive calculations would not otherwise indicate.

The absorption of solar energy by a building surface raises its temperature above that of the surrounding air by some degree, depending on surface color, radiation intensity, and the countervailing rate at which its temperature is reduced by the wind and radiant heat loss. The effect of these variables is expressed as a fictitious air temperature which would produce the same rate of heat exchange that occurs with the actual combination of incident solar radiation, radiant heat exchange with the sky and other surroundings, and convective heat exchange with the air. This is called the *sol-air temperature*, and it is dependent on season, time of day, latitude, orientation, and wind speed.

Radiant barriers are now marketed for new and retrofit construction which significantly reduce radiant heat gain in buildings in the summer. Cold climates benefit less from radiant barriers because summer cooling loads are less, and the barrier blocks desirable winter heat gain. Sometimes referred to as *reflective insulation,* radiant barriers are available in sheets and rolls of single- or multilayer construction and in preformed shapes with integral air spaces. The bright-finish aluminized materials reflect as much as 95 to 97% of the incident radiant heat which strikes their surfaces. The same effect can be accomplished with a foil-faced insulation material. New glazing systems with low-emissivity films also help control radiant heat loss through windows during the winter. Radiant barriers are particularly effective in attic or ceiling spaces in hot climates where the majority of heat gain through the roof is due to radiant heat. In a retrofit application, the barrier should be attached to the bottom of the roof rafters. In new construction a radiant barrier film can be draped across the top of the rafters, directly under the sheathing (*Fig. 3.5*). To work effectively, there must be a minimum ¾ in. air space above the barrier.

TABLE 3.2 Reflectance, Absorptance, and Emittance of Various Surfaces

Material	Solar reflectance	Solar absorptance	Thermal emittance, ε
Aluminum, clear finish	0.40	0.60	0.08–0.12
Aluminum paint	0.60	0.40	0.50–0.54
Mineral board, natural	0.25	0.75	
Mineral board, white	0.39	0.61	
Brick, light buff	0.30–0.50	0.50–0.70	0.95
Brick, red	0.15–0.35	0.65–0.85	0.95
Brick, white	0.50–0.75	0.25–0.50	0.95
Concrete, natural	0.35	0.65	0.92–0.95
Copper, tarnished	0.20	0.80	0.20
Copper, patina	0.35	0.65	0.15
Galvanized steel	0.10	0.90	0.90
Galvanized steel, white	0.74	0.26	0.90
Glass, clear (¼")	0.5–0.7	0.93–0.95	0.84
Glass, tinted (¼")	0.4–0.6	0.94–0.96	
Glass, clear, low-e	0.6–0.9	0.91–0.94	0.20
Glass, reflective (¼")	0.8–0.36	0.64–0.92	
Marble, white	0.42	0.58	0.85
Surface color			
Black	0.50	0.95	0.95
Dark gray	0.20	0.80	0.93
Light gray	0.35	0.65	0.90
White	0.55	0.45	0.90
Tinned surface	0.95	0.50	0.88
Wood, smooth	0.22	0.78	0.91–0.93

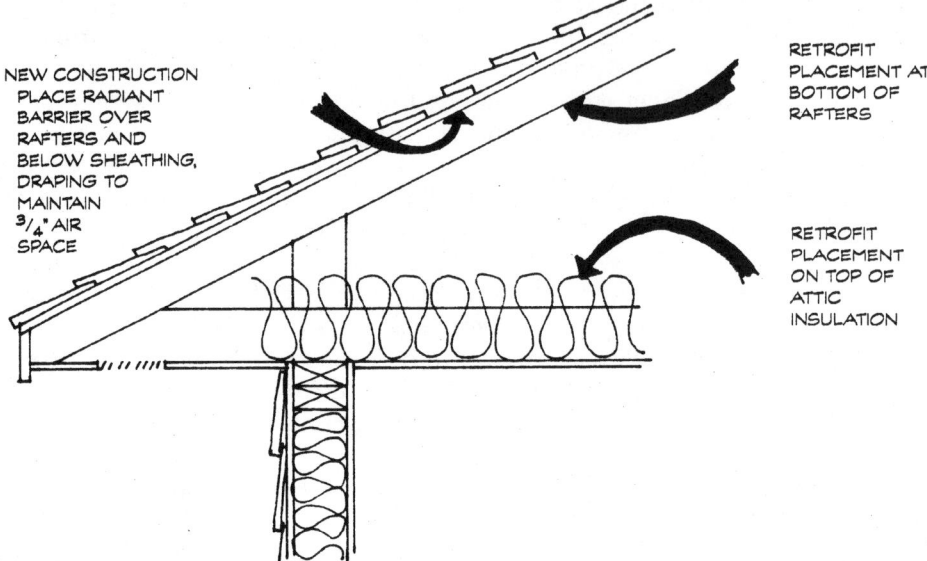

Figure 3.5 Alternative locations of radiant barrier. (*From Parsec Radiant Barrier Company.*)

Radiant barriers are an effective means of reducing summer cooling costs in residential, commercial, and industrial applications.

Radiant heat gain is one of the primary factors contributing to the phenomenon called the *urban heat island.* The warmer temperatures associated with urban areas compared to rural areas are attributable to reductions in the natural tree canopy and the extensive use of dark, heat-absorbing materials in buildings, streets, and parking lots. As part of a Department of Energy research program aimed at reducing energy consumption for heating and cooling buildings, a study performed by the Lawrence Berkeley Laboratory in California rated the solar reflectance of various exterior building and paving materials. *Table 3.3* lists the solar reflectance of such surfaces. While the color of a material indicates its reflectance of visible radiation, it is not possible to determine by visual observation whether the material is reflective of ultraviolet and infrared radiation. For this reason, two surfaces which appear to be the same color may have different solar reflectances. All colors tend to diminish in intensity with age. Dark colors fade and become more reflective, with an attendant rise in reflectance, while light colors become soiled and lose reflectance.

The American Society for Testing and Materials (ASTM) Committee E6 on Building Performance has a task group that is developing a field test method for measuring solar reflectance and a standard practice for calculating a solar reflectance index. Solar reflectance index (SRI) is defined as the relative summer surface temperature of a material with respect to standard white (SRI=100) and standard black (SRI=0), under specified solar and ambient conditions. Work is also under way to develop an extensive listing of materials and their solar reflectance indexes so that a labeling system can be generated to readily identify the characteristics of the materials for specifiers and consumers.

The urban heat island effect, and the cost of cooling buildings in warm climates, can be reduced by providing trees to shade building surfaces and by using lighter-colored materials for exterior building envelopes. In climates where energy costs are primarily for heating, increased insulation R-value is the best method of conserving energy dollars, but in climates where energy costs are primarily for cooling, heat-reflective materials are very effective in reducing operating expenses. *Figure 3.6* shows solar absorptance and surface temperature of horizontal surfaces adjusted to noon on a clear, windless summer day in Austin, Texas. The average city is shown to be about 6°F warmer than rural areas. The hypothetical "green city" incorporates white roofs, light-colored streets and parking lots, and urban vegetation for shade. Strategically planted deciduous shade trees can provide a summer cooling effect while allowing winter heat gain. The hypothetical green city is

TABLE 3.3 Solar Reflectance of Typical Urban Materials and Areas

Surface	Solar reflectance*
Streets	
Asphalt (fresh 0.05, aged 0.20)	0.05–0.2
Walls	
Concrete	0.10–0.35
Brick/stone	0.20–0.40
Roofs	
Smooth-surface asphalt (weathered)	0.07
Tar and gravel	0.08–0.18
Tile	0.10–0.35
Slate	0.10
Thatch	0.25–0.20
Corrugated iron	0.10–0.16
Highly reflective roof after weathering	0.6–0.7
Paints	
White, whitewash	0.50–0.90
Red, brown, green	0.20–0.35
Black	0.02–0.15
Urban areas†	
Range	0.10–0.27
Average	0.15

*Solar reflectance in the fraction of solar radiation reflected by a surface.
†Based on midlatitude cities in snow-free conditions.
SOURCE: Downey, 1996, pp. 10–12.

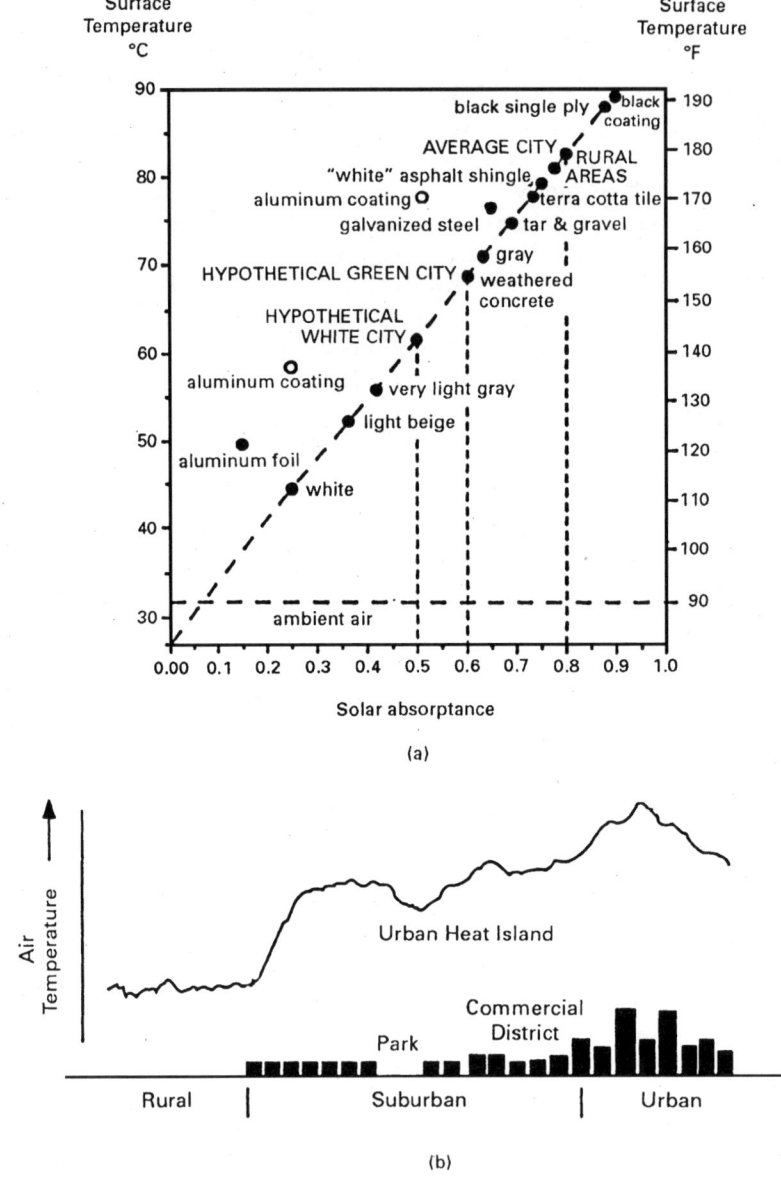

Figure 3.6 Urban heat island effect and the hypothetical "green city." (*From Downey, "Cool Construction Materials," Interface: Journal of the Roof Consultants Institute, Sept. 1996.*)

shown to be an estimated 20 to 25°F cooler than the rural areas because urban settings can contain more areas with highly reflective surfaces than rural settings. Lawrence Berkeley Laboratory has estimated, for example, that the surface area of the city of Sacramento is made up of 28% rooftops, 16% streets, and 14% parking lots, sidewalks, school yards and driveways.

The theoretical "white city" represents urban developments in arid regions where shade trees are not indigenous. The absence of such vegetation allows for increased areas of high-reflectance materials, resulting in an estimated temperature nearly 15°F cooler than the green city and 35 to 40°F cooler than the average city. These studies indicate that a combination of thermal resistance insulation and materials with high solar reflectance will contribute significantly to reduced energy consumption, lower building operating costs, and increased occupant comfort in warm climates.

3.2 Thermal Resistance

Thermal *conductance* C is the time rate of heat flow through one square foot of area at a temperature difference of 1°F, for a specified thickness of material. *Conductivity* k of a homogeneous material is its conductance for a standard unit thickness. For building materials, this standard thickness is 1 in., and for soil and snow, the standard thickness is 1 ft. Material density affects the rate of conductive heat flow. The conductance of a homogeneous material is found by dividing its conductivity by its thickness:

$$C = \frac{k}{n} \tag{3.1}$$

where C = thermal conductance, Btu/h · ft²·°F
 k = thermal conductivity, Btu/h · ft² · °F · in.
 n = thickness, in.

The reciprocal of thermal conductance is thermal *resistance*. Thermal resistance is expressed as the temperature difference, in degrees Fahrenheit, required to cause heat to flow through a square foot of area of a building material or component at the rate of 1 Btu/h. Thermal resistance is generally referred to as R-value:

$$R = \frac{1}{C} \tag{3.2}$$

where R=thermal resistance, °F/Btu/h·ft², and C=thermal conductance, Btu/h·ft²·°F

Thermal resistances are additive, so the total thermal resistance of a building section can be obtained by adding the resistances of its component layers. The thermal efficiency of a building's component materials or assemblies are normally judged by their thermal resistance, or R-value. Thermal resistance depends on the density of the material—as density increases, thermal resistance decreases. Materials in which heat flow is identical in all directions are considered thermally homogeneous. Materials that are not isotropic with respect to heat transmission (such as hollow masonry units, metal studs, and concrete tees) are considered thermally nonhomogeneous, or heterogeneous. For layered construction with heat flow paths in series, the total thermal resistance of the wall is obtained by adding the thermal resistances of each layer ($R = R_1 + R_2 \ldots + R_n$).

The reciprocal of the total resistance, including the resistance of boundary air layers and air spaces, is called the *overall conductance coefficient* or, more commonly, the *coefficient of thermal transmittance* or the *overall heat-transfer coefficient* or the *U*-value:

$$U = \frac{1}{R_T} \qquad (3.3)$$

where U=thermal transmittance, Btu/h·ft²·°F, and R_T=total thermal resistance, °F/Btu/h·ft².

To find the overall thermal performance of a building envelope that includes different kinds of construction assemblies in parallel heat flow paths, we must use the *U*-value. The *U*-value can be used to calculate total thermal transmission of a wall or roof area for a specific difference between indoor and outdoor temperature. The *U*-value is used to judge the overall thermal efficiency of building walls or roofs including both opaque areas and fenestration.

Some building energy codes give allowable *U*-values based on a weighted average for the coefficients for opaque walls, windows and doors. This weighted average is designated as U_0, and is found by multiplying the *U*-value times the area for each component, and dividing by the total area:

$$U_0 = \frac{(U_w \times A_w) + (U_f \times A_f) + (U_d \times A_d)}{A_w + A_f + A_d} \qquad (3.4)$$

where U_w = *U*-value of opaque wall
U_f = *U*-value of fenestration
U_d = *U*-value of doors
A_w = area of opaque wall
A_f = area of fenestration
A_d = area of doors

The overall heat loss or heat gain for each square foot of gross wall area is then obtained by multiplying the total temperature difference between inside and outside by this weighted *U*-value.

The thermal resistance of nonhomogeneous materials, or those materials that are not of uniform thickness, must be obtained experimentally, and cannot be adjusted for different thicknesses. Manufacturers usually supply this type of information. Thermal resistance values for typical building materials are listed in *Table 3.4*.

3.2.1 Effect of air spaces

Air spaces have an effect on both conductive and convective heat flow. With thermal conduction, the longer the heat flow path, the smaller the quantity of heat transferred, so a wide air space will reduce the conductive heat transfer. To reduce convective heat transfer, a narrow air space is required. If the air space is less than ¼ in. wide, though, there is diminishing benefit from

TABLE 3.4 Thermal Design Values of Common Building and Insulating Materials

Description	Density, lb/ft³	Conductivity k	Conductance (C)	Resistance R Per inch of thickness $(1/k)$	Resistance R For thickness listed $(1/C)$
Board products					
Gypsum boards, ⅜ in.	50		3.10		0.32
Gypsum board, ½ in.	50		2.22		0.45
Gypsum board, ⅝ in.	50		1.78		0.56
Plywood	34	0.80		1.25	
Plywood, ½ in.	34		1.60		0.62
Plywood or wood panels, ¾ in.	34		1.07		0.93
Fiber board sheathing, ½ in.	18		0.76		1.32
Hardboard, medium density	50	0.73		1.37	
Particleboard					
Low density	37	0.71		1.41	
Medium density	50	0.94		1.06	
High density	62.5	1.18		0.85	
Underlayment, ⅝ in.	40		1.12		0.82
Waferboard	37	0.63		1.59	
Wood subfloor, ¾ in.			1.06		0.94
Building membrane					
Felt, vapor-permeable			16.7		0.06
Plastic film, vapor diffusion retarder			8.35		Negligible
Finish flooring materials					
Carpet and rubber pad			0.48		1.23
Terrazzo, 1 in.			12.50		0.08
Hardwood, ¾ in.			1.47		0.68
Tile, vinyl or rubber			20.00		0.05
Roofing					
Asphalt roll roofing	70		6.50		0.15
Asphalt shingles	70		2.27		0.44
Built-up roofing	70		3.00		0.33
Slate shingles			20.00		0.05
Wood shingles			1.06		0.94
Single-ply membrane, 45 mil	83		2.0		0.50

TABLE 3.4 Thermal Design Values of Common Building and Insulating Materials (Continued)

Description	Density, lb/ft³	Conductivity k	Conductance (C)	Resistance R Per inch of thickness (1/k)	Resistance R For thickness listed (1/C)
Soil					
Wet		1.00		1.0/ft	
Dry		0.50		2.0/ft	
Insulating materials					
Batt or blanket insulation, glass fiber or rock wool,					
3–4 in.	0.3–2.0		0.091		11.0
3½ in.	0.3–2.0		0.077		13.0
5½–6½ in.	0.3–2.0		0.053		19.0
6–7½ in.	0.3–2.0		0.045		22.0
9–10 in.	0.3–2.0		0.033		30.0
12–13 in.	0.3–2.0		0.026		38.0
Board insulation					
Cellular glass	8.5	0.35		2.86	
Glass fiber, organic bonded	4.0–9.0	0.25		4.00	
Perlite, expanded, organic bonded	1.0	0.36		2.78	
Polystyrene, extruded	1.8–3.5	0.20		5.40 at 40°F	
				5.00 at 75°F	
Polystyrene, expanded (molded beads)	1.0	0.26		4.17 at 40°F	
				3.85 at 75°F	
Polyurethane/polyisocyanurate, unfaced	1.5	0.16–0.18		6.25–5.56	
Polyurethane/polyisocyanurate, vapor permeable facers	1.5–2.5	0.16–0.18		6.25–5.56	
Polyurethane/polyisocyanurate, vapor impermeable facers	2.0	0.14		7.20	
Phenolic, closed cell	3.0	0.12		8.20	
Phenolic, open cell	1.8–2.2	0.23		4.40	
Mineral fiber insulation	16–17	0.34		2.94	
Mineral fiber acoustical tile	18.0	0.35		2.86	
Wood fiber acoustical tile, ½ in.			0.80		1.25
Fiber cement board	25.0–27.0	0.50–0.53		2.0–1.89	
Loose fill					
Cellulose	2.3–3.2	0.27–0.32		3.70–3.13	
Perlite, expanded	2.0–4.1	0.27–0.31		3.7–3.3	
	4.1–7.4	0.31–0.36		3.3–2.8	

TABLE 3.4 Thermal Design Values of Common Building and Insulating Materials (Continued)

Description	Density, lb/ft³	Conductivity k	Conductance (C)	Resistance R Per inch of thickness (1/k)	Resistance R For thickness listed (1/C)
Insulating materials					
Loose fill					
Mineral fiber, glass or rock wool					
3¾–5 in.	0.6–2.0				11.0
6½–8¾ in.	0.6–2.0				19.0
7½–10 in.	0.6–2.0				22.0
10¼–13¾ in.	0.6–2.0				30.0
Mineral fiber, glass or rock wool,					
3½ in. closed sidewall application	2.0–3.5				12.0–14.0
Vermiculite, exfoliated	7.0–8.2	0.47		2.13	
	4.0–6.0	0.44		2.27	
Concrete					
Lightweight aggregate concrete and cellular concrete	120	5.5–11.0		0.18–0.09	
	100	3.7–5.9		0.27–0.17	
	80	2.5–3.5		0.40–0.29	
	60	1.6–1.8		0.63–0.56	
	40	0.93–1.11		1.08–0.90	
Sand and gravel or crushed stone aggregate concrete	140	10.0–20.0		0.10–0.05	
Masonry					
Granite and marble	150–175	20		0.05	
Limestone and sandstone		12.5		0.08	
Brick, common	80	2.2–3.2		0.45–0.31	
	90	2.7–3.7		0.37–0.27	
	100	3.3–4.3		0.30–0.23	
	110	3.5–5.5		0.29–0.18	
	120	4.4–6.4		0.23–0.16	
	130	5.4–9.0		0.19–0.11	
Brick, face	130	5.4–9.0		0.19–0.11	
Concrete block					
6 in., normal-weight aggregate	125+				1.6–2.0
Same with perlite-filled cores	125+				2.2–3.7
Same with vermiculite-filled cores	125+				2.2–3.5

TABLE 3.4 Thermal Design Values of Common Building and Insulating Materials (Continued)

Description	Density, lb/ft^3	Conductivity k	Conductance (C)	Resistance R Per inch of thickness $(1/k)$	Resistance R For thickness listed $(1/C)$
Masonry					
Concrete block					
8 in., normal-weight aggregate	125+				1.7–2.2
Same with perlite-filled cores	125+				2.7–4.8
Same with vermiculite-filled cores	125+				2.7–4.6
12 in., normal-weight aggregate	125+				1.9–2.3
Same with perlite-filled cores	125+				3.6–6.8
Same with vermiculite-filled cores	125+				3.6–6.5
6 in., medium-weight aggregate	105–124				1.8–2.2
Same with perlite-filled cores	105–124				3.0–4.8
Same with vermiculite-filled cores	105–124				2.9–4.5
8 in., medium-weight aggregate	105–124				2.0–2.4
Same with perlite-filled cores	105–124				3.8–6.3
Same with vermiculite-filled cores	105–124				3.7–5.9
12 in., medium-weight aggregate	105–124				2.1–2.6
Same with perlite-filled cores	105–124				5.2–9.1
Same with vermiculite-filled cores	105–124				5.1–8.5
6 in., lightweight aggregate	85–104				2.1–2.5
Same with perlite-filled cores	85–104				4.1–6.1
Same with vermiculite-filled cores	85–104				3.9–5.6
8 in., lightweight aggregate	85–104				2.3–2.7
Same with perlite-filled cores	85–104				5.3–8.2
Same with vermiculite-filled cores	85–104				5.0–7.5
12 in., lightweight aggregate	85–104				2.4–3.0
Same with perlite filled cores	85–104				7.6–12.1
Same with vermiculite filled cores	85–104				7.2–11.0

TABLE 3.4 Thermal Design Values of Common Building and Insulating Materials (Continued)

Description	Density, lb/ft³	Conductivity k	Conductance (C)	Resistance R Per inch of thickness $(1/k)$	Resistance R For thickness listed $(1/C)$
Plaster and stucco					
Cement plaster, sand aggregate stucco	116	5.0		0.20	
⅝ in.	116		13.3		0.08
¾ in.	116		6.66		0.15
Gypsum plaster					
Lightweight aggregate, ½ in.	45		3.12		0.32
Lightweight aggregate, ⅝ in.	45		2.67		0.39
Lightweight aggregate on metal lath, ¾ in.			2.13		0.47
Perlite aggregate	45	1.5		0.67	
Sand aggregate	105	5.6		0.18	
Sand aggregate, ½ in.	105		11.10		0.09
Sand aggregate, ⅝ in.	105		9.10		0.11
Sand aggregate on metal lath, ¾ in.			7.70		0.13
Vermiculite aggregate	45	1.7		0.59	
Siding materials					
Wood shingles, 16 in., 7½ in. exp.			1.15		0.87
Hardboard siding, ⁷⁄₁₆ in. thick			0.49		0.67
Wood drop siding, 1×8			1.27		0.79
Wood bevel siding, ½ ×8, lapped			1.23		0.81
Wood bevel siding, ¾ ×8, lapped			0.95		1.05
Plywood, ⅜ in.			1.59		0.59
Aluminum over					
⅜-in. insulating sheathing			0.55		1.82
⅜-in. foil-faced insulating sheathing			0.34		2.96
Woods (12% moisture content)					
Hardwoods					
Oak	41.2–46.8	1.12–1.25		0.89–0.80	
Birch	42.6–45.4	1.16–1.22		0.87–0.82	
Ash	38.4–41.9	1.06–1.14		0.94–0.88	

TABLE 3.4 Thermal Design Values of Common Building and Insulating Materials (Continued)

Description	Density, lb/ft³	Conductivity k	Conductance (C)	Resistance R Per inch of thickness (1/k)	Resistance R For thickness listed (1/C)
Woods (12% moisture content)					
Softwoods					
Southern pine	35.6–41.2	1.00–1.12		1.00–0.89	
Douglas fir, larch	33.5–36.3	0.95–1.01		1.06–0.99	
Hemlock, spruce-pine-fir	24.5–31.4	0.74–0.90		1.35–1.11	
West Coast cedar	21.7–31.4	0.68–0.90		1.48–1.11	
California redwood	24.5–28.0	0.74–0.82		1.35–1.22	

Air surfaces	Heat flow direction	R Nonreflective surface	R Reflective aluminum surface	R Highly reflective foil surface
Still air (position of surface)				
Horizontal	Up	0.61	1.10	1.32
45° slope	Up	0.62	1.14	1.37
Vertical	Horizontal	0.68	1.35	1.70
45° slope	Down	0.76	1.67	2.22
Horizontal	Down	0.92	2.70	4.55
Moving air (any position)				
15-mph wind (winter)	Any	0.17		
7½-mph wind (summer)	Any	0.25		

Air spaces, position of air space and thickness	Heat flow direction	R Both surfaces nonreflective, ε=0.82	R One surface nonreflective, one aluminum-coated, ε=0.20	R One surface nonreflective, one highly reflective foil, ε=0.03
Horizontal				
½ in., winter	Up	0.91	1.45	1.73
½ in., summer	Up	0.73	1.51	2.13
1½ in., winter	Up	0.97	1.63	2.01
1½ in., summer	Up	0.77	1.71	2.55
3½ in., winter	Up	1.03	1.79	2.25
3½ in., summer	Up	0.80	1.83	2.84

TABLE 3.4 Thermal Design Values of Common Building and Insulating Materials (Continued)

Air spaces, position of air space and thickness	Heat flow direction	R		
		Both surfaces nonreflective, $\varepsilon=0.82$	One surface nonreflective, one aluminum-coated, $\varepsilon=0.20$	One surface nonreflective, one highly reflective foil, $\varepsilon=0.03$
45° slope				
½ in., winter	Up	1.02	1.76	2.20
½ in., summer	Up	0.76	1.65	2.44
1½ in., winter	Up	1.04	1.82	2.30
1½ in., summer	Up	0.80	1.86	2.92
3½ in., winter	Up	1.06	1.90	2.42
3½ in., summer	Up	0.82	1.97	3.18
Vertical				
½ in., winter	Horizontal	1.13	2.14	2.82
½ in., summer	Horizontal	0.77	1.67	2.47
1½ in., winter	Horizontal	1.12	2.10	2.76
1½ in., summer	Horizontal	0.87	2.25	3.99
3½ in., winter	Horizontal	1.14	2.17	2.88
3½ in., summer	Horizontal	0.85	2.15	3.69
45° slope				
½ in., winter	Down	1.15	2.19	2.91
½ in., summer	Down	0.77	1.67	2.48
1½ in., winter	Down	1.27	2.68	3.85
1½ in., summer	Down	0.91	2.56	5.07
3½ in., winter	Down	1.27	2.66	3.81
3½ in., summer	Down	0.90	2.49	4.81
Horizontal				
½ in., winter	Down	1.15	2.20	2.94
½ in., summer	Down	0.77	1.67	2.48
1½ in., winter	Down	1.49	3.91	7.03
1½ in., summer	Down	0.94	2.79	6.09
3½ in., winter	Down	1.62	4.87	10.90
3½ in., summer	Down	1.00	3.41	10.07

reduced convective heat flow, and there is an increase in heat flow by conduction. As the air space width is increased above ½ in., the heat lost by conduction is reduced, but convection begins to play an increasingly important role. The minimum total heat transfer under the combined effects of conduction and convection is obtained at a width of about ⅝ in. Over ⅝ in., the increase in convection more than offsets the reduced thermal conduction. When the air space width exceeds 1¼ in., the combined heat transfer is independent of width (*Fig. 3.7*).

Radiant heat transfer across an air space is independent of the air space width, but is affected by the emissivities of the boundary surfaces. Polished metal or metallic surfaces have lower emissivities than most other building materials, so reflective surfaces can be used as a barrier to reduce radiant heat transfer across an air space. Values for effective emittance of various building materials are given in *Table 3.5*. Effective emittance E is based on the combined effect of emittances ε from both boundary surfaces when they are parallel and of a dimension much larger than the distance between them. Thermal resistance for plane air spaces ½ to 3½ in. wide, and based on effective emittance, position of the air space (vertical, horizontal, or sloping), and direction of heat flow (up, down, or horizontal), are given in *Table 3.6*. The effective thermal resistance of attic spaces is dependent on surface temperatures, ventilation rates, and the resistance of the ceiling construction (Table 3.7). Surface temperature, which is easier to calculate, can be substituted for sol-air temperature in the table if 0.25 is subtracted from the attic resistance shown.

The layer of air immediately adjacent to a surface adds some thermal resistance to building envelope assemblies, depending on position (vertical, horizontal, or sloping), direction of heat flow (up, down, or horizontal), and air movement (still inside, moving outside). Typical values for surface heat transfer coefficients h and thermal resistances (R-values) are given in *Table 3.8*. Whenever an opaque wall assembly is analyzed, it should include both the

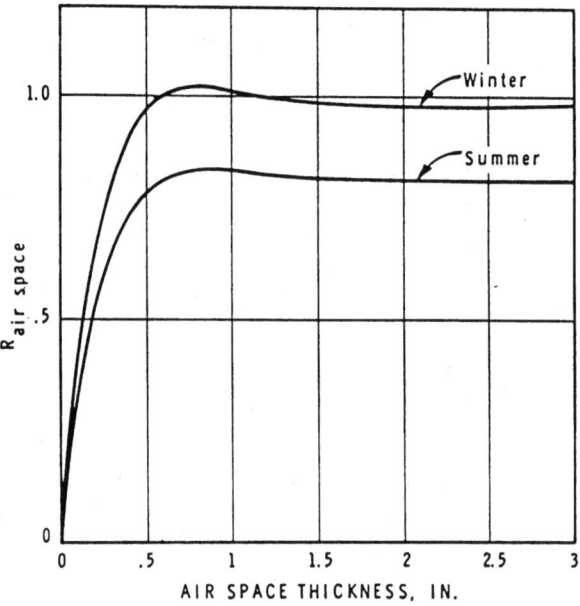

Figure 3.7 Variation in thermal resistance of air spaces with width of space. (*From Latta, Walls, Windows and Roofs for the Canadian Climate.*)

TABLE 3.5 Emittance Values of Various Surfaces and Effective Emittances of Air Spaces*

Surface	Average emittance ε	Effective emittance E of air space	
		One surface emittance ε; the other 0.9	Both surfaces emittance ε
Aluminum foil, bright	0.05	0.05	0.03
Aluminum foil, with condensate just visible (>0.7 gr/ft^2)	0.30†	0.29	
Aluminum foil, with condensate clearly visible (>2.9 gr/ft^2)	0.70†	0.65	
Aluminum sheet	0.12	0.12	0.06
Aluminum coated paper, polished	0.20	0.20	0.11
Steel, galvanized, bright	0.25	0.24	0.15
Aluminum paint	0.50	0.47	0.35
Building materials: wood, paper, masonry, nonmetallic paints	0.90	0.82	0.82
Regular glass	0.84	0.77	0.72

*These values apply in the 4- to 40-μm range of the electromagnetic spectrum.
†Values are based on data presented by Bassett and Trethowen (1984).
SOURCE: *ASHRAE Handbook of Fundamentals.*

inside and outside air surfaces because these surfaces affect convection and conduction of heat. The inclusion of these air surfaces makes all opaque wall assemblies "layered" construction.

3.2.2 Effect of mass

Heat flow through solid materials is not instantaneous. The time delay involving absorption of the heat is called *thermal lag*. As the temperature rises on one side of a wall, heat begins to migrate toward the cooler side. Before heat transfer from one side to the other can be completed, though, the wall itself must undergo a temperature increase. The amount of thermal energy necessary to produce this increase is directly proportional to the weight of the material. Masonry and concrete are heavy, so they can absorb and store a significant amount of heat and substantially retard its migration. This characteristic is called *thermal storage capacity*. Thermal storage capacity affects the rate of conductive heat transfer and therefore the thermal performance of the building envelope, and is a critical consideration in passive solar heating and cooling strategies, including off-peak load shifting.

Most building materials absorb and store at least some heat, but higher density and greater mass cause slower absorption and longer retention. So although density decreases thermal resistance, or R-value, it increases thermal absorption and storage capacity and delays heat flow through the material. The speed with which a particular material will heat up or cool down depends on its density, thickness, specific heat, and thermal conductivity.

Table 3.6 Thermal resistance of plane air spaces, °F·ft²·hr/Btu (From ASHRAE Handbook of Fundamentals).

Position of Airspace	Direction of Heat Flow	Airspace Mean Temp.[d], °F	Airspace Temp. Diff.[d], °F	0.5-in. Airspace[c] Effective Emittance, $E^{d,e}$ 0.03	0.05	0.2	0.5	0.82	0.75-in. Airspace[c] Effective Emittance, $E^{d,e}$ 0.03	0.05	0.2	0.5	0.82	1.5-in. Airspace[c] Effective Emittance, $E^{d,e}$ 0.03	0.05	0.2	0.5	0.82	3.5-in. Airspace[c] Effective Emittance, $E^{d,e}$ 0.03	0.05	0.2	0.5	0.82
		90	10	2.13	2.03	1.51	0.99	0.73	2.34	2.22	1.61	1.04	0.75	2.55	2.41	1.71	1.08	0.77	2.84	2.66	1.83	1.13	0.80
		50	30	1.62	1.57	1.29	0.96	0.75	1.71	1.66	1.35	0.99	0.77	1.87	1.81	1.45	1.04	0.80	2.09	2.01	1.58	1.10	0.84
		50	10	2.13	2.05	1.60	1.11	0.84	2.30	2.21	1.70	1.16	0.87	2.50	2.40	1.81	1.21	0.89	2.80	2.66	1.95	1.28	0.93
Horiz.	Up ↑	0	20	1.73	1.70	1.45	1.12	0.91	1.83	1.79	1.52	1.16	0.93	2.01	1.95	1.63	1.23	0.97	2.25	2.18	1.79	1.32	1.03
		0	10	2.10	2.04	1.70	1.27	1.00	2.23	2.16	1.78	1.31	1.02	2.43	2.35	1.90	1.38	1.06	2.71	2.62	2.07	1.47	1.12
		−50	20	1.69	1.66	1.49	1.23	1.04	1.77	1.74	1.55	1.27	1.07	1.94	1.91	1.68	1.36	1.13	2.19	2.14	1.86	1.47	1.20
		−50	10	2.04	2.00	1.75	1.40	1.16	2.16	2.11	1.84	1.46	1.20	2.37	2.31	1.99	1.55	1.26	2.65	2.58	2.18	1.67	1.33
		90	10	2.44	2.31	1.65	1.06	0.76	2.96	2.78	1.88	1.15	0.81	2.92	2.73	1.86	1.14	0.80	3.18	2.96	1.97	1.18	0.82
		50	30	2.06	1.98	1.56	1.10	0.83	1.99	1.92	1.52	1.08	0.82	2.14	2.06	1.61	1.12	0.84	2.26	2.17	1.67	1.15	0.86
45° Slope		50	10	2.55	2.44	1.83	1.22	0.90	2.90	2.75	2.00	1.29	0.94	2.88	2.74	1.99	1.29	0.94	3.12	2.95	2.10	1.34	0.96
	Up ↗	0	20	2.20	2.14	1.76	1.30	1.02	2.13	2.07	1.72	1.28	1.00	2.30	2.23	1.82	1.34	1.04	2.42	2.35	1.90	1.38	1.06
		0	10	2.63	2.54	2.03	1.44	1.10	2.72	2.62	2.08	1.47	1.12	2.79	2.69	2.12	1.49	1.13	2.98	2.87	2.23	1.54	1.16
		−50	20	2.08	2.04	1.78	1.42	1.17	2.05	2.01	1.76	1.41	1.16	2.22	2.17	1.88	1.49	1.21	2.34	2.29	1.97	1.54	1.25
		−50	10	2.62	2.56	2.17	1.66	1.33	2.53	2.47	2.10	1.62	1.30	2.71	2.64	2.23	1.69	1.35	2.87	2.79	2.33	1.75	1.39
		90	10	2.47	2.34	1.67	1.06	0.77	3.50	3.24	2.08	1.22	0.84	3.99	3.66	2.25	1.27	0.87	3.69	3.40	2.15	1.24	0.85
		50	30	2.57	2.46	1.84	1.23	0.90	2.91	2.77	2.01	1.30	0.94	2.58	2.46	1.84	1.23	0.90	2.67	2.55	1.89	1.25	0.91
		50	10	2.66	2.54	1.88	1.24	0.91	3.70	3.46	2.35	1.43	1.01	3.79	3.55	2.39	1.45	1.02	3.63	3.40	2.32	1.42	1.01
Vertical	Horiz. →	0	20	2.82	2.72	2.14	1.50	1.13	3.14	3.02	2.32	1.58	1.18	2.76	2.66	2.10	1.48	1.12	2.88	2.78	2.17	1.51	1.14
		0	10	2.93	2.82	2.20	1.53	1.15	3.77	3.59	2.64	1.73	1.26	3.51	3.35	2.51	1.67	1.23	3.49	3.33	2.50	1.67	1.23
		−50	20	2.90	2.82	2.35	1.76	1.39	2.90	2.83	2.36	1.77	1.39	2.64	2.58	2.18	1.66	1.33	2.82	2.75	2.30	1.73	1.37
		−50	10	3.20	3.10	2.54	1.87	1.46	3.72	3.60	2.87	2.04	1.56	3.31	3.21	2.62	1.91	1.48	3.40	3.30	2.67	1.94	1.50
		90	10	2.48	2.34	1.67	1.06	0.77	3.53	3.27	2.10	1.22	0.84	5.07	4.55	2.56	1.36	0.91	4.81	4.33	2.49	1.34	0.90
		50	30	2.64	2.52	1.87	1.24	0.91	3.43	3.23	2.24	1.39	0.99	3.58	3.36	2.31	1.42	1.00	3.51	3.30	2.28	1.40	1.00
45° Slope		50	10	2.67	2.55	1.89	1.25	0.92	3.81	3.57	2.40	1.45	1.02	5.10	4.66	2.85	1.60	1.09	4.74	4.36	2.73	1.57	1.08
	Down ↘	0	20	2.91	2.80	2.19	1.52	1.15	3.75	3.57	2.63	1.72	1.26	3.85	3.66	2.68	1.74	1.27	3.81	3.63	2.66	1.74	1.27
		0	10	2.94	2.83	2.21	1.53	1.15	4.12	3.91	2.81	1.80	1.30	4.92	4.62	3.16	1.94	1.37	4.59	4.32	3.02	1.88	1.34
		−50	20	3.16	3.07	2.52	1.86	1.45	3.78	3.65	2.90	2.05	1.57	3.62	3.50	2.80	2.01	1.54	3.77	3.64	2.90	2.05	1.57
		−50	10	3.26	3.16	2.58	1.89	1.47	4.35	4.18	3.22	2.21	1.66	4.67	4.47	3.40	2.29	1.70	4.50	4.32	3.31	2.25	1.68
		90	10	2.48	2.34	1.67	1.06	0.77	3.55	3.29	2.10	1.22	0.85	6.09	5.35	2.79	1.43	0.94	10.07	8.19	3.41	1.57	1.00
		50	30	2.66	2.54	1.88	1.24	0.91	3.77	3.52	2.38	1.44	1.02	6.27	5.63	3.18	1.70	1.14	9.60	8.17	3.86	1.88	1.22
		50	10	2.67	2.55	1.89	1.25	0.92	3.84	3.59	2.41	1.45	1.02	6.61	5.90	3.27	1.73	1.15	11.15	9.27	4.09	1.93	1.24
Horiz.	Down ↓	0	20	2.94	2.83	2.20	1.53	1.15	4.18	3.96	2.83	1.81	1.30	7.03	6.43	3.91	2.19	1.49	10.90	9.52	4.87	2.47	1.62
		0	10	2.96	2.85	2.22	1.53	1.16	4.25	4.02	2.87	1.82	1.31	7.31	6.66	4.00	2.22	1.51	11.97	10.32	5.08	2.52	1.64
		−50	20	3.25	3.15	2.58	1.89	1.47	4.60	4.41	3.36	2.28	1.69	7.73	7.20	4.77	2.85	1.99	11.64	10.49	6.02	3.25	2.18
		−50	10	3.28	3.18	2.60	1.90	1.47	4.71	4.51	3.42	2.30	1.71	8.09	7.52	4.91	2.89	2.01	12.98	11.56	6.36	3.34	2.22

Thermal resistance values were determined from the relation, $R = 1/C$, where $C = h_c + Eh_r$; h_c is the conduction-convection coefficient, Eh_r is the radiation coefficient $\cong 0.00686E[(t_m + 460)/100]^3$, and t_m is the mean temperature of the airspace. Values for h_c were determined from data developed by Robinson *et al.* (1954). Equations 5 through 7 in Yarbrough (1983) show the data in Table 2 in analytic form. For extrapolation from Table 2 to airspaces less than 0.5 in. (as in insulating window glass), assume

$$h_c = 0.159(1 + 0.0016\, t_m)/l$$

where l is the airspace thickness in in., and h_c is heat transfer through the airspace only.

[b]Values are based on data presented by Robinson *et al.* (1954). Values apply for ideal conditions, *i.e.*, airspaces of uniform thickness bounded by plane, smooth, parallel surfaces with no air leakage to or from the space. When accurate values are required, use overall U-factors determined through calibrated hot box (ASTM C 976) or guarded hot box (ASTM C 236) testing. Thermal resistance values for multiple airspaces must be based on careful estimates of mean temperature differences for each airspace.

[c]A single resistance value cannot account for multiple airspaces; each airspace requires a separate resistance calculation that applies only for the established boundary conditions. Resistances of horizontal spaces with heat flow downward are substantially independent of temperature difference.

[d]Interpolation is permissible for other values of mean temperature, temperature difference, and effective emittance E. Interpolation and moderate extrapolation for airspaces greater than 3.5 in. are also permissible.

[e]Effective emittance E of the airspace is given by $1/E = \epsilon_1 + 1/\epsilon_2 - 1$ where ϵ_1 and ϵ_2 are the emittances of the surfaces of the airspace (see Table 3).

TABLE 3.7 Effective Thermal Resistance of Ventilated Attics[a] (Summer Condition)

		No ventilation[b]		Natural ventilation				Power ventilation[c]			
		\multicolumn{10}{c}{Ventilation rate, cfm/ft^2}
		0		0.1[d]		0.5		1.0		1.5	
		\multicolumn{10}{c}{Ceiling resistance[e] R, °F · ft^2 · h/Btu}
Ventilation air temp., °F	Sol-air[f] temp., °F	10	20	10	20	10	20	10	20	10	20
\multicolumn{12}{c}{Part A. Nonreflective Surfaces}
	120	1.9	1.9	2.8	3.4	6.3	9.3	9.6	16	11	20
80	140	1.9	1.9	2.8	3.5	6.5	10	9.8	17	12	21
	160	1.9	1.9	2.8	3.6	6.7	11	10	18	13	22
	120	1.9	1.9	2.5	2.8	4.6	6.7	6.1	10	6.9	13
90	140	1.9	1.9	2.6	3.1	5.2	7.9	7.6	12	8.6	15
	160	1.9	1.9	2.7	3.4	5.8	9	8.5	14	10	17
	120	1.9	1.9	2.2	2.3	3.3	4.4	4	6	4.1	6.9
100	140	1.9	1.9	2.4	2.7	4.2	6.1	5.8	8.7	6.5	10
	160	1.9	1.9	2.6	3.2	5.0	7.6	7.2	11	8.3	13
\multicolumn{12}{c}{Part B. Reflective Surfaces[g]}
	120	6.5	6.5	8.1	8.8	13	17	17	25	19	30
80	140	6.5	6.5	8.2	9.0	14	18	18	26	20	31
	160	6.5	6.5	8.3	9.2	15	18	19	27	21	32
	120	6.5	6.5	7.5	8.0	10	13	12	17	13	19
90	140	6.5	6.5	7.7	8.3	12	15	14	20	16	22
	160	6.5	6.5	7.9	8.6	13	16	16	22	18	25
	120	6.5	6.5	7.0	7.4	8	10	8.5	12	8.8	12
100	140	6.5	6.5	7.3	7.8	10	12	11	15	12	16
	160	6.5	6.5	7.6	8.2	11	14	13	18	15	20

[a]Although the term *effective resistance* is commonly used when there is attic ventilation, this table includes values for situations with no ventilation. The effective resistance of the attic, added to the resistance ($1/U$) of the ceiling yields the effective resistance of this combination based on sol-air and room temperatures. These values apply to wood frame construction with a roof deck and roofing that has a conductance of 1.0 Btu/h · ft^2 · °F.
[b]This condition cannot be achieved in the field unless extreme measures are taken to tightly seal the attic.
[c]Based on air discharging outward from attic.
[d]When attic ventilation meets the requirements stated in Chap. 23 of the source document, 0.1 cfm/ft^2 is assumed as the natural summer ventilation rate for design purposes.
[e]When determining ceiling resistance, do not add the effect of a reflective surface facing the attic, as it is accounted for in Part B.
[f]Roof surface temperature rather than sol-air temperature can be used if 0.25 is subtracted from the attic resistance shown.
[g]Surfaces with effective emittance E of 0.05 between ceiling joists facing the attic space.
SOURCE: ASHRAE *Handbook of Fundamentals*.

TABLE 3.8 Surface Heat Transfer Coefficients h, Btu/h · ft² · °F, and Resistances R, °F · ft² · h/Btu, for Air*†‡§

Position of surface	Direction of heat flow	Non-reflective, $\varepsilon=0.90$		Reflective $\varepsilon=0.20$		$\varepsilon=0.05$	
		h_i	R	h_i	R	h_i	R
Still air							
Horizontal	Upward	1.63	0.61	0.91	1.10	0.76	1.32
Sloping—45°	Upward	1.60	0.62	0.88	1.14	0.73	1.37
Vertical	Horizontal	1.46	0.68	0.74	1.35	0.59	1.70
Sloping—45°	Downward	1.32	0.76	0.60	1.67	0.45	2.22
Horizontal	Downward	1.08	0.92	0.37	2.70	0.22	4.55
Moving air (Any position)		h_O	R	h_O	R	h_O	R
15-mph wind (for winter)	Any	6.00	0.17				
7.5-mph wind (for summer)	Any	4.00	0.25				

*No surface has both an air space resistance value and a surface resistance value. No air space value exists for any surface facing an air space of less than 0.5 in.

†For ventilated attics or spaces above ceilings under summer conditions (heat flow down), see Table 3.7.

‡Conductances are for surfaces of the stated emittance facing virtual blackbody surroundings at the same temperature as the ambient air. Values are based on a surface-air temperature difference of 10°F and for surface temperature of 70°F.

§See Chap. 3 of source document for more detailed information, especially Tables 5 and 6, and see Figure 1 for additional data.

¶Condensate can have a significant impact on surface emittance.

SOURCE: ASHRAE *Handbook of Fundamentals.*

3.2.3 Effect of insulation

Thermal insulation materials are very low density, and therefore have very low conductive heat flow and high thermal resistance or R-value. Insulation can be used to improve the thermal performance of building walls and roofs by reducing both conductive heat flow through the section and convective heat flow in air spaces. When coupled with radiant barrier foil facings, insulation materials provide a highly effective means of maintaining interior conditions within desirable and affordable comfort zones. The proper selection of insulating materials, though, depends on more than just thermal resistance. Important properties and characteristics include

- Stable R-value over time
- Dimensional stability under varying temperature and moisture conditions
- Resistance to deterioration

In addition, wall insulation must have a means of secure attachment, and loose fill materials must resist settlement. Some roofing insulations must also have good compressive strength, chemical compatibility with roofing membranes, and the ability to withstand the high temperatures of hot-applied bitumens. Different properties will be important for different applications, so product selection should be based on specific performance requirements and conditions of use.

3.2.4 Effect of thermal bridges

A *thermal bridge* occurs when a material or object of relatively high thermal conductivity penetrates a material of relatively low thermal conductivity, increasing the rate of heat flow at the penetration. For example, the heat flow is increased where a metal stud penetrates a layer of wall insulation, and the insulation is "bridged."

The net effect of thermal bridging depends on the location of the insulation. A thermal bridge will be relatively warm if its exposure on the warm side of the insulation is greater than its exposure on the cold side, and vice versa. If exposure on the cold side is increased and exposure on the warm side decreased by moving the insulation, the temperature of the bridge itself will be lower (*Fig. 3.8*). Depending upon whether heat loss or heat gain is the predominant problem, heating climates and cooling climates may have slightly different optimum locations for insulation within the envelope. These requirements, of course, must be balanced with those for controlling condensation. Many walls contain thermal bridges, which must be taken into account. Methods of calculating the effect of thermal bridges are discussed below. ASHRAE Standard 90.1, *Energy Efficient Design of New Buildings Except New Low-Rise Residential Buildings,* gives correction factors for metal stud walls that are based on size, gauge and spacing of the studs (*Table 3.9*).

Thermal bridging is also a problem with hollow-core masonry units. Many concrete masonry units are designed with proprietary insulation inserts which increase the overall R-value, but do not compensate for the thermal bridges. According to the Expanded Shale, Clay and Slate Institute (ESCSI), hollow-core units made from lightweight concrete have higher R-values than normal weight units with insulation inserts because the lower thermal conductance reduces the effect of the thermal bridges (*Table 3.10*). One proprietary block

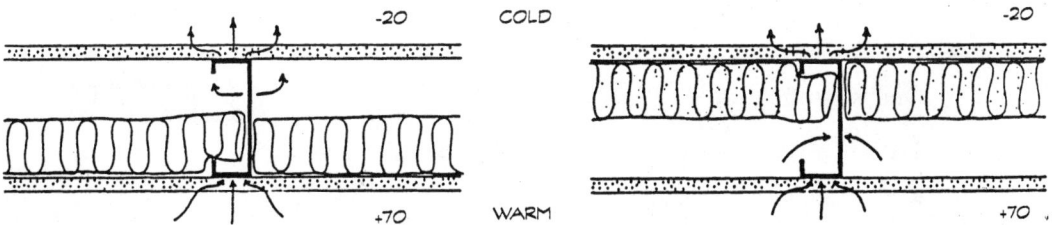

Figure 3.8 Thermal bridge formed by a metal stud. (*From Latta, Walls, Windows and Roofs for the Canadian Climate.*)

TABLE 3.9 Correction Factors* for Metal Stud Walls Based on Size, Gauge, and Spacing of Studs

Size of members	Gauge of stud†	Spacing of framing, in. on centers	Cavity insulation R-value	Correction factor	Effective framing/cavity R-value
2×4	18–16	16	11	0.50	5.5
			13	0.46	6.0
			15	0.43	6.4
2×4	18–16	24	11	0.60	6.6
			13	0.55	7.2
			15	0.52	7.8
2×6	18–16	16	19	0.37	7.1
			21	0.35	7.4
2×6	18–16	24	19	0.45	8.6
			21	0.43	9.0
2×8	18–16	16	25	0.31	7.8
2×8	18–16	24	25	0.38	9.6

*This table was developed for C-channel metal studs.
†These factors can be applied to metal studs of this gauge or thinner.

unit marketed in recent years features a design that incorporates continuous insulation inserts that eliminate thermal bridges completely (*Fig. 3.9*).

Thermal bridges occur at the exterior corners of wood stud buildings (*Fig. 3.10A*), at roof/wall intersections (*Fig. 3.10B*), at wall/floor intersections (*Fig. 3.10C*) and at foundation perimeters. Mechanical fasteners through the insulation layer of a roofing assembly can also create thermal bridges (*Table 3.11*).

Thermal bridges are common locations for mold and mildew growth in heating climates because of the higher relative humidities found near cold surfaces. Dust can also accumulate at these cold spots. This phenomenon is caused by *brownian motion*—the vibration of particles due to temperature. The lower the temperature of the particles, the slower they vibrate and the more likely they are to adhere to these cold surfaces, "dust marking" the thermal bridges. The higher humidity near the surface also collects more dust particles.

3.3 Types of Insulation

One way of classifying insulation materials is according to physical form:

- Loose-fill insulation, including poured and blown
- Flexible and semirigid insulation
- Rigid board insulation
- Formed-in-place insulation

3.3.1 Loose-fill insulation

Loose-fill insulations are made of fibers, granules, or chips that are poured into wall cavities or blown into attics. Most loose-fill insulations are made from either inert minerals or organic cellulose fibers.

Mineral wool insulation includes chopped fiber materials made from *rock, slag,* and *glass wool.* Glass wool is often referred to separately as *fiberglass.* Mineral wool is both moisture- and vermin-resistant, and loose mineral wool is used most commonly to fill between ceiling joists in attic spaces where it can be poured from bags or blown into place with mechanical blowers. Loose mineral wool is not effective in wall cavities because clumps form at nails and other obstructions, and settlement over time creates voids and gaps in coverage. Mineral wool insulation is quick and economical, but does not provide the most efficient thermal resistance for its thickness.

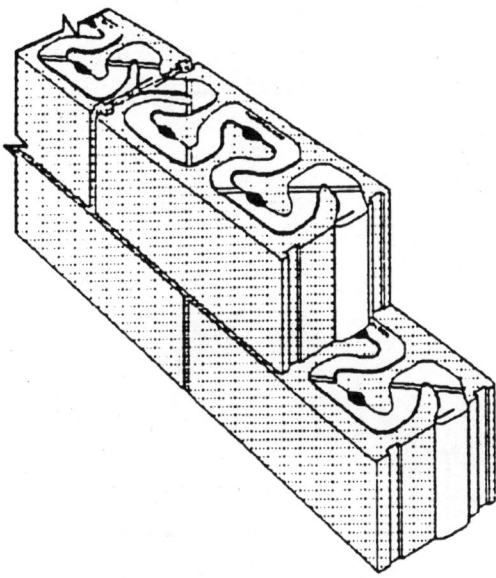

Figure 3.9 A proprietary concrete masonry unit with continuous insulation insert which eliminates thermal bridges. (*From Thermalock Products.*)

TABLE 3.10 Thermal Resistance of Concrete Masonry Units Made with Lightweight versus Normal-Weight Aggregate

CMU size, in.	Lightweight, 90-lb/ft^3 uninsulated	Normal weight 135-lb/ft^3 core insert insulated
8	2.9	2.7
10	3.0	2.8
12	3.1	2.8

SOURCE: Expanded Shale, Clay, and Slate Institute (ESCSI).

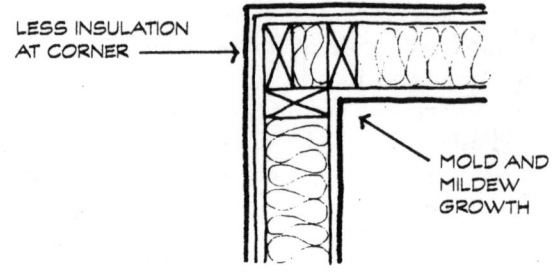

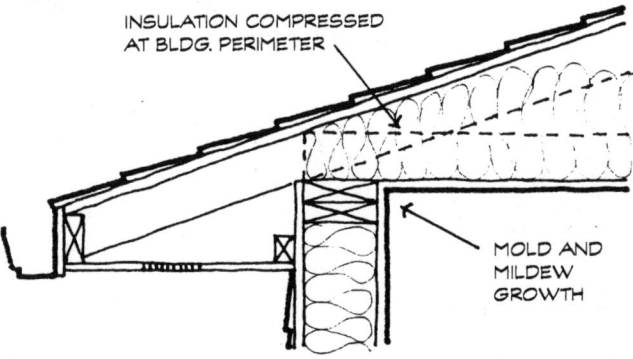

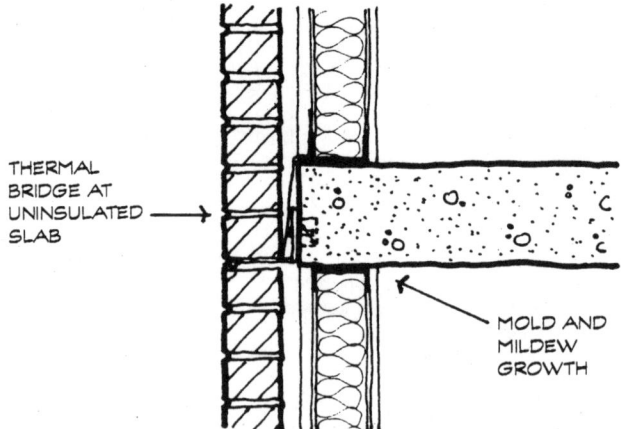

Figure 3.10 Common thermal bridges. (*From Lstiburek and Carmody, Moisture Control Handbook.*)

TABLE 3.11 Thermal Bridges at Roofing Fasteners

System design	Fasteners @ 1 per 4 sq. ft	Fasteners @ 1 per 2 sq. ft	Fasteners @ 1 per 1 sq. ft
Insulation type	Extruded	Extruded	Extruded
Fastener type	Deck screw, ¼ in. dia.	Deck screw, ¼ in. dia.	Deck Screw, ¼ in. dia.
Fastener conductivity	365	365	365
Fastener area	490.87 sq. in.	981.75 sq. in.	1963.50 sq. in.
Roof area	5,760,000 sq. in.	5,760,000 sq. in.	5,760,000 sq. in.
Insulation conductivity	0.2	0.2	0.26
Deck conductivity	365	365	365
Open deck area	0.00 sq. in.	0.00 sq. in.	0.00 sq. in.
Surface resistance	0.85	0.85	0.85
Deck thickness	0.14 in.	0.14 in.	0.14 in.

Comparison Results: Assumed System Resistance and U-value versus Actual System Resistance and U-value

Insulation thickness, in.	R-value Assumed	R-value Calculated	Loss, %	R-value Assumed	R-value Calculated	Loss, %	R-value Assumed	R-value Calculated	Loss, %
1.0	5.85	5.18	11.5	5.85	4.66	20.27	4.7	3.45	26.49
1.5	8.35	7.34	12.08	8.35	6.57	21.3	6.62	4.75	28.19
2.0	10.85	9.51	12.4	10.85	8.48	21.86	8.54	6.05	29.13
3.0	15.85	13.83	12.73	15.85	12.29	22.44	12.39	8.66	30.13
4.0	20.85	18.16	12.9	20.85	16.11	22.75	16.24	11.26	30.66
5.0	25.85	22.49	13.01	25.85	19.92	22.94	20.08	13.86	30.98
6.0	30.85	26.81	13.08	30.85	23.74	23.06	23.93	16.46	31.2
7.0	35.85	31.14	13.13	35.85	27.55	23.15	27.77	19.06	31.36
8.0	40.85	35.47	13.17	40.85	31.36	23.22	31.62	21.67	31.48

ASSUMPTIONS: Roof connection information based on typical manufacturer specifications with assumed ¼-in. deck screw size. Fastener requirements for membrane attachments dictated by wind uplift requirements.
SOURCE: Watts (1996).

Vermiculite is an inert, lightweight insulating material made from aluminum silicate expanded into cellular granules about 15 times their original size. *Perlite* is a white, inert, lightweight granular insulating material made from volcanic siliceous rock expanded up to 20 times its original volume. These granular fill materials are used primarily in single-wythe hollow masonry walls and in masonry cavity walls. Chemical treatments are used to impart a degree of water repellency, but the fill must be free-draining to avoid trapping moisture in the wall cavity. Settlement should not exceed 0.5% or a thermal gap will be created at the top of the wall. Open weep holes at the base of the wall must be screened to prevent the granules from spilling out.

Cellulose insulation is composed of recycled newspaper, wood chips, and other organic fibers. It can be installed by blowing or pouring into cavities, or can be combined with water and adhesives for spray application. The fibers are chemically treated to provide a degree of fire resistance. Cellulose is gaining in popularity as an environmentally friendly material because it uses recycled materials, because its manufacture is a relatively clean process, and because toxic content is usually limited to ink residues.

3.3.2 Flexible and semirigid insulation

Flexible and semirigid insulations includes *batt* and *blanket* materials of mineral wool or fiberglass. Minimum property requirements are covered in ASTM C665 *Specification for Mineral Fiber Blanket Thermal Insulation*. Batts are manufactured in 48-in. lengths, while blankets are longer and packaged in large rolls. These insulations inhibit conductive heat flow by trapping small pockets of air between the fibers. R-values increase with thickness and, up to a point, with density. If the fibers are packed too tightly, though, or if the insulation is compressed, less air is trapped and the R-value drops.

Batt and blanket insulations are typically manufactured unfaced or with coverings or facings on one or both sides. These facings serve as reinforcement, vapor retarders, reflective surfaces, or finishes, and most commonly include combinations of laminated foil, plastic, and paper. Standard widths are designed to fit between framing members spaced either 16 in. or 24 in. on center. Standard thicknesses include $3\frac{1}{2}$ in. and $5\frac{1}{2}$ in. to fill the cavities in 2×4 and 2×6 walls respectively, and $8\frac{1}{2}$ in. for attics and 2×10 cathedral ceilings where an air space is required for ventilation below the roof deck.

3.3.3 Rigid insulation

Rigid insulations include wood fiber, plastic foam, perlite, fiberglass, and cellular glass materials formed into board stock. *Wood fiberboard* insulation is usually manufactured in multiple plies and impregnated with asphalt to improve its moisture resistance. It is sturdy and economical, but also combustible, susceptible to moisture damage, and dimensionally unstable. Wood fiberboard insulation has different physical properties from wood fiber sheathing, and should meet the minimum requirements of ASTM C208 *Specification for Cellulosic Fiber Insulating Board*.

Rigid fiberglass insulation, or glass fiberboard, is made from fine, high-density glass fibers with a resinous binder, sandwiched between facings of fiber-reinforced asphalt and kraft paper. Fiberglass roof insulation resists warping, buckling, and shrinking, and provides a base for built-up roofing applications. It has low compressive strength and high water absorption. Minimum property requirements are covered in ASTM C612 *Specification for Mineral Fiber Block or Board Thermal Insulation*.

Mineral fiber board is similar to rigid fiberglass in physical characteristics and performance. ASTM C726 *Specification for Mineral Fiber Roof Insulation Board* covers minimum property requirements.

Perlite board is a mineral aggregate board composed of particles of expanded perlite combined with cellulose binders and sizing agents. It has good compressive strength and moisture resistance. Perlite board insulation is used primarily in roofing applications. Minimum property requirements are covered in ASTM C728 *Specification for Perlite Thermal Insulation Board*.

Cellular glass, also known as foam glass insulation, is made of heat-fused, closed glass cells. It is impermeable to moisture except at the joints, where both liquid and vapor can migrate. Cellular glass insulation is unfaced, so freeze-thaw cycles of free water at the surface can cause physical deterioration. It is non-absorptive, non-combustible, dimensionally stable, and has high compressive strength, but is friable and will crush under concentrated point loads. Cellular glass insulation is used only in roofing applications, and is available in 24-×48-in. boards as well as 12-×18-in. and 18-×24-in. blocks ranging in thickness from 1½ to 4 in. in ½-in. increments. Tapered boards are also available to provide slope for roof drainage. Cellular glass insulation is compatible with roofing bitumen, is moisture-resistant and dimensionally stable, and has a stable R-value. Minimum property requirements are covered in ASTM C552 *Specification for Cellular Glass Thermal Insulation.*

There are several types of rigid plastic foam insulation. Polystyrene insulation includes both molded and extruded products. What is commonly called *expanded polystyrene* (EPS) is more properly referred to as *molded expanded polystyrene* (MEPS). It is a white rigid board sometimes referred to as "bead board," manufactured in a range of low densities, and popular as a backing material for exterior insulation and finish systems. It is moisture- and vapor-resistant, and is the least expensive of the plastic foam insulations.

Extruded expanded polystyrene (XEPS) is a rigid board, typically manufactured in a density of about 2 lb/ft^3, and has a higher R-value per inch than molded expanded polystyrene board. It has a closed cell structure and will not absorb water, so it is particularly effective in below-grade applications and protected membrane roofing systems. Minimum properties for both molded and extruded polystyrene are covered in ASTM C578 *Specification for Preformed, Cellular Polystyrene Thermal Insulation.*

Polyisocyanurate foam is a closed cell rigid board insulation usually faced with asphalt-saturated felt facers or glass fiber mat facers for roofing applications and aluminum foil facers for walls.

Polyurethane insulation is very similar to polyisocyanurate, with only slightly different physical properties. Polyisocyanurate and polyurethane foam insulations are so similar that the terms are often used interchangeably. Polyurethane once dominated the rigid insulation board market, but is no longer manufactured in the United States in board form. Polyurethane board insulation has been largely replaced by polyisocyanurate board insulation. Polyisocyanurate has a high R-value per inch of thickness, but is subject to "thermal drift" or loss of resistance over time when the facers are vapor-permeable. Most manufacturers now cite "aged" R-values to account for this phenomenon. The R-value of polyisocyanurate insulation with vapor-impermeable facers is stable. Minimum property requirements are covered in

ASTM C591 *Specification for Preformed Cellular Polyisocyanurate Thermal Insulation* and ASTM C1289 *Specification for Faced Rigid Cellular Polyisocyanurate Thermal Insulation Board.*

Phenolic foam has a very small, tight cell structure and has the highest R-value per inch of thickness. The closed cells tend to open with time, though, decreasing the long-term R-value, and phenolic insulation absorbs moisture more readily than the other foam plastic insulations. Because of problems associated with board shrinkage and accelerated corrosion of metal decks, domestic production of phenolic foam board ceased in 1992, and this insulation is no longer commonly used in the United States.

Composite insulation boards are composed of two or more components of different types of insulation laminated together in the factory. Typically, the primary insulation is polyisocyanurate or polystyrene that is laminated to a perlite or wood fiberboard which serves as secondary insulation and as a substrate for application of bituminous roofing membranes. Other composites include polyisocyanurate laminated to a nailable roofing substrate such as oriented strand board (OSB), waferboard, or plywood. ASTM C984 *Specification for Perlite Board and Rigid Cellular Polyurethane Composite Roof Insulation* and ASTM C1050 *Specification for Cellular Polystyrene-Cellulosic Fiber Composite Board Insulation* are the only two industry standards covering composite roof insulations.

All of the "poly-" foam insulations are heat-sensitive and combustible, and cannot be used as a substrate for hot-applied membrane roofing assemblies unless they have an overlayment of fiberboard, fiberglass, or perlite board. Both isocyanurate and urethane insulations require thermal barriers in wall or roof assemblies to protect against accidental ignition, and both are capable of emitting potentially lethal gases. Ultraviolet (UV) resistance is provided by facings of asphaltic impregnated paper or foil to protect the materials from degradation prior to encapsulation in a wall or roof.

3.3.4 Formed-in-place insulation

Formed-in-place insulations include both polyurethane and polyisocyanurate materials. *Urethane* foam insulations have been around the longest, and offer the highest R-value per inch. They are used in floor and wall cavities where a high R-value must be achieved in limited space. Urethane foam hardens with a rigid skin that has good moisture resistance. The foam should not completely fill the cavity at the time of application, however, because it expands to its maximum volume after injection. Sprayed polyurethane foam is also used in roofing applications where it can be easily applied to curved surfaces that would otherwise be difficult to insulate. The largest installation of sprayed polyurethane foam roofing is on the SuperDome in New Orleans, which is over an acre in area. Sprayed polyurethane can be applied over almost any type of roof deck and in varying thicknesses. The foam is sprayed as a liquid in which the isocyanurate component reacts with the resin component to form a solid with a surface skin that is water-resistant. Most manufacturers offer different seasonal grades of chemical mixtures formulated according to the anticipated weather conditions at the project site and the ambient and surface temperatures anticipated during application. Sprayed polyurethane is UV-sensitive, and in roofing applications must have a protective coating to prevent degradation and mechanical damage. Minimum property requirements for materials are covered in ASTM C1079 *Specification for Spray-Applied Rigid Cellular Polyurethane Thermal Insulation.*

Cellulose insulation may be mixed with water and adhesives for a wet spray application. Sprayed cellulose is typically used for walls, and dry cellulose for horizontal areas such as attics. Sprayed cellulose insulation must be allowed to dry thoroughly before being enclosed in a wall, or mildew problems could later develop.

3.3.5 Insulation requirements

The properties and characteristics of the most commonly used building insulation materials are listed in *Table 3.12*. State and local energy codes often dictate requirements for minimum thermal resistance of building components. The map in *Fig. 3.11* gives recommendations for different climates in the continental United States as listed in the ninth edition of *Architectural Graphic Standards.* The most widely accepted method of calculating minimum required insulation for various building envelope components is the American Society of Heating, Refrigerating and Air Conditioning Engineers (ASHRAE) Standard 90.1 *Energy Efficient Design of New Buildings Except Low-Rise Residential Buildings.* Although the ASHRAE method is referenced by each of the three model building codes [i.e., Building Officials and Code Administrators (BOCA) *National Building Code,* Southern Building Code Congress (SBCC) *Standard Building Code,* and International Conference of Building Officials (ICBO) *Uniform Building Code*], some local jurisdictions impose more stringent requirements.

3.3.6 Loss of thermal resistance

All insulation materials have an acceptable level of moisture, below which they are considered "dry" and above which they are considered "wet." Acceptable moisture levels and "wet" insulation are often incorrectly defined in terms of equilibrium moisture content (EMC) which is based on the moisture in an insulation material when it is in a laboratory atmosphere of 75°F at 90% relative humidity. This is actually a very small amount of moisture, and does not have any measurable effect on the thermal resistance of the material. A better indicator of the performance of insulation is the thermal resistance ratio (TRR).

Thermal resistance ratio is calculated as the ratio of a material's wet thermal resistivity to its dry thermal resistivity, expressed as a percentage. A dry specimen

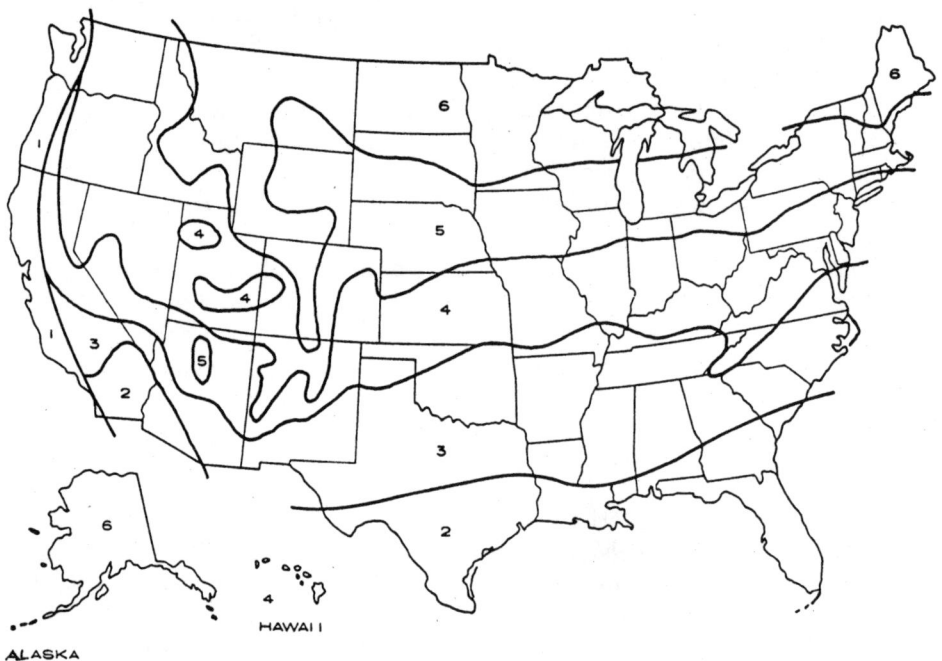

Figure 3.11 Recommended minimum thermal resistances for ceilings, walls and floors. (*From Architectural Graphic Standards, 9th ed.*)

has a TRR of 100%. As moisture accumulates in an insulation and its thermal resistance decreases, its TRR also decreases. The roofing industry has established that all roof insulation must maintain a TRR of 80% as a pass/fail criteria for in-service conditions. Insulation with a TRR of 80% or less is considered wet. *Table 3.13* shows the relationship of equilibrium moisture content (EMC) to moisture content by percent of dry weight and percent of volume for a thermal resistance ratio (TRR) of 80%. The wide variation in weight and density among various types of insulation gives the moisture content percentages a wide range also, but each material is considered to provide acceptable thermal performance up to the stated moisture content. *Figure 3.12* shows the relationship between TRR and moisture content for different insulating materials.

3.4 Calculating Thermal Gradients

The temperature variation through the thickness of a material or assembly of materials is called the *temperature gradient*. The outside surface temperature of a building varies with the seasons, with changes in the weather, with time of day, and with cloud cover or pollutants which obscure or filter solar radiation. The precise calculation of a temperature gradient through the building envelope under these constantly fluctuating conditions is difficult, and is made more so by the heat storage capacity of the various materials. A high degree of precision, however, is not usually required. For most purposes, it is adequate to calculate the temperature gradient by assuming steady-state parallel heat flow conditions. If we assume that the heat flow is straight through the building enclosure and is not deflected sideways, then the rate of heat flow is directly proportional to the magnitude of the temperature difference, and inversely proportional to the thermal resistance. The temperature drop through each component of a wall or roof is proportional to its thermal resistance.

Since the temperature drop through a building section is proportional to the resistance of the components, the total temperature drop can be apportioned to the various components in the ratio of their thermal resistance to

TABLE 3.12 Properties of Common Thermal Insulation Materials

Insulation	Density, lb/ft^3	R-value per inch thickness	Water vapor permeability, perm in.	Water absorption, % by weight	Dimensional stability
Glass fiber					
Batts/blankets	1.5–4.0	3.14	100	2	No change
Rigid boards	4.0–9.0	3.8–4.8	100	10	No change
Rock or slag wool	1.5–2.5	2.9–3.7	100	2	No change
Cellulose (loose)	2.0–3.0	2.8–3.7	100	15	Settles 0–20%
Molded polystyrene (boards)	0.9–1.8	3.6–4.4	1.2–5.0	2–3*	No change
Extruded polystyrene (boards)	1.6–3.0	4.0–6.0	0.3–0.9	1–4	No change
Polyurethane	1.7–4.0	5.8–6.2†	2.0–3.0	Negligible	0–12% change
Polyisocyanurate	1.7–4.0	5.8–7.8†	2.5–3.0	Negligible	0–12% change
Perlite (loose)	5.0–8.0	2.63	100	Low	Settles 0–10%
Vermiculite (loose)	4.0–10.0	2.4–3.0	100	None	Settles 0–10%

*By volume.
†Unfaced.
SOURCE: *Architectural Graphic Standards,* 9th ed., John Wiley & Sons, New York, 1994.

TABLE 3.13 Equilibrium Moisture Content and Moisture Content at 80% Thermal Resistance Ratio

Insulation	Equilibrium moisture content (% dry weight) at 75°F and 90% RH	Moisture content (% dry weight) at 80% TRR	Moisture content (% of volume) at 80% TRR
Fiberboard	8.5–10.0	15	4.4
Gypsum	1.0–2.5	8	7.0
Cellular glass	0.2	23	3.1
Expanded polystyrene, 1 psf	2.0	383	6.1
Expanded polystyrene, 2 psf		248	7.2
Expanded polystyrene, 3 psf		82	4.3
Extruded polystyrene	0.8	185	5.9
Urethane	6.0	262	8.8
Isocyanurate	3.0	262	8.8
Lightweight concrete, 23 pcf	5.0	10	3.7
Lightweight concrete, 37 pcf	6.0	9	5.3
Perlite board	5.0	17	2.7
Phenolic	23.4	25	1.0
Cork		39	9.9
Formed-in-place urethane		130	6.5
Fibrous glass	1.1	42	6.2

SOURCE: U.S. Department of Commerce, National Bureau of Standards, NBS Technical Note 965, and Tobiasson et. al., Cold Regions Research and Engineering Laboratory, *New Wetting Curves for Common Roof Insulations.*

the total thermal resistance. The temperature drop can be arithmetically calculated (see *Table 3.14*), and the resulting gradient then plotted on a scale drawing of the section (*Fig. 3.13A*). Using an alternative graphical method with each component of the section drawn to a scale proportional to its thermal resistance, a straight line joining inside and outside temperature will automatically distribute the temperature gradient in proportion to the thermal resistance of each component (*Fig. 3.13B*). The arithmetic calculation is not a lengthy one. It is probably the easier method to use if a wall is being designed to meet fixed internal and external temperature conditions and the components of the wall are selected to suit. On the other hand, if a tentative wall design is chosen and the effects of varying temperature conditions need to be studied, the graphical method may be more convenient.

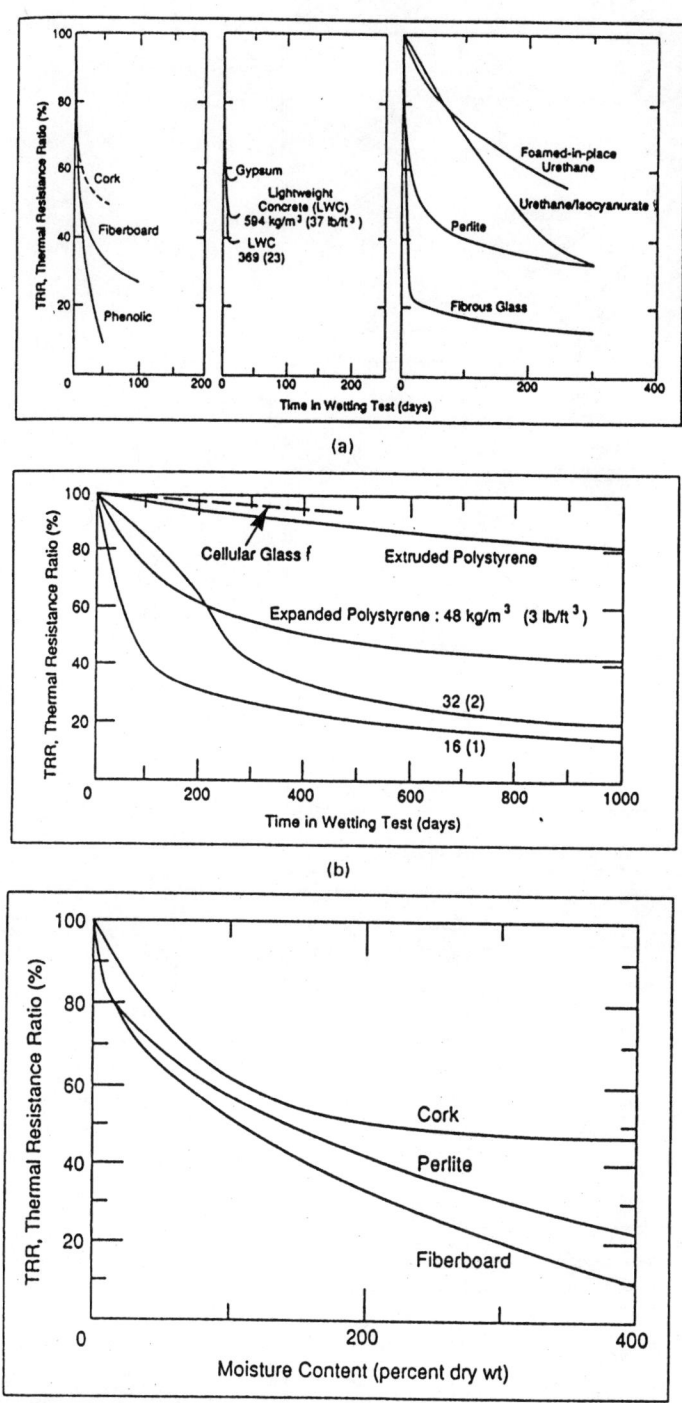

Figure 3.12 Thermal resistance and moisture content of insulation. (*From Tobiasson, New Wetting Curves for Common Roof Insulations.*)

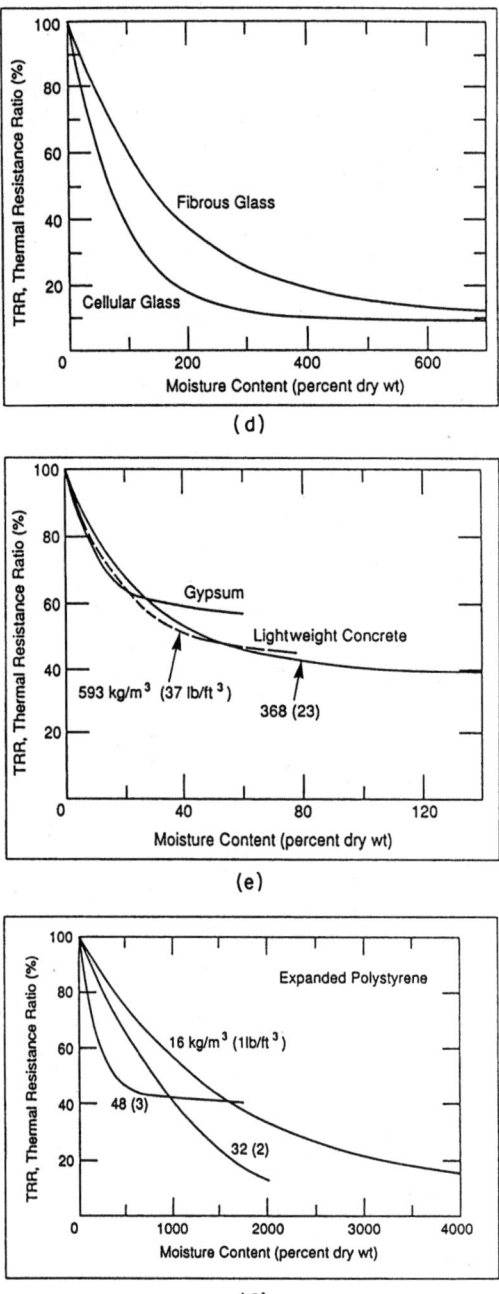

Figure 3.12 (*Continued*)

TABLE 3.14 **Arithmetic Determination of Temperature Gradient**

Component	Thickness, in.	R-value	Temperature drop, °F	Interface temperature, °F
Internal air film (still air)		0.68	3	70
Gypsum plaster (sand aggregate)	⅝	0.11	1	67
Concrete block (lightweight aggregate)	8	2.0	9	66
				57
Cement mortar	¼	0.05	0	
				57
Foamed plastic insulation	2	8.33	39	
Air space	2	1.14	5	18
Face brick	4	0.44	2	13
External air film (15-mph wind)		0.17	1	11
				10
Total		12.92	60	

SOURCE: Latta (1973).

3.4.1 Calculating surface temperatures

The *interior* surface temperature of a wall, ceiling, or floor can be calculated from the following formulas. For horizontal heat flow (through walls):

$$T_s = [0.68 \times U \times (t_i - t_o)] \tag{3.5}$$

T_s = surface temperature where
t_i = interior air temperature
t_o = exterior air temperature
U = U-value

For vertical heat flow upward (through ceilings in winter):

$$T_s = [0.61 \times U \times (t_i - t_o)] \tag{3.6}$$

For vertical heat flow downward (through ceilings in summer or floors in winter):

$$T_s = [0.92 \times U \times (t_i - t_o)] \tag{3.7}$$

The effect on interior surface temperature of increasing or decreasing thermal resistance can be calculated with these formulas. The difference between surface temperature and air temperature can be important in determining the potential for interior surface condensation. If the thermal resistance of the enclosure is

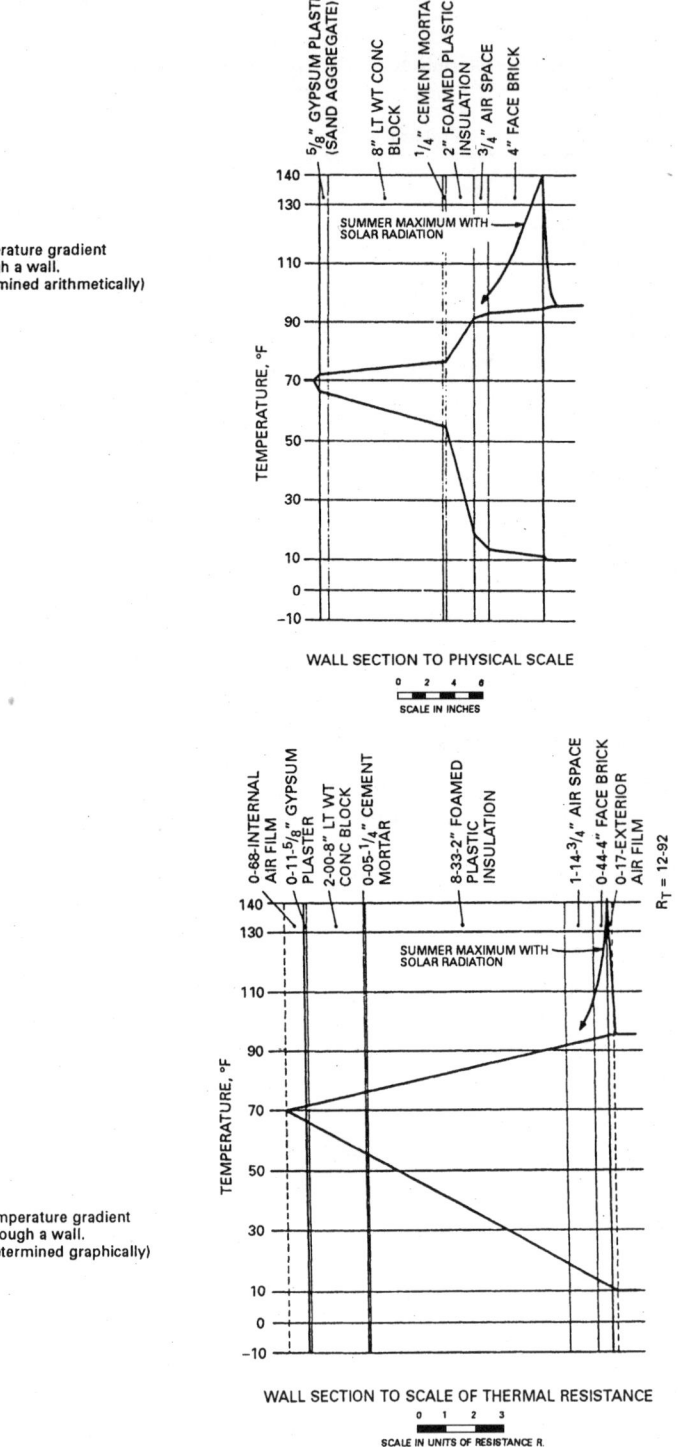

Figure 3.13 Temperature gradients determined arithmetically and graphically. (*From Latta, Walls, Windows and Roofs for the Canadian Climate.*)

increased by adding more insulation, the difference between interior surface temperature and interior air temperatures decreases.

It may sometimes also be necessary to calculate *exterior* wall surface temperatures. Extreme *winter* wall surface temperatures are usually established simply from ASHRAE winter design dry bulb air temperatures, since the wall surface will usually be within a few degrees of ambient, depending on the amount of insulation present. *Summer* wall surface temperatures, however, are affected by both ambient temperature and solar radiation, so radiative heat gain must be considered:

$$T_s = T_a + XS \tag{3.8}$$

where T_s = surface temperature
 T_a = summer air temperature (dry bulb)
 X = constant for thermal storage capacity of material (*Table 3.15*)
 S = solar absorption coefficient of material (*Table 3.16*).

The extreme surface temperature range or total temperature differential to which a wall or roof section will be exposed can be found by subtracting winter surface temperature from summer surface temperature.

3.4.2 Calculating the effect of thermal bridges

Heat flow and thermal gradients through thermal bridges can be calculated in the same way as any other component of a building envelope. *Figure 3.14* shows graphically and arithmetically the difference in temperature through a stud and the adjacent wall section between studs. An insulating sheathing on the outside of the studs would mitigate some of the effects of the thermal bridging. The relative area of thermal bridges to total surface area also has an effect on overall performance. Changing stud spacing from 16 in. to 24 in. on center, for example, decreases the effective bridged area from 10% to 6.2% of the total wall area.

To fully account for the effects of metal bridges, some tributary area must be assigned to the relative thermal resistances, depending on the size, shape, and location of the bridge. The wall in *Fig. 3.15* has thermal bridges where the wood studs interrupt the layer of insulation. The parallel *path* method is used to calculate the effect of nonmetallic thermal bridges, where the width of the stud is path A, and the path at the insulation is path B. The calculations in this example show that the average U-value is actually 6% higher than the U-value at the insulated section alone would indicate. So unless the bridging effect is taken into account, the U-value of a "typical" wall section would be misleading as to the actual thermal performance of the building.

The wall in *Fig. 3.16* shows an insulated double-wythe masonry cavity wall with a thermal bridge at the metal tie. Metallic bridges are analyzed by the parallel *zone* method, where a slightly larger area is assumed to be affected than just the actual area of the metal itself (zone A). The ASHRAE *Handbook of Fundamentals* prescribes a method for determining the size and shape of each zone. In the case of a metal beam, the surface shape of zone A would be

TABLE 3.15 Constant for Thermal Storage Capacity of Material (X)

$X=$	100	Low heat capacity materials*
or		
$X=$	130	Solar radiation reflected on low heat capacity materials†
$X=$	75	High heat capacity materials*
or		
$X=$	100	Solar radiation reflected on high heat capacity materials†

*Materials such as exterior insulation and finish systems and well-insulated metal panel curtain walls have low thermal storage capacity. Materials such as precast concrete panels and masonry walls, on the other hand, have high thermal storage capacity.
†If the wall surface receives reflected as well as direct radiation, use the larger constant. Reflected radiation can come from adjacent wall surfaces, roofs, and paving.
SOURCE: O'Connor (1990).

a strip of width W centered on the beam. Since the metal tie in *Fig. 3.16* is of circular wire, zone A is a circle of diameter W, which is calculated from the equation

$$W = m + 2d \tag{3.9}$$

where W = width or diameter of the zone
m = width or diameter of the metal path
d = distance from the panel surface to the metal, in. (but not less than 0.5 in.)

In this example, assuming one masonry tie for every $4\frac{1}{2}$ sq ft of wall area, the thermal bridge increases the average U-value of the wall by 3.19%. In an uninsulated cavity wall (*Fig. 3.17*), the effect of the thermal bridge is considerably less (0.26%) because there is less difference between the thermal conductance of the wall section and the metal tie.

As the distance between the exterior face of the wall and the edge of the metal bridge increases, the calculated size of the affected zone also increases. *Figure 3.18* illustrates this phenomenon. The web thickness (0.0359 in.) rather than the flange width of the metal stud is used in calculating the area of the zone. The diameter of the bridge zone from the stud through each veneer anchor (W) is calculated as the distance from the stud flange to the exterior surface, including the thickness of the brick veneer (3.75 in.), plus the air space (1 in.), plus the sheathing (0.50 in.), so

$$W = m + 2d = 0.0359 + 2(3.75 + 1.0 + 0.5) = 10.536 \text{ in.}$$

With the zone of the thermal bridge at each veneer anchor this large, the calculated average U-value is 88.89% higher than the calculated U-value of the wall without the bridge. Using an insulating sheathing in the cavity instead of batt insulation between the studs would prevent this thermal heat loss.

3.4.3 Effect of insulation location

Neither the amount of heat that flows through a wall nor the interior surface temperature are affected by insulation location, but other things are. In addition to affecting the rate of heat loss through thermal bridges, the location of insulation within a building section also affects thermal expansion and contraction of envelope materials, and condensation on and within the envelope.

TABLE 3.16 Solar Absorption Coefficients (S)

Material	Coefficient	Material	Coefficient
Optical flat black paint	0.98	Red oil paint	0.74
Glass, tinted (1/4")	0.94–0.96	Brick, light buff (yellow)	0.50–0.70
Flat black paint	0.95	Surface color, light gray	0.65
Glass, clear (1/4")	0.93–0.95	Concrete, natural	0.65
Glass, clear (low-e)	0.91–0.94	Mineral board, white	0.61
Glass, reflective	0.64–0.92	Aluminum, clear finish	0.60
Black lacquer	0.92	Medium light buff bricks	0.60
Dark gray paint	0.91	Medium dull green paint	0.59
Black concrete	0.91	Medium orange paint	0.58
Dark blue lacquer	0.91	Marble, white	0.58
Galvanized steel, unfinished	0.90	Medium yellow paint	0.57
Black oil paint	0.90	Glass, tinted (1/4")	0.48–0.53
Stafford blue bricks	0.89	Medium blue paint	0.51
Dark olive drab paint	0.89	Medium kelly green paint	0.51
Dark brown paint	0.88	Brick, white	0.25–0.50
Dark blue-gray paint	0.88	Light green paint	0.47
Blue or dark green lacquer	0.88	Surface color, white	0.45
Brown concrete	0.85	Polished brass, copper	0.40
Brick, red	0.65–0.85	Aluminum paint	0.40
Medium brown paint	0.84	White semigloss paint	0.30
Glass, reflective (1/4")	0.60–0.83	Galvanized steel, white	0.26
Surface color, dark gray	0.80	White gloss paint	0.25
Copper, tarnished	0.80	Silver paint	0.25
Medium light brown paint	0.80	White lacquer	0.21
Brown or green lacquer	0.79	Polished alum., chrome	0.20
Medium rust paint	0.78	Polished alum. reflector	0.12
Wood, smooth	0.78	Aluminized mylar film	0.10
Mineral board, natural color	0.75	Tinned surface	0.05
Light gray oil paint	0.75	Vapor-deposited coatings	0.02

SOURCE: O'Connor (1990).

Component	R-value	Temperature Drop, °F	Interface Temperature, °F
Through Wall Cavity			
Inside Air Film	0.68	5	70
			65
1/2" Gypsum Board	0.45	4	
			61
2" Insulation	8.00	66	
			-5
1-5/8" Air Space	1.26	10	
			-15
1/2" Gypsum Board	0.45	4	
			-19
Outside Air Film	0.17	1	-20
TOTAL	11.01	90	
Through Stud			
Inside Air Film	0.68	10	70
			60
1/2" Gypsum Board	0.45	6	
			54
3-5/8" Wood Stud	4.55	65	
			-11
1/2" Gypsum Board	0.45	6	
			-17
Outside Air Film	0.17	3	-20
TOTAL	6.30	90	

Note: Temperature drop through any component = (Total inside/outside temperature difference ÷ Total R-value) x (R-value of component)

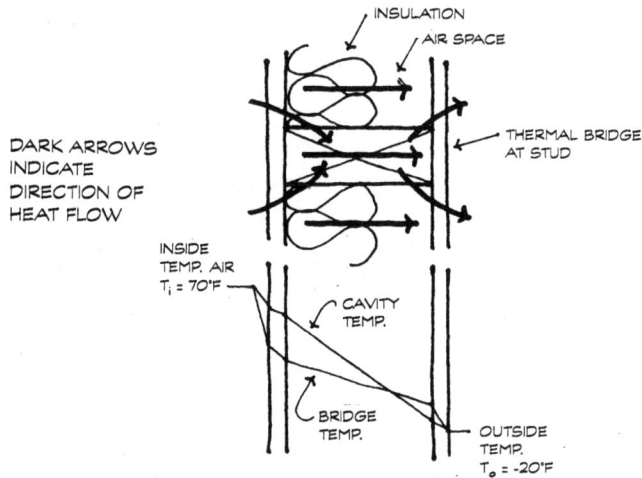

Figure 3.14 Temperature drop through wall cavity and through stud. (*From Latta, Walls, Windows and Roofs for the Canadian Climate.*)

Heat Flow and Insulation   105

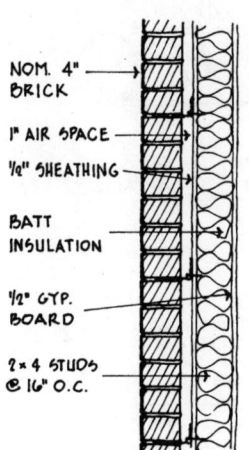
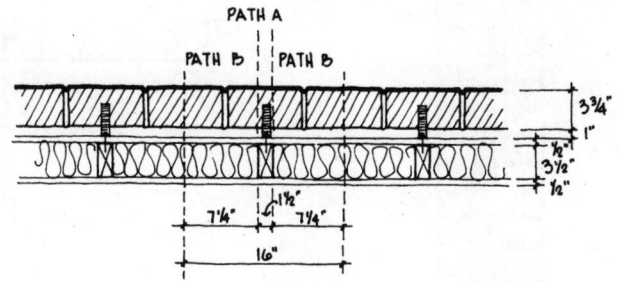

NOM. 4" BRICK
1" AIR SPACE
1/2" SHEATHING
BATT INSULATION
1/2" GYP. BOARD
2 × 4 STUDS @ 16" O.C.

Stud spacing = 16.00 in. o.c., or 1.33 ft o.c.
Height of the section = 12.00 in. or 1.00 ft
Total area, A_t = 1.33 × 1.00 = 1.33 sq ft
Width of Path A = 1.50 in. or 0.125 ft
Area of Path A, A_A = 0.125 × 1.00 = 0.125 sq ft
Width of Path B = 16.00 − 1.50 = 14.50 in. or 1.208 ft
Area of Path B, A_B = 1.208 × 1.000 = 1.208 sq ft

Section	C (Btu /(hr·°F·sq ft))	K ((Btu·in.) /(hr·°F·sq ft))	x (in.)	C_x (Btu /(hr·°F·sq ft))	Path A $1/C_x$ ((hr·°F·sq ft) /Btu)	Path B $1/C_x$ ((hr·°F·sq ft) /Btu)
Outside air surface	6.000			6.000	0.17	0.17
4-in. nominal face brick		9.000	3.75	2.400	0.42	0.42
1-in. airspace	1.030			1.030	0.97	0.97
Exterior fiberboard sheathing	0.760			0.760	1.32	1.32
2-in. × 4-in. wood stud		0.800	3.50	0.229	4.37	
3½-in. batt insulation						11.00
½-in. gypsum wallboard	2.250			2.250	0.45	0.45
Inside air surface	1.470			1.470	0.68	0.68
					R_A = 8.38	R_B = 15.01
						U_B = 0.067

R_A/A_A = 67.04 R_B/A_B = 12.43
$1/(R_A/A_A)$ = 0.015 $1/(R_B/A_B)$ = 0.080
U_{avg} = [1/(R_A/A_A) + 1/(R_B/A_B)]/($A_A + A_B$) = (0.015 + 0.080)/(0.125 + 1.208) = 0.071 Btu/(hr·°F·sq ft)

$$\frac{U_{avg} - U_B}{U_B} \times 100\% = \frac{0.071 - 0.067}{0.067} \times 100\% = 6.0\%$$

Figure 3.15 Thermal calculations for brick veneer/wood stud wall. (*From BIA Tech Note 4 Revised, Brick Institute of America, 11490 Commerce Park Drive, Reston, VA.*)

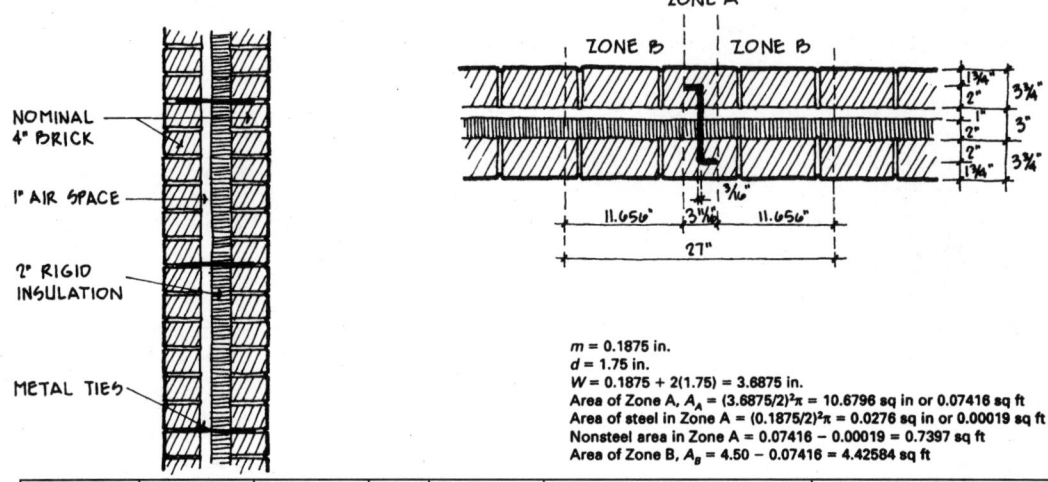

$m = 0.1875$ in.
$d = 1.75$ in.
$W = 0.1875 + 2(1.75) = 3.6875$ in.
Area of Zone A, $A_A = (3.6875/2)^2\pi = 10.6796$ sq in or 0.07416 sq ft
Area of steel in Zone A = $(0.1875/2)^2\pi = 0.0276$ sq in or 0.00019 sq ft
Nonsteel area in Zone A = 0.07416 − 0.00019 = 0.7397 sq ft
Area of Zone B, $A_B = 4.50 − 0.07416 = 4.42584$ sq ft

					Zone A			Zone B		
Section	C (Btu /(hr·°F·sq ft))	K ((Btu·in.) /(hr·°F·sq ft))	x (in.)	C_x (Btu /(hr·°F·sq ft))	A (sq ft)	$C_x \cdot A$ (Btu /(hr·°F))	$\frac{1}{C_x \cdot A} = \frac{R}{A}$ ((hr·°F) /Btu)	A (sq ft)	$C_x \cdot A$ (Btu /(hr·°F))	$\frac{1}{C_x \cdot A} = \frac{R}{A}$ ((hr·°F) /Btu)
Outside air surface	6.000			6.000	0.07416	0.445	2.25	4.42584	26.555	0.04
4-in. nominal face brick		9.000	3.75	2.400				4.42584	10.622	0.09
Brick		9.000	1.75	5.143	0.07416	0.381	2.62			
Brick		9.000	2.00	4.500	0.07397	0.333				
Steel		314.000	2.00	157.000	0.00019	0.030				
					Subtotal	0.363	2.75			
1-in. airspace	1.030			1.030	0.07397	0.076		4.42584	4.559	0.22
Steel		314.000	1.00	314.000	0.00019	0.060				
					Subtotal	0.136	7.35			
2-in polystyrene rigid board insulation		0.250	2.00	0.125	0.07397	0.009		4.42584	0.553	1.81
Steel		314.000	2.00	157.000	0.00019	0.030				
					Subtotal	0.039	25.64			
Brick		9.000	2.00	4.500	0.07397	0.333				
Steel		314.000	2.00	157.000	0.00019	0.030				
					Subtotal	0.363	2.75			
Brick		9.000	1.75	5.143	0.07416	0.381	2.62			
4-in. nominal face brick		9.000	3.75	2.400				4.42584	10.622	0.09
Inside air surface	1.470			1.470	0.07416	0.109	9.17	4.42584	6.506	0.15
					$R_A/A_A = 55.15$			$R_B/A_B = 2.40$		
					$1/(R_A/A_A) = 0.018$			$1/(R_B/A_B) = 0.417$		

$U_{avg} = [1/(R_A/A_A) + 1/(R_B/A_B)]/(A_A + A_B) = (0.018 + 0.417)/(0.07416 + 4.42584) = 0.097$ Btu/(hr·°F·sq ft)

$U_B = [1/(R_B/A_B)]/A_B = 0.417/4.42584 = 0.094$ Btu/(hr·°F·sq ft) $\frac{U_{avg} - U_B}{U_B} \times 100\% = \frac{0.097 - 0.094}{0.094} \times 100\% = 3.19\%$

Figure 3.16 Thermal calculations for insulated brick masonry cavity wall. (*From BIA Tech Note 4 Revised, Brick Institute of America, 11490 Commerce Park Drive, Reston, VA.*)

Heat Flow and Insulation   107

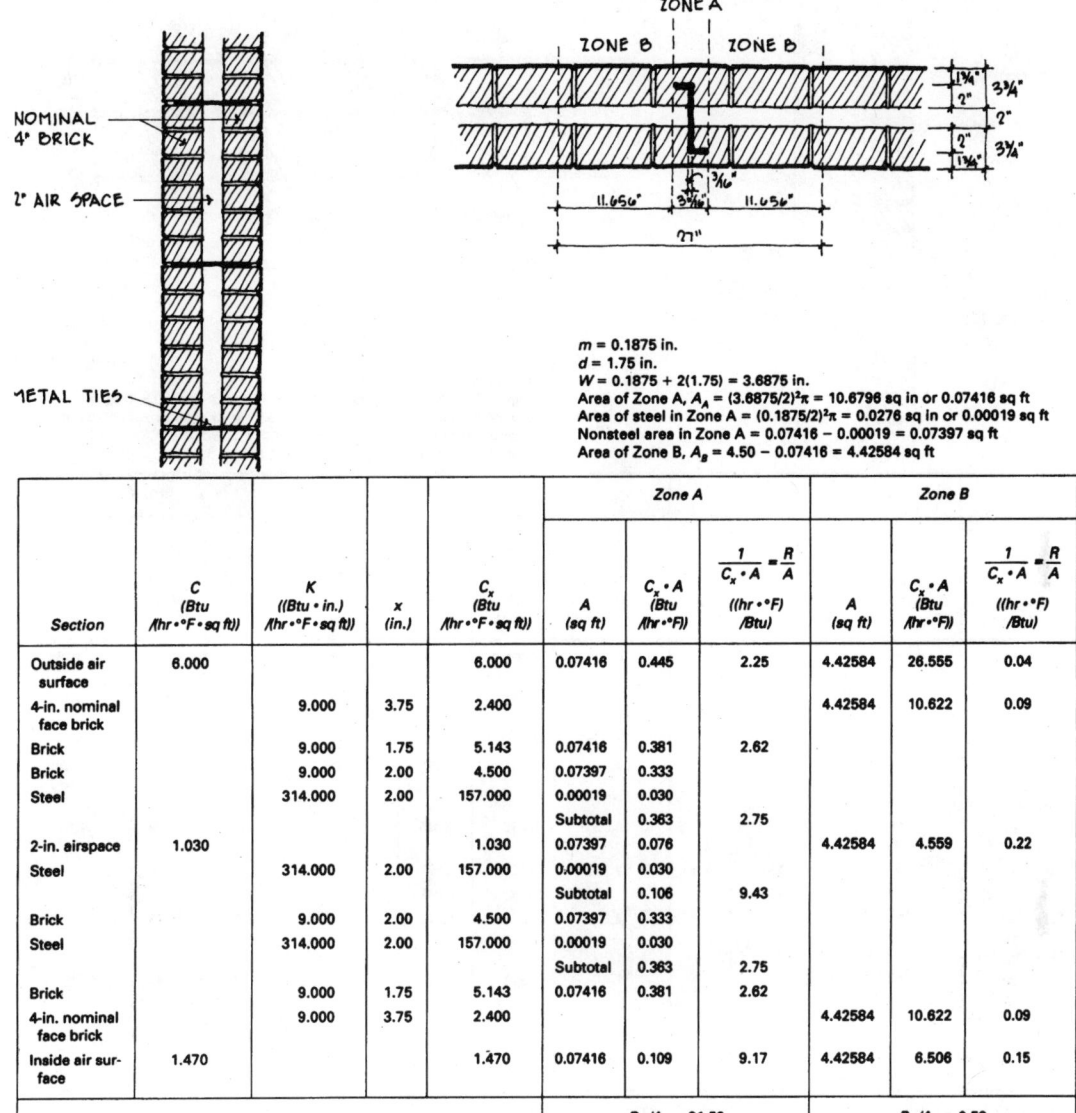

Figure 3.17 Thermal calculations for uninsulated brick masonry cavity wall. (*From BIA Tech Note 4 Revised, Brick Institute of America, 11490 Commerce Park Drive, Reston, VA.*)

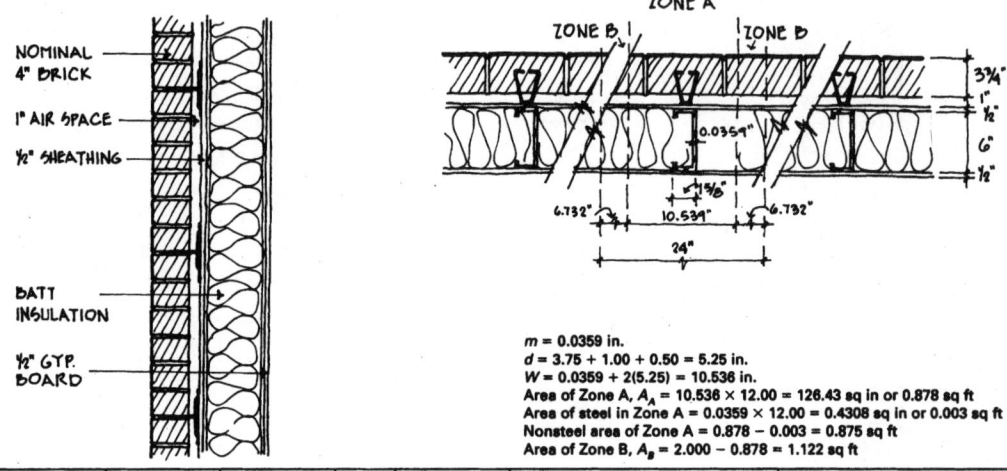

$U_{avg} = [1/(R_A/A_A) + 1/(R_B/A_B)]/(A_A + A_B) = (0.118 + 0.051)/(0.878 + 1.122) = 0.085 \text{ Btu/(hr} \cdot {}^\circ\text{F} \cdot \text{sq ft)}$

$U_B = [1/(R_B/A_B)]/A_B = 0.051/1.122 = 0.045 \text{ Btu/(hr} \cdot {}^\circ\text{F} \cdot \text{sq ft)} \qquad \dfrac{U_{avg} - U_B}{U_B} \times 100\% = \dfrac{0.085 - 0.045}{0.045} \times 100\% = 88.89\%$

Figure 3.18 Thermal calculations for brick veneer/metal stud wall. (*From BIA Tech Note 4 Revised, Brick Institute of America, 11490 Commerce Park Drive, Reston, VA.*)

Insulation reduces the rate of heat flow. If an infinite amount of insulation could be used, heat could not escape from the interior, so building envelope components outside the insulation would be exactly the same temperature in winter as the outside air. Since an infinite amount of insulation cannot be used, there is still some measurable heat flow through the envelope. Depending on the amount and the location of the insulation, the outer envelope surface will always be at least slightly warmer than the outside air temperature in winter. In hot weather, the situation in an air-conditioned building will be reversed—the outer envelope surface will be cooler than the outside air temperature as long as the surface is not exposed to direct or reflected solar radiation. The inner envelope surface will be warmer than room temperature in summer and cooler than room temperature in winter. Since room temperature variations are usually modest, the outer components of the building envelope are subjected to a much wider range of temperatures than the inner components. Changing the location of the insulation within the building section changes the range of temperatures to which various component are subjected, affecting relative expansion and contraction and the potential for condensation.

The effects of insulation location can be illustrated by calculating thermal gradients through a simple precast concrete wall section. The uninsulated wall consists of the inside and outside air films, and a 4-in. thickness of concrete. The temperature gradient can be calculated by the arithmetic method and then plotted graphically as shown in *Fig. 3.19*. With an inside temperature of 70°F and an outside temperature of 10°F, the inside wall surface is only slightly above freezing at 36°F. At this low temperature, condensation on the wall surface during the winter is very likely even with only modest indoor relative humidity. (See Chap. 5 for a discussion of dew points and condensation.)

The thermal performance of the wall is improved by adding 1 in. of insulation to the inside face of the concrete panel. To protect the insulation, plaster or gypsumboard must be added. The calculation for this revised section (*Fig. 3.20*) shows the inside surface of the plaster now at 63°F. At this significantly higher temperature, surface condensation is unlikely at typical interior humidity levels.

Although insulation located on the inside of the wall reduces the rate of heat loss, it can create other problems. If the concrete panel is exposed to a relatively wide range of exterior temperatures, it will experience much greater thermal expansion and contraction than the interior plaster. The differential movement between the two must be accommodated by flexible connections or the plaster will crack. If the concrete panel is an infill between structural frame members, then the structural members form thermal bridges through the insulation. Under the same temperature conditions as above, the frame surfaces exposed to the room will be cold enough to produce winter surface condensation. Differential movement between frame members and infill panels also makes it difficult to maintain a weathertight seal in the perimeter joints. If the panels are installed outside the structural frame, there is often a space between the frame and the panel in which it is difficult to install insulation but in which air can circulate. Any indoor air which escapes into this cavity will condense unless its relative humidity is very low. Because the panel temperature is well below freezing, the condensation will form as frost and be released in accumulated quantities when warmer temperatures cause melting.

While the problem of interior surface condensation is solved by adding insulation on the inside of the concrete panel, others are created. Most of

	Thickness n. in.	Conductivity, k	Conductance $C = \frac{k}{n}$	Resistance $R = \frac{one}{C}$	Temperature Drop, °F	Interface Temperature, °F
						70° Inside Air
Inside Air Film	-	-	1.46	0.68	34	
						36
Precast Concrete	4	12.0	3.00	0.33	17	
						19
Outside Air Film	-	-	6.00	0.17	9	
						10° Outside Air
TOTAL THERMAL RESISTANCE				1.18		

Drop in temperature through any component

$= \dfrac{\text{Difference between outside and inside air temperatures}}{\text{Total Thermal Resistance of Wall}} \times$ Thermal Resistance of Component

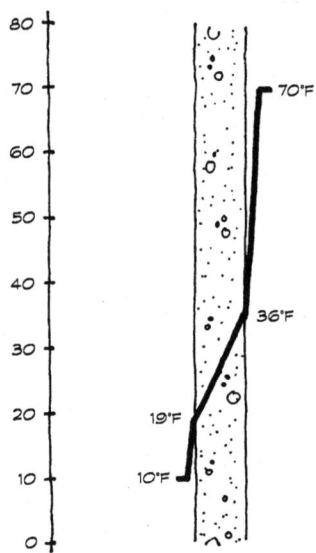

Figure 3.19 Calculation of temperature gradient for an uninsulated concrete wall. (*From Latta, Walls, Windows and Roofs for the Canadian Climate.*)

	Thickness n, in.	Conductivity k	Conductance $C = \frac{k}{n}$	Resistance $R = \frac{one}{c}$	Temperature Drop, °F	Interface Temperature, °F
						70° Inside Air
Inside Air Film	–	–	1.46	0.68	7	
						63
Plaster (sand aggregate)	3/4	–	7.70	0.13	1	
						62
Foamed Plastic Insulation	1	0.24	0.24	4.17	46	
						16
Precast Concrete	4	12.0	3.00	0.33	4	
						12
Outside Air Film	–	–	6.00	0.17	2	
						10° Outside Air

TOTAL THERMAL RESISTANCE   5.48

Drop in temperature through any component

$$= \frac{\text{Difference between outside and inside air temperatures}}{\text{Total Thermal Resistance of Wall}} \times \text{Thermal Resistance of Component}$$

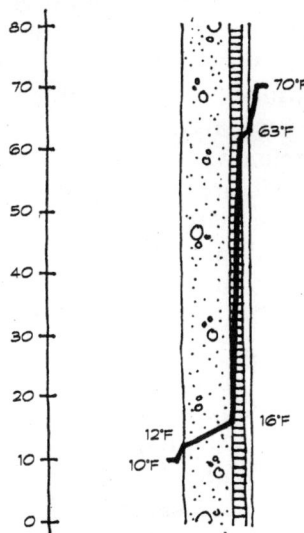

Figure 3.20 Calculation of temperature gradient for a concrete wall insulated on the inside. (*From Latta, Walls, Windows and Roofs for the Canadian Climate.*)

the problems arise either because the panel itself is colder than before, or because its wider temperature fluctuations cause increased expansion and contraction. If the insulation is moved to the outside of the panel and covered with stucco, the thermal gradient is calculated as shown in *Fig. 3.21*. The inside surface temperature is still 63°F, but with the insulation on the outside, both the structural frame and the panels are in a more stable thermal environment. Differential movement between the panels and the frame, and between adjacent panels, is reduced, and room air which infiltrates the space between the structural frame and the panels cannot condense because the surface temperatures are too warm.

3.4.4 Slabs and below-grade walls

Soil is not a good insulating material, but it does have thermal mass which minimizes fluctuations in temperature. *Daily* temperature fluctuations affect only the top $1\frac{1}{2}$ to 2 ft of soil. *Annual* temperature fluctuations above and below the "steady-state" temperature affect the first 20 to 30 ft of soil (*Fig. 3.22*). Below this depth, the soil temperature is constant. Approximate steady-state soil temperatures for the 48 contiguous states are given in *Fig. 3.23*. Winter frost depths are shown in *Fig. 3.24*.

Since average ground temperatures for most of the United States are below comfortable room temperatures, basements and below-grade structures continuously lose some heat to the soil. In summer, this provides a source of cooling for interior spaces. Uninsulated slabs in northern climates do create small heating loads on mechanical equipment, but they are usually considered negligible. In climates or occupancies with high humidity levels, however, the temperature of an uninsulated floor slab may fall below the dew point, requiring either the addition of continuous underfloor insulation or the dehumidification of indoor air. Insulation may also be required when groundwater levels are high. Even though high groundwater levels are often seasonal and usually lowest in winter, increased soil moisture content because of the capillary rise of water can increase the soil's conductivity. Continuous underslab insulation may then be desirable, but underslab drainage must also receive special attention. Refer to Chap. 7 for more information.

The thermal resistance of soil is generally estimated at R-1 to R-2 per foot of thickness. At an average of R-1.25, it takes 4 ft of soil to equal the insulating value of 1 in. of extruded polystyrene insulation. Because heat flow from floor slabs and below-grade walls follows a radial path (*Fig. 3.25*), however, the effective insulating value of soil is greater than would be initially apparent from R-values alone. This radial path of heat flow means that the perimeter of a slab-on-grade is subject to much greater heat loss than the interior floor. *Figure 3.26* shows the heat flow from the perimeter of a floor slab to a cold exterior ground surface as a series of nearly concentric radial paths. As the length of the heat flow path increases, the effective insulating value of the soil increases, so thermal insulation is generally required only at the perimeter of the slab. Placing this insulation vertically on the outside of the foundation provides the

	Thickness n, in.	Conductivity k	Conductance $C = \frac{k}{n}$	Resistance $R = \frac{one}{C}$	Temperature Drop °F	Interface Temperature °F
						70° Inside Air
Inside Air Film	-	-	1.46	0.68	7	
						63
Precast Concrete	4	12.0	3.00	0.33	4	
						59
Foamed Plastic Insulation	1	0.24	0.24	4.17	45	
						14
Stucco	3/4	5.0	6.67	0.15	2	
						12
Outside Air Film	-	-	6.00	0.17	2	
						10° Outside Air

TOTAL THERMAL RESISTANCE   5.50

Drop in temperature through any component

= $\dfrac{\text{Difference between outside and inside air temperatures}}{\text{Total Thermal Resistance of Wall}}$ x Thermal Resistance of Component

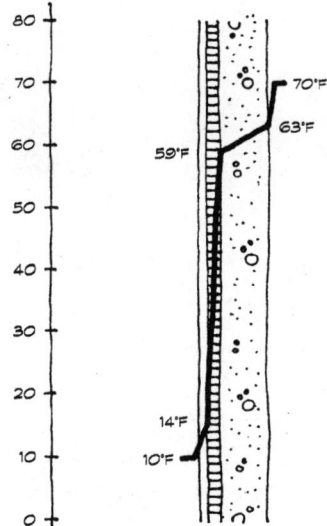

Figure 3.21 Calculation of temperature gradient for a concrete wall insulated on the outside. (*From Latta, Walls, Windows and Roofs for the Canadian Climate.*)

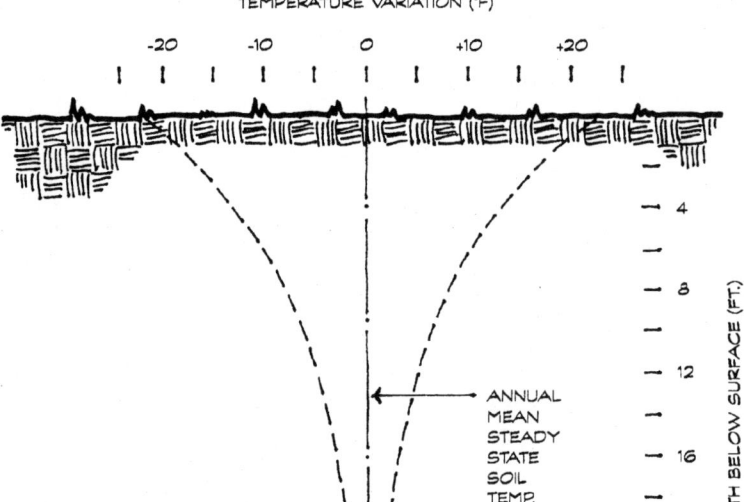

Figure 3.22 Annual temperature fluctuations in the soil. (*From Watson and Labs, Climatic Building Design.*)

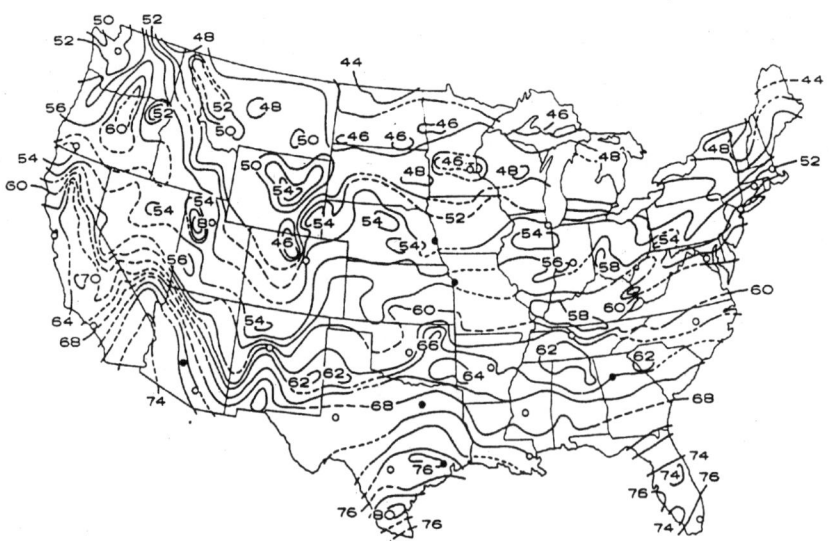

Figure 3.23 Approximate steady-state soil temperatures in the United States. (*From Architectural Graphic Standards, 9th ed.*)

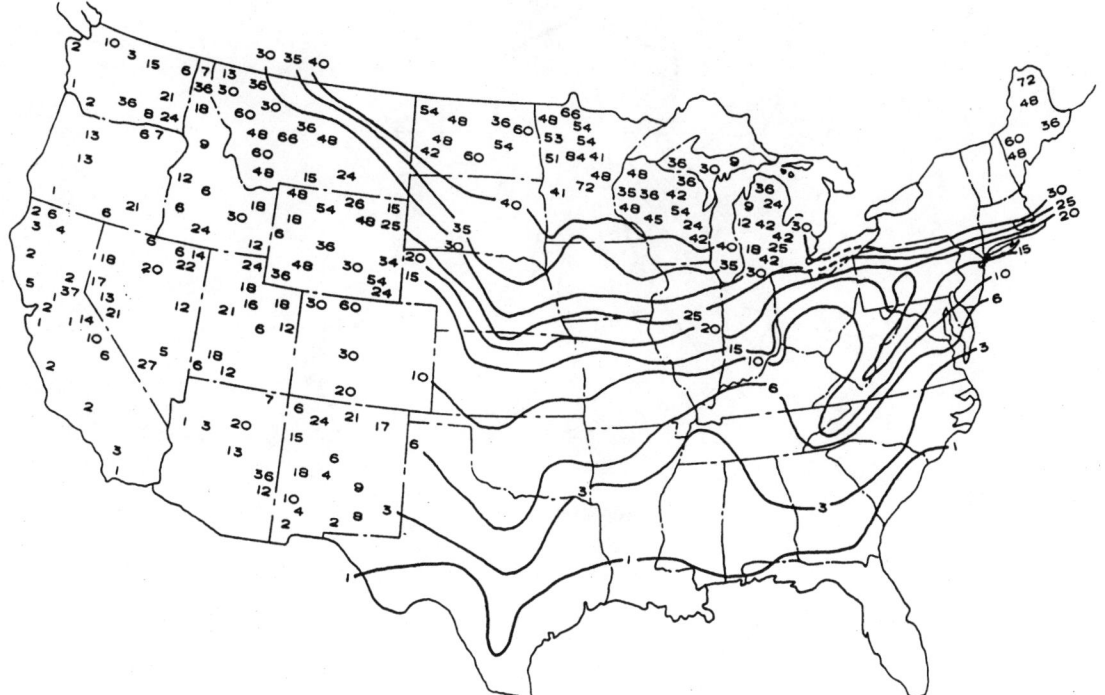

Figure 3.24 Winter frost depths in the United States. (*From Architectural Graphic Standards, 9th ed.*)

greatest protection from freeze-thaw stresses, but placing continuous horizontal insulation below the slab isolates the floor from all slab-to-ground interactions. When the groundwater level is less than 2 ft below the underside of the slab, horizontal insulation is recommended. When the groundwater level is more than 4 ft below grade, vertical placement is preferred. ASHRAE recommended R-values for perimeter insulation are shown in *Fig. 3.27*.

The heat flow paths shown in *Fig. 3.25* are affected by seasonal changes in ground temperature, so the drawing is only a simplified representation of actual conditions. It does, however, show that the center of the floor slab exchanges heat more directly with the deep soil that is more stable in temperature. With time, a relatively stable thermal gradient is established between the slab and the soil below it, and heat exchange is relatively small. The stability of this heat flow can help minimize internal temperature changes as heating or cooling equipment cycles off and on.

Heat flow from below-grade walls forms a radial path to the surface as shown in *Fig. 3.28*. Near-surface underground heat transfer can be calculated using the formula

$$R_{\text{eff}} = R_{\text{soil}} \times 2\pi r \left(\frac{90° - m°}{360°} \right) \quad (3.10)$$

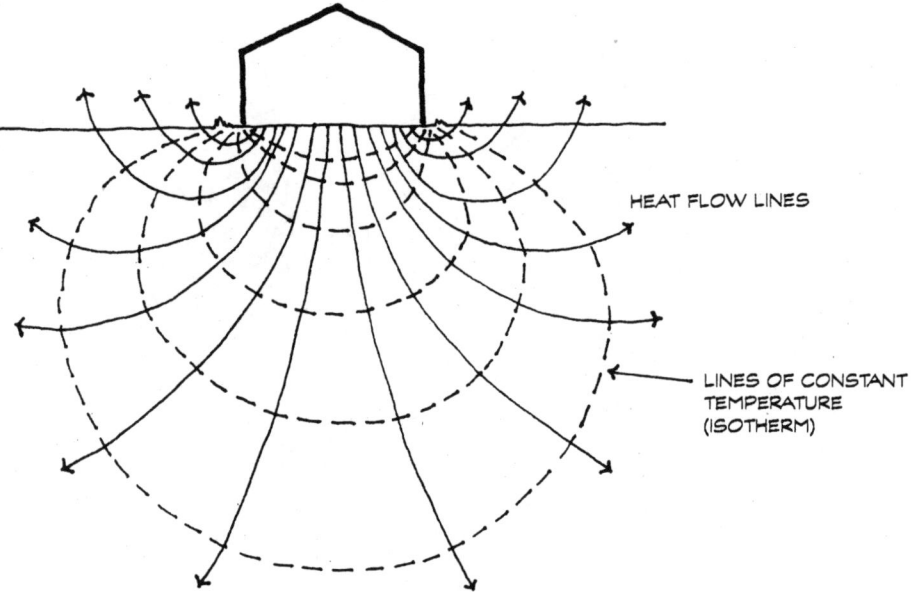

Figure 3.25 Heat flow paths from building slabs. (*From Watson and Labs, Climatic Building Design.*)

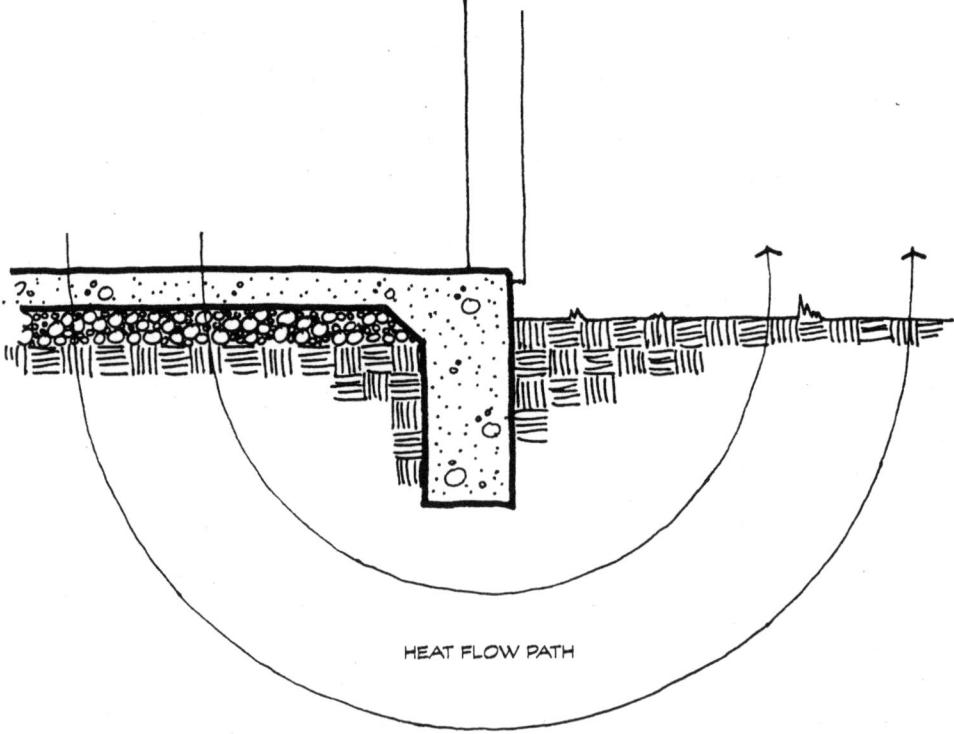

Figure 3.26 Heat flow path at slab perimeter. (*From Watson and Labs, Climatic Building Design.*)

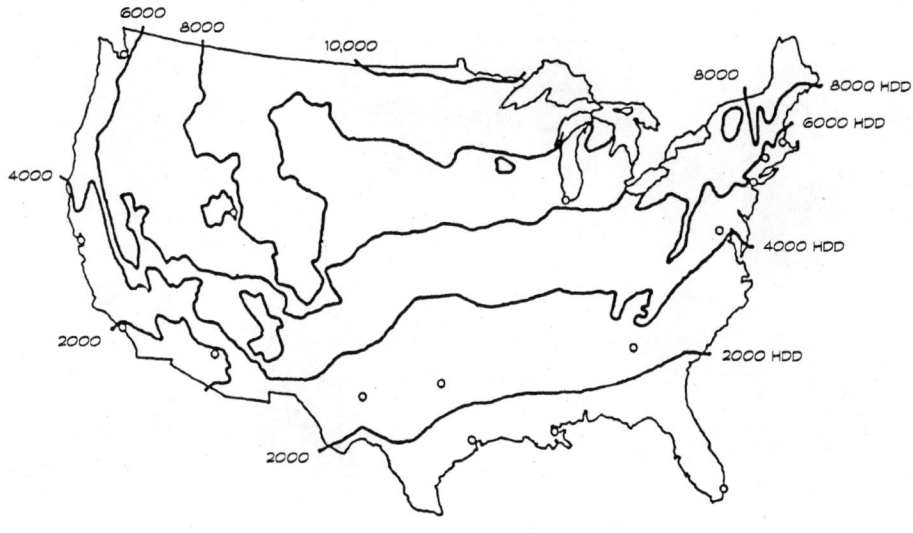

Heating Degree Days (HDD)	Recommended R-value
10,000	7.5
9,000	6.8
8,000	6.2
7,000	5.5
6,000	4.8
5,000	4.2
4,000	3.5
3,000	2.8
2,000	None required

Figure 3.27 ASHRAE recommended R-values of foundation based on number of heating degree days. (*From Watson and Labs, Climatic Building Design.*)

where R_{eff} = effective thermal resistance
R_{soil} = thermal resistance of soil (assume R-1.25)
r = radius of heat flow path, ft
m = slope of grade, degrees

Heat loss from a basement includes that which takes place through the wall above grade, and that which takes place through the wall and floor below grade. In addition, there is heat loss in the movement of air. There is a potential path of significant air leakage through the joint between the top of the base-

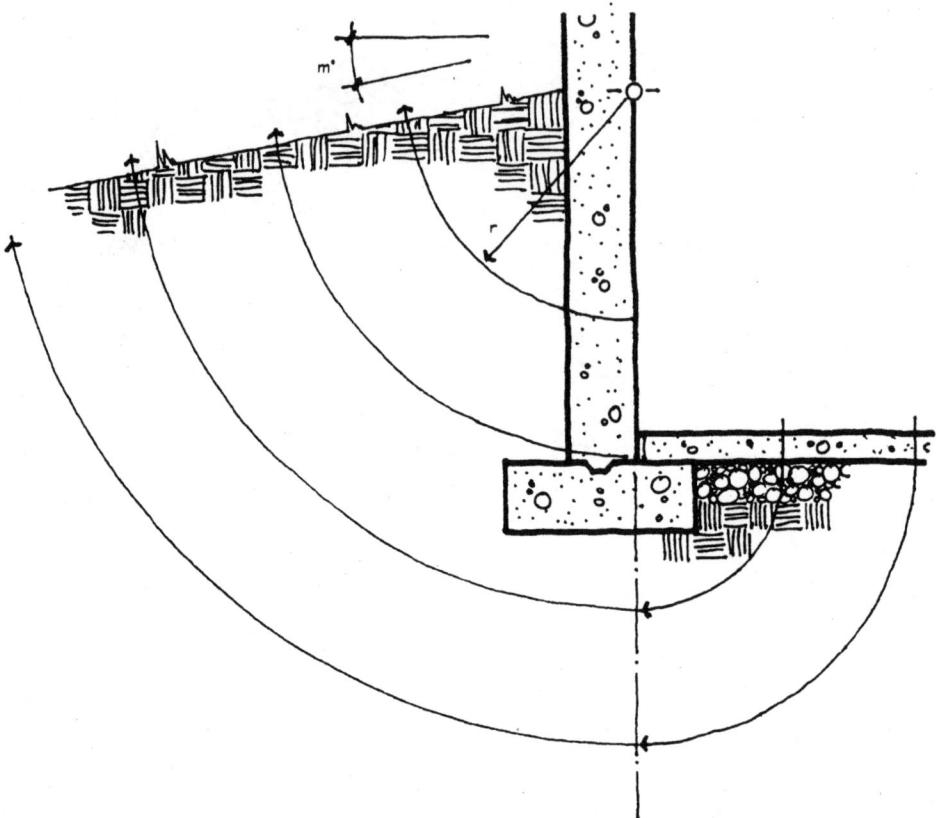

Figure 3.28 Heat flow from below-grade walls. (*From Watson and Labs, Climatic Building Design.*)

ment wall and the sill plate of the superstructure. Adding insulation to the inside face of a below-grade wall makes both the wall and the adjacent soil colder. The lower soil temperature may cause frost heave problems in some climates. With a hollow concrete block wall, part of which is exposed above grade, air in the block cores is cooled and sinks by convection, displacing warmer air in the lower parts of the wall. This causes additional heat loss from the basement as the lower portions of the wall are cooled. Insulating the outside of the wall will minimize this effect, and grouting the wall will eliminate the convective air spaces. Except in extreme northern climates where the ground temperature is colder, it is usually necessary to insulate only the first 3 to 6 ft of below-grade walls. Below this level, the cumulative thermal resistance of the soil is sufficient to prevent serious heat loss.

Building codes generally require that foundations be placed below the frost depth to prevent winter heave. In northern climates, this additional foundation depth adds significantly to the cost of construction. Home builders in the United States, however, are now using techniques first developed in Scandinavia which allow the construction of much shallower foundations than would typically be required, known as *frost-protected shallow foundations* (*Fig. 3.29*). Model building codes are beginning to change code language to allow alternative foundation designs such as these to be widely adopted.

A 1- to 2-in. layer of extruded polystyrene insulation applied to the vertical stem of the foundation wall, and a horizontal "wing" of insulation placed outside the perimeter of the building effectively blocks the natural heat flow lines and artificially raises the frost depth. Heat loss from the interior spaces raises the soil temperature above its winter norm, eventually stabilizing any temperature fluctuations since the heat cannot escape to the exterior ground surface. Continuous insulation below the slab is recommended only in northern climates with high summer humidity levels or a high water table (*Fig. 3.30*). Insulating the slab in high humidity areas prevents surface condensation on slabs by keeping the concrete temperature above the dew point of the air. In areas with a high water table, the groundwater has much the same effect as a wind chill, creating a continuous loss of heat into the soil. When high groundwater levels are present, careful attention must also be paid to subsurface drainage and vapor diffusion (see Chap. 7). For unheated spaces in cold climates such as residential garages and commercial warehouses, continuous underslab insulation is recommended to prevent heat loss to the soil through the slab.

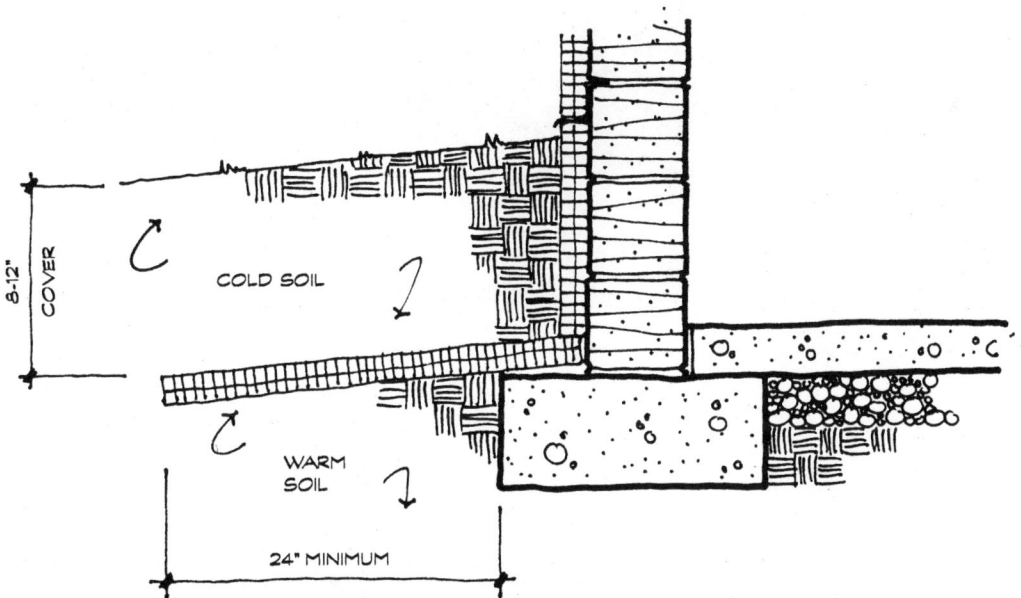

Figure 3.29 Frost-protected shallow foundation. (*From Watson and Labs, Climatic Building Design.*)

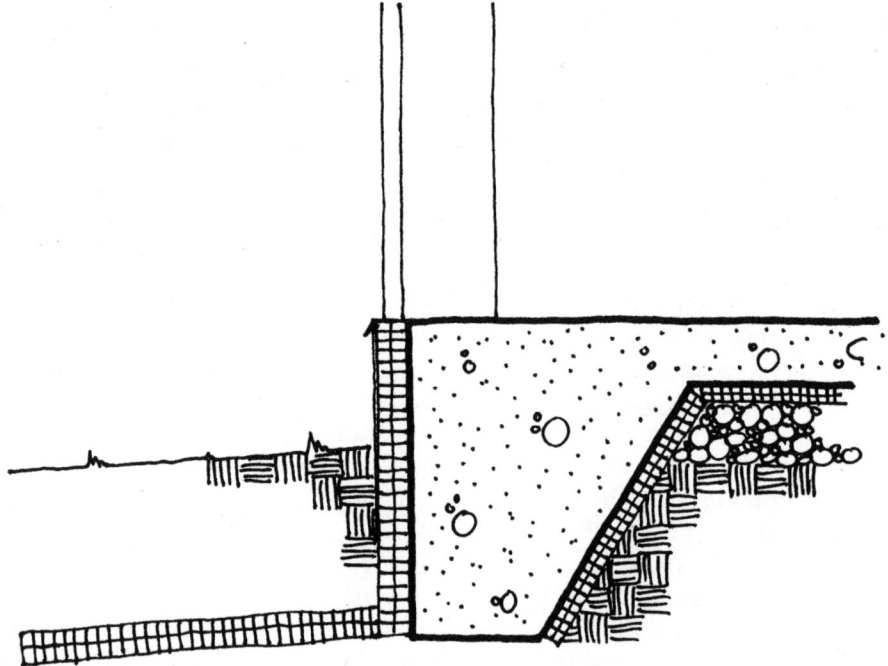

Figure 3.30 Continuous insulation below slab. (*From Lstiburek and Carmody, Moisture Control Handbook.*)

Chapter 4

Water Penetration

Moisture can occur as a liquid, solid, or gas. In the design of buildings, we are concerned with moisture in the form of rain, snow, ice, and water vapor as well as the transformation of ice and snow to water, rain to ice, and water vapor to a condensed liquid, or condensed frost.

4.1 Rain and Groundwater

Liquid water penetration primarily involves rain, melting snow, and groundwater. Rain impinges on roofs, walls, windows, doors, balconies, decks, and all exposed above-grade building elements. Rainfall amounts vary with geographic location, and exposure risk is related to prevailing winds and microclimatic conditions. Snow that accumulates on horizontal or sloping surfaces, or drifts against walls melts and seeps into porous materials, cracks, and defective joints. Groundwater pushes against below-grade walls and slabs. The highest elevation of groundwater in a given location is called the *water table*. Water tables rise and fall with seasonal moisture conditions, as well as surface contour. In low-lying areas, the water table can be very close to the ground surface, while at higher elevations, it can be quite deep. Most structures, whether they include basement space or not, are affected by the presence of groundwater.

4.2 Water+Opening+Force

Water will move from one location to another when acted upon by a force. Some forces move relatively large quantities of water over surfaces or through clearly defined passages, and other forces redistribute moisture within a material. Water penetration through a building enclosure depends on the simultaneous occurrence of three things:

- The presence of water
- An opening through which water can enter
- A physical force to move the water

This is true for water penetration under any driving force. The size of opening which will permit water penetration varies with the nature of the force. There are several forces which can cause water penetration through a building enclosure:

- Gravity
- Kinetic energy (momentum)
- Surface tension
- Capillary suction
- Air currents
- Air pressure
- Hydrostatic pressure

Rain is pulled down the face of a wall by gravitational forces, but it can also be driven into cracks and openings by air currents and by the kinetic force or momentum of the raindrops themselves (*Fig. 4.1*). Rain can also enter a building through capillary suction, surface tension, and air pressure differentials. Below-grade walls may also be subjected to hydrostatic pressure.

4.2.1 Gravity

Joints and openings in horizontal surfaces are particularly vulnerable to water penetration caused by gravity flow (*Fig. 4.1A*). Flat roofs are far more likely to develop leaks than sloped roofs, and the membranes used to protect them must form a complete barrier to water entry. Cracks or imperfections in the joints of a flat masonry wall coping, metal cap flashing, or window sill can allow water to flow directly into the wall.

4.2.2 Air currents

Rain occurring in the absence of wind, or on the leeward side of a building, strikes only the horizontal and sloping surfaces of a building, while the walls stay dry. Rain driven by wind, though, is carried at an angle by the air currents, and strikes the windward walls as well as the horizontal and sloping surfaces. Wind-driven rain not only wets building surfaces that might remain dry in the absence of the wind, it also blows into openings (*Fig. 4.1B*). Even relatively light winds and air currents can blow snow in through an unprotected gable vent in an attic. When it later melts, the accumulated snow can cause considerable damage.

Wind patterns create swirling forces at the tops and corners of buildings (refer to *Fig. 2.13* in Chap. 2), increasing the wetted area at these locations

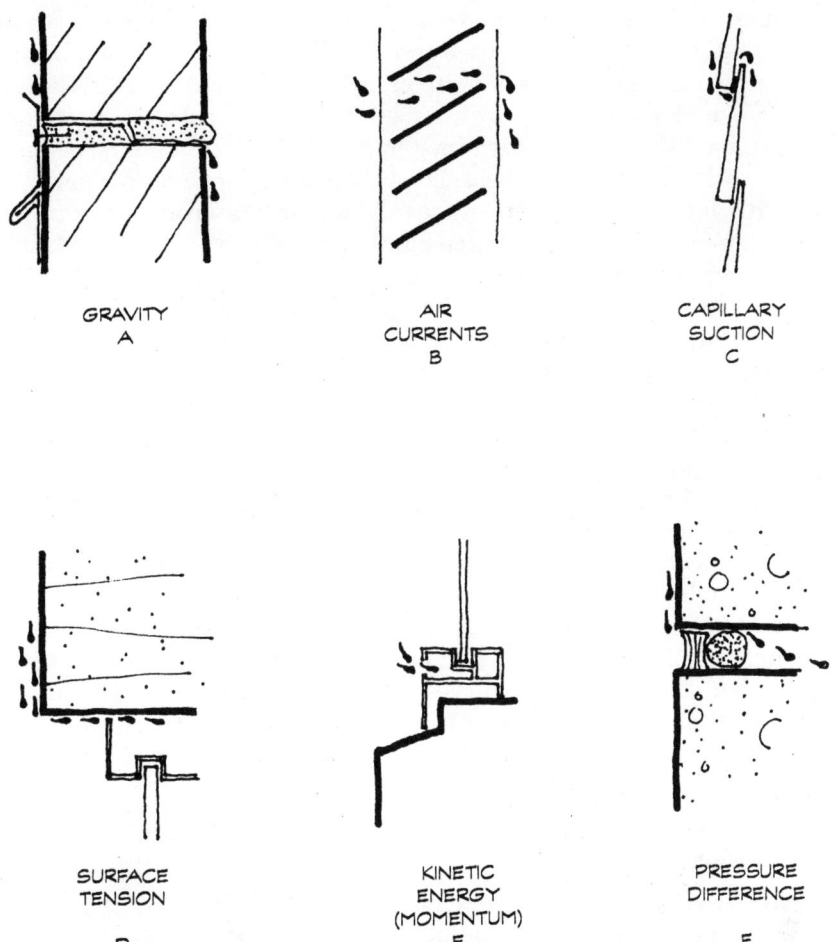

Figure 4.1 Forces causing water penetration through building enclosures.

(*Fig. 4.2*). Research programs in Canada show that the tops and edges of a building may be subjected to 20 to 30 times the rain wetting as the central portion of a facade. In other words, the upper portion of the leeward walls and the upper corners of a building will become wet even if the rain is blowing from the opposite direction. This means that parapet walls are at a higher risk of wind-driven rain exposure than any other element of a building. Not only are they wetted by rain blowing from any direction, but they are wetted on both sides of the wall. Parapets and parapet copings therefore require the most attention to detailing which will prevent rain penetration.

4.2.3 Capillary suction

Small capillary passages (less than $1/8$ in.) in and between exterior building materials can create a suction force which draws in water (*Fig. 4.1C*). The smaller the passage, the greater the force and the higher the capillary potential. Capillary force also works to distribute water uniformly throughout porous materials until a state of equilibrium is reached. Where two different porous materials are in contact with one another, the amount of water retained in each is determined by their relative capillary potential. A material with fine capillary passages will attract and retain more water than one with larger capillary passages. A porous material separating two environments with different relative humidities will have a different moisture content on each side. The face of the material on the side with high humidity will have a higher moisture content than on the side with lower humidity, and a capillary flow of moisture will take place between the two surfaces until a state of equilibrium is reached.

Some of the rain which strikes a wall is absorbed by the capillary passages in porous cladding materials and in the small joints and cracks between cladding panels. Water penetrates the wall until the capillary potential of the material is satisfied, or until it reaches a barrier which it cannot cross. Many historic buildings have solid masonry walls without any kind of capillary barrier, but they are unlikely to have serious rain penetration problems because of capillary suction alone. The storage capacity of the wall mass is so great that before it can be satisfied, the rain will have stopped and evaporative drying begins. If water penetrates the wall for other reasons, though, capillary action can spread it to interior finishes causing staining and deterioration. With less massive walls with smaller capacity for moisture storage, an uninterrupted capillary passage has a greater possibility of transmitting water to the interior finishes.

4.2.4 Surface tension

Water clings to a surface by surface tension. Water that is flowing down the face of a vertical surface will turn and follow the profile of the surface, flowing horizontally onto soffits and into cracks, joints and openings (*Fig. 4.1D*).

4.2.5 Kinetic energy (momentum)

The momentum of blowing rain will carry it directly through open windows, doors and joints. If the openings are small, or if the trajectory of the rain is such that it cannot enter directly, water can splash into openings as rain drops will splash from a sill through an open window or an attic vent or splash into a horizontal joint (*Fig. 4.1E*).

4.2.6 Air pressure

In addition to the force of the wind driving rain against a surface and blowing it into openings, wind also creates air pressure differentials across the building envelope. The exact pressure differential at any given point on a wall or roof will vary depending on the location with respect to the wind direction and the shape of the building (refer to *Fig. 2.15* in Chap. 2). Positive pressures on the

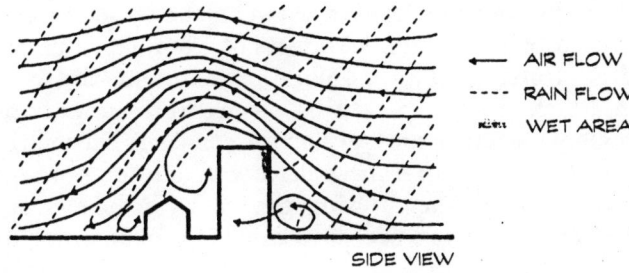

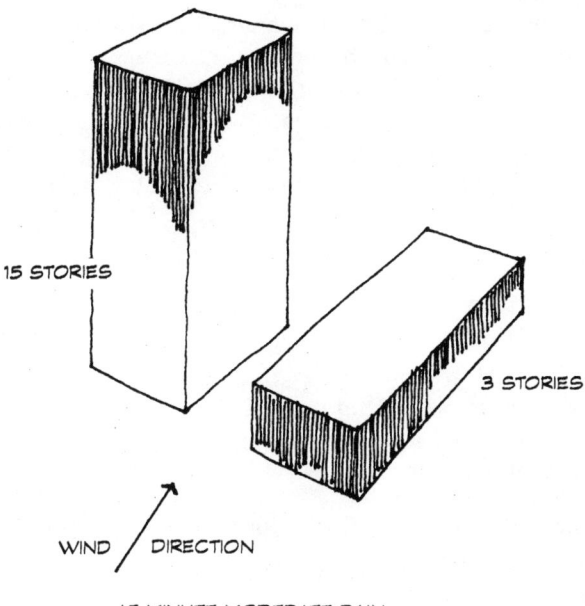

Figure 4.2 Wind increases wetting at building tops and corners. (*From Quirouette, "Rain Penetration Control."*)

windward face of a building are highest in the center, and diminish toward the perimeter. Suction pressures on the leeward facade are highest at the perimeter, and diminish toward the center. The pressure on the side walls parallel to the wind is normally negative, but can rapidly change in magnitude, or even change from negative to positive pressure as the wind direction changes. Because of the suction on roofs and side walls, the windward wall itself will have a steep pressure gradient at the top and corners. This maximum pressure gradient coincides with the areas of a building facade that are subjected to the most frequent and most intense wetting from a wind-driven rain.

The total air pressure difference between the outside and the inside of a wall is produced by the combined effects of wind, stack effect, and mechanical ventilation (refer to Chap. 2). Wind pressure has the most effect on rain penetration. Stack effect and ventilation pressures have a greater effect on vapor movement because they are more constant rather than fluctuating (see Chap. 5). Pressure differences on a wall can force water through even microscopic openings in a building envelope through a sort of pumping action (*Fig. 4.1F*), and wind pressure can even drive water uphill.

4.2.7 Hydrostatic pressure

The hydrostatic pressure of a fluid at any point is caused by the weight of the fluid above it. Standing water and groundwater can both exert hydrostatic pressure against a building envelope. Groundwater in below-grade soil strata exerts hydrostatic pressure against a basement waterproofing membrane. The deeper the structure, the higher the pressure exerted. Ponded water exerts gravitational pressure on a roof membrane, but also exerts hydrostatic pressure on the vertical perimeter flashings and joints. Continuous hydrostatic pressure places stringent requirements on waterproofing membranes and is very unforgiving of small flaws or discontinuities (refer to Chap. 7).

4.3 Preventing Water Problems

What constitutes a leak? Is water that has been absorbed into a porous cladding material considered a leak? Is water that has penetrated a weeped glazing system considered a leak? How about water that has entered the drainage cavity behind a masonry veneer? Water that has migrated into the occupied space of a building would be considered a leak by almost any analysis, but perhaps some definitions are in order.

- *Water penetration.* Process in which water enters a material or system through an exposed surface, joint, or opening.
- *Water absorption.* Process in which water enters a material or system through capillary pores and interstices and is retained without transmission.
- *Water permeation.* Process in which water enters, flows within, and spreads throughout a material or system.
- *Water saturation.* The maximum amount of water a material or system can retain without discharge or transmission.
- *Water infiltration.* Process in which water passes through a material or system and reaches an area that is not directly or intentionally exposed to the water source.
- *Water leakage.* Water infiltration that is unintended; uncontrolled; exceeds the resistance, retention, or discharge capacity of the system; or causes damage or accelerated deterioration.

All of these processes might be involved as water passes from the outside surface of a roof or cladding system into and through a building envelope. In addition to a defined "leak," some of the other conditions listed may constitute serious water problems under certain circumstances. For example, a saturated masonry veneer can be seriously damaged by freeze-thaw cycling. Water that has infiltrated the framing of a weeped glazing system constitutes a problem only if it is unable to drain freely to the outside or if it exceeds the drainage capacity of the system. Ultimately, water problems are defined by the circumstances under which they occur. Preventing water problems in buildings of any kind involves both good design and proper installation of materials and systems.

Remember that water penetration through a building enclosure depends on the simultaneous occurrence of three things: *water,* an *opening* through which water can enter, and a physical *force* to move the water. If any one of the three conditions is eliminated, water cannot penetrate the enclosure. It is not possible to eliminate rain, snow, and groundwater or the forces that move them, and in many building systems it is impossible or impractical to eliminate all of the openings and penetrations which occur intentionally or unintentionally in a building enclosure. It is possible, though, to mitigate all three factors which contribute to water penetration by developing designs which:

- Limit water penetration into a building or building enclosure with
  - Barriers such as membranes and joint sealants
  - Diversions such as sloping surfaces and gutters
  - Screens such as projections and baffles

- Prevent water accumulation by providing
  - Drainage
  - Drying/evaporation
  - Ventilation

- Neutralize the physical forces that transport water with
  - Capillary breaks
  - Drips
  - Protected openings
  - Rain screens

4.3.1 Limit water penetration

To limit water penetration, cracks and openings in exterior surfaces must be minimized or eliminated. Continuous membranes such as built-up and single-ply roofing and below-grade waterproofing are designed to act as barriers to water penetration. The effectiveness of barrier systems depends on the integrity of seams and joints, as well as properly detailed flashing at perimeters and penetrations. Joint sealants are used to form a barrier against water penetration into joints. For primary seals on an exterior weather barrier system, joint design and sealant installation must be near perfect for effective performance.

Water penetration can also be limited by diverting the flow of water across the building enclosure. Sloping surfaces are far more effective than flat surfaces in draining water quickly from a roof, coping, or sill. Flat building surfaces seldom begin and certainly do not remain dead level. Considering the inaccuracy of the construction process and the anticipated tolerances in erecting and assembling building components, most flat surfaces actually slope in one direction or another. Unfortunately, the unintended slope can easily be in a direction which collects rather than sheds water. So-called "flat" roofs should be sloped to interior or perimeter drains so that water cannot pond in the low spots created by natural roof deflections. Sills should be sloped away from windows and copings should be sloped to drain onto roof surfaces.

Water penetration can also be limited by reducing the amount of water that strikes building surfaces and joints. Overhangs and projections can protect walls and openings from rain and staining, louvers can shield an attic vent, baffles or batten strips can block joints, and gutters and downspouts can direct water away from foundation walls (*Fig. 4.3*).

4.3.2 Prevent water accumulation

Water that penetrates a building enclosure may not necessarily cause damage if it can be drained or dried quickly. Flashing and weep holes are used in many different kinds of building systems to collect and drain water that has penetrated the exterior surface (*Fig. 4.4*). Porous materials can be protected from excessive surface absorption with coatings or water repellent treatments, and drying can be expedited by providing air circulation behind the cladding.

4.3.3 Neutralize forces

To resist water penetration caused by *gravity* flow, particular attention must be paid to horizontal surfaces. Wall caps or copings should be sloped to shed water, and the joints sealed with a high-performance sealant (*Fig. 4.5A*).

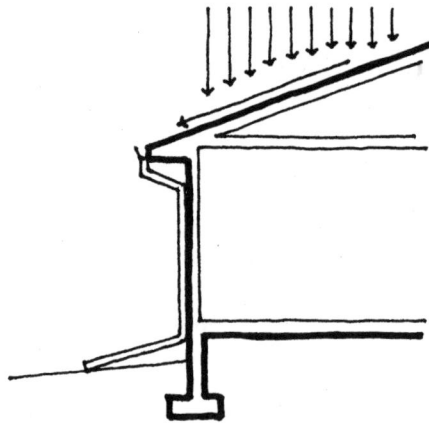

Figure 4.3 Overhangs, gutters and downspouts keep water away from foundation perimeter. (*From Lstiburek and Carmody, Moisture Control Handbook.*)

Water Penetration 129

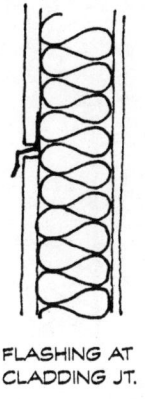

FLASHING AT
CLADDING JT.

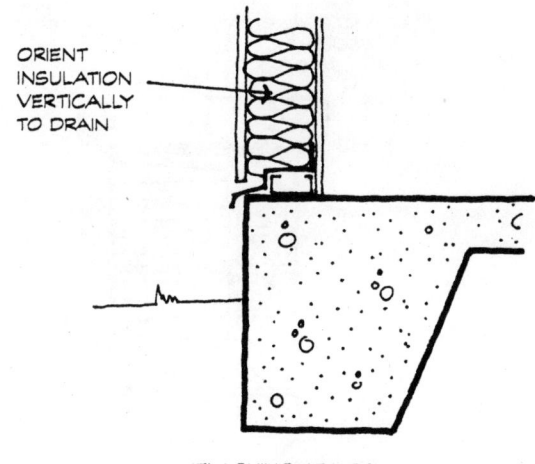

FLASHING AT METAL
BLDG. PANEL

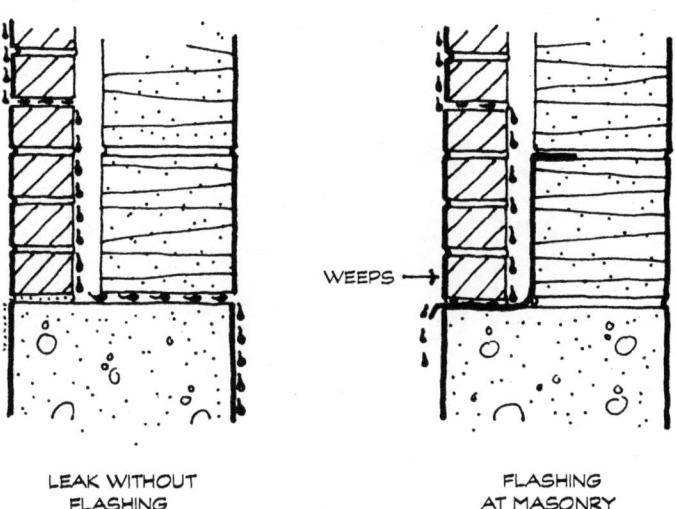

LEAK WITHOUT
FLASHING

FLASHING
AT MASONRY

Figure 4.4 Flashing and weeps prevent moisture accumulation. (*From Lstiburek and Carmody, Moisture Control Handbook.*)

Window sills and other projections should be sloped to the outside so that water will run off. A minimum 15° slope is generally recommended for sills (*Fig. 4.5B*). Drainage flashing and weep holes can provide a second line of defense to collect and drain any water which may penetrate through sills or copings. On vertical surfaces, cladding materials resist the gravitational force of downward flowing water best if the panels or sections are applied overlapping from top to bottom like shingles or clapboard siding (*Fig. 4.5C*).

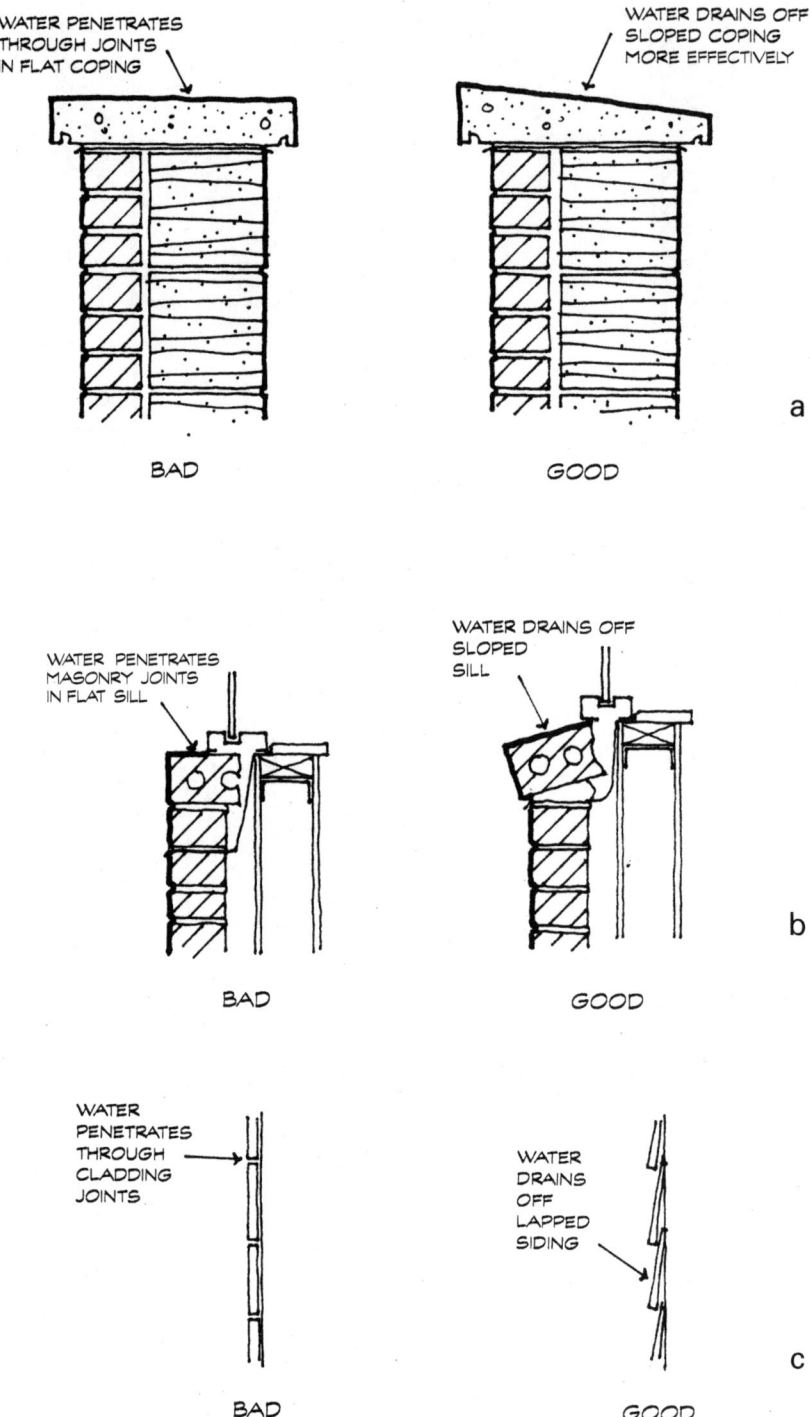

Figure 4.5 Sloped surfaces shed water.

To prevent rain penetration caused by the *kinetic energy* or momentum of the rain, roof overhangs and projections such as balconies and porches provide much the same protection as an umbrella (*Fig. 4.6A*), intercepting the rain drops before they strike the wall surface (*Fig. 4.6B*). Similarly, a shield in front of an opening will stop the rain. A shield can be incorporated in the joints between components of the cladding such as a batten strip on vertical wood siding, or the joint itself can be shaped or baffled to avoid a straight-through opening (*Fig. 4.6C and D*). It is not necessary for these devices to be sealed tightly to prevent the momentum of the rain from carrying it through the opening, so small imperfections are of no consequence.

If an exterior joint is recessed, a shielding device may not be necessary. The number of raindrops which will be aimed accurately enough to penetrate to the back of such recesses will be very small. Others may strike the sides of the recess and splash inward to some extent, but it is unlikely that much water will penetrate a recessed joint simply from the momentum of the falling rain.

To prevent the penetration of water caused by *surface tension,* the surface profile should be designed to break the surface tension and force water to drip off. Since water cannot flow upward without some force being applied, a sharp change in direction to another vertical surface will cause the water to collect until its weight is sufficient that gravity overcomes the surface tension, and the water drips downward. *Figure 4.7* shows several examples of drip profiles or "drips" that can be incorporated in masonry, precast, exterior insulation and finish systems (EIFS), and stucco systems as well as doors and sills.

Capillary suction can be eliminated by interrupting the passages with a sheet material such as metal flashing or a waterproofing membrane. Capillary suction can also be eliminated by introducing an air gap into the moisture path (*Fig. 4.8*). Air gaps, drainage cavities, and gravel beds are effective capillary barriers commonly used in drainage-type wall systems such as masonry cavity walls, under floor slabs, and adjacent to basement walls. Air gaps in vertical walls or joints can also serve as drainage channels to direct penetrated water back to the outside. Sheet membrane barriers are commonly used as protection against capillary suction or rising damp in walls of porous materials which penetrate below grade (*Fig. 4.9*).

To prevent the penetration of rain blown against a building by *air currents,* openings in the enclosure must be shielded in a manner similar to that discussed under kinetic energy forces. Roof overhangs protect the tops of walls, balcony and porch projections shield door openings, and awning-type windows allow natural ventilation while shielding the opening from blowing rain (*Fig. 4.10*). Similarly, a shield incorporated in the joints between components of the cladding as baffles, or splines will prevent rain from blowing into the opening.

Equalizing the pressure differentials on two opposite sides of a wall stops rain penetration caused by *air pressure*. Pressure equalization requires an air chamber between the inside and outside face of the enclosure that can be sealed against air leakage at the sides and at the inside plane (*Fig. 4.11*). Pressure equalization principles were first developed for use in the curtain wall industry, but have been adapted with varying degrees of success to other cladding systems.

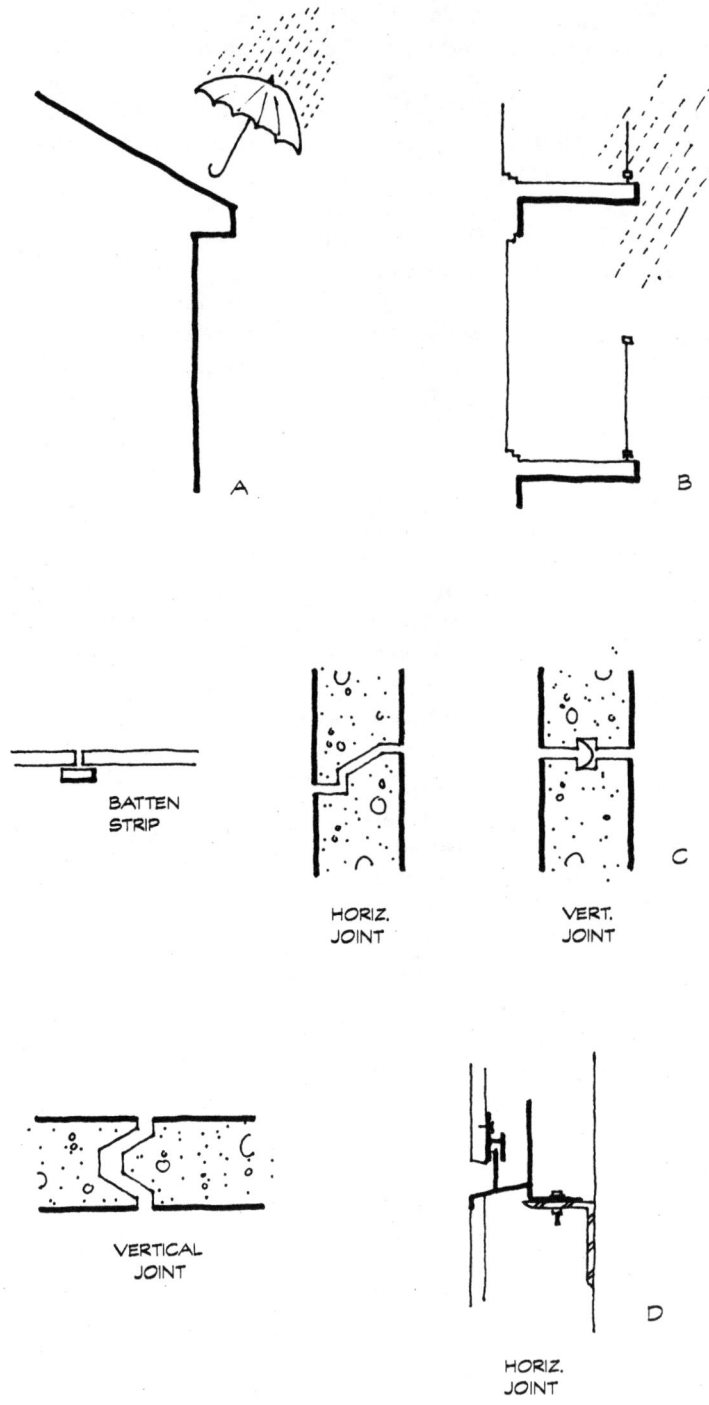

Figure 4.6 Overhangs, shields, and labyrinth joints.

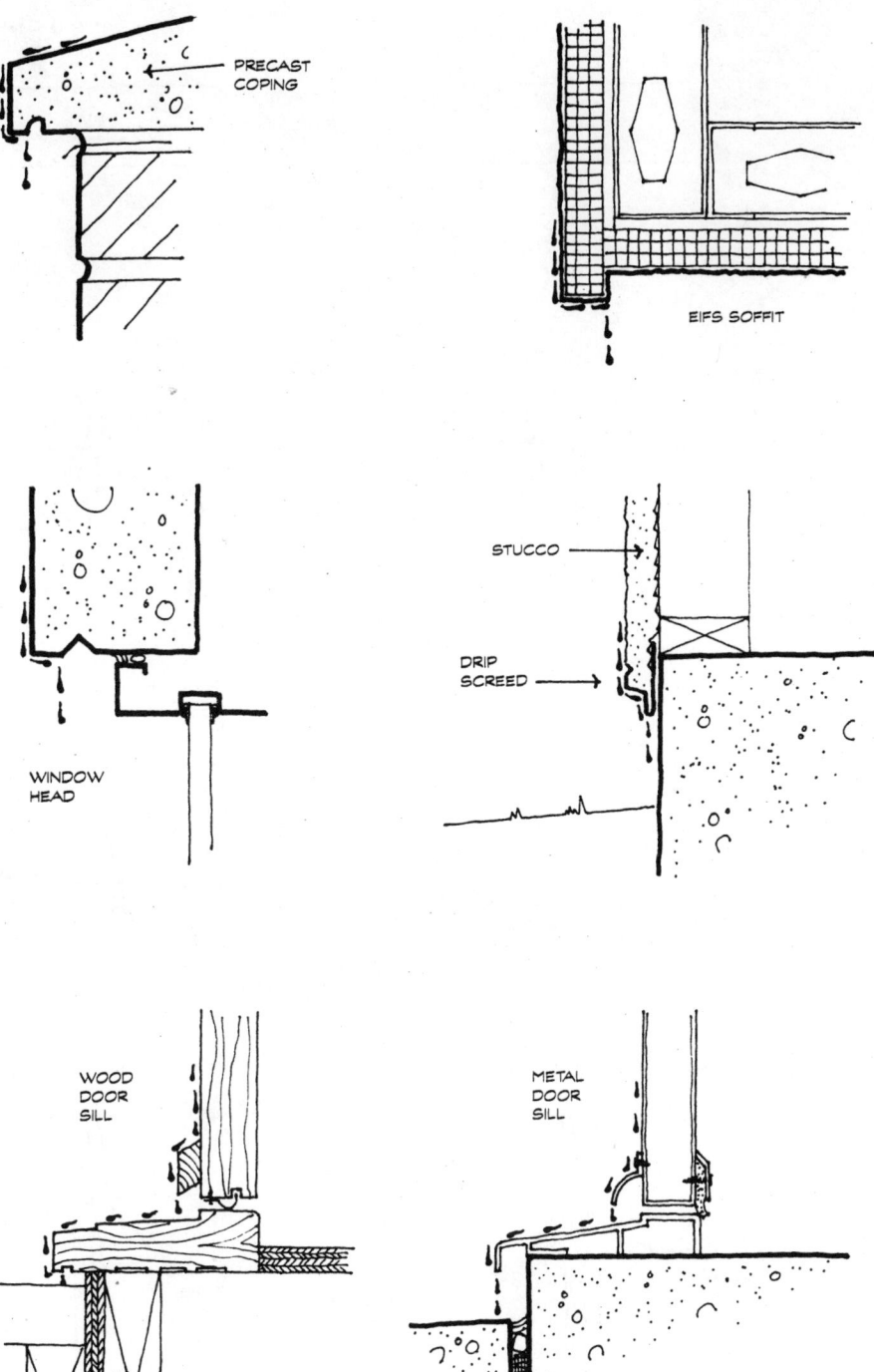

Figure 4.7 Drips break surface tension.

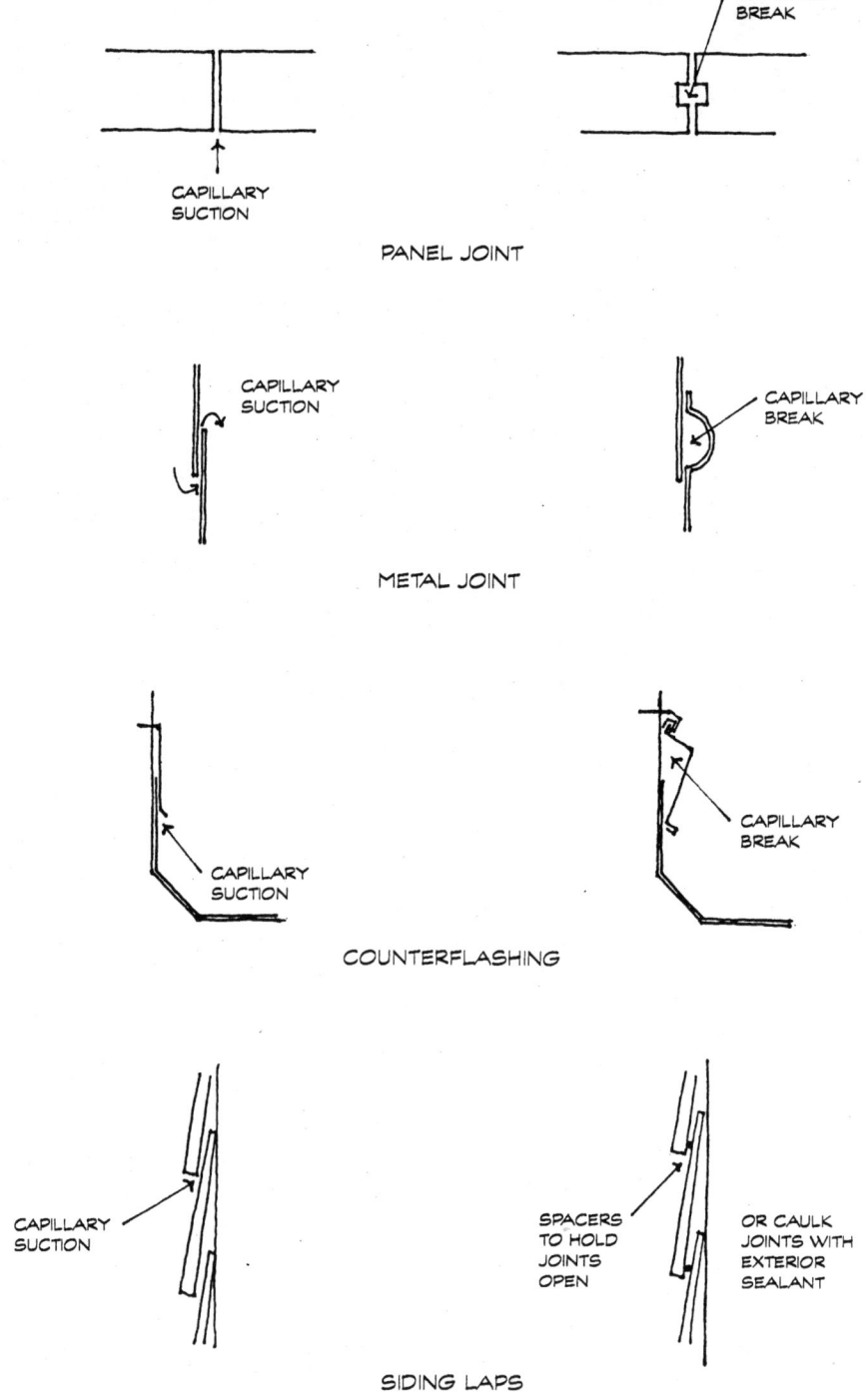

Figure 4.8 Capillary suction and capillary breaks.

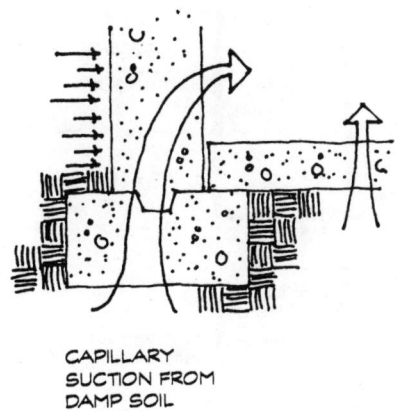

Figure 4.9 Capillary moisture penetration from soil.

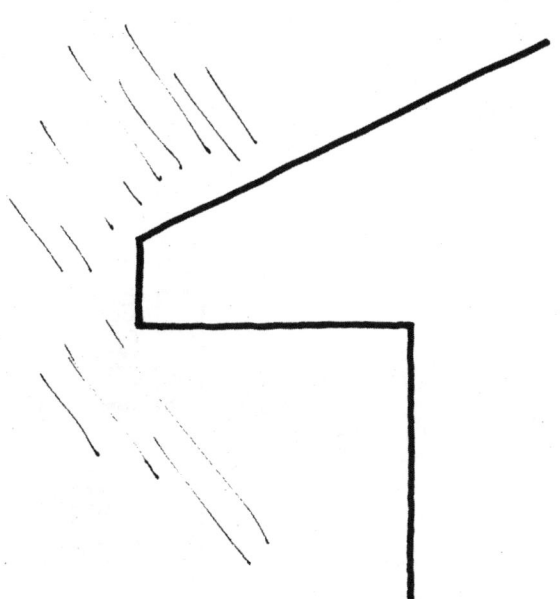
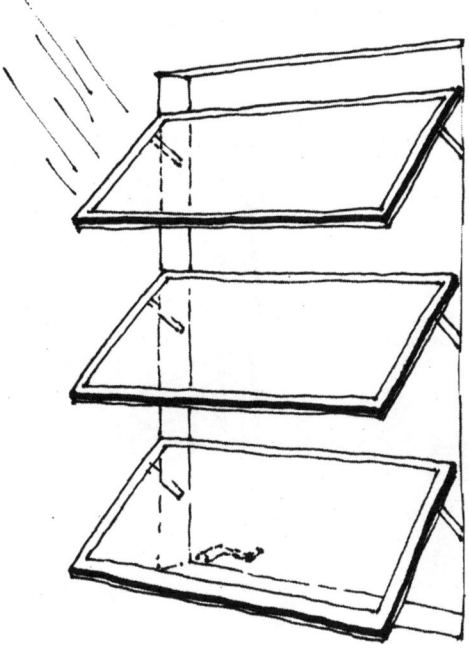

Figure 4.10 Overhangs and awnings.

Wind pressure can also drive water through openings and up a vertical surface. Where wind pressures cannot be equalized, a back dam or flashing overlap must be of a certain height to contain this water, depending on the expected wind speed and pressure (*Fig. 4.12*).

Hydrostatic pressure leaks can be prevented by a waterproof barrier membrane and below-grade drainage. The hydrostatic pressure can be relieved with a permeable drainage mat or gravel layer adjacent to basement walls and below slabs (*Fig. 4.13*).

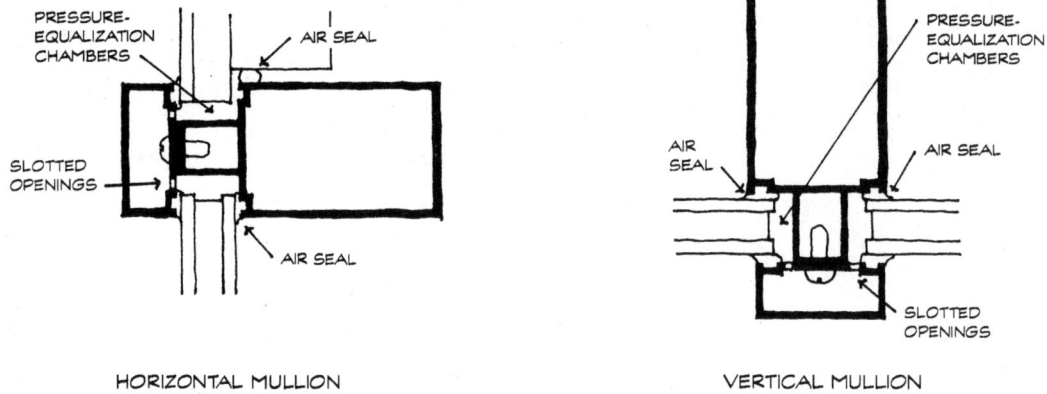
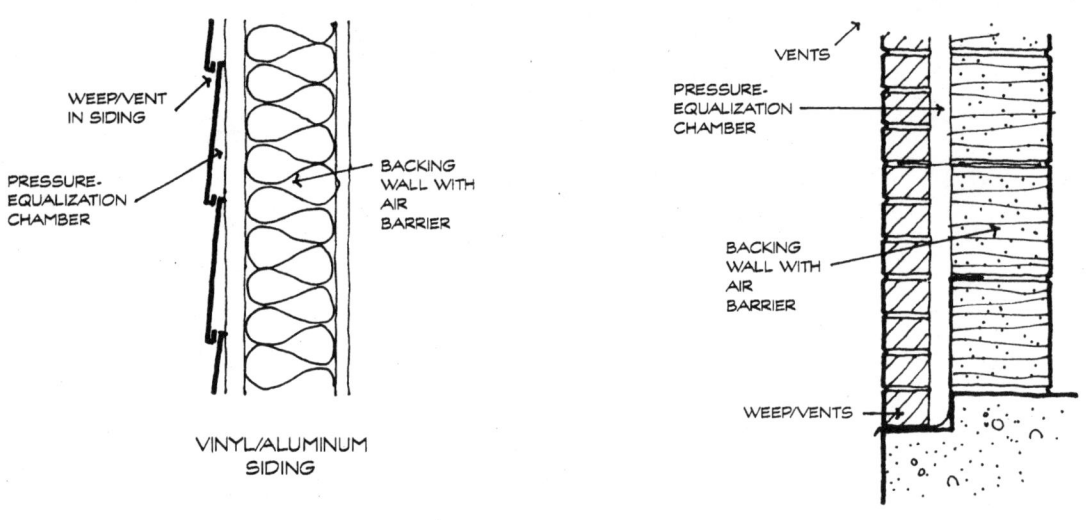

Figure 4.11 Pressure equalization. (*From AAMA Aluminum Curtain Wall Design Guide Manual; Lstiburek and Carmody, Moisture Control Handbook; and Beall, Masonry Design and Detailing, 4th ed.*)

Required Height of Vertical Dams and Overlaps

Wind Speed, mph	Wind Pressure, psf	Required Height, inches (H)
45	5	1
60	10	2
90	20	4
110	30	6
125	40	8
140	50	10

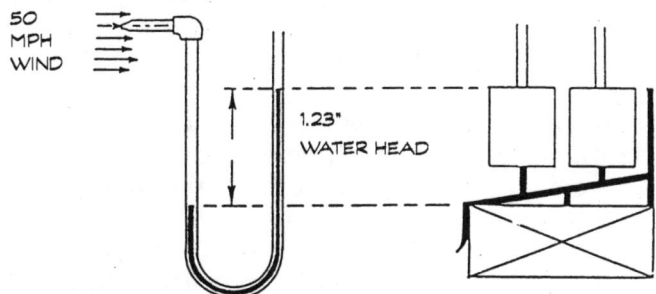

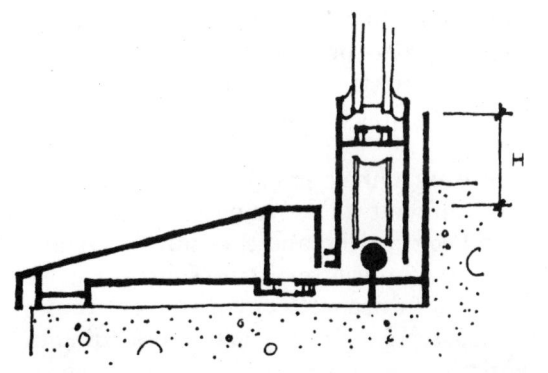
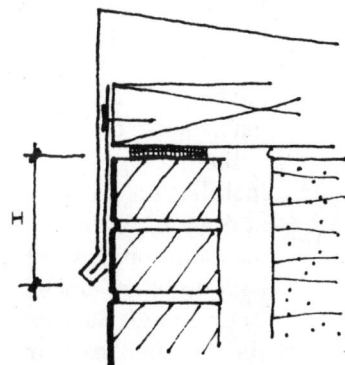

Figure 4.12 Required height of vertical dams and overlaps.

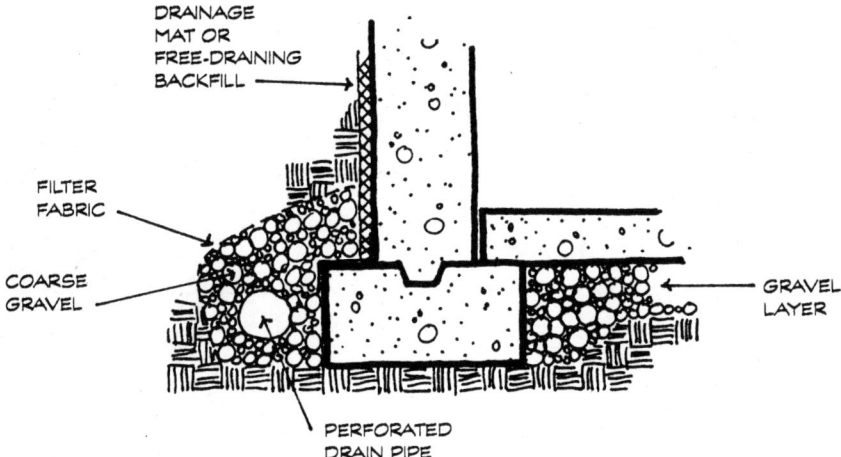

Figure 4.13 Relief of hydrostatic pressure below grade. (*From Lstiburek and Carmody, Moisture Control Handbook.*)

4.4 Barriers, Drains, and Rain Screens

The three basic design concepts which incorporate one or more of these moisture protection principles are *barrier* systems, *drainage* systems, and *rain screen* systems. Barrier systems attempt to limit the number of openings which occur in a building envelope, drainage systems limit the amount of water that can accumulate by controlling its flow, and rain screens neutralize some of the primary forces which move water through the enclosure.

4.4.1 Barrier systems

Barrier systems are also sometimes referred to as *face seal* or *prime seal* *methods* of moisture protection. Barrier systems are used in roofing, below-grade waterproofing, exterior insulation and finish systems (EIFS), most architectural precast cladding, and most metal building systems (*Fig. 4.14*). Barrier systems rely exclusively on an air and water seals at the *exterior* weathering surface. There is no secondary line of defense, so all exterior joints and openings must be perfectly sealed or water will enter and be trapped inside the wall. To expect such perfection is unrealistic. A building envelope that relies on joint sealants as its first and only line of defense will leak sooner or later.

Barrier systems are unforgiving of even minor errors in application and installation. Perfect exterior seals are difficult, if not impossible, to achieve and maintain because of expansion and contraction and exposure to ultraviolet radiation. Interstitial condensation from interior moisture sources can also be a problem, since barrier systems do not incorporate any method of venting or drainage. Barrier systems are the least expensive method of providing initial weather resistance, but they require frequent maintenance of jointing systems to assure satisfactory performance.

4.4.2 Drainage systems

Cavity drainage systems were first developed for masonry walls when their primary usage changed from loadbearing walls to veneers. Drainage systems have also been adapted to precast concrete and other cladding systems. The capillary suction of porous surface materials and any water paths that penetrate the cladding are interrupted by an air space between the cladding and backing. The cavity is then drained of penetrated water by flashing and weep holes (*Fig. 4.15*). This multilayer approach provides redundant protection where the exterior skin stops most of the incident rain but does not have to be a perfect barrier. The internal collection and drainage system provides a second line of defense and prevents water from penetrating to the interior. Cavity drainage systems are generally more expensive than barrier systems, but when properly designed and constructed, they require less maintenance, have a longer expected service life, and provide a higher level of protection against water penetration.

Cavity drainage systems are more forgiving than barrier walls, but can still experience substantial water penetration and leakage problems. If the interior wythe or backing wall is not airtight, pressure differentials between the outside atmosphere and the cavity space can draw large quantities of water through exterior cladding joints, hairline cracks, and defects. Penetrated water can cross the cavity on metal ties, joint reinforcement, and other connectors and either damage or penetrate the backing wall. If the flashing and weep holes are not adequately designed and installed, excessive water can accumulate within the wall, causing efflorescence, freeze-thaw spalling, corrosion of metal components, and deterioration of thermal insulation. The absence of an air barrier can also allow exfiltration of moist interior air and infiltration of humid outdoor air, leading respectively to the potential for winter or summer condensation problems.

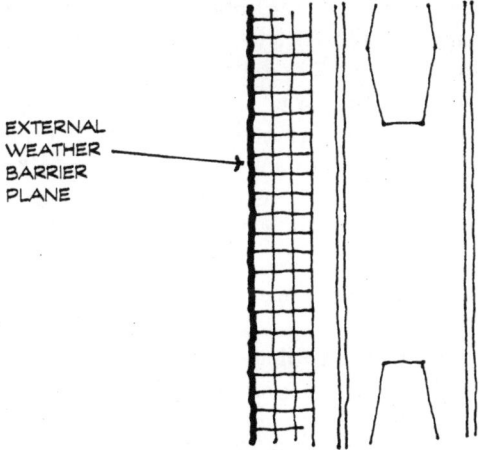

Figure 4.14 Rain barrier concept.

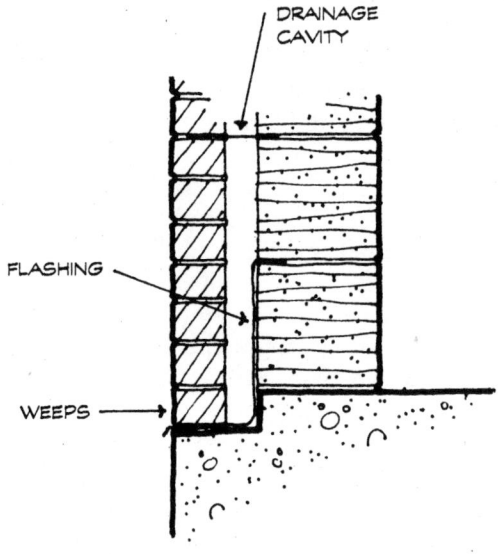

Figure 4.15 Drainage wall concept.

4.4.3 Rain screen systems

Rain screen systems create a *pressure-equalized* cavity behind the cladding and must include a structurally supported air barrier, a sealed and compartmented air chamber, adequate drains, and external vents. Wind pressure is transmitted to the cavity through drain and vent holes, while an air barrier and compartment seals confine the cavity pressure and prevent air from moving to the building interior or around corners (*Fig. 4.16*). As a rule of thumb, intentional openings in the outer wall should be 10 times the area of the unintentional openings in the air barrier. At this ratio, 1% of the pressure difference across a building envelope occurs between the exterior and the air chamber and 99% occurs between the air chamber and the interior, so for practical purposes pressure equalization has been achieved. The force producing rain penetration is reduced to less than 1% of what it would have been, and the quantity of water penetrating to about one-tenth of what it would have been. The air chamber may also serve as a drainage cavity for collecting penetrated water and directing it back to the outside. Pressure-equalized rain screen principles can be applied to different cladding systems with varying degrees of success.

Rain screens are high-performance wall systems. They provide multiple layers of protection by (1) limiting water penetration, (2) collecting and draining penetrated water, and (3) moving the primary air seal from the weathering surface of the wall to a protected location within the wall. Rain screen systems are more expensive than barrier and drainage wall systems, but also have the lowest maintenance and the highest safety factor in protecting against water infiltration.

So-called modified rain screens are really no more than imperfect rain screens. While they may provide improved performance over simple cavity drainage systems, they seldom provide true pressure equalization. Increased venting of the exterior cladding will help minimize pressure differentials, but without the air barrier and a compartmented air chamber, water penetration can still be substantial. In an effective rain screen design, pressure equalization occurs in fractions of a second when the air chamber is properly divided into small compartments. If a single large air chamber extends between two locations which are at different pressures, effective pressure equalization is impossible. Where large pressure differences occur over relatively short distances such as at building corners, compartmenting the air chamber is essential. Since pressure gradients are greatest at the tops and edges of a facade, compartment sizes may generally be larger in the middle of the building.

Canada Mortgage and Housing Corporation (CMHC) has developed a computer program called *RAIN* to predict pressure-equalization performance for a given wall design. The program requires inputting the area of vent openings in the cladding, an assumed amount of unintentional leakage in the air barrier, the flexibility of both the air barrier and the cladding, and the volume of the cavity.

4.4.4 Combining protective strategies

In reality, many enclosure systems use a combination of methods to resist water penetration. A building might consist of barrier membrane waterproofing below

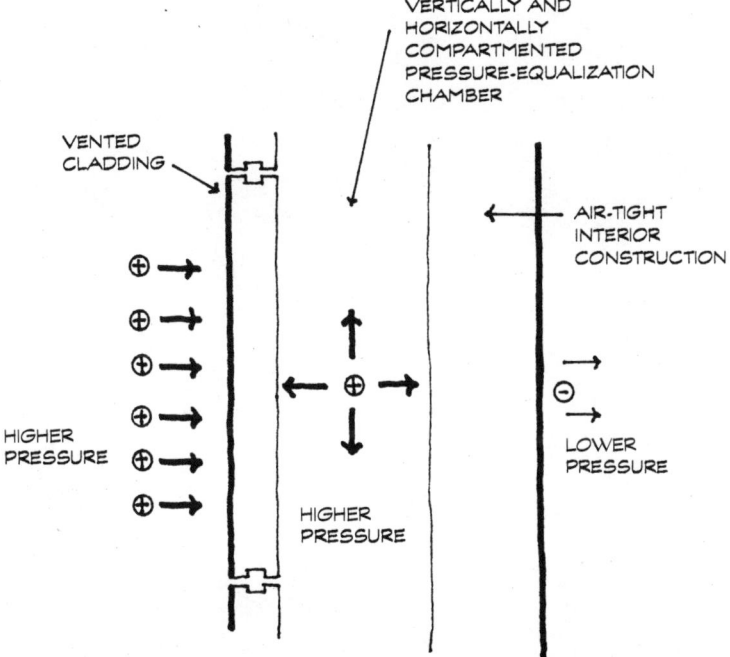

Figure 4.16 Rain screen concept.

grade with a drainage medium provided as a capillary break to hydrostatic pressure, a sloping roof to shed snow and rain, a pressure-equalized aluminum curtain wall system, and a masonry cavity wall with flashing and weep holes for drainage. A combination of protective strategies can be selected to suit the type of building and its occupancy requirements, the severity of macro- and micro-environmental exposure, and appropriateness for the type of building system under consideration.

The application of these various methods of resisting water penetration in roofing, waterproofing, and cladding systems are discussed in detail in Chaps. 6 through 8, including industry standards and recommended details.

Chapter 5

Vapor Condensation

Increased thermal insulation, tighter enclosures, and more efficient heating and cooling systems have improved the comfort and energy performance of buildings, but have also significantly increased the incidence of condensation and moisture accumulation. Condensation can occur on exposed surfaces or be concealed in the walls or roof of a building, where it may accumulate unnoticed until deterioration of the envelope materials is extensive. If we understand the physical forces that affect the movement of water vapor and the flow of heat, it is possible to anticipate how conditions within buildings and within a wall, floor, or roof section will be affected by the design and its environment.

5.1 Psychrometrics

Psychrometry is the branch of physics dealing with the measurement or determination of atmospheric conditions, and particularly the study of the behavior of moist air under various temperature and humidity conditions.

The air that we breathe is actually a mixture of dry air and water vapor. The air and vapor are not combined at normal atmospheric temperatures and pressures, but merely share a given volume of space. The vapor can move without the air moving with it, but the air cannot move without taking the vapor along. There is a maximum amount of vapor which can be retained within a given volume of air at any given temperature. When the air contains the maximum amount of vapor it can hold at a given temperature, it is saturated. The temperature which corresponds to 100% saturation is called the *dew point*. The maximum amount of water vapor that can be contained in a given volume of air is governed by the air temperature. Warm air can hold more vapor than cool air. The vapor exerts a pressure in proportion to its volume of the air mass. Warm air with more vapor is at a higher vapor pressure than cool air with less vapor. In seeking a natural state of equilibrium, vapor will always move from a location of higher vapor pressure to a location of lower vapor pressure.

As warm air is cooled, its relative humidity increases until saturation (100% RH) is reached at the dew point temperature. If the temperature is reduced below the dew point, vapor will begin to condense out of the air. If the dew point is above freezing, the vapor will condense as a liquid. If the dew point is below freezing, the vapor will condense as frost. If the temperature of any surface in an area is below the dew point of the air that surrounds it, the air in contact with the surface will be cooled sufficiently that condensation will form on the surface. As moisture condenses out of the air, its vapor pressure is reduced. The adjacent warmer air with higher pressure moves to equalize the difference and is, in turn, cooled by the surface and its vapor condenses to liquid or frost. Because of the pressure gradient caused by the cooling and condensation, there is a constant flow of moist air toward the condensing surface.

Humidity can be expressed in absolute terms as the weight of water vapor per unit weight of dry air. This is called the *absolute humidity* or *humidity ratio,* and is usually given in pounds or grains of water vapor per pound of dry air. There are 455 grains to an ounce and 7000 grains in a pound. At 20°F, it takes only 16 grains of vapor to saturate one pound of dry air, while at 80°F, it takes 156 grains. Humidity can also be expressed as a measure of the amount of water vapor in the air relative to the maximum amount the air can contain at the same temperature. This is called the *relative humidity* (RH), and is expressed as a percentage. Saturated air has 100% relative humidity. Air that is at 50% relative humidity contains only one-half the maximum amount it can hold at that particular temperature. Another way of expressing relative humidity is the ratio between the *actual vapor pressure* of the air and the *saturation vapor pressure* at the same temperature. If the temperature and relative humidity are known, the actual vapor pressure can be calculated by multiplying the relative humidity by the saturation vapor pressure. Saturation vapor pressures for different air temperatures are given in *Fig. 5.1*.

A *psychrometric chart* is a graphical representation of air/vapor conditions and relationships. *Figure 5.2* is a simplified psychrometric chart. The chart can be used to find the dew point of air at a given temperature and relative humidity, to find the humidity ratio, and other properties of moist air masses. Using the chart is easier than it looks. Dry bulb air temperatures are listed along the bottom, moisture content or humidity ratio is on the right, relative humidity is represented by the curved lines, and dew point temperatures are along the left-hand curved boundary line. If two properties of the air are known, all other properties can be found from the chart. For example, to find the dew point of air at 85°F and 40% relative humidity, follow the vertical line from the bottom of the chart at 85°F until it intersects the 40% curved relative humidity line. At the point of intersection, follow a horizontal line to the left to find the dew point temperature (58°F) along the left-hand boundary curve (*Fig.* 5.3). To find the actual moisture content of the air (0.010 lb or 70 grains of vapor per lb of dry air), follow a horizontal line to the right from the same intersection of the temperature and relative humidity lines.

Vapor Condensation

Temp. (°F)	Vapor pressure (in. Hg)	Temp. (°F)	Vapor pressure (in. Hg)
−20	0.01259	41	0.25748
−19	0.01333	42	0.26763
−18	0.01411	43	0.27813
−17	0.01493	44	0.28899
−16	0.01579	45	0.30023
−15	0.01671	46	0.31185
−14	0.01767	47	0.32386
−13	0.01868	48	0.33629
−12	0.01974	49	0.34913
−11	0.02086	50	0.36240
−10	0.02203	51	0.37611
−9	0.02327	52	0.39028
−8	0.02457	53	0.40492
−7	0.02594	54	0.42004
−6	0.02738	55	0.43565
−5	0.02889	56	0.45176
−4	0.03047	57	0.46840
−3	0.03214	58	0.48558
−2	0.03388	59	0.50330
−1	0.03572	60	0.52159
0	0.03764	61	0.54047
1	0.03967	62	0.55994
2	0.04178	63	0.58002
3	0.04401	64	0.60073
4	0.04634	65	0.62209
5	0.04878	66	0.64411
6	0.05134	67	0.66681
7	0.05402	68	0.69019
8	0.05683	69	0.71430
9	0.05978	70	0.73915
10	0.06286	71	0.76475
11	0.06608	72	0.79112
12	0.06946	73	0.81828
13	0.07300	74	0.84624
14	0.07670	75	0.87504
15	0.08056	76	0.90470
16	0.08461	77	0.93523
17	0.08884	78	0.96665
18	0.09327	79	0.99899
19	0.09789	80	1.0323
20	0.10272	81	1.0665
21	0.10777	82	1.1017
22	0.11305	83	1.1379
23	0.11856	84	1.1752
24	0.12431	85	1.2135
25	0.13032	86	1.2529
26	0.13659	87	1.2934
27	0.14313	88	1.3351
28	0.14966	89	1.3779
29	0.15709	90	1.4219
30	0.16452	91	1.4671
31	0.17227	92	1.5135
32	0.18035	93	1.5612
33	0.18778	94	1.6102
34	0.19456	95	1.6606
35	0.20342	96	1.7123
36	0.21166	97	1.7654
37	0.22020	98	1.8199
38	0.22904	99	1.8759
39	0.23819	100	1.9333
40	0.24767		

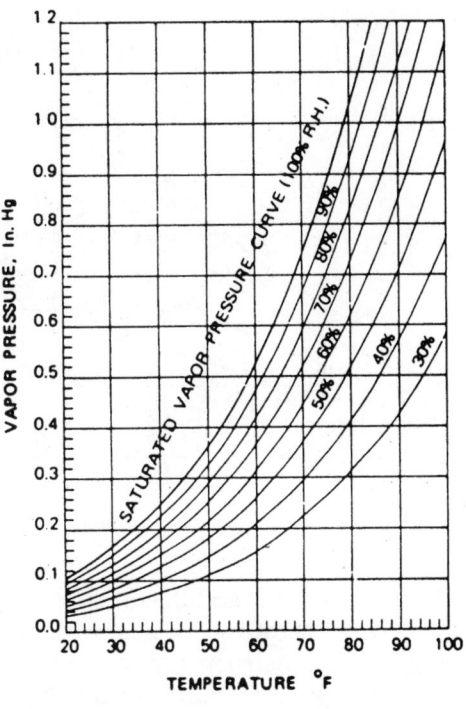

Figure 5.1 Saturation vapor pressures for various temperatures.

146 Chapter Five

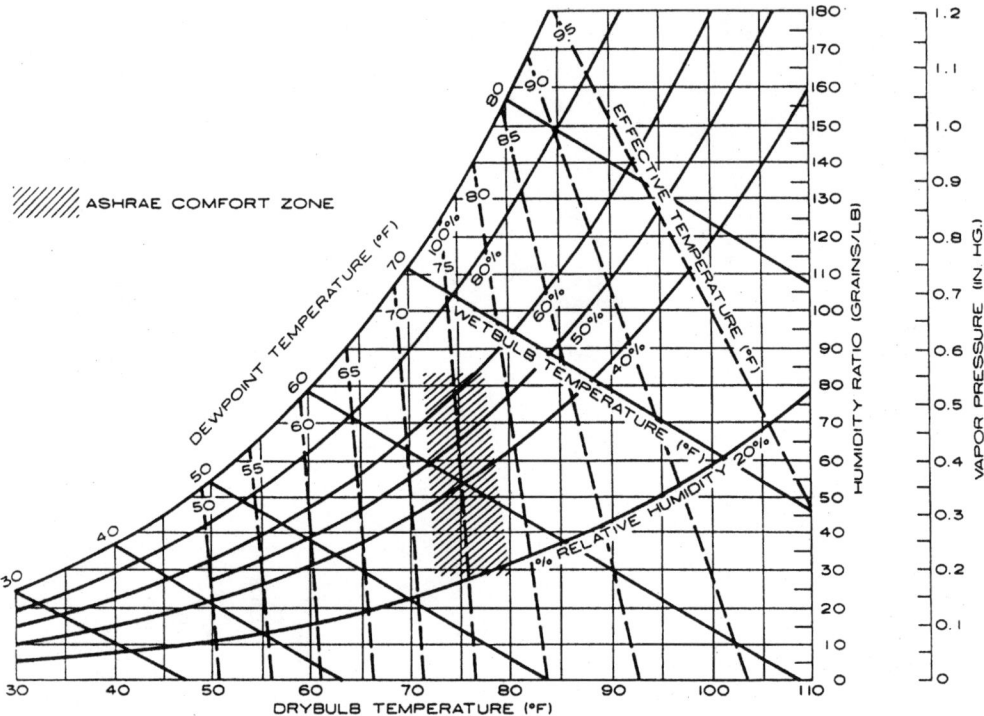

Figure 5.2 Simplified psychrometric chart based on the ASHRAE *Handbook of Fundamentals*.

For architects, the most commonly sought information is the dew point temperature. To simplify access to this information, dew point tables have been constructed for a range of temperatures from 30 to 110°F, and a range of relative humidities from 22 to 60% (*Table 5.1*).

5.2 Condensation in Buildings

Winter condensation problems are most frequent in cold climates in insulated buildings with high indoor humidity. Summer condensation problems are most frequent in warm, humid climates in air-conditioned buildings.

5.2.1 Humidity, comfort, and health

The comfort zone for relative humidity for persons at rest or doing light work can range from 20 to 60% as long as the air temperature is less than 78°F. A humidity range of 25 to 40% is generally considered to be most healthy. Some industrial facilities require relative humidities higher than 60% because of the equipment used, the manufacturing process, or the types of products stored. At the other extreme, some manufacturing facilities require relative humidities less than 20%. High humidity retards natural human heat loss by evaporative cooling (i.e., sweating) and by respiration. Low humidity tends to dry throat and nasal passages. In buildings, low humidity can also cause loosened furniture joints, cracked book bindings, etc. In hospitals, the static electricity that accompanies low humidity creates a risk of spark and explosion with oxygen and other combustible gases. Relative humidity levels in hospitals, computer facilities, museums, and other buildings are often maintained at 50% or more, putting them at high risk of winter condensation problems.

Winter humidification devices can introduce moisture directly into the air stream of mechanical systems to improve the comfort of dry winter air, and summer dehumidification for comfort is a normal part of the mechanical cooling process. The cooler the temperature, the higher the humidity level needed for comfort. Indoor winter humidification levels, however, have an effect on exterior envelope components. In cold climates, for example, with low outdoor winter design temperatures, condensation will form on single-glazed windows even at low indoor humidities. To minimize the potential for condensation and still provide comfortable conditions, indoor relative humidity should be maintained at a maximum of 25% when outdoor temperatures are 0°F or below. When the outdoor temperature is 10°F, indoor humidity can be safely raised to

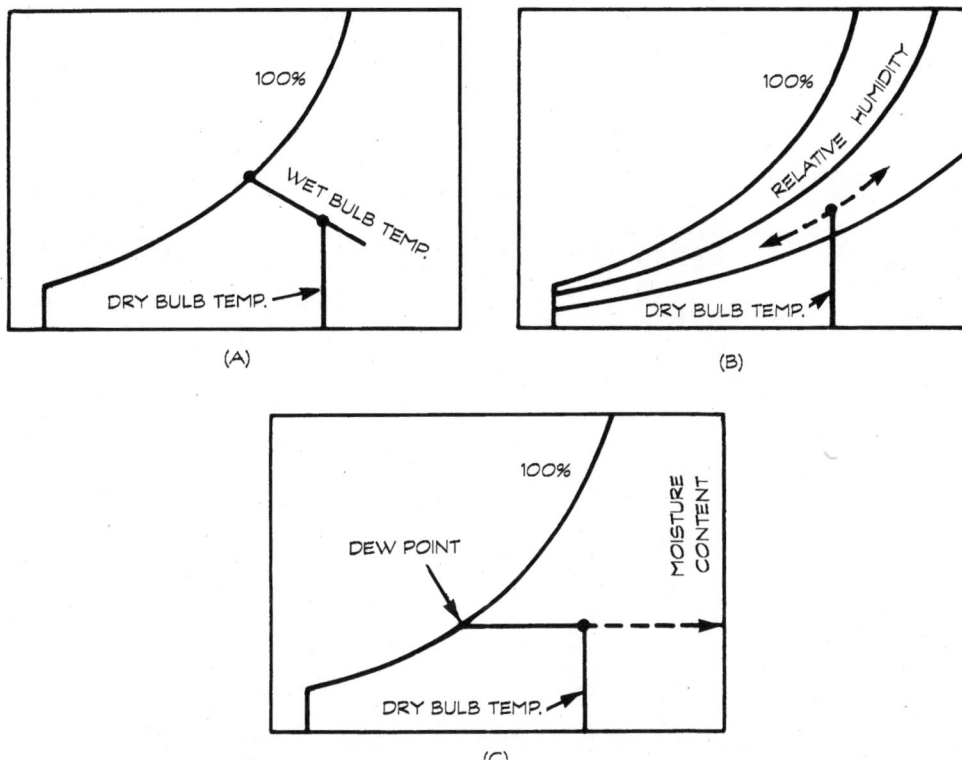

Figure 5.3 If two properties of air are known, all other properties can be found from a psychrometric chart. (*From PCA, How to Calculate Heat Transmission Coefficients and Vapor Condensation Temperatures of Concrete Masonry Walls.*)

TABLE 5.1 Dew Point Tables

Dry Bulb Temp. (°F)	RELATIVE HUMIDITY (%) Dew-point Temperature (°F)															
	22	24	26	28	30	32	34	36	38	40						
30	1.0	0.7	2.2	3.7	5.0	6.3	7.5	8.6	9.7	10.7						
35	3.1	4.8	6.4	7.9	9.2	10.5	11.7	12.9	14.0	15.0						
40	7.0	8.7	10.3	11.8	13.2	14.5	15.7	16.9	18.1	19.1						
45	10.8	12.6	14.2	15.7	17.2	18.5	19.8	21.0	22.1	23.2						
50	14.6	16.3	18.1	19.6	21.1	22.4	23.7	25.0	26.1	27.2						
55	18.4	20.3	21.9	23.5	25.0	26.4	27.7	29.0	30.1	31.3						
60	22.2	24.1	25.8	27.4	28.9	30.3	31.7	33.0	34.4	35.7						
61	23.0	24.8	26.6	28.2	29.7	31.1	32.5	33.9	35.3	36.6						
62	23.7	25.6	27.3	29.0	30.5	31.9	33.4	34.8	36.2	37.5						
63	24.5	26.4	28.1	29.7	31.2	32.7	34.2	35.7	37.0	38.3						
64	25.2	27.1	28.9	30.5	32.0	33.6	35.1	36.6	37.9	39.2						
65	26.0	27.9	29.6	31.3	32.9	34.5	36.0	37.4	38.8	40.1						
66	26.7	28.6	30.4	32.0	33.7	35.2	36.9	38.3	39.7	41.0						
67	27.5	29.4	31.2	32.9	34.6	36.2	37.8	39.2	40.6	41.9						
68	28.3	30.2	31.9	33.7	35.5	37.1	38.6	40.1	41.5	42.8						
69	29.0	30.9	32.8	34.6	36.3	38.0	39.5	41.0	42.4	43.7						
70	29.8	31.7	33.6	35.5	37.2	38.8	40.4	41.9	43.3	44.6						
71	30.5	32.5	34.5	36.3	38.1	39.7	41.3	42.8	44.2	45.5						
72	31.2	33.3	35.3	37.2	38.9	40.6	42.1	43.6	45.1	46.4						
73	32.0	34.1	36.2	38.0	39.8	41.5	43.0	44.5	45.9	47.3						
74	32.8	35.0	37.0	38.9	40.7	42.3	43.9	45.4	46.8	48.2						
75	33.7	35.8	37.9	39.7	41.5	43.2	44.8	46.3	47.7	49.1						
76	34.5	36.7	38.7	40.6	42.4	44.1	45.7	47.2	48.6	50.0						
77	35.3	37.5	39.6	41.5	43.2	44.9	46.5	48.0	49.5	50.9						
78	36.1	38.3	40.4	42.3	44.1	45.8	47.4	48.9	50.4	51.8						
79	37.0	39.2	41.2	43.2	45.0	46.7	48.3	49.8	51.3	52.7						
80	37.8	40.0	42.1	44.0	45.8	47.5	49.2	50.7	52.2	53.5						
81	38.6	40.9	42.9	44.9	46.7	48.4	50.0	51.6	53.0	54.4						
82	39.5	41.7	43.8	45.7	47.6	49.3	50.9	52.4	53.9	55.3						
83	40.3	42.5	44.6	46.6	48.4	50.1	51.8	53.3	54.8	56.2						
84	41.1	43.4	45.5	47.4	49.3	51.0	52.7	54.2	55.7	57.1						
85	42.0	44.2	46.3	48.3	50.1	51.9	53.5	55.1	56.6	58.0						
86	42.8	45.1	47.2	49.1	51.0	52.7	54.4	56.0	57.5	58.9						
87	43.6	45.9	48.0	50.0	51.9	53.6	55.3	56.9	58.3	59.8						
88	44.4	46.7	48.9	50.8	52.7	54.5	56.1	57.7	59.2	60.7						
89	45.3	47.6	49.7	51.7	53.6	55.3	57.0	58.6	60.1	61.6						
90	46.1	48.4	50.5	52.6	54.4	56.2	57.9	59.5	61.0	62.5						
92	47.8	50.1	52.2	54.3	56.2	57.9	59.6	61.2	62.8	64.2						
94	49.4	51.7	53.9	56.0	57.9	59.7	61.4	63.0	64.5	66.0						
96	51.1	53.4	55.6	57.7	59.6	61.4	63.1	64.7	66.3	67.8						
98	52.7	55.1	57.3	59.4	61.3	63.1	64.9	66.5	68.1	69.6						
100	54.4	56.7	59.0	61.1	63.0	64.9	66.6	68.2	69.8	71.3						
102	56.0	58.4	60.7	62.7	64.7	66.6	68.3	70.0	71.6	73.1						
104	57.6	60.1	62.3	64.4	66.4	68.3	70.1	71.7	73.3	74.9						
106	59.3	61.7	64.0	66.1	68.1	70.0	71.8	73.5	75.1	76.6						
108	60.9	63.4	65.7	67.8	69.8	71.7	73.5	75.2	76.9	78.4						
110	62.6	65.1	67.4	69.5	71.5	73.5	75.3	77.0	78.6	80.2						

Dry Bulb Temp. (°F)	RELATIVE HUMIDITY (%) Dew-point Temperature (°F)										
	42	44	46	48	50	52	54	56	58	60	
30	11.7	12.7	13.6	14.4	15.3	16.1	16.9	17.6	18.3	19.1	
35	16.0	17.0	17.9	18.8	19.7	20.5	21.3	22.1	22.8	23.5	
40	20.1	21.1	22.1	23.0	23.8	24.7	25.5	26.3	27.0	27.8	
45	24.2	25.2	26.1	27.1	28.0	28.8	29.6	30.4	31.2	32.0	
50	28.3	29.3	30.3	31.2	32.1	33.1	34.1	35.0	35.8	36.7	
55	32.4	33.6	34.7	35.7	36.7	37.7	38.6	39.6	40.5	41.4	
60	36.9	38.1	39.2	40.3	41.3	42.3	43.3	44.3	45.2	46.1	
61	37.8	39.0	40.1	41.2	42.2	43.3	44.3	45.2	46.2	47.1	
62	38.7	39.9	41.0	42.1	43.2	44.2	45.2	46.2	47.1	48.0	
63	39.5	40.7	41.9	43.1	44.1	45.1	46.1	47.1	48.0	48.9	
64	40.5	41.7	42.9	44.0	45.1	46.1	47.1	48.0	49.0	49.9	
65	41.4	42.6	43.8	44.9	45.9	47.0	48.0	49.0	49.9	50.8	
66	42.3	43.5	44.6	45.7	46.8	47.9	48.9	49.9	50.8	51.8	
67	43.2	44.4	45.6	46.7	47.8	48.8	49.8	50.8	51.8	52.7	
68	44.1	45.3	46.5	47.6	48.7	49.7	50.8	51.8	52.7	53.7	
69	45.0	46.2	47.4	48.5	49.6	50.7	51.7	52.7	53.7	54.6	
70	45.9	47.2	48.3	49.4	50.5	51.6	52.6	53.6	54.6	55.5	
71	46.8	48.0	49.2	50.3	51.4	52.5	53.5	54.5	55.5	56.4	
72	47.7	48.9	50.1	51.3	52.4	53.5	54.5	55.4	56.4	57.4	
73	48.6	49.8	51.0	52.2	53.3	54.4	55.4	56.4	57.4	58.3	
74	49.5	50.8	52.0	53.1	54.2	55.3	56.3	57.3	58.3	59.2	
75	50.4	51.6	52.9	54.0	55.1	56.2	57.3	58.3	59.2	60.2	
76	51.3	52.5	53.8	54.9	56.0	57.1	58.2	59.2	60.2	61.1	
77	52.2	53.5	54.7	55.8	57.0	58.1	59.1	60.1	61.1	62.1	
78	53.1	54.4	55.6	56.7	57.9	59.0	60.1	61.1	62.0	63.0	
79	54.0	55.3	56.5	57.7	58.8	59.9	61.0	62.0	63.0	63.9	
80	54.9	56.2	57.4	58.6	59.7	60.8	61.9	62.9	63.9	64.9	
81	55.8	57.1	58.3	59.5	60.6	61.7	62.8	63.8	64.8	65.8	
82	56.7	58.0	59.2	60.4	61.5	62.7	63.7	64.8	65.8	66.8	
83	57.5	58.8	60.1	61.3	62.5	63.6	64.6	65.7	66.7	67.7	
84	58.5	59.8	61.0	62.2	63.4	64.5	65.6	66.6	67.6	68.6	
85	59.3	60.7	61.9	63.1	64.3	65.4	66.5	67.5	68.6	69.6	
86	60.3	61.6	62.8	64.0	65.2	66.3	67.4	68.5	69.5	70.5	
87	61.2	62.5	63.7	65.0	66.1	67.2	68.3	69.4	70.4	71.4	
88	62.0	63.4	64.6	65.9	67.0	68.2	69.3	70.3	71.4	72.4	
89	62.9	64.3	65.6	66.8	67.9	69.1	70.2	71.3	72.3	73.3	
90	63.9	65.2	66.5	67.7	68.9	70.0	71.1	72.2	73.2	74.2	
92	65.6	67.0	68.3	69.5	70.7	71.9	73.0	74.1	75.1	76.1	
94	67.4	68.8	70.1	71.3	72.5	73.6	74.7	75.9	77.0	78.0	
96	69.2	70.6	71.9	73.1	74.3	75.5	76.7	77.8	78.8	79.9	
98	71.0	72.4	73.7	75.0	76.2	77.3	78.5	79.6	80.7	81.7	
100	72.8	74.2	75.5	76.8	78.0	79.2	80.3	81.5	82.6	83.6	
102	74.6	75.9	77.3	78.6	79.8	81.0	82.2	83.3	84.4	85.5	
104	76.3	77.8	79.1	80.4	81.6	82.9	84.0	85.2	86.3	87.3	
106	78.1	79.5	80.9	82.2	83.5	84.7	85.9	87.0	88.1	89.2	
108	79.9	81.3	82.7	84.0	85.3	86.5	87.7	88.9	90.0	91.1	
110	81.7	83.1	84.5	85.8	87.1	88.4	89.6	90.7	91.8	92.9	

30%. At 20°F, the humidity can be raised to 35%, and to 40% at 30°F (*Fig. 5.4*). Alternatively, double or triple glazing should be used when indoor humidity must be above the levels indicated.

5.2.2 Sources of moisture

Very dry desert climates may occasionally have outdoor relative humidities between 5 and 10% but, more commonly, dry regions have relative humidities between 10 and 30%. Moist regions (tropical and subtropical) have relative humidities from 60 to 80%. In any climate, whenever it is raining or there is fog, the relative humidity is 100%.

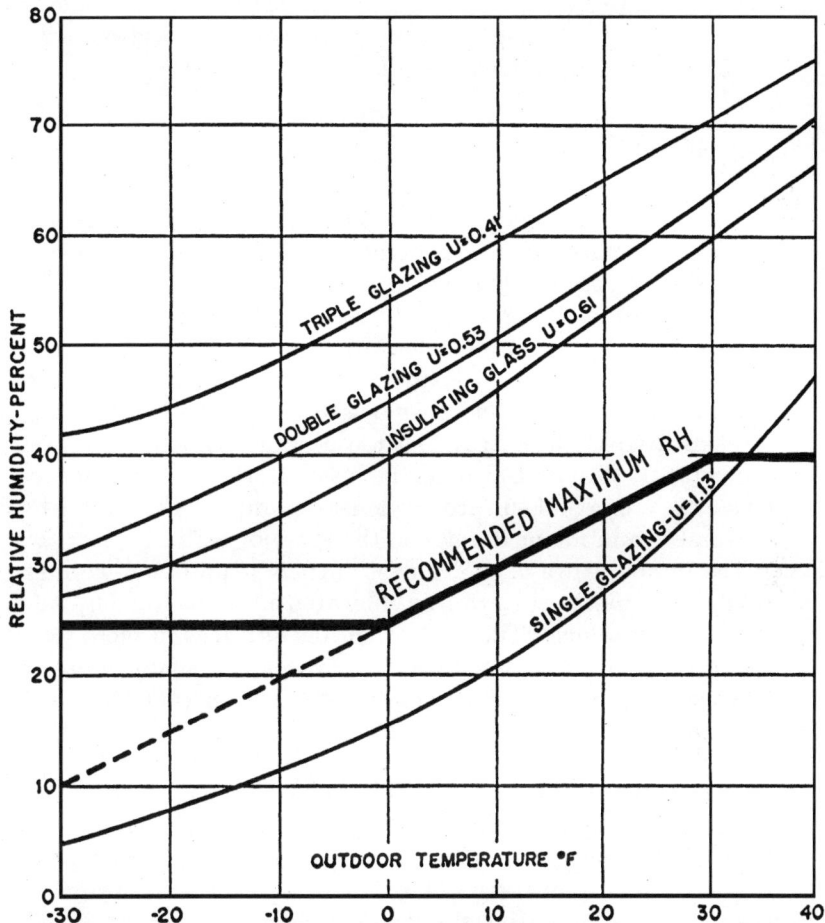

* Below 0°F, RH values maintained at 25% to provide acceptable degree of occupant comfort (solid line); however, to minimize condensation on single glazing, use RH values represented by dashed line.

Figure 5.4 Recommended maximum interior humidity to prevent surface condensation. (*From Olin, Construction Principles, Materials and Methods, 5th ed.*)

People generate moisture by breathing and perspiring. One and a half to two gallons of water vapor per day are generated in a house with four occupants under normal living conditions, but there are other significant sources of indoor moisture. One of the largest sources of moisture, regardless of climate or season, is the migration of moisture from the surrounding soil into foundations and conditioned spaces. It is estimated that as much as 80% of the moisture entering a structure originates from the soil. It moves into the structure as a liquid by capillary action, or as a vapor by diffusion through below-grade walls and floors. Leaky basements and the evaporation of moisture from exposed soil in crawl spaces contributes significantly to moisture-related problems. The evaporation of construction moisture in newly constructed buildings can contribute as much as 10 pints of water vapor per day for the first year or two. *Table 5.2* lists the most common sources of indoor moisture vapor and their estimated contributions.

5.2.3 Surface condensation

When the inside surface temperatures of exterior walls are substantially below indoor air temperatures, visible condensation may occur when air strikes the cold surface. The higher the indoor humidity, the less the temperature differential needed for surface condensation to form. For a 68°F indoor air temperature, the graph in *Fig. 5.5* shows the surface temperature differentials which will cause condensation at various indoor humidity levels. For example, if the indoor air is at 40% relative humidity, an interior wall surface temperature that is 42 degrees below the air temperature will cause surface condensation to form. At indoor temperatures higher than 68°F, the required temperature differential would be less. Interior wall surface temperatures can be elevated to avoid surface condensation during the winter by increasing the thermal insulation or R-value of the exterior wall.

The temperature of the coldest surface usually determines the maximum relative humidity that can be tolerated in a room before surface condensation begins to form. The coldest winter surfaces in most rooms are the windows. The coldest summer surfaces are the air conditioning ducts. To operate a space at higher and more comfortable winter relative humidities, the temperature of the coldest surface must be kept above the dew point of the indoor air. If room air at 75°F with a relative humidity of 40% is cooled by contact with a cold window, it is forced to deposit some of its moisture on the window. To prevent surface condensation of this sort, one solution would be to limit the moisture content of the air so that the dew point is lower than the temperature of the window surfaces. Another solution would be to keep the moist air away from the window by circulating dry air against it. If this circulating air is also warm, it can raise the surface temperature of the window above the dew point of the room air. Even normal convective air movement, when it is not blocked by draperies or other window treatments, can elevate the surface temperature of the window (*Fig. 5.6*). Another way of raising the surface temperature of the window would be to use insulated

TABLE 5.2 Most Common Sources of Indoor Moisture

Moisture source by type	Estimated moisture amount, pints
Household Produced	
Aquariums	Replacement of evaporative loss
Bathing:	
Tub (excludes towels and spillage)	0.12/standard size bath
Shower (excludes towels and spillage)	0.52/5-minute shower
Clothes washing (automatic, lid closed, standpipe discharge)	0+/load (usually nil)
Clothes drying:	
Vented outdoors	0+/load (usually nil)
Not vented outdoors or indoor drying line	4.68 to 6.18/load (more if gas dryer)
Combustion (unvented kerosene space heater)	7.6/gal of kerosene burned
Cooking:	
Breakfast (family of four, average)	0.35 (plus 0.58 if gas cooking)
Lunch (family of four, average)	0.53 (plus 0.68 if gas cooking)
Dinner (family of four, average)	1.22 (plus 1.58 if gas cooking)
Simmer at 203°F, 10 min, 6-in. pan (plus gas)	less than 0.01 if covered, 0.13 if uncovered
Boil 10 min, 6-in. pan (plus gas)	0.48 if covered, 0.57 if uncovered
Dishwashing (by hand):	
Breakfast (family of four, average)	0.21
Lunch (family of four, average)	0.16
Dinner (family of four, average)	0.68
Firewood storage indoors (cord of green firewood)	400 to 800/6 months
Floor mopping	0.03/sq ft
Gas range pilot light (each)	0.37 or less/day
House plants (5 to 7 average plants)	0.86 to 0.96/day
Humidifiers	0 to 120+/day (2.08 average/h)
Pets	Fraction of human adult weight
Respiration and perspiration (family of four, average)	0.44/hour (family of four, average)
Refrigerator defrost	1.03/day (average)
Saunas, steam baths, and whirlpools	0 to 2.7+/h
Vegetable storage (large-scale storage is significant)	0+ (not estimated)
Non-Household Produced	
Combustion exhaust gas backdrafting or spillage	0 to 6720+/year
Desorption of materials:	
Seasonal	6.33 to 16.91/average day
New construction	10+/average day
Ground moisture migration	0 to 105/day
Plumbing leaks	0+ (not estimated)
Rain or snowmelt penetration	0+ (not estimated)
Seasonal high outdoor absolute humidity	64 to 249+/day

SOURCE: W. Angell and W. Olson, Cold Climate Housing Information Center, University of Minnesota, and Lstiburek and Carmody, *Moisture Control Handbook.*

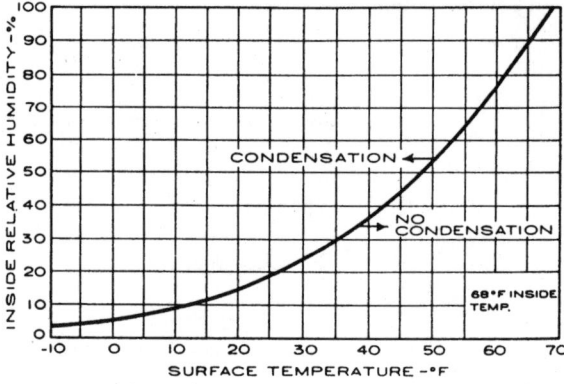

Figure 5.5 Conditions causing surface condensation at 68°F. (*From AAMA Aluminum Curtain Wall Design Guide Manual.*)

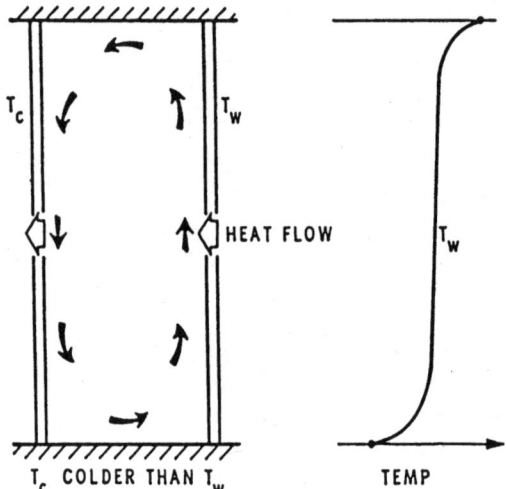

Figure 5.6 Effect of convection in an air space on the warm side surface temperature. (*From Latta, Walls, Windows and Roofs for the Canadian Climate.*)

TABLE 5.3 Maximum Relative Humidities at 70°F at Which No Condensation Will Occur at the Center of a Pane of Glass

Outdoor temperature, °F	Relative humidities at 70°F, %			
	Single window		Double window	
	Wind	No wind	Wind	No wind
+20	24	41	53	61
0	12	27	41	49
−20	6	17	32	39
−40	2	10	23	31

SOURCE: Latta, *Walls, Windows and Roofs for the Canadian Climate.*

glass in a frame with higher thermal resistance. *Table 5.3* shows the relative humidity that can be maintained in a room before surface condensation takes place on the center of a window pane with different outdoor conditions and glazing systems.

With winter humidification for comfort in cold climates, the need to raise window surface temperatures to control condensation led to the development of higher performance glazing and framing systems. Ironically, this has caused a higher incidence of concealed moisture problems in cold-climate enclosures because visible surface condensation no longer alerts occupants to the problem.

5.2.4 Concealed condensation

Although the most common and the most visible areas of condensation are the windows, condensation can make more insidious attacks inside the walls, beneath flooring, and within roofs. When vapor passes through a wall that is warm on one side and cold on the other, it may reach its dew point and condense into water or frost. The temperature drop through a wall is directly proportional to the thermal resistance of the various components. Thermal bridges through studs and connectors or at uninsulated expansion joints may also result in cold contact surfaces within the wall where concealed condensation can form. If the temperature at any point within a wall falls below the dew point temperature of the air, condensation will take place at that point.

The use of thermal insulation in wall cavities increases interior surface temperatures in cold climates and therefore reduces the likelihood of interior surface condensation. However, thermal insulation also lowers the winter temperature of the outer portions of walls and roofs by reducing heat flow from the conditioned space. In a well-insulated wall or roof with little heat loss from the interior, the outside surface temperatures will be very close to the ambient outdoor air temperature. These lower winter wall temperatures increase the likelihood of concealed condensation within the wall.

Concealed condensation can be controlled by reducing the transport of vapor into the wall or roof cavity or by elevating interstitial temperatures. Interstitial temperatures can be elevated by locating insulation toward the cold side of the section. When insulation is added to a wall or roof section, it elevates the temperature of everything on the warm side, and lowers the temperature of everything on the cold side. In cold climates, exterior insulation keeps the wall or roof components warm during the winter when condensation is most likely. In hot humid climates, interior insulation keeps the wall or roof components warm during the summer when condensation is most likely.

In very cold weather, concealed condensation can form as frost, or condensed water may turn to ice. Thawing of accumulated condensation can release large quantities of water very rapidly, leading to the misdiagnosis of a moisture problem as a leak rather than condensation. Freezing temperatures can also cause the formation of ice crystals within the body of porous materials that are saturated with absorbed condensate, resulting in physical damage or disintegration from the mechanical force of expansion. In hot weather, accumulated condensation can encourage mold and mildew growth, insect infestation, and wood rot.

5.3 Vapor Movement

Water vapor moves through building enclosures in two ways—it diffuses through solid materials, and is carried by air currents through cracks and openings. Warm air has a higher vapor pressure than cool air. If separated by a wall, the higher-pressure vapor will migrate through the wall toward the lower-pressure atmosphere. In cold climates, the predominant vapor diffusion drive is in winter from the warm, heated inside of the building toward the cooler outside (*Fig. 5.7A*). In warm humid climates, the predominant vapor diffusion drive is in summer from the outside inward toward the cool, air-conditioned interior (*Fig. 5.7B*). An easy way to determine the direction of vapor diffusion is to use the graph in *Fig. 5.7C*. If the ordinate for outdoor ambient conditions is above the ordinate for indoor conditions, vapor flow is *inward*. If the ordinate for outdoor ambient conditions is below the ordinate for indoor conditions, vapor flow is *outward*. Seasonal weather changes can cause a reversal of vapor flow. The behavior and performance of each building is governed by its unique set of internal and external conditions.

5.3.1 Diffusion

Vapor diffusion directly through the envelope materials occurs independent of air flow, but usually accounts for much less moisture transfer than airborne vapor. When there is a difference in vapor pressure between two points, vapor will flow from the point of higher vapor pressure to the point of lower vapor pressure, even without any corresponding flow of air. Diffusion is the process by which water vapor migrates *through* a material because of vapor pressure differentials (*Fig. 5.8*). Even though vapor diffusion transports less moisture through a building envelope than the movement of water vapor by air flow, an analysis of the vapor diffusion characteristics of an enclosure will show if conditions are conducive to condensation, whether the moisture will accumulate, or if the wall or roof has the ability to dry out.

When a vapor pressure difference exists between two sides of a wall or roof section, the water vapor will diffuse at a rate that is determined by

- The vapor pressure difference between inside and outside
- The length of the flow path
- The vapor diffusion resistance of the materials

Most building materials are permeable to vapor, and conventional construction is made up of a variety of materials of different permeance. The unit of *permeance* is the *perm*, which is defined as one grain of water passing through one square foot in one hour under a vapor pressure difference of one inch of

Vapor Condensation 155

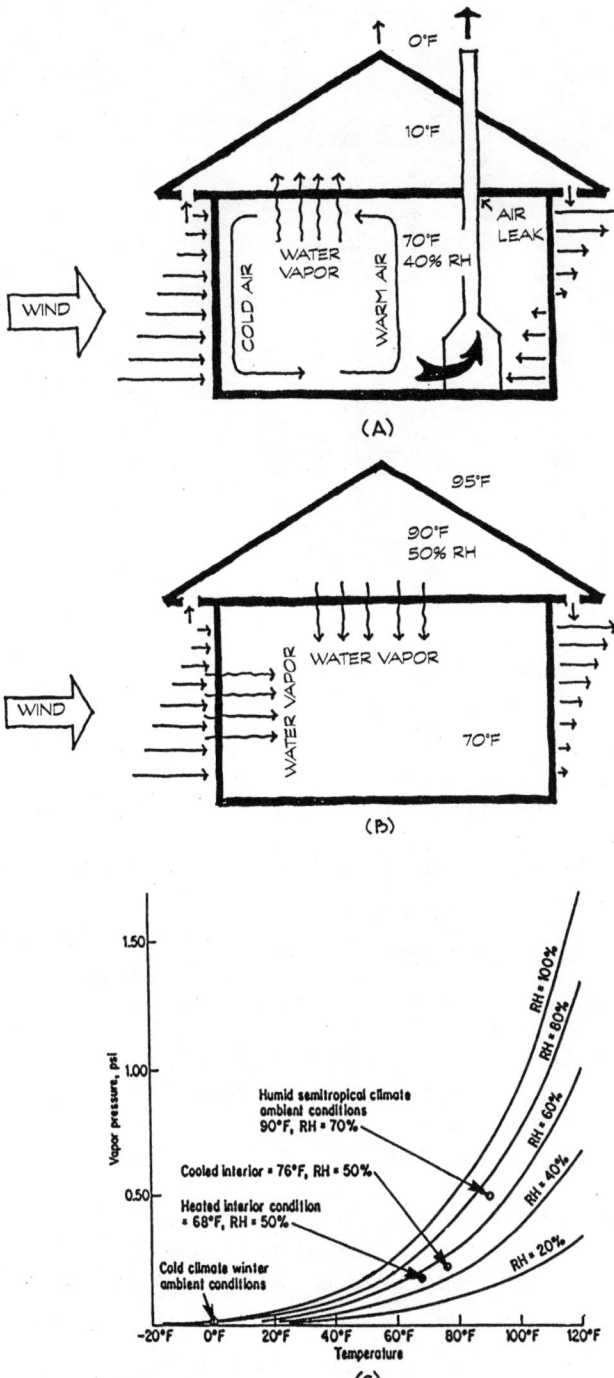

Figure 5.7 Vapor drive. (*Graph from Griffin, Manual of Low Slope Roof Systems, 3d ed.*)

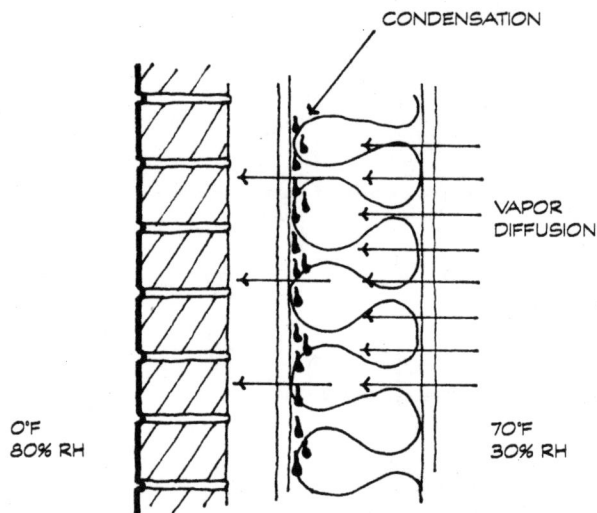

Figure 5.8 Diffusion is the process by which vapor migrates through a material. (*From Quirouette, The Difference Between a Vapor Barrier and an Air Barrier.*)

mercury. *Permeability* is the permeance of a one inch thickness of a homogeneous material. The unit of permeability is the *perm-inch*. The vapor flow resistance of a material is the reciprocal of its permeance. The unit of *resistance* is a *rep*, which is the reciprocal of perm (rep=1/perm). Values for the vapor permeance and vapor permeability of building materials are given in *Table 5.4*.

Vapor diffusion calculation is similar to heat flow calculation, using the formula

$$\text{Vapor diffusion} = \frac{\text{vapor pressure difference}}{\text{vapor diffusion resistance}}$$

One important difference is the ability of the vapor to condense. In a steady-state heat flow calculation, it is assumed that all the heat which enters the wall at one side emerges on the other. A vapor flow calculation starts with the same assumption of continuity of flow. If it is found that the assumption is invalid, and that on its passage through the building enclosure the vapor is cooled below the dewpoint, condensation will occur and the basis of the calculation is upset. Even so, once the plane of condensation has been established, the same method can be use to calculate the flow of vapor to and away from this plane. The difference between the two gives the rate of accumulation of water within the enclosure.

5.3.2 Air flow

Moving air carries water vapor with it. As long as the moisture stays in the air and keeps moving, it does no harm, but if it condenses to water or frost within

TABLE 5.4 Vapor Permeance of Building Materials*

Material	Permeance, perms, gr/h·ft^2·in. Hg	Permeability, perm-inches, gr/h·ft^2·(in. HG/in.)$^{-1}$
Common Building Materials		
Concrete, uncracked		3.2
Concrete, uncracked, 6 in. thick	0.5	
Brick masonry, 4 in. thick	0.8	
Concrete block, cored, limestone aggregate, 8 in. thick	2.4	
Tile or masonry, glazed	0.12	
Cement board	4.0–8.0	
Plaster on metal lath	15.0	
Gypsum wall board, $3/8$ in.	50.0	
Gypsum sheathing, asphalt impregnated		20.0
Hardboard, $1/8$ in.	11.0	
Metal roof deck, not considering laps and joints	0.7 0.0	
Metal roof deck, considering laps and joints	+1.0	
Plywood, exterior glue, $1/4$ in.	0.7	
Plywood, interior glue, $1/4$ in.	1.9	
Wood plank	Varies	
Common Roofing Materials		
Asphalt built-up membrane, hot applied	0.0	
Asphalt, hot applied, 2 lb/100 ft^2	0.5	
Asphalt, hot applied, 3.5 lb/100 ft^2	0.1	
No. 15 asphalt-saturated felt	1.0	
No. 15 tarred felt	1.0	
No. 30 asphalt-saturated felt	0.5	
No. 43 asphalt-saturated and coated felt	0.3	
Roll roofing, saturated and coated	0.05	
45-mil EPDM	0.04	
60-mil EPDM	0.03	
Plastic and Metal Films and Foils		
Aluminum foil, 1 mil	0.0	
Asphalt-laminated Kraft paper	0.3	
4-mil polyethylene sheet	0.8–1.4	
6-mil polyethylene sheet	0.06	
4-mil polyvinyl chloride (PVC)	1.2	
Vinyl wall covering	0.08–0.13	

TABLE 5.4 Vapor Permeance of Building Materials* (Continued)

Material	Permeance, perms, gr/h·ft^2·in. Hg	Permeability, perm-inches, gr/h·ft^2·(in. HG/in.)$^{-1}$
Common Insulation Materials		
Air, still		120.0
Cellular glass		0.0
Polyurethane, 1 in. thick		0.4–1.6
Polystyrene, extruded		1.2
Polystyrene, expanded		2.0–5.8
Mineral wool, unprotected		116.0
Phenolic foam, covering removed		26.0
Foil-faced insulation	0.5	
Liquid-Applied Coatings		
Commercial latex paint:		
Vapor retarder paint, 3.1 mil dry film thickness (dft)	0.45	
Primer sealer, 1.2 mil dft	6.28	
Vinyl acetate/acrylic primer, 2.0 mil dft	7.42	
Vinyl acrylic primer, 1.6 mil dft	8.62	
Semigloss vinyl acrylic enamel, 2.4 mil dft	6.61	
Exterior acrylic house and trim paint, 1.7 mil dft	5.47	
Two-coat paint application:		
Asphalt paint on plywood	0.4	
Aluminum varnish on wood	0.3–0.5	
Enamel on smooth plaster	0.5–1.5	
Primers and sealers on interior gypsum board	0.9–2.1	
Various primers plus one coat flat oil paint on plaster	1.6–3.0	
Flat paint on interior gypsum board	4.0	
Water emulsion paint on interior gypsum board	30.0–85.0	
Styrene-butadiene latex coating	11.0	
Polyvinyl acetate latex coating	5.5	
Chlorosulfonated polyethylene mastic	1.7	
Asphalt cut-back mastic		
$1/16$ in. dry	0.14	
$3/16$ in. dry	0.0	

*This table is for comparison purposes only. Exact values for permeance and permeability should be obtained from the material manufacturer. Values shown indicate variations among mean values for materials that are similar but of different density, orientation, lot, or source.

the building envelope, the potential for damage is great. Water will condense if the air containing it is cooled sufficiently as it moves from a warm location to a cold one. The infiltration and exfiltration of air through cracks and openings in a wall or roof can move much larger quantities of water vapor than diffusion. Because building construction is generally more airtight than it once was, air that leaks through the exterior envelope is usually funneled through small openings at relatively high volumes. Any condensation which occurs is thus often concentrated at limited locations, but with an attendant increase in severity. The uncontrolled movement of air through or within a building envelope can lead to serious problems of moisture leakage, condensation, and undesirable heat loss or heat gain.

Openings through a building envelope can occur at cracks or joints between infill components and structural elements, at poor connections between walls and floors or roofs, and at window and utility penetrations. Air can also pass through porous materials such as concrete block and batt insulation. Some openings follow direct paths as in an uncapped steel deck at the perimeter of a floor or roof. Other openings may present a more circuitous path such as through an outlet box, between the joints of a sheathing material, into a wall cavity, and through the weeps of the exterior cladding. Some cracks or openings may develop after construction because of thermal expansion and contraction, differential movement, or creep and frame shortening, and others may result from poorly designed or installed sealant joints.

Three different types of forces can induce air flow through openings in a building enclosure:

- Wind pressure
- Stack effect
- Mechanical ventilation

The net air pressure difference across a wall or roof may be a combination of all three forces, and the magnitudes of the differentials vary from one part of a building to another (*Fig. 5.9*). The combined pressure difference caused by wind load, stack effect, and fan pressurization can range from about 0.2 lb/sq ft in a small house to as much as 42 lb/sq ft in a high-rise building. Air leakage accounts for at least 6 or 7 times the volume of vapor movement as diffusion, and may account for up to 100 times as much vapor movement. The Division of Building Research of the National Research Council of Canada estimates that over one heating season in Ottawa, a small hole at an electrical box or other penetration in a 1-m^2 wall area accounts for 30 L of moisture transported by air leakage compared to only $^1/_3$ L of moisture transported by diffusion over the same 1-m^2 wall area (*Fig. 5.10*).

Wind pressure. Wind blowing on the outside of a building creates a pressure differential between the outside and inside air masses. If the building were

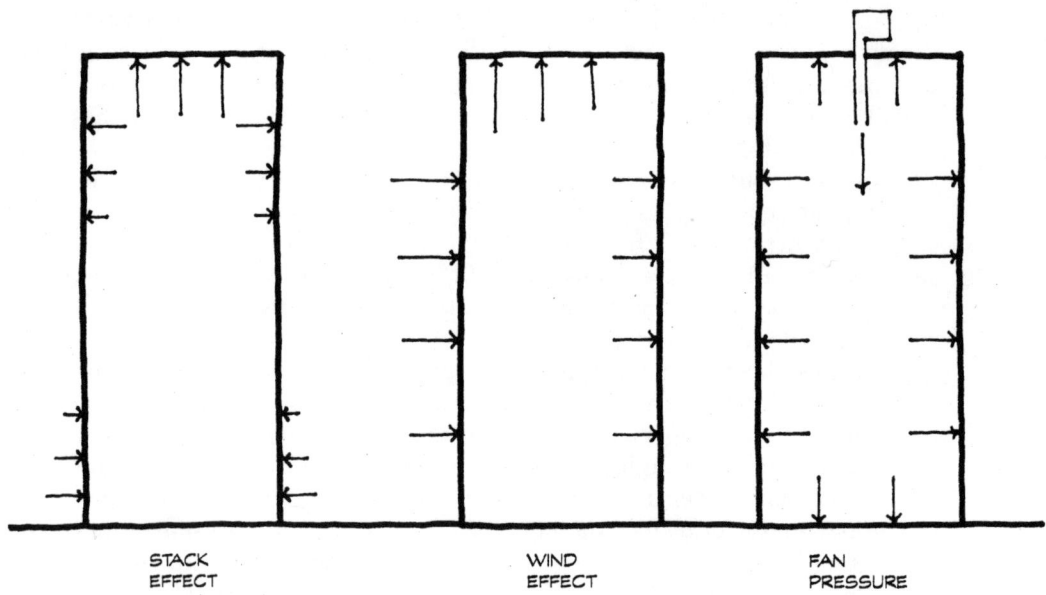

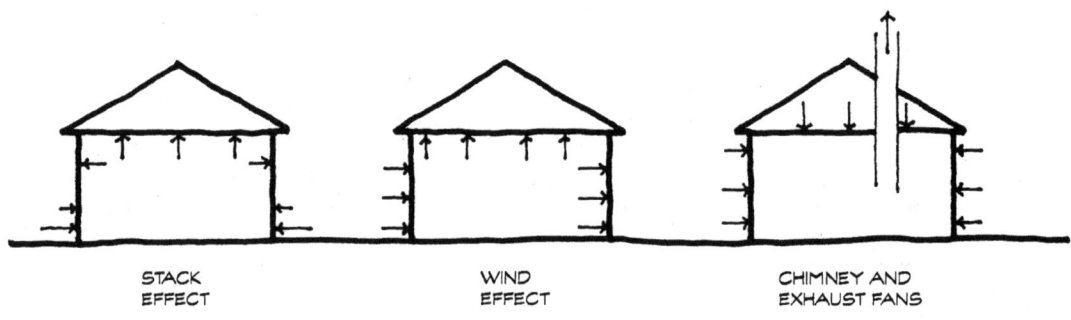

Figure 5.9 Air pressure differentials. (*Adapted from Quirouette, The Difference Between a Vapor Barrier and an Air Barrier.*)

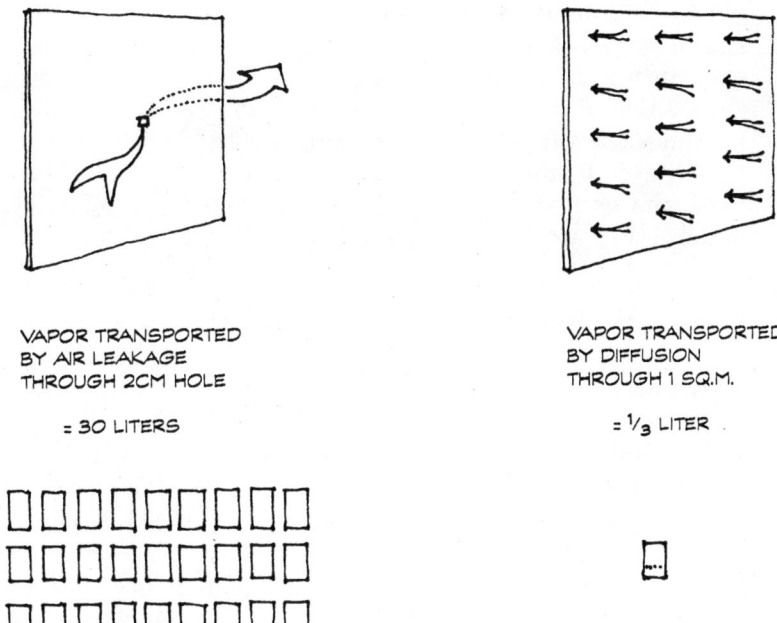

Figure 5.10 Vapor movement caused by air leakage can be 90 times that caused by diffusion. (*From Canadian Home Builders Association, Builders' Manual.*)

tightly sealed the inside pressure would remain unchanged, but all buildings have openings in the envelope through which air can flow in a natural effort to equalize the pressures. As air is forced into or out of the building by wind pressure, the interior air pressure changes. The magnitude of change is affected by the size, number, and location of openings. If a window is opened on the windward side of a house, the interior pressure will rise and become almost equal to the pressure on the exterior windward wall. The interior walls and ceiling, however, undergo a substantial increase in pressure difference, which in turn will increase air exfiltration through any openings in the ceiling and the leeward wall of the building.

Wind speed is seldom constant either with height or with time, nor does it produce anything approaching a uniform pressure over a building. Wind pressure can change from positive to negative in a relatively short distance around a corner, or from one side of a parapet to the other. Wind causes the infiltration of air through cracks and openings on the windward side of a building and exfiltration through cracks and openings on the leeward side and on the sides parallel to the wind. A flat roof will also generally experience air exfiltration because of wind. Since wind velocity increases with height, the difference in pressure across the building envelope also increases with height. The top floors of a high-rise building will experience the greatest pressure differentials.

The maximum wind pressure that is exerted on a building—when roofs are ripped off, trees uprooted, and windows blown in—will probably occur only once during the life of a building. Many buildings may never be subjected to the maximum possible wind pressure that could occur. On the other hand, all buildings are subjected almost continuously to the smaller pressure differentials caused by typical wind pressures. It is these pressure differentials that move the greater mass of air and the moisture vapor it contains, simply because they act for a much longer time.

Stack effect. Not all air flow through buildings is caused by wind. Convection currents caused by differences in temperature can also set up patterns of air movement. When air is heated it expands, so each cubic foot of heated air is less dense than the same volume of unheated air. The lighter air rises and is replaced by colder air at lower levels. This is analogous to the draft produced in a chimney stack. During the winter, a similar convective air movement occurs in heated buildings, even though the temperature difference is much less than that in a chimney. In one- and two-story buildings, winter stack effect can increase air leakage, and in tall buildings it can cause significant pressure differences and air movement through exterior walls. This air movement is accompanied by vapor movement and the potential for condensation as moist air moves back and forth through a building enclosure that is warm on one side and cold on the other. During the summer when outside air temperatures are higher than those inside, the pattern of pressure difference and air flow can be reversed, but pressure differentials are much less in summer because of the smaller temperature difference between indoor and outdoor air.

Winter stack effect creates a slight positive (outward) pressure at the top of a building, and a slight negative (inward) pressure at the base which causes air to infiltrate at the lower floors and exfiltrate at the upper levels. The pressure from stack effect is directly proportional to the effective height of the building. Such air pressure differences are the reason outward swinging doors are often difficult to open at the lobby level of high rise buildings. Inward opening exterior doors at roof level can also be difficult to open. In order to reduce the force required to open lobby level entrance doors, and to reduce air infiltration, vestibules or revolving door entrances are commonly used in tall buildings. A vestibule divides the pressure difference through the entrance so that each set of doors resists only half the total.

In multi-story buildings, air moving upward on stack-effect convection currents flows mainly through vertical shafts such as elevators, stairwells, and service shafts that penetrate the floors. The total pressure difference from inside to outside is a function of the temperature difference and the height of the building. The neutral pressure plane at which pressure changes from negative to positive is a function of the relative airtightness of the various components. For a heated building with an elevator shaft and with a uniform distribution of openings in the enclosure, through each floor, and into the elevator shaft at each floor, the pattern of pressure distribution caused by stack effect is shown in *Fig. 5-11*. The openings in a real building envelope, though, are seldom uniformly distributed. Whatever the distribution of openings, air inflow must equal air outflow. The total pressure difference for a heated building, the sum of the inward pressure at the bottom and the outward pressure at the top, can be estimated from *Fig. 5.12*. With a neutral pressure plane at midheight for a simple building shape, this total will be divided equally between the top and bottom of the building.

In tall buildings, air leakage and vapor migration rates are significantly increased by the chimney or stack effect—the inward movement of cool air at

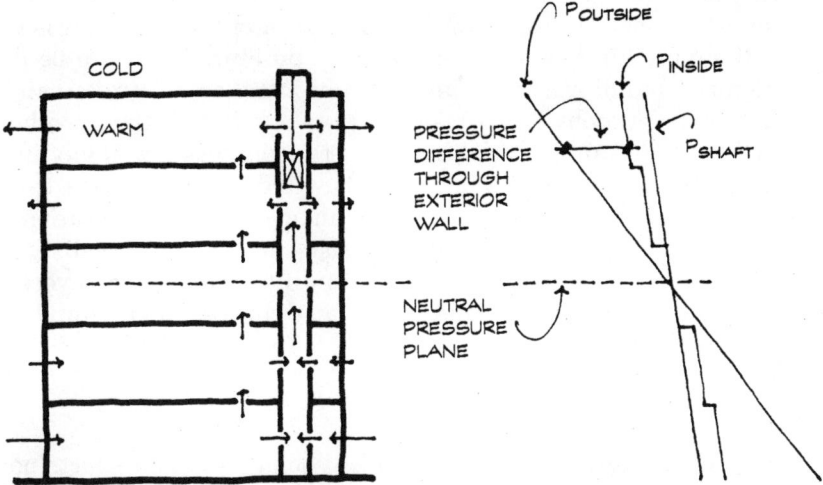

Figure 5.11 Example of pressure distribution in a building. (*From Latta, Walls, Windows and Roofs for the Canadian Climate.*)

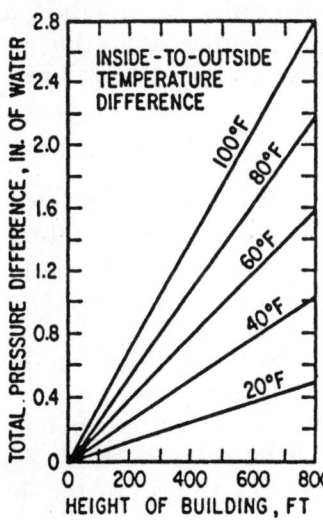

Figure 5.12 Total pressure difference between inside and outside of a heated building. (*From Latta, Walls, Windows and Roofs for the Canadian Climate.*)

lower stories and outward movement of warm air at upper stories due to pressure differentials (*Fig. 5.13*). A seasonal pattern of efflorescence or dampness at the top of a building can be a telltale sign of exfiltration caused by stack effect.

Mechanical ventilation. Ventilation air for contemporary buildings is usually provided by fans rather than operable windows. These may be small residential bathroom fans or large commercial or industrial systems. Hotels may have many small fans serving each guest room. Regardless of size, these fans are used to introduce or exhaust air from a building. An excess of exhaust over supply prevents exfiltration of moist interior air into cold roof or wall cavities. This is desirable in buildings such as those for swimming pools or industrial facilities where the moisture content of indoor air is very high.

If supply air exceeds exhaust air, a building is said to be *pressurized*. In high-rise buildings, this minimizes uncontrolled infiltration and the pressure difference at lobby levels caused by stack effect. If the hypothetical building in *Fig. 5.11* is pressurized by introducing excess air supply uniformly at all levels, the infiltration at lower levels will decrease, and the exfiltration at all levels will increase. Pressurization can aggravate condensation problems that result from the exfiltration of moist air into cold wall or roof cavities. Pressurized buildings that are humidified can suffer severe condensation problems in the winter unless the airtightness of the building envelope is increased.

5.3.3 Radiant solar heating

Movement of water vapor by incident solar radiation affects porous cladding materials. It occurs when a wet or saturated material is heated on one side, driving the absorbed moisture from the warmer to the cooler side. For example, when a porous cladding has absorbed rainwater and is later warmed by solar heating to very high surface temperatures, the absorbed rain water will be partially vaporized and driven deeper into the wall by the temperature differential (*Fig. 5.14*). As the cladding dries out, moisture will move back to the outside surface and evaporate. If vapor is driven into an air cavity that is well ventilated to the outside, drying will be more rapid because it occurs at both the inside and outside surfaces of the cladding.

Solar heating of wet cladding materials can create an inward vapor drive in both hot and cold climates. In the summer, if the interior spaces are air-conditioned, this solar-driven vapor may reach its dew point near the inside of the wall. If trapped there by an interior vapor retarder, condensation can cause a seasonal accumulation of moisture. To avoid such problems in the United Kingdom, the Building Research Establishment recommends that walls facing east-southeast to west-southwest incorporate a drainage cavity ventilated to the exterior to promote drying. In hot climates, inward vapor drives created by solar heating of wet cladding materials during the summer can be many times greater than the vapor

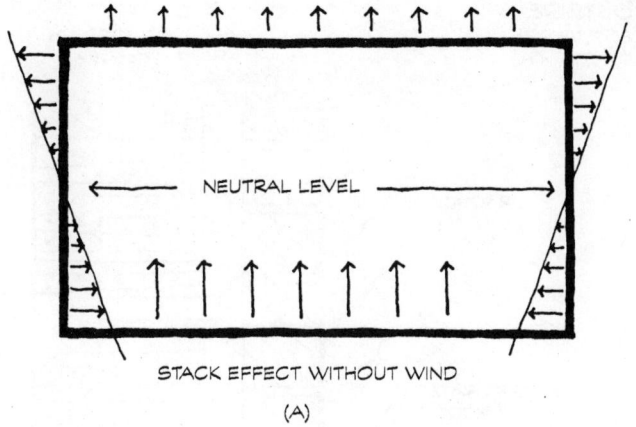

STACK EFFECT WITHOUT WIND

(A)

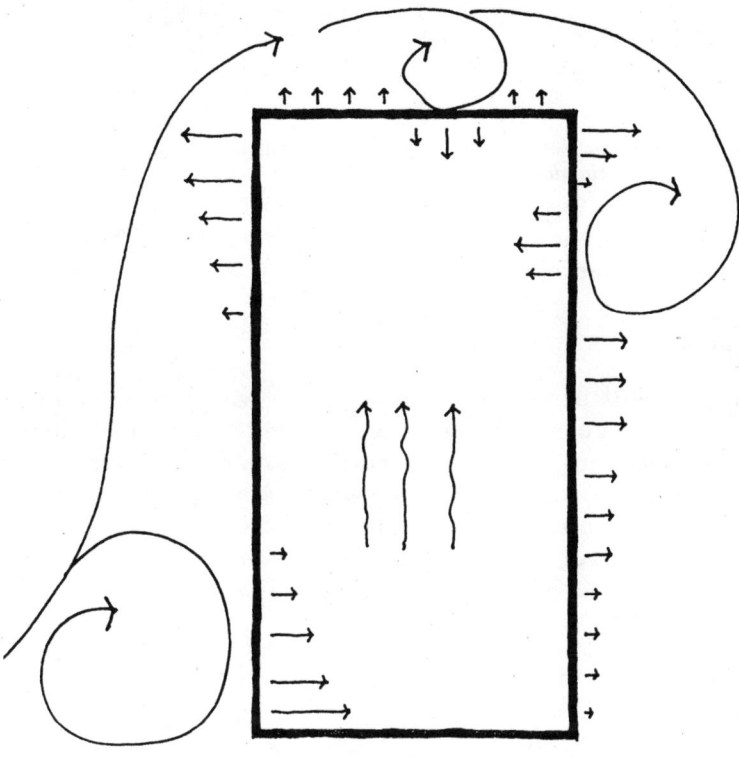

STACK EFFECT COMPLICATED BY
BUILDING HEIGHT AND WIND

(B)

Figure 5.13 Stack effect complicated by wind pressure differentials.

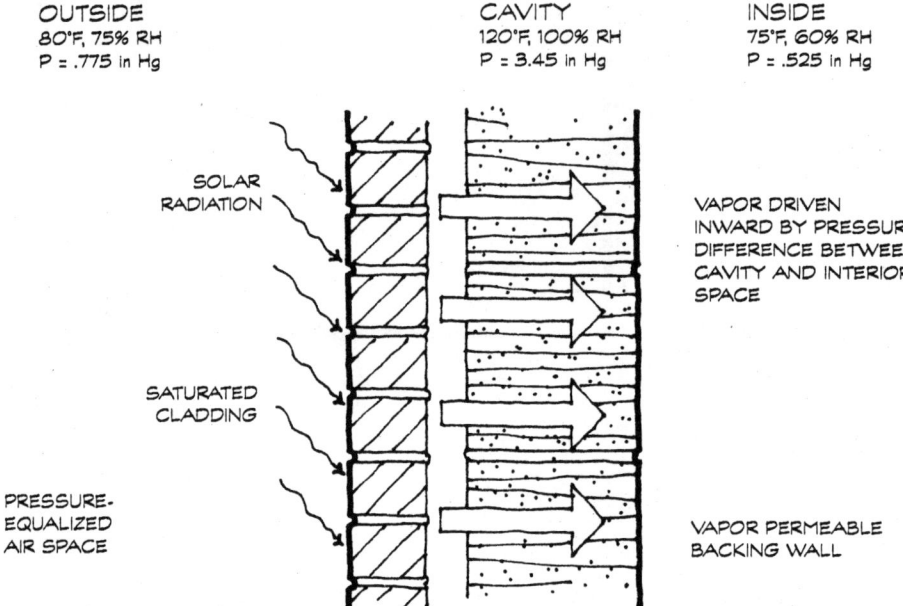

Figure 5.14 Effect of radiant solar heating on vapor drive. (*From Lstiburek and Carmody, Moisture Control Handbook.*)

pressure differences driving moisture outward from heated buildings in the harshest winter climates.

5.4 Vapor Resistance

The materials used in construction vary widely in their resistance to heat flow and to vapor diffusion. The relative thermal and vapor resistances of each material influence vapor movement and the potential for condensation on or within a wall or roof section. It is a relatively simple and straightforward process to calculate vapor diffusion and determine whether or not condensation will occur in a particular enclosure assembly under a given set of environmental conditions. In order to analyze the diffusion characteristics we must know

- Indoor and outdoor temperatures and relative humidities
- Thermal resistance of each material in the assembly
- Vapor resistance of each material in the assembly

Indoor design temperature and relative humidity information can be supplied by the mechanical engineer. Summer and winter outdoor design temperatures for a given location can be obtained from ASHRAE standards and other publications. *Table 5.5* is an excerpt from the 1989 *ASHRAE Handbook of Fundamentals.* The design dry bulb and design wet bulb temperatures given in the table can be used to find outdoor humidity from a psychrometric chart,

TABLE 5.5 Design Temperatures for U.S. Cities

City	Winter design dry bulb temp., °F	Summer design dry bulb temp. and mean coincident wet bulb temp., °F
Alabama		
Birmingham	21	94/75
Huntsville	16	93/74
Mobile	29	93/77
Tuscaloosa		
Alaska		
Anchorage	−18	68/58
Fairbanks	−47	78/60
Juneau	1	70/58
Nome	−27	62/55
Arizona		
Flagstaff	4	82/55
Phoenix	34	107/71
Tucson	32	102/66
Yuma	39	109/72
Arkansas		
Fayetteville	12	94/73
Fort Smith	17	98/76
Hot Springs	23	97/77
Little Rock	20	96/77
California		
Bakersfield	32	104/70
Burbank	39	91/68
Los Angeles	43	80/68
Modesto	30	98/68
Palm Springs	35	110/70
San Diego	44	80/69
San Francisco	38	77/63
Santa Barbara	36	77/66
Colorado		
Colorado Springs	2	88/57
Denver	1	91/59
Durango	4	87/59
Greeley	−5	96/60

TABLE 5.5 Design Temperatures for U.S. Cities (Continued)

City	Winter design dry bulb temp., °F	Summer design dry bulb temp. and mean coincident wet bulb temp., °F
Connecticut		
Hartford	7	88/73
Delaware		
Wilmington	14	89/74
Dist. of Columbia		
Washington	17	91/74
Florida		
Gainesville	31	93/77
Key West	57	90/78
Miami	47	90/77
Tampa	40	91/77
Georgia		
Atlanta	22	92/74
Augusta	23	95/76
Macon	25	93/76
Savannah	27	93/77
Hawaii		
Hilo	62	83/72
Honolulu	63	86/73
Kaneohe Bay	66	83/74
Wahiawa	59	84/72
Idaho		
Boise	10	94/64
Coeur d'Alene	−1	86/61
Idaho Falls	−6	87/61
Twin Falls	2	95/61
Illinois		
Champaign	2	92/74
Chicago	−4	89/74
Joliet	0	90/74
Springfield	2	92/74
Indiana		
Bloomington	5	92/75
Evansville	9	93/75

TABLE 5.5 Design Temperatures for U.S. Cities (Continued)

City	Winter design dry bulb temp., °F	Summer design dry bulb temp. and mean coincident wet bulb temp., °F
Indiana		
Indianapolis	2	90/74
South Bend	1	89/73
Iowa		
Cedar Rapids	−5	88/75
Des Moines	−5	91/74
Dubuque	−7	88/73
Keokuk	0	92/75
Kansas		
Atchison	2	93/76
Hutchinson	8	99/72
Topeka	4	96/75
Wichita	7	98/73
Kentucky		
Bowling Green	10	92/75
Covington	6	90/72
Lexington	8	91/73
Louisville	10	93/74
Louisiana		
Baton Rouge	29	93/77
Lake Charles	31	93/77
New Orleans	33	92/78
Shreveport	25	96/76
Maine		
Augusta	−3	85/70
Bangor	−6	83/68
Caribou	−13	81/67
Portland	−1	84/71
Maryland		
Baltimore	13	91/75
Frederick	12	91/75
Hagerstown	12	91/74
Salisbury	16	91/75

TABLE 5.5 Design Temperatures for U.S. Cities (Continued)

City	Winter design dry bulb temp., °F	Summer design dry bulb temp. and mean coincident wet bulb temp., °F
Massachusetts		
Boston	9	88/71
Gloucester	5	86/71
Lowell	1	88/72
Worcester	4	84/70
Michigan		
Detroit	6	88/72
Kalamazoo	5	88/72
Lansing	1	87/72
Sault Ste. Marie	−8	81/69
Minnesota		
Duluth	−16	82/68
International Falls	−25	83/68
Minneapolis	−12	89/73
St. Cloud	−11	88/72
Mississippi		
Biloxi	31	92/79
Jackson	25	95/76
Tupelo	19	94/77
Vicksburg	26	95/78
Missouri		
Cape Girardeau	13	95/75
Hannibal	3	93/76
St. Louis	6	94/75
Springfield	9	93/74
Montana		
Billings	−10	91/64
Cut Bank	−20	85/61
Helena	−16	88/60
Missoula	−6	88/61
Nebraska		
Grand Island	−3	94/71
Lincoln	−2	95/74
North Platte	−4	94/69
Omaha	−3	91/75

TABLE 5.5 Design Temperatures for U.S. Cities (Continued)

City	Winter design dry bulb temp., °F	Summer design dry bulb temp. and mean coincident wet bulb temp., °F
Nevada		
Carson City	9	91/59
Elko	−2	92/59
Las Vegas	28	106/65
Reno	10	92/60
New Hampshire		
Berlin	−9	84/69
Concord	−3	87/70
Manchester	−3	88/71
Portsmouth	2	85/71
New Jersey		
Atlantic City	13	89/74
Newark	14	91/73
New Brunswick	10	89/73
New Mexico		
Albuquerque	16	94/61
Carlsbad	19	100/67
Los Alamos	9	87/60
New York		
Albany	−1	88/72
Buffalo	6	85/70
New York City	15	89/73
Utica	−6	85/71
North Carolina		
Asheville	14	87/72
Charlotte	22	93/74
Greensboro	18	91/73
New Bern	24	90/78
North Dakota		
Bismark	−19	91/68
Fargo	−18	89/71
Grand Forks	−22	87/70
Minot	−20	89/67

TABLE 5.5 Design Temperatures for U.S. Cities (Continued)

City	Winter design dry bulb temp., °F	Summer design dry bulb temp. and mean coincident wet bulb temp., °F
Ohio		
Akron	6	86/71
Cincinnati	6	90/72
Columbus	5	90/73
Toledo	1	88/73
Oklahoma		
Bartlesville	10	98/74
Lawton	16	99/74
Oklahoma City	13	97/74
Tulsa	13	98/75
Oregon		
Eugene	22	89/66
Klamath Falls	9	87/60
Pendleton	5	93/64
Portland	23	85/67
Pennsylvania		
Erie	9	85/72
Lancaster	8	90/74
Philadelphia	14	90/74
Pittsburgh	5	86/71
Rhode Island		
Newport	9	85/72
Providence	9	86/72
South Carolina		
Charleston	28	92/78
Columbia	24	95/75
Greenville	22	91/74
Orangeburg	24	95/75
South Dakota		
Aberdeen	−15	91/72
Pierre	−10	95/71
Rapid City	−7	92/65
Sioux Falls	−11	91/72

TABLE 5.5 Design Temperatures for U.S. Cities (Continued)

City	Winter design dry bulb temp., °F	Summer design dry bulb temp. and mean coincident wet bulb temp., °F
Tennessee		
Chattanooga	18	93/74
Dyersburg	15	94/77
Memphis	18	95/76
Nashville	14	94/74
Texas		
Amarillo	11	95/67
Austin	28	98/74
Dallas	22	100/75
El Paso	24	98/64
Houston	32	94/77
Midland	21	98/69
San Antonio	30	97/73
Wichita Falls	18	101/73
Utah		
Moab	11	98/60
Ogden	5	91/61
Salt Lake City	8	95/62
Vermont		
Barre	−11	81/69
Burlington	−7	85/70
Rutland	−8	84/70
Virginia		
Charlottesville	18	91/74
Norfolk	22	91/76
Richmond	17	92/76
Roanoke	16	91/72
Washington		
Olympia	22	83/65
Seattle	26	80/64
Spokane	2	90/63
Yakima	5	93/65

TABLE 5.5 Design Temperatures for U.S. Cities (Continued)

City	Winter design dry bulb temp., °F	Summer design dry bulb temp. and mean coincident wet bulb temp., °F
West Virginia		
Clarksburg	10	90/73
Martinsburg	10	90/74
Morgantown	8	87/73
Wheeling	5	86/71
Wisconsin		
Green Bay	−9	85/72
Madison	−7	88/73
Milwaukee	−4	87/73
Wausau	−12	88/72
Wyoming		
Casper	−5	90/57
Cheyene	−1	86/58
Cody	−13	86/60
Laramie	−6	81/56

Excerpted from ASHRAE *Handbook of Fundamentals.*

or average humidity conditions can be obtained from weather service data (see App. C, Bibliography, under National Oceanic and Atmospheric Administration). Thermal resistance values for various building materials are given in *Table 3.4* in Chap. 3. Vapor resistance ratings can be derived from *Table 5.4* as the reciprocal of permeance. With this information, the thermal gradient and the vapor pressure gradient through the wall can be calculated to determine whether the conditions will cause summer or winter condensation.

5.4.1 Calculating the thermal gradient

The temperature drop through a wall or roof section is directly proportional to the thermal resistance of the various components. The total temperature drop can be apportioned to each component in the ratio of its thermal resistance to the total thermal resistance. The temperature drop can be arithmetically calculated as described in Chap. 3, and the resulting gradient plotted on a scale drawing of the section (*Fig. 5.15*). From this tabulation and drawing, it is easy to determine the temperature at each plane.

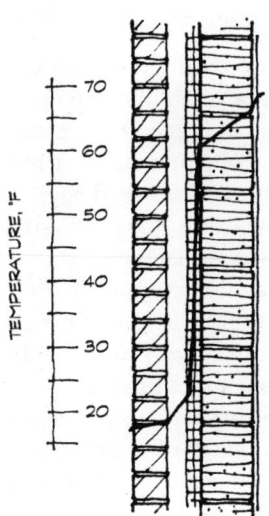

Material	Thermal Resistance, (R)	% of Total Resistance	Temp. Difference, (°F)	Temp., (°F)
Inside temperature		--	--	72
Inside still air film	0.68	5	3	69
1/2" gypsum board	0.45	4	2	67
6" concrete block	1.53	11	6	61
1-1/2" expanded polystyrene insulation	9.38	69	38	23
1" air space	0.97	7	4	19
4" brick	0.44	3	1	18
outside air film, 15 mph	0.17	1	1	17
outside temperature		--	--	17
Totals	13.62	100%	55	

Figure 5.15 Thermal gradient calculation.

5.4.2 Calculating the vapor pressure gradient

The vapor pressure drop through a wall or roof section is directly proportional to the vapor resistance of the various components. The temperatures found at each plane in the thermal gradient calculation are listed along with the corresponding saturated vapor pressures from *Figure 5.1*. The actual vapor pressures for the indoor and outdoor design temperatures are calculated as a percentage of the saturated vapor pressure of those temperatures based on given relative humidity conditions for this example of of 73% RH outdoors and 50% RH indoors. Saturated vapor pressure (SVP) at a given temperature times relative humidity (RH) at the same temperature=actual vapor pressure (AVP) required for continuity of flow. The difference between the actual indoor and actual outdoor vapor pressures is the total vapor pressure difference. The total vapor pressure difference is apportioned among the various components of the enclosure in the ratio of each material's vapor resistance to the total vapor resistance, and the resulting gradient plotted on a scale drawing of the section (*Fig. 5.16*). The saturated vapor pressures are plotted on the same drawing. At any location where the actual vapor pressure is higher than the saturated vapor pressure, condensation will occur. In this example, the dew point of the moist indoor air is reached at the plane of the concrete block and insulation interface. If the joints of the insulation are taped and the face of the insulation has little or no permeability, the condensed vapor will be trapped in the insulation layer. If vapor can migrate through the insulation, it will condense on the next cold surface, which is the cavity side of the brick veneer. If the wall is designed with flashing and weep holes in the cavity, the condensed moisture can be effectively drained and no damage will occur.

5.5 Analyzing the Risk

This steady-state calculation method considers only vapor diffusion and does not take into account the large quantities of vapor that may move through the envelope with air leaks. It does, however, give us the opportunity to find the location of the dew point within the section, to calculate whether or not the wall will be able to dry out under normal conditions or whether condensation will accumulate, and to decide whether or not a vapor diffusion retarder is required and where it should be located. If vapor diffusion calculations show that condensation is likely to occur, air leakage paths through the enclosure will increase the quantity of vapor and the severity of the condensation.

In assessing the risk of condensation, the vapor diffusion analysis can provide several pieces of information. The first is whether the occurrence of condensation is possible, probable, or definite. The graphic analysis shows this best. If the SVP and AVP lines just barely touch, condensation will occur only under the most severe of weather conditions. If the lines cross, but the area between them is small, condensation is probable under typical or average seasonal conditions. If the lines cross and the area between them is large, condensation will likely occur even under relatively mild conditions. The analysis also shows

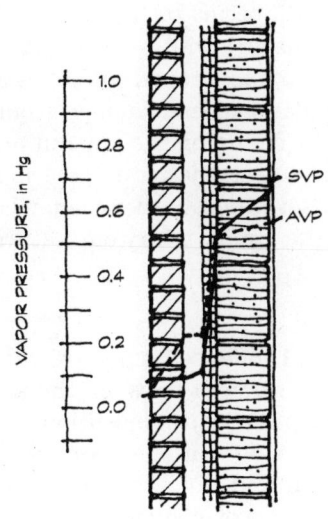

Temp., °F	Saturated Vapor Pressure, SVP	Vapor Resistance, Rep	% of Total Resistance	Vapor Pressure Difference	Actual Vapor Pressure, AVP
72	0.791	--	--	--	0.577
69	0.714	0.008	--	--	0.577
67	0.667	0.027	1	0.005	0.572
61	0.540	0.300	7	0.037	0.535
23	0.119	2.34	59	0.314	0.221
19	0.098	0.008	--	--	0.221
18	0.093	1.3	33	0.176	0.045
17	0.089	--	--	--	0.045
Totals		3.983	100	0.532	

Figure 5.16 Vapor pressure gradient calculation.

where condensation will occur. In the example given, condensation within the drainage cavity of a masonry wall is not likely to be of concern because it will be collected by the flashing and weeped out of the wall. If condensation were shown to occur at another plane where damage would be likely, adjustments to the design can be made by changing the amount or location of the insulation, or by adding a vapor diffusion retarder on the warm side of the section. The analysis can also be used to determine whether condensation will accumulate or whether the wall or roof has the ability to dry out naturally on its own.

5.5.1 Moisture accumulation

To determine whether condensed moisture will accumulate in the section, the actual vapor pressure gradient between the inside and the plane of condensation, and between the plane of condensation and the outside can be determined by calculation using the saturation vapor pressure of 0.119 in. Hg at the plane between the insulation and the air cavity. The vapor pressure drops between the inside and the plane of condensation (0.577−0.119=0.458 in. Hg), and between the plane of condensation and the outside (0.119−0.045=0.74 in. Hg) are distributed among the various components in proportion to their resistance to vapor diffusion. The vapor flow to the plane of condensation is

$$\frac{0.577-0.119}{0.008+0.027+0.300+2.34} =$$

$$\frac{0.458}{2.675} = 0.17 \text{ gr/ft}^2 \cdot \text{h}$$

and from the plane of condensation

$$\frac{0.119-0.045}{0.008+1.3} =$$

$$\frac{0.74}{1.308} = 0.57 \text{ gr/ft}^2 \cdot \text{h}$$

The flow of vapor from the condensation plane to the outside (0.57 gr/ft²·h) is greater than the flow of vapor from the inside to the condensation plane (0.17 gr/ft²·h), so the wall can easily dry out and there is no moisture accumulation (0.17−0.57=−0.40 gr/ft²·h). If the situation were reversed and vapor flow to the condensation plane exceeded vapor flow away from the condensation plane, then moisture would accumulate at a rate equal to the difference between the two. If condensation may occur frequently (i.e., under moderate conditions), then the enclosure may build up a high moisture content, since the vapor can get into the wall section more easily than it can get out. Under such circumstances, the wall or roof design must be adjusted by changing the amount or location of the insulation, or by adding a vapor diffusion retarder on the warm side of the section. Every wall or roof must be analyzed individually, because changes in materials or in indoor temperature and humidity conditions change the location of the dew point within the section.

5.5.2 Wetting and drying cycles

Occasional or limited condensation that is effectively drained from walls or vented from roofs is not generally a problem. However, repeated or excessive condensation can accumulate in porous materials, causing wood to warp or decay; metal to corrode; insulation to lose its thermal resistance; and masonry to shrink or expand, effloresce, or suffer damage from freeze-thaw cycles.

Seasonal wetting and drying patterns can be analyzed for any given location. The saturated vapor pressures (SVP) for average outdoor monthly temperatures are plotted, along with the average actual outdoor vapor pressure (VP_{avg}) based on average humidity conditions (*Fig. 5.17*). A horizontal line is drawn at the indoor vapor pressure based on indoor design temperature and humidity. If seasonal drying potential exceeds seasonal wetting potential, there should be no long-term accumulation of moisture, even if calculations show that there is a short-term accumulation. If seasonal wetting potential exceeds seasonal drying potential, then a vapor diffusion retarder is needed to prevent long-term moisture accumulation. Average

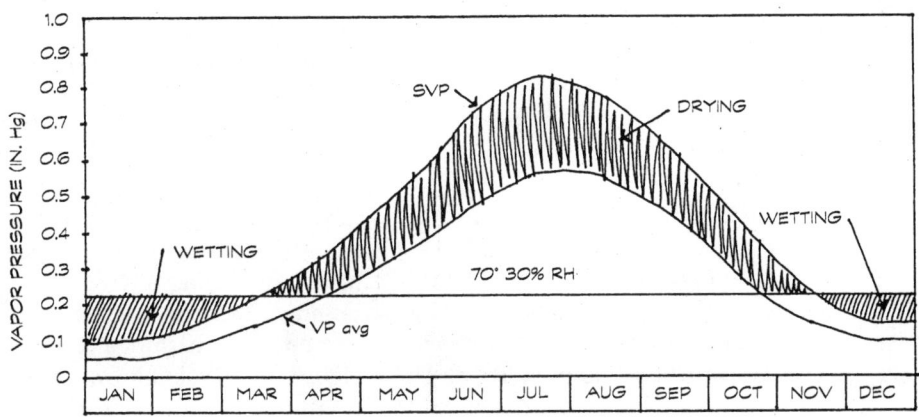

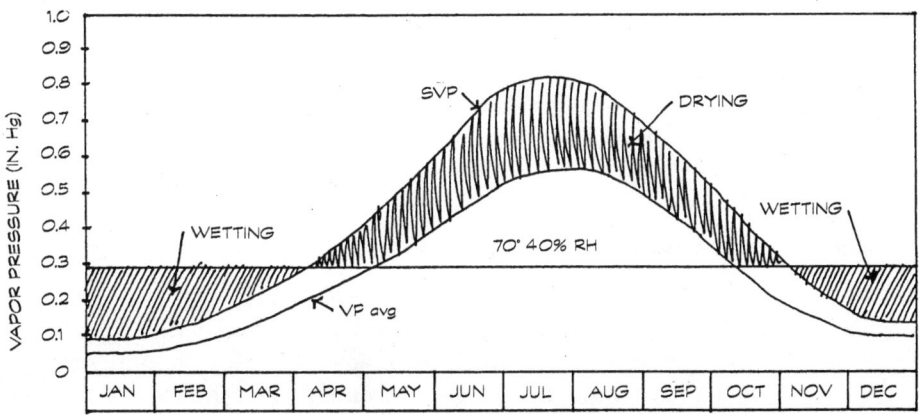

Figure 5.17 Seasonal drying can prevent progressive accumulation of moisture.

outdoor temperature and humidity information can be obtained from weather service data (see App. C, Bibliography, under National Oceanic and Atmospheric Administration).

5.5.3 Cold climates

In cold climates, condensation problems generally occur during the winter months when warm, moist indoor air is driven through walls and roofs toward the cooler, drier outside atmosphere. *Figures 5.18* and *5.19* show a sample vapor analysis for a precast concrete cladding over a metal stud wall. The diffusion calculations show that condensation will occur in the cavity behind the precast, and that moisture will accumulate at the rate of 3.23 gr/ft^2·h. The seasonal analysis, however, shows that drying potential exceeds wetting potential for the indoor design conditions, so if flashing and weeps are provided to drain the condensed water, there should be no long-term accumulation of moisture in the wall.

5.5.4 Hot, humid climates

In hot, humid climates, condensation problems can occur during the summer months when warm, moist outdoor air is driven through walls and roofs toward cooler, drier interior spaces. Air conditioners reduce indoor vapor pressure by condensing moisture from the air as it is cooled (dehumidification). When this occurs, a vapor drive develops than can cause humid outdoor air to diffuse through a building envelope toward the cooler indoor atmosphere. If the moisture content of the outdoor air is high enough and temperatures within the building enclosure are cool enough, condensation can occur.

Figures 5.20 and *5.21* show a sample vapor analysis for a brick veneer over a metal stud backing wall. The diffusion calculations show that condensation will occur behind the interior gypsum board, but that moisture will accumulate at only 0.0009 gr/ft^2·h. The seasonal analysis shows that drying potential exceeds wetting potential for the indoor design conditions, so there is no long-term accumulation of moisture. However, mold and mildew growth are likely under the warm summer conditions, and subsequent drying of the wall will not kill the mold spores. Adjusting the design to change the amount or location of the insulation could eliminate the potential for condensation, or a vapor diffusion retarder may be needed on the summer warm side of the insulation. If vinyl wall coverings are used on the inside surface, they act as a vapor diffusion retarder, but are located on the wrong side of the insulation and may aggravate condensation or mold and mildew growth.

5.6 Control Strategies

To prevent condensation from occurring in a wall or roof section, the actual vapor pressure must be kept below the saturated vapor pressure. As previously mentioned, one way this can be achieved is by changing the amount or location of the insulation to raise the temperature and the corresponding saturated

Vapor Condensation

WINTER: Outdoor: 14 °F (design) 66% RH (avg.)
Indoor: 68°F (measured) 45% RH (meas.)

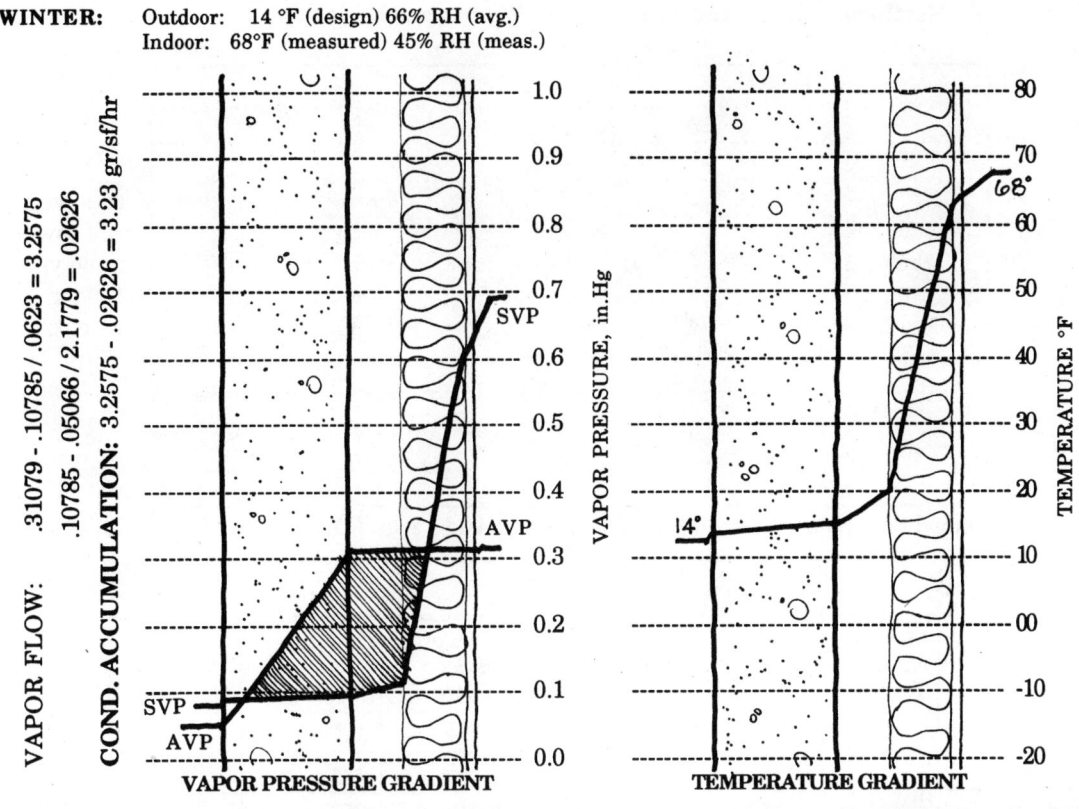

MATERIAL	Thermal Analysis				Vapor Pressure Analysis				
	R	%	ΔT	°F	SVP	VR	%	VPD	AVP
inside temperature				68	.69065				.31079
inside air film	0.68	4.0	2.0			.0083	0.34	.00088	
				66	.64452				.30991
paint finish	--	--	--			.0120	0.50	.00130	
				66	.64452				.30861
5/8" gypsum board	0.56	4.0	2.0			.0334	1.50	.00390	
				64	.60112				.30471
3-1/2" fiberglass insul.	13.0	80	43			.0086	0.33	.00086	
				21	.10785				.30385*
3" air space	1.12	7.0	4.0			.0079	0.33	.00086	
				17	.08891				.30299*
7" precast concrete	0.70	4.0	2.0			2.170	97.0	.25233	
				15	.08062				.05066
outside air film	0.17	1.0	1.0						
outside temperature				14	.07675				.05066
TOTAL	16.23	100	54			2.2402	100	.26013	

Figure 5.18 Condensation analysis for a cold climate.

182   Chapter Five

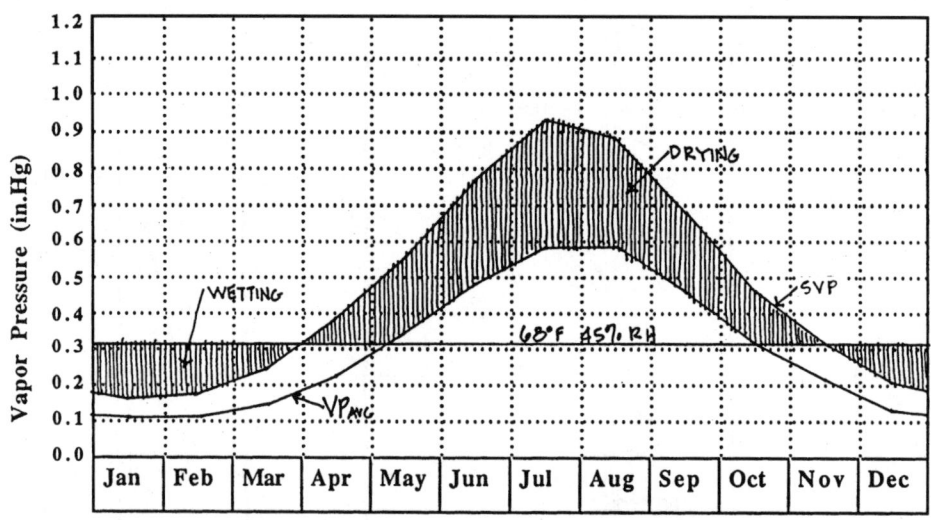

Month	Mean Outdoor Temp., °F	SVP in.Hg	Average Outdoor RH, %	Average Outdoor VP, in.Hg	Wetting	Drying
January	31.3	.17481	66.5	.11625	.13598	
February	32.8	.18643	63.75	.11885	.12436	
March	41.2	.25968	60.5	.15712	.05111	
April	52.1	.39200	58.0	.22736		.08121
May	62.3	.56632	62.75	.35537		.25553
June	71.5	.77845	63.25	.49237		.46766
July	76.8	.92976	63.75	.59272		.61897
August	75.5	.89048	66.5	.59217		.57969
September	68.2	.69548	68.5	.47640		.37921
October	57.2	.47213	67.75	.31987		.16134
November	46.5	.31806	67.5	.21469		.00727
December	35.5	.20768	67.25	.13967	.10311	
				Total	.41456	2.55088

Indoor temperature: 68°F   Wetting / Drying ratio: .2
Indoor humidity: 45%
Indoor vapor pressure: .31079

Figure 5.19 Condensation accumulation analysis for a cold climate.

Vapor Condensation

SUMMER: Outdoor: 94°F (design) 75% RH (avg)
Indoor: 73°F / 74% RH (measured)

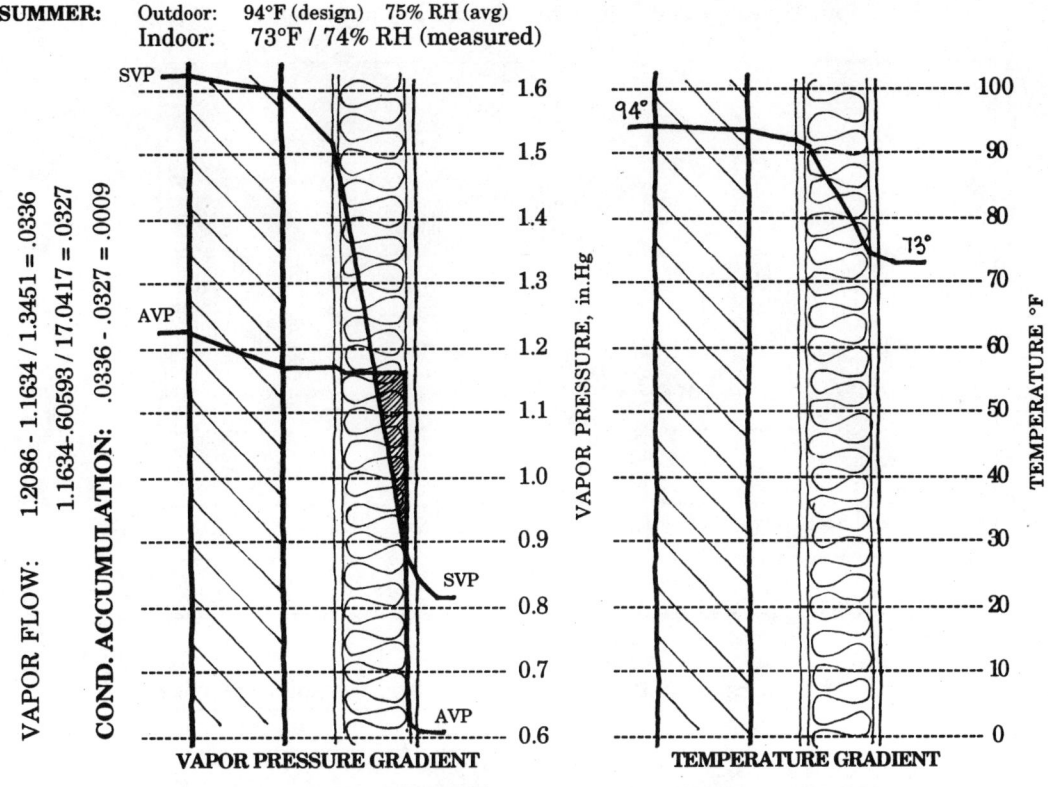

VAPOR FLOW: 1.2086 - 1.1634 / 1.3451 = .0336
1.1634-.60593 / 17.0417 = .0327
COND. ACCUMULATION: .0336 - .0327 = .0009

	Thermal Analysis					Vapor Pressure Analysis			
MATERIAL	R	%	ΔT	°F	SVP	VR	%	VPD	AVP
inside temperature				73	.81883				.60593
inside air film	0.68	5	1.0			.0083	---	---	
				74	.84682				.60593
5/8" gypsum wallboard	0.56	4	1.0			.0334	0.5	.00301	
				75	.87564				.60894
6 mil polyethylene	---	---	---			17.00	92.0	.55446	
				75	.87564				1.1634*
unfaced fibergalss insul.	11.0	75	16			.0086	---	---	
				91	1.4682				1.1634
1/2" GP Dens-Glass sheath.	0.56	4	1.0			.0286	0.5	.00301	
				92	1.5147				1.1664
2" air space	1.12	8	1.5			.0079	---	---	
				93.5	1.5870				1.1664
brick veneer	0.44	3	0.5			1.300	7.0	.04219	
				94	1.6115				1.2086
outside air film	0.25	1	---						
outside temperature				94	1.6115				1.2086
TOTAL	14.61	100	21			18.3868	100	.60267	

Figure 5.20 Condensation analysis for a hot, humid climate.

184 Chapter Five

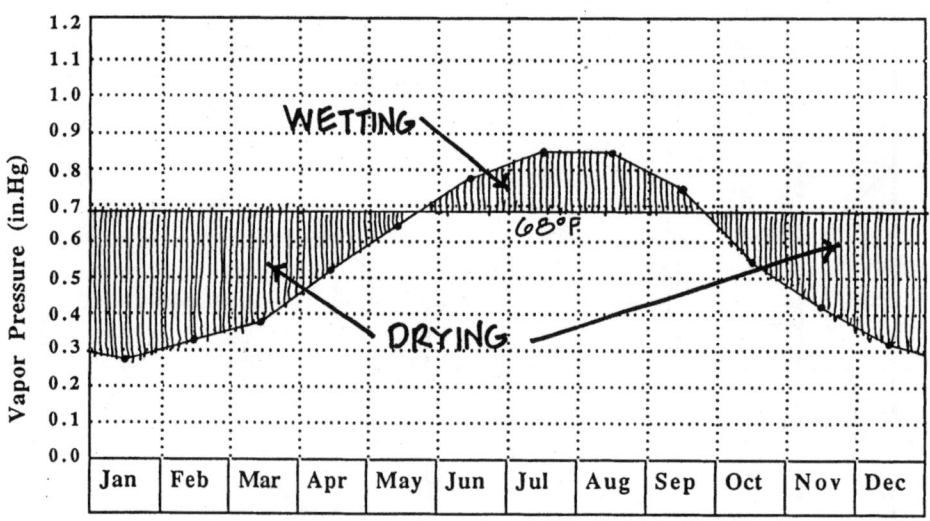

Month	Mean Outdoor Temp., °F	SVP in.Hg	Average Outdoor RH, %	Average Outdoor VP, in.Hg	Wetting	Drying
January	51.4	0.38203	74.25	0.28366		0.40699
February	54.5	0.42812	72.75	0.31146		0.37919
March	61.0	0.54081	72.25	0.39074		0.29991
April	68.7	0.70754	72.75	0.51474		0.17591
May	74.9	0.87276	75.00	0.65457		0.03608
June	80.6	1.05354	75.00	0.79016	0.09951	
July	83.1	1.14250	74.50	0.85116	0.16051	
August	82.6	1.12426	75.25	0.84601	0.15536	
September	78.4	0.98026	77.25	0.75725	0.06660	
October	69.7	0.73218	75.75	0.55463		0.13602
November	60.1	0.52381	76.50	0.40072		0.28993
December	54.0	0.42031	75.00	0.31523		0.37542
				Total	0.48198	2.09945

Indoor design temperature: 68°F   Wetting / Drying ratio: 0.23
Indoor saturation pressure: 0.69065

Figure 5.21 Condensation accumulation analysis for a hot, humid climate.

pressure values, by changing to warm side materials with higher vapor flow resistances so that the actual vapor pressure is reduced faster, or by using a combination of these methods. Condensation control can also be achieved by control of moisture sources, attention to indoor ventilation, and the judicious use of vapor diffusion retarders and air barriers.

5.6.1 Limiting moisture sources

In residences, cooking, laundering, and showering, which are significant contributors to airborne moisture, cannot be eliminated. However, ground moisture can be controlled by using vapor retarders and capillary breaks below slabs on grade and by covering the soil in crawl spaces with a vapor resistant film. Latent construction moisture can be driven off more rapidly by operating heating systems before occupancy. Storing firewood outdoors rather than indoors will eliminate this source of moisture. Other common sources can be controlled by ventilation and dehumidification. When the moisture source cannot be minimized or controlled through ventilation, vapor retarders and air barriers can be incorporated in the design to impede vapor movement.

5.6.2 Ventilation and dehumidification

Moisture levels inside buildings can be reduced by direct ventilation to the exterior of bathrooms, clothes dryers, and gas appliances such as stoves and water heaters. During the winter, moisture levels can be diluted by the exchange of indoor moisture-laden air with dry outdoor air. As long as the outdoor air is drier than the indoor air, the greater the air change, the greater the dilution of indoor airborne moisture levels. Dilution usually occurs in older buildings through natural air changes (uncontrolled infiltration and exfiltration), but in newer construction, it occurs through mechanical ventilation by fans or blowers. Dilution by air change is possible only where the outdoor air is drier than the indoor air. In cooling climates or during cooling periods, this is often not the case, so dilution of indoor airborne moisture levels is limited to heating climates and during heating seasons.

Dehumidification involves the removal of moisture from a space and usually involves the cooling of warm, moisture-laden air to reduce its ability to hold moisture, thereby forcing the moisture to condense. Dehumidification is often coupled with air conditioning and is common in cooling climates or during the cooling season.

Air circulation at the inside surface of exterior walls can also help control surface condensation on walls and windows. When air movement is obstructed by draperies or other window treatments, condensation is more likely to occur on walls or windows with low thermal resistance. Attic ventilation can be a critical element in preventing condensation in cold climates, and in reducing summer heat gain. Ventilation of low slope roofs can also be important in many applications. Chapter 6 covers the why and how of these roofing design issues.

5.6.3 Vapor retarders and air barriers

Water vapor moves through building enclosures in two ways—it diffuses through solid materials, and is carried by air currents through cracks and openings (*Fig. 5.22*). Vapor retarders are intended to impede the flow of vapor by diffusion. Air barriers are intended to stop the airborne transport of water vapor and to reduce heat loss and energy consumption caused by air leaks. Research has consistently shown that more vapor transport occurs because of air leaks than because of diffusion, and the emphasis has been placed in recent years on the use of air barriers rather than vapor retarders to prevent condensation problems. If uncontrolled air movement is stopped, so is the large majority of water vapor movement.

All buildings need protection against air leakage, and some may need protection against vapor diffusion as well. A vapor diffusion analysis can determine whether an air barrier should be vapor permeable or vapor resistant. A vapor permeable air barrier can be located anywhere within the wall section, regardless of vapor diffusion flow characteristics. If vapor diffusion resistance is also needed, then a vapor-retardant air barrier can perform both functions,

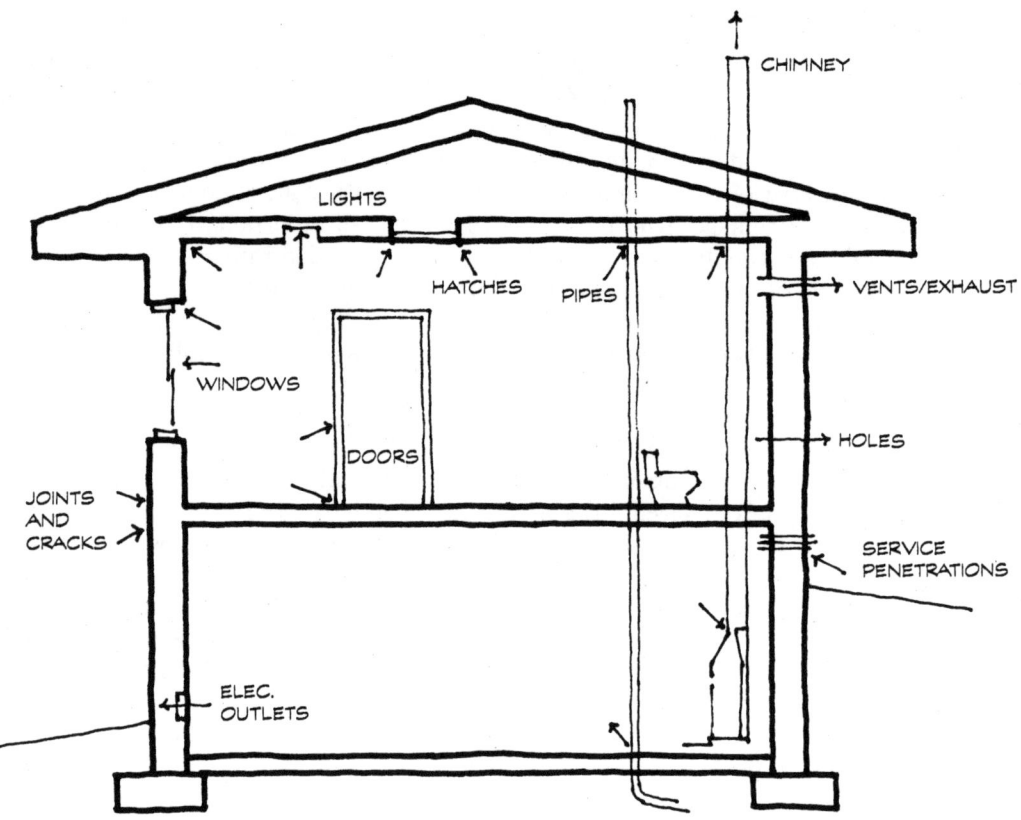

Figure 5.22 Air and vapor leakage paths. (*From Canadian Home Builders Association, Builders' Manual.*)

but its optimum location must be dictated by the predominant direction of vapor diffusion flow. Vapor retarders and air barriers can also be installed as separate elements. If air leakage is minimized, though, vapor diffusion is usually of secondary concern. It is important to understand the difference between air barriers and vapor retarders because their function, location, performance, and installation requirements are different.

Vapor retarders. Most building materials have some permeance to vapor diffusion. Some materials are simply more permeable than others. Water vapor migrates through air, polyethylene film, fiberboard, concrete, insulation, and many other materials, but at very different rates. A vapor retarder is a material with lower permeance to water vapor diffusion than most other materials. In order to be classified as a vapor retarder, vapor permeance must be less than 1 perm. In roofing applications, the criterion is a perm rating less than 0.5. The Canadian construction industry distinguishes between Type I vapor retarders with a permeance of 0.25 perm or less and Type II vapor retarders with a permeance of 1.0 or less.

Where vapor diffusion is a potential problem, the flow of vapor must be impeded on the warm side of the insulation so that it cannot reach a point at which the temperature is low enough to cause condensation. Not all climates and not all circumstances require a vapor retarder. In cold climates, where the predominant vapor flow is from inside to outside, vapor retarders should be on the inside. In hot, humid climates where the predominant vapor flow is from outside to inside, vapor retarders should be on the outside. In moderate climates where neither inward nor outward vapor flow predominates, vapor retarders generally are not needed and, in fact, can create problems by trapping moisture in a wall or roof assembly (*Fig. 5.23*).

Six-mil polyethylene film is probably the material most commonly used for vapor retarders in walls, but other materials such as aluminum foil, vinyl wall coverings, mastics, metal, glass, some types of closed-cell insulation, some paints, and even concrete of sufficient thickness can function as vapor retarders. Foil-faced insulation and foil-backed gypsum board are also popular vapor retarders. It is not possible to seal the joints in foil-backed gypsum board, but contrary to popular belief, a vapor retarder need not be perfectly continuous to be effective in retarding vapor diffusion. Unsealed laps, gaps, or penetrations, occasional pinholes and minor cuts do not increase the overall moisture diffusion rate into a wall or roof cavity appreciably because total vapor diffusion is a function of surface area. A vapor retarder that covers 90% of the enclosure is 90% effective in retarding vapor diffusion. Discontinuities, though, may create air leakage paths. If a vapor retarder is also intended to stop air leakage and airborne vapor movement, it must be selected, detailed, and installed as a continuous, vapor-retardant air barrier with all laps, penetrations, and edges sealed.

The use of vapor retarders in low-slope roofing applications has always been controversial, and industry recommendations are changing based on recent

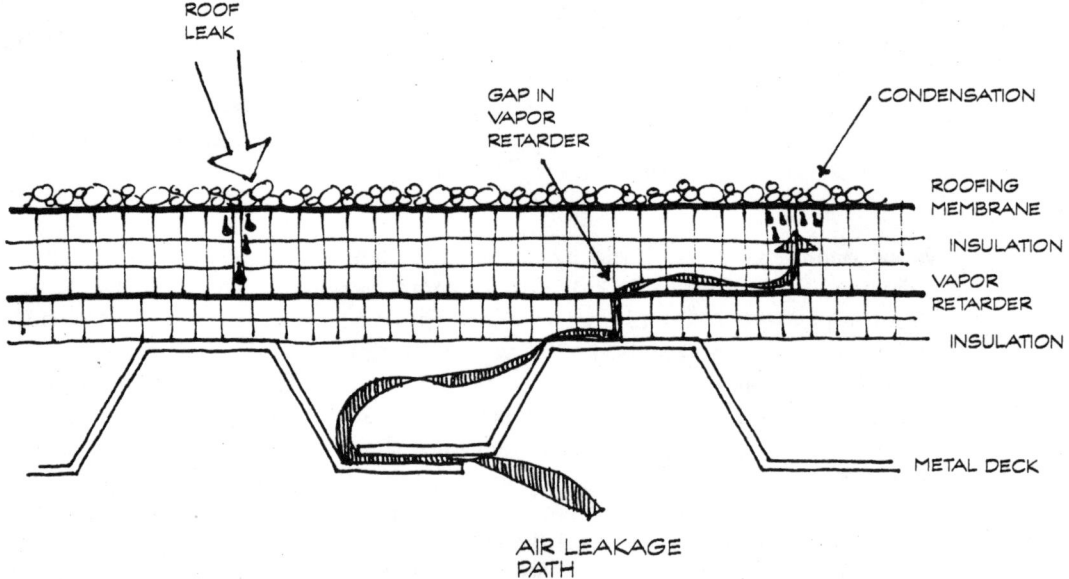

Figure 5.23 Vapor retarders can trap moisture if improperly used. (*From National Research Council of Canada, Roofs That Work.*)

research. A detailed discussion of vapor retarders in roofing systems is included in Chap. 6.

Traditional recommendations for the use of vapor retarders to prevent winter condensation have been confusing because of the various criteria used to determine which climates need protection, and under what circumstances. The National Roofing Contractors Association (NRCA) has generally used the criteria of an average 40°F January temperature *and* 45% indoor relative humidity. ASHRAE defines three winter condensation zones bounded by winter design temperatures of about 20°F, 0°F, and −20°F. The most recent and most detailed research has been done by the Cold Regions Research and Engineering Laboratory (CRREL). Tobiasson and Harrington (1985) have drawn vapor drive maps of the United States derived from National Weather Service and U.S. Air Force Weather Service climatic data. The isolines on the map in *Fig. 5.24* represent the relative humidity of 68°F indoor air above which vapor retarders are needed to prevent winter condensation. For example, in northern Alaska, the indoor relative humidity for buildings without vapor retarders is limited to 20%. In the northern tier of states, vapor retarders are needed in buildings with indoor relative humidities higher than 35%. Further south, indoor relative humidity can increase to much higher levels before vapor retarders are needed. For indoor air temperatures other than 68°F, the authors developed a graph for correcting the mapped values for allowable indoor relative humidity. Where winter condensation is not a potential problem, vapor permeable air barriers may be needed instead of vapor retarders to control heat loss caused by air leaks, or to provide pressure equalization for rain screen walls.

Vapor Condensation   189

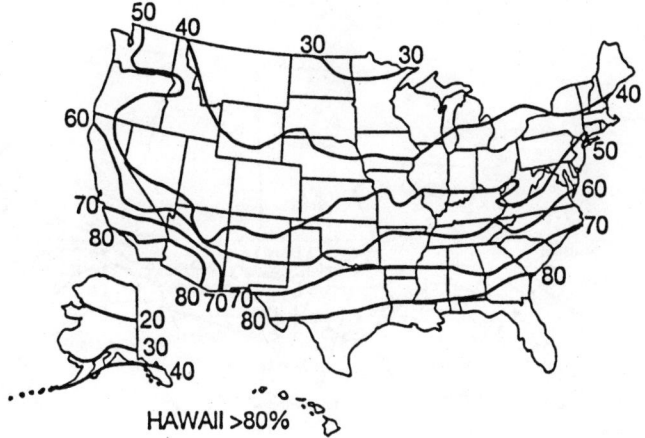

INDOOR RELATIVE HUMIDITIES ABOVE WHICH BUILDINGS WITH AN INDOOR TEMPERATURE OF 68°F NEED VAPOR RETARDERS TO PREVENT WINTER WETTING OF THE BUILDING ENVELOPE

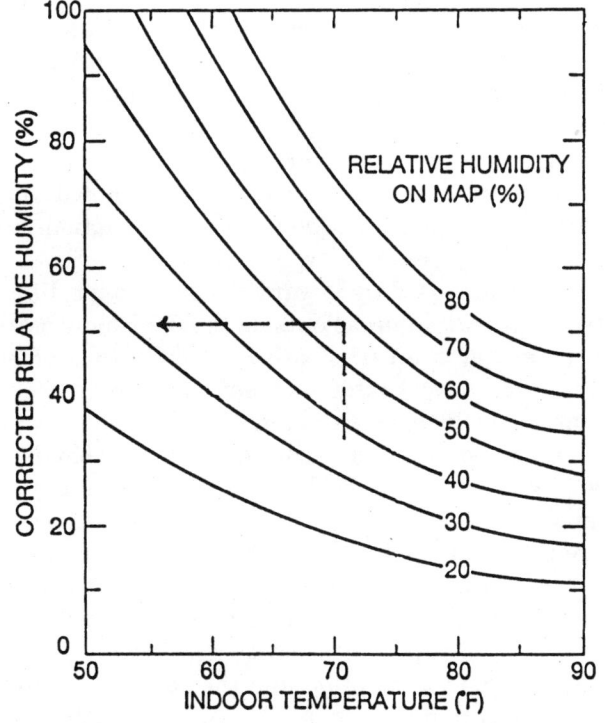

MAPPED RELATIVE HUMIDITY CORRECTION GRAPH FOR INDOOR TEMPERATURES OTHER THAN 68°F

Figure 5.24 Recommendations for use of vapor retarders. (*From Tobiasson and Harrington, Vapor Drive Maps of the U.S.*)

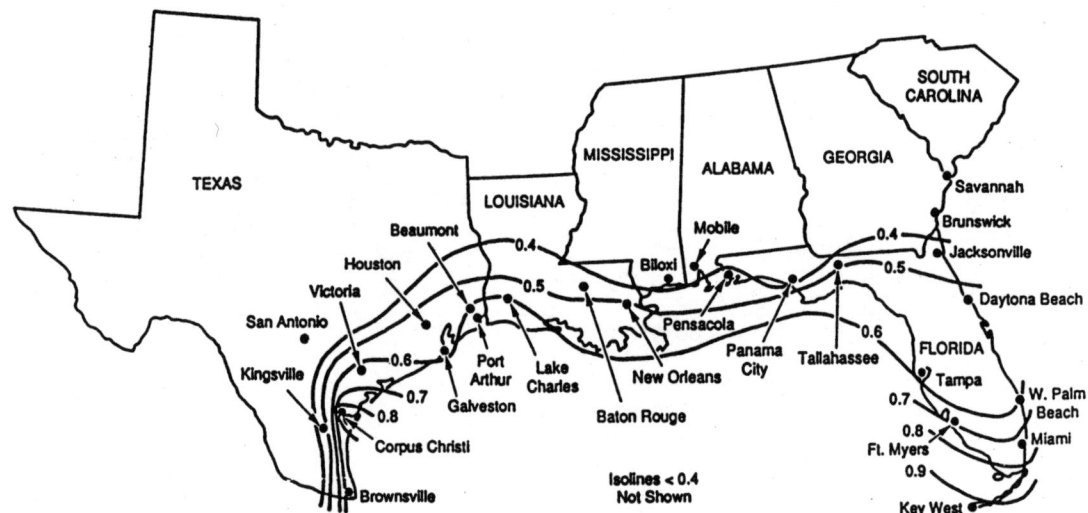

Figure 5.25 Vapor drive map for hot, humid climates. (*From Tobiasson, Vapor Retarders to Control Summer Condensation.*)

In hot, humid climates, summer condensation can be a problem in air-conditioned buildings. Tobiasson has developed a vapor drive map for the U.S. Gulf Coast and South Atlantic Coast to provide guidelines on where vapor retarders are needed. The map in *Fig. 5.25* is based on indoor temperatures of 68°F and the assumption that a vapor retarder is present on the inside surface in the form of a vinyl wall covering or a paint film with low permeability. I question the general applicability of these assumptions, but agree with Tobiasson that they represent a "worst case" scenario. According to the research, condensation caused by inward summer vapor diffusion presents a seasonal wetting potential in areas outside the 0.6 isoline on the map. Indoor air temperatures higher than 68°F are less likely to develop condensation problems, and interior surfaces that are relatively permeable will have smaller accumulations of moisture. In areas inside the 0.6 isoline where there is less summer condensation potential, a vapor-permeable air barrier may be needed instead of a vapor retarder to control heat loss or heat gain caused by air leaks, or to provide pressure equalization for rain screen walls.

Both the summer and winter vapor drive maps are based on diffusion caused by vapor pressure differences across the building envelope. They do not take into account vapor movement caused by air leakage. If the potential for condensation exists from vapor diffusion alone, then air leakage will increase the volume of moisture accumulation and the severity of any resulting damage. Vapor retarders in this case should be combined with air barriers to provide effective protection.

Air barriers. An air barrier is a system or network of materials which prevents air movement through a building enclosure. In doing so, it also blocks the movement of vapor contained in that air. An air barrier is not necessarily a vapor retarder. The spunbonded polyolefin and woven polypropylene "house wrap" materials that are now popular are intended to shed water and resist air flow, but they have high vapor permeability to allow walls to breathe. Since air leakage has been determined to be the primary cause of concealed vapor condensation within building envelopes, the function of vapor retardance, when needed, should be combined with that of air movement control. In much of the construction industry the term vapor retarder has been replaced with the term air barrier/vapor retarder. To make matters even more confusing, the terms moisture barrier and weather barrier have also crept into the vocabulary. Not all buildings require protection against vapor diffusion, so some air barriers can be vapor permeable and still perform effectively. Until standardized terms are widely adopted, perhaps the most descriptive and least confusing terminology would be

- *Vapor retarder:* A material with a vapor permeance rating of 1.0 or less
- *Vapor-permeable air barrier:* A material or system of materials sealed to prevent air movement, but with a vapor permeance rating of more than 1.0
- *Vapor-retardant air barrier:* A material or system of materials with a vapor permeance rating of 1.0 or less which is also sealed to prevent air movement

In cold climates, a vapor-permeable air barrier such as polyolefin house wrap can be installed in the conventional manner near the outside of the envelope to stop air movement and airborne vapor flow, in combination with a separate vapor retarder installed in the conventional manner near the inside of the envelope to stop vapor diffusion (*Fig. 5.26*).

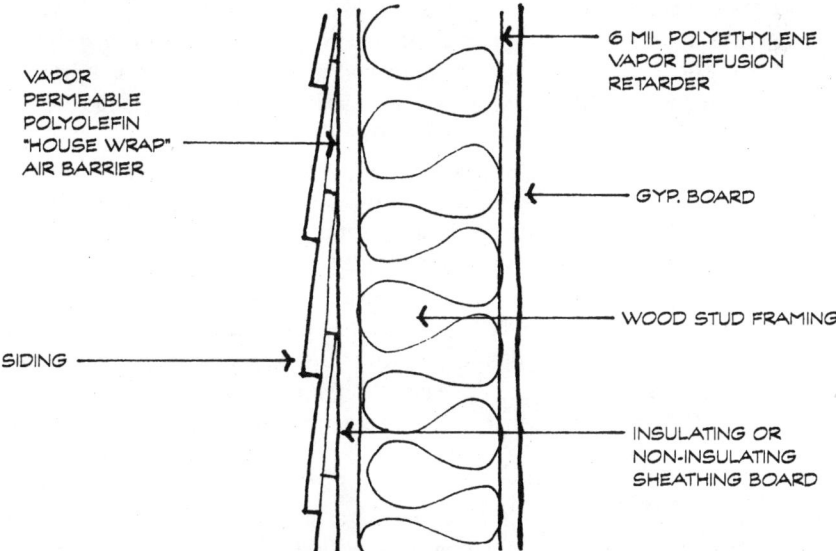

Figure 5.26 Air barrier and separate vapor diffusion retarder for cold climate.

Unlike vapor retarders, air barriers must be sealed and continuous to be effective (*Fig. 5.27*). Even though most of the area may be protected, gaps in coverage, unsealed laps, penetrations, punctures and edges will experience higher velocity air flow because of the funneling effect, and damage from condensation can be localized and severe. When air barriers are used with pressure-equalized rain screens, continuity and structural support against displacement are critical (see "Rigid air barriers" below).

Rigid air barriers. If an air pressure difference cannot move air, it will act to displace the materials that prevent the air from flowing. If an air barrier is effective in stopping air movement through the envelope, it is subjected to the loads induced by the air pressure difference. If the air pressure difference exceeds the capability of the air barrier system to resist the load, then airtightness will be destroyed and increased air leakage will result.

Each membrane or assembly of materials intended to support a differential air pressure load must be designed and constructed to carry that load, or it must receive the necessary support from other elements of the wall. If the air barrier system is made of flexible membrane materials, then it must be supported on *both* sides by materials capable of resisting the peak air pressure loads; or it must be made of self-supporting materials, such as board products adequately fastened to the structure. In *Fig. 5.28,* a polyethylene air barrier is installed between the stud framing and the interior gypsum board. The gypsum board does not continue above the ceiling line, so the air barrier is unsupported for a distance and may be dislodged by wind pressure. In *Fig. 5.29,* an air barrier is installed in a parapet wall between rigid insulation board and the interior wythe of masonry. If the insulation board is not adequately attached to the wall, negative wind pressure may dislodge both the insulation and the air barrier. Some membrane products can be satisfactorily adhered to a solid substrate to form a composite air barrier system (membrane+substrate). For example, a thick mastic or rubberized asphalt membrane applied to the surface of a masonry infill wall forms an effective air barrier if the perimeter joints are also sealed with an elastomeric joint sealant.

To form an effective air barrier, particularly in pressure-equalized rain screen walls, air permeance should be minimal. *Table 5.6* lists measured air leakage rates for a variety of common building materials. Even when constructed of materials impermeable to air flow, air barriers are effective only if the joints, seams, edges and penetrations are sealed as well. Typical air leakage values at 0.30 in H_2O for commercial buildings are 0.10, 0.30, and 0.60 cfm/ft^2 for tight, average and leaky walls, respectively. In residential buildings, 35% of air leakage typically occurs through walls, 15% through doors and windows, 18% through ceilings and heating systems, and 12% through fireplaces.

5.7 Installing Air Barriers

There are several different approaches to achieving airtightness in both residential and commercial building envelopes. Membrane air barriers, rigid air

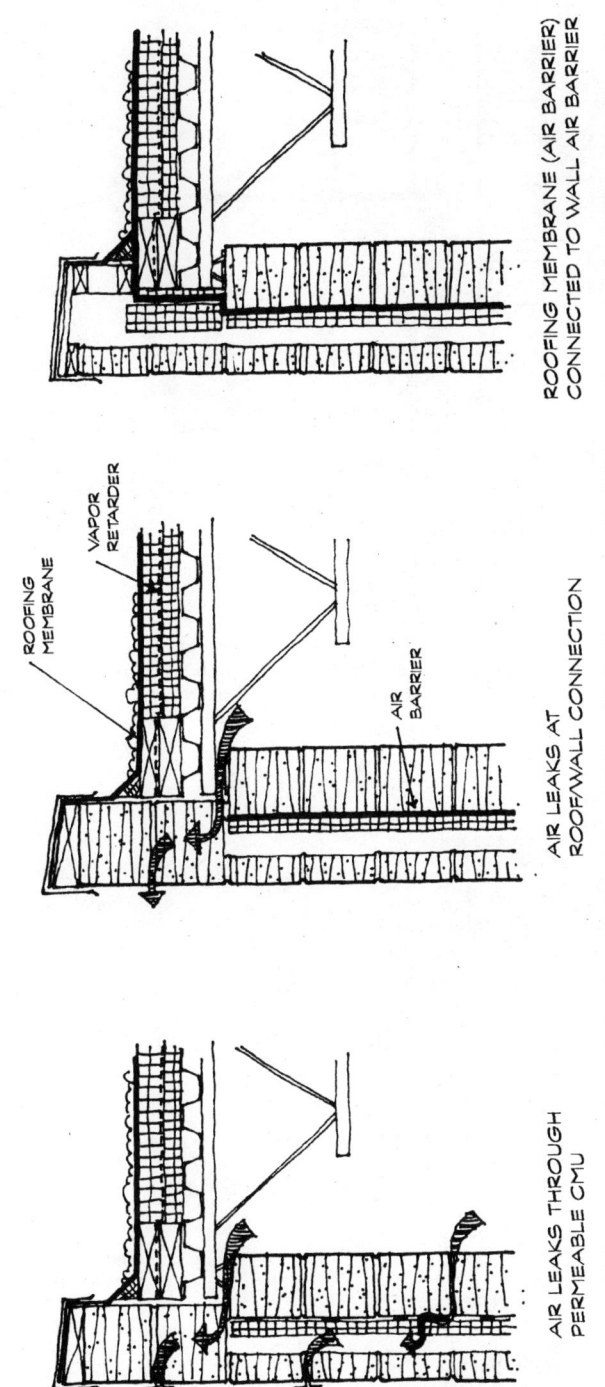

Figure 5.27 Air barriers must be sealed and continuous to be effective. (*From Quirouette, The Difference Between a Vapour Barrier and an Air Barrier.*)

194   Chapter Five

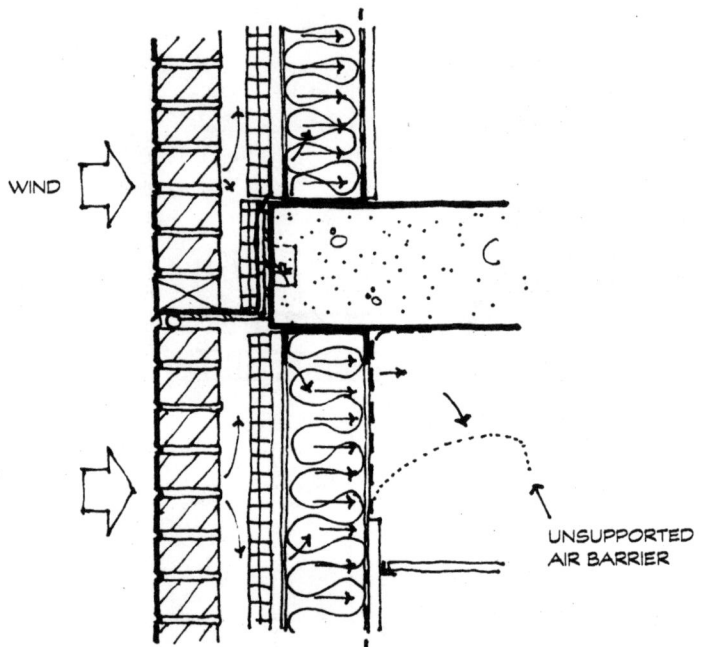

Figure 5.28 Unsupported air barrier displaced by air pressure differentials. (*From Quirouette, The Difference Between a Vapour Barrier and an Air Barrier.*)

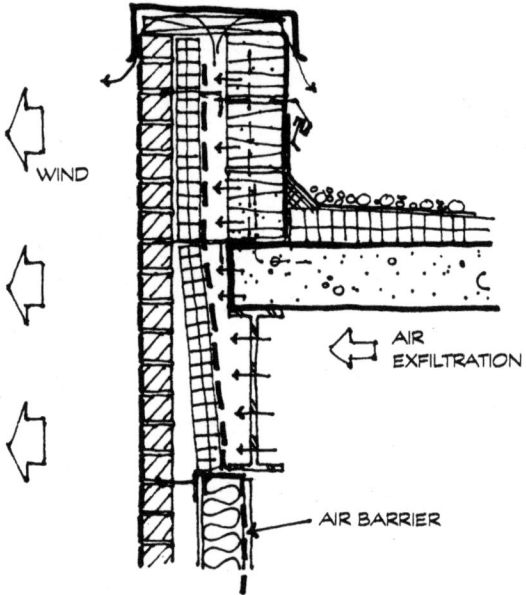

Figure 5.29 Unsupported air barrier displaced by air pressure differentials. (*From Quirouette, The Difference Between a Vapour Barrier and an Air Barrier.*)

TABLE 5.6 Air Permeance of Common Building Materials

Material	Measured leakage at 75 Pa (0.30 in H_2O), $L/s \cdot m^2$ (cfm/ft^2)
Materials showing a non-measurable air flow	
6-mil polyethylene	0.00
$1/16$-in. smooth surface roofing membrane	0.00
$3/32$-in. modified bituminous torch on grade membrane (glass fiber mat)	0.00
1-mil aluminum foil	0.00
$1/16$-in. modified bituminous self-adhesive membrane	0.00
$3/32$-in. modified bituminous torch on grade membrane (polyester reinforced mat)	0.00
$3/4$-in. plywood sheathing	0.00
1.5-in. extruded polystyrene	0.00
1-in. foil back urethane insulation	0.00
1-in. phenolic insulation board	0.00
2-in. phenolic insulation board	0.00
$1/2$-in. cement board	0.00
$1/2$-in. foil back gypsum board	0.00
Glass	0.00
Metal	0.00
Materials having a measurable air flow	
$3/8$-in. plywood sheathing	0.0067 (0.15)
$5/8$-in. oriented strand board	0.0069 (0.16)
$1/2$-in. gypsum board (M/R)	0.0091 (0.21)
$3/4$-in. oriented strand board	0.0108 (0.25)
$1/2$-in. particle board	0.0155 (0.35)
Reinforced non-perforated polyolefin	0.0195 (0.44)
$1/2$-in. gypsum board	0.0196 (0.45)
$5/8$-in. particle board	0.0260 (0.59)
$1/8$-in. hardboard	0.0274 (0.68)
1-in. expanded polystyrene - type II	0.1187 (2.71)
Spunbonded polyolefin film	0.1776 (4.05)
30-lb. roofing felt	0.1873 (4.27)
15-lb. non-perforated asphalt felt	0.2706 (6.17)
15-lb. perforated asphalt felt	0.3962 (9.04)
Glass fiber rigid insulation board with spunbonded polyolefin film on one face	0.4880 (11.13)

TABLE 5.6 Air Permeance of Common Building Materials (Continued)

Material	Measured leakage at 75 Pa (0.30 in H_2O), $L/s \cdot m^2$ (cfm/ft^2)
Materials having a measurable air flow	
$7/16$-in. plain fiberboard	0.8223 (18.76)
$7/16$-in. asphalt impregnated fiberboard	0.8285 (18.90)
Spunbonded polypropylene film	3.2186 (73.41)
4-mil type II perforated polyethylene	3.2307 (73.69)
4-mil type I perforated polyethylene	4.0320 (91.96)
1-in. expanded polystyrene—type I	12.2372 (279.11)
Tongue-and-groove planks	19.1165 (436.01)
6-in. glass fiber or wool insulation	36.7327 (837.81)
Vermiculite insulation	70.4926 (1607.81)
Cellulose insulation (spray-on)	86.9457 (1983.08)

SOURCE: Canadian Home Builders Association, *Builders' Manual.*

barriers, and exterior air barriers can all be effective if installed and sealed to provide continuity from floor to floor in wall construction and between walls and roofs.

5.7.1 Membrane air barriers

Interior membranes are installed on the winter warm side of the insulation, and may be made of polyethylene sheet or elastomeric materials.

Polyethylene membranes are used most commonly as interior air barriers in cold climates. One advantage is that they also provide resistance to winter vapor diffusion. Seams in the membrane should occur over solid backing, be overlapped and sealed with a flexible, non-drying sealant. To provide maximum structural rigidity to resist air pressure differences, polyethylene membranes should be sandwiched between two layers of rigid materials such as gypsum drywall. Exterior air barrier house-wrap membranes are often used in conjunction with polyethylene to provide continuity between floors and at heads and sills (*Fig. 5.30*). Detailing at penetrations such as electrical outlets requires special attention. Recommended practices are

- Ensure that membrane is located on warm side of wall assembly.
- Use UV stabilized, 6-mil polyethylene.
- Ensure compatibility of seam lap sealants with polyethylene.
- Protect membrane from long-term exposure to ultraviolet light.
- Minimize number of seams by using largest possible sheets.

- Overlap seams a minimum of 6 in.
- Ensure that lap seams are located over rigid backing such as framing members (*Fig. 5.31*).
- Caulk lap seams with flexible, non-hardening sealant such as acoustical sealant, and staple through sealant into solid backing.
- Avoid stapling at locations other than seams and edges.
- Sandwich membrane between two solid materials such as two layers of drywall.

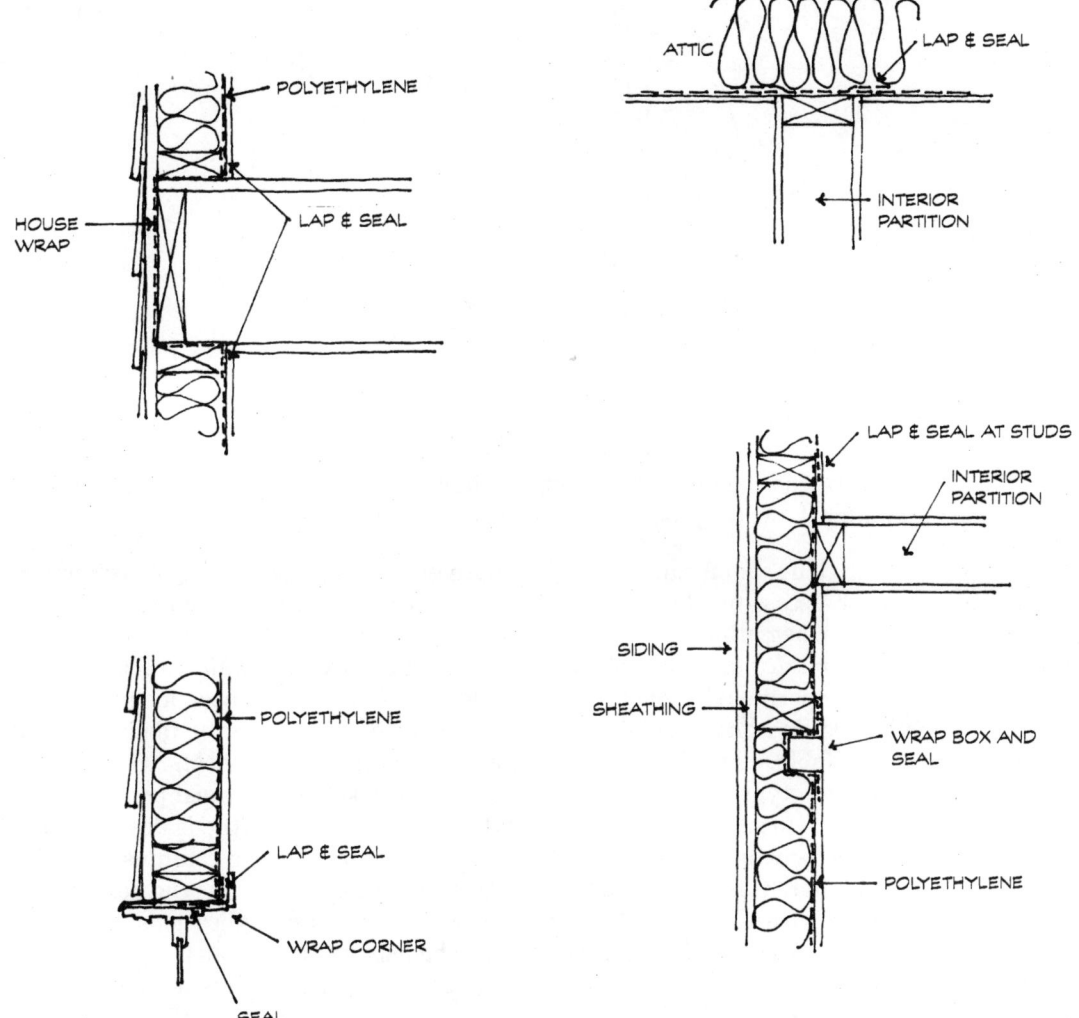

Figure 5.30 Air barrier sealing and continuity at openings and intersections. (*From Lstiburek and Carmody, Moisture Control Handbook.*)

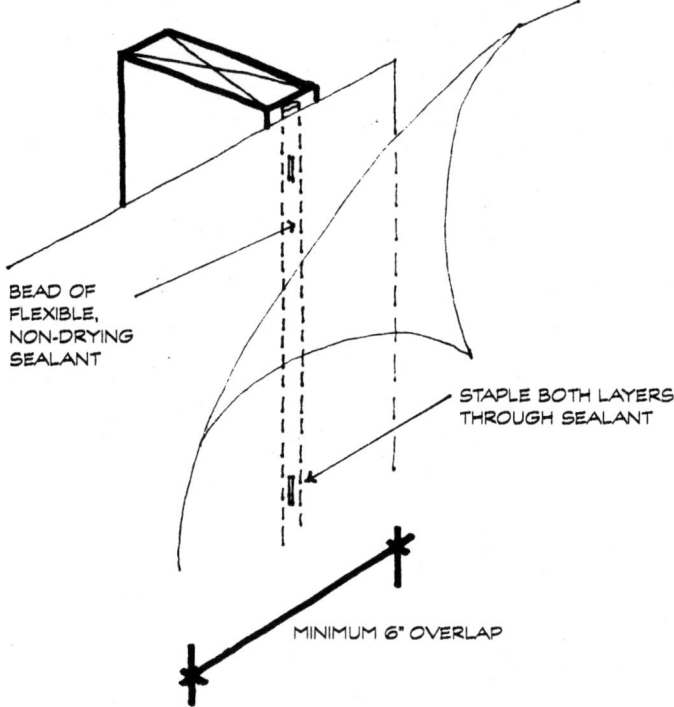

Figure 5.31 Locate seam laps over rigid backing. (*From Canadian Home Builders' Association, Builders' Manual.*)

- Ensure that membrane is not in direct contact with hot surfaces such as chimneys, water pipes, or baseboard heaters.

Polyethylene membrane air barriers are easily damaged during construction, are not accessible for repairs after construction, and are difficult to seal around penetrations.

Elastomeric membranes can be used in masonry cavity walls to seal the face of an air-permeable concrete block backing (*Fig. 5.32*). Sheet membranes of SBS modified bitumen bonded on both sides of a polyester or fiberglass mat are available, as well as self-adhering rubberized asphalt sheets. The application of both types is complicated by the masonry ties which must penetrate the membrane to secure the exterior wythe. If the exterior wythe is designed as a reinforced curtain wall spanning between floors or between columns, then the intermediate ties can be eliminated. If the exterior wythe must be designed as a veneer with ties to the backing wythe, some mastic applications, if applied carefully and in sufficient thickness, can be used to achieve airtightness. In all cases, care must be taken to assure that the perimeter joints are sealed with a compatible joint sealant.

Spunbonded polyolefin and spunbonded polypropylene *house wrap* materials are commonly used to provide air barriers behind wood siding and masonry veneers (*Fig. 5.33*). Since they are vapor permeable, their location on the outside of the wall does not interfere with wall performance when used in conjunction with interior polyethylene vapor diffusion retarders, as is common in cold climates. They have the advantage of fewer electrical penetrations at the exterior plane than on the interior plane making installation simpler. When used in conjunction with an interior vapor diffusion retarder, they allow less stringent installation of that membrane because the air sealing function has been separated from it. Seams in house-wrap membranes are lapped and sealed with a compatible tape. The air barrier must be sealed to the sill plate, top plate, windows and doors for continuity. As noted in *Table 5.6,* these membranes do have some air permeance, so they are typically installed over sheathing board to improve resistance to air flow. To provide structural resistance to inward- and outward-acting wind pressures, the membrane must be sandwiched between two layers of sheathing, or between the sheathing and siding.

5.7.2 Gypsum drywall air barriers

Gypsum drywall can be assembled to function as part of an air barrier. By sealing the drywall at all junctions and by incorporating subfloors and framing members as part of the system, an effective air barrier is relatively easy to achieve and relatively economical to install (*Fig. 5.34*). The Canadian Home Builders' Association *Builders' Manual* is a comprehensive guide to detailing and building wood frame construction with gypsum drywall air barriers. A few of those details are reproduced here to illustrate methods of sealing penetrations (*Figs. 5.35* through *5.37*). Many of the details are adaptable to metal stud construction as well. The following practices are recommended:

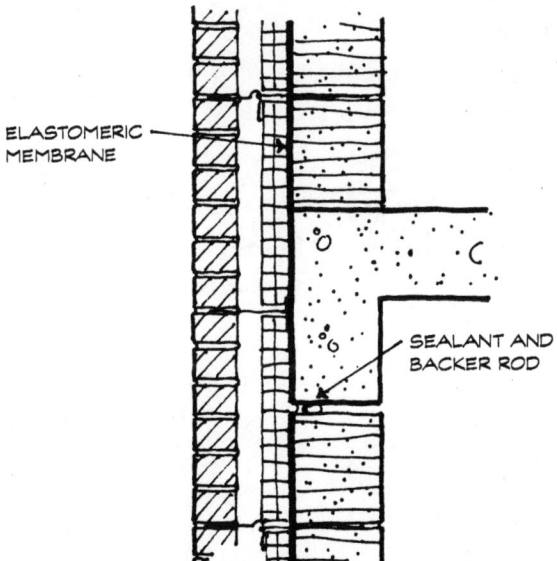

Figure 5.32 Air seal at permeable CMU backing wall. (*From Quirouette, The Difference Between a Vapor Barrier and an Air Barrier.*)

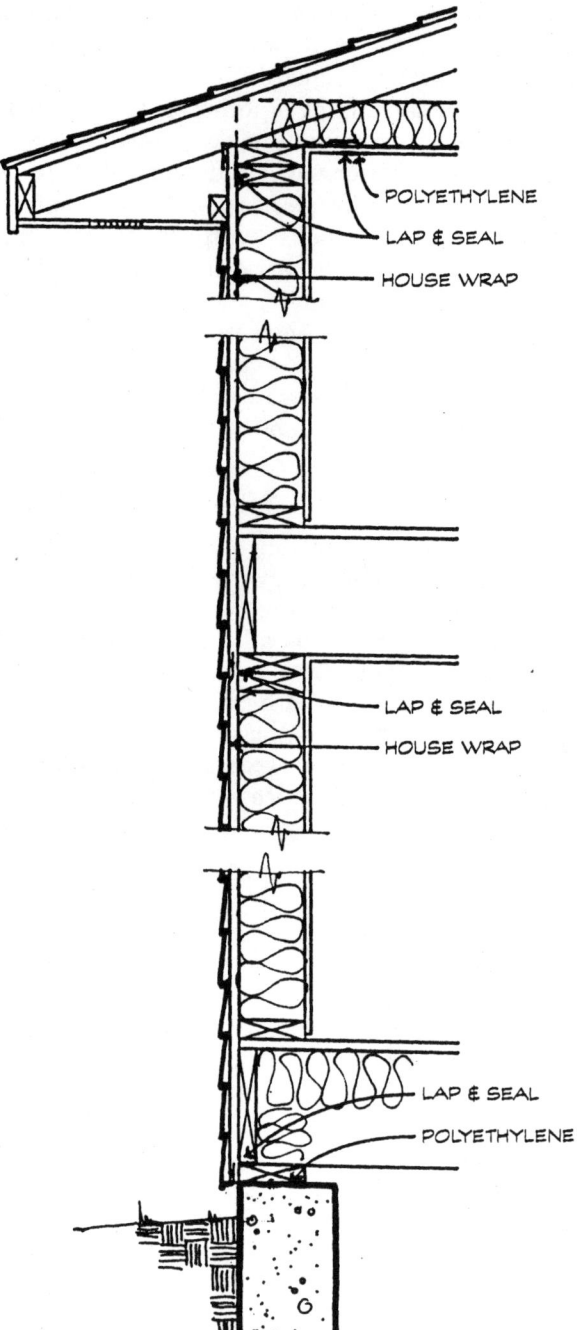

Figure 5.33 Typical "house wrap" air barrier. (*From Lstiburek and Carmody, Moisture Control Handbook.*)

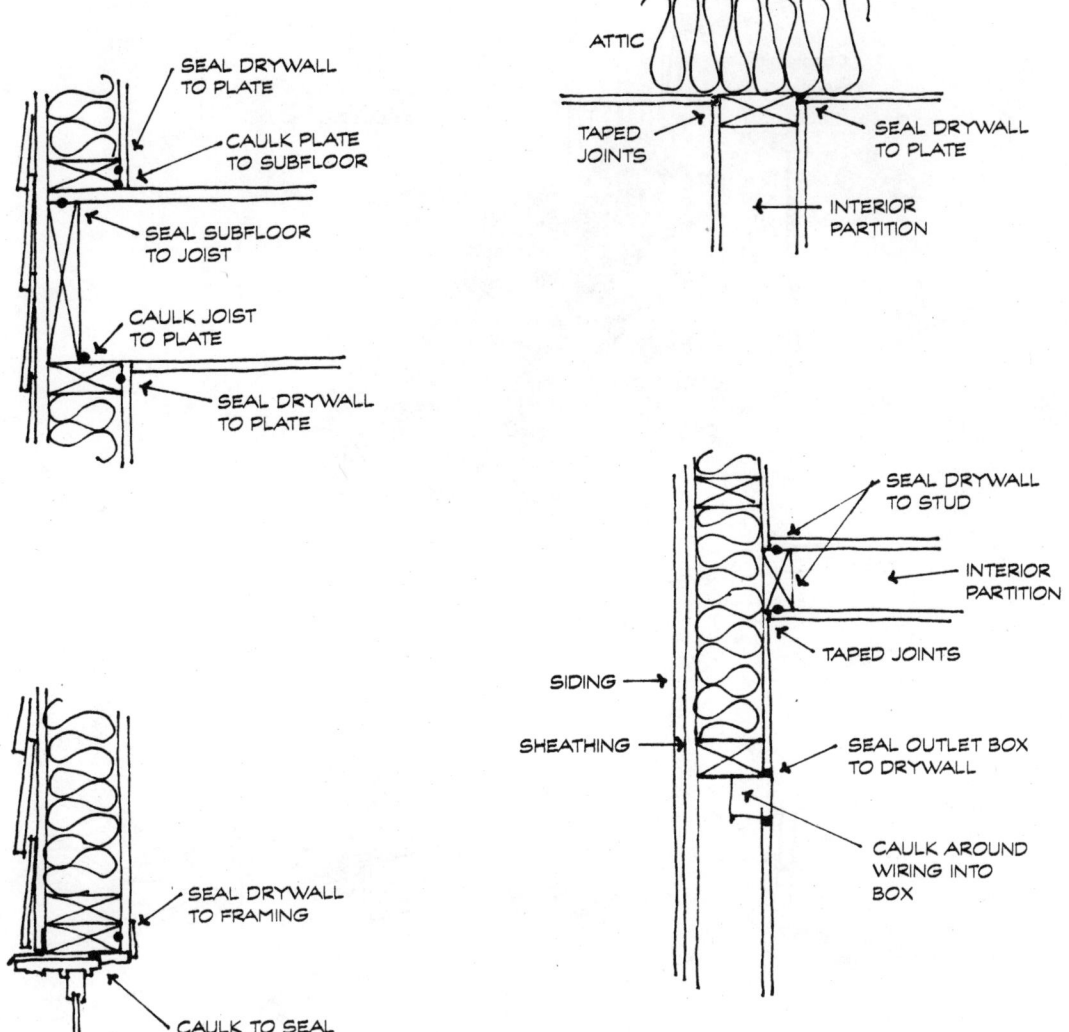

Figure 5.34 Drywall air barrier. (*From Lstiburek and Carmody, Moisture Control Handbook.*)

- Seams between all components of the air barrier must be adequately and continuously sealed.
- Joints between wallboards are finished in the customary manner with tape and joint compound.
- Drywall is sealed to the framing with adhesive-backed foam tape. The tape must be soft and resilient so that it will accommodate movement and fill all gaps. Most commonly used is a $1/2$-×$3/16$-in. low-density, closed-cell PVC foam tape which must be stapled to the framing to ensure that it will stay in place during the drywall installation. The drywall must be nailed or screwed at 8 in. on center over the tape to provide an airtight seal.

202   Chapter Five

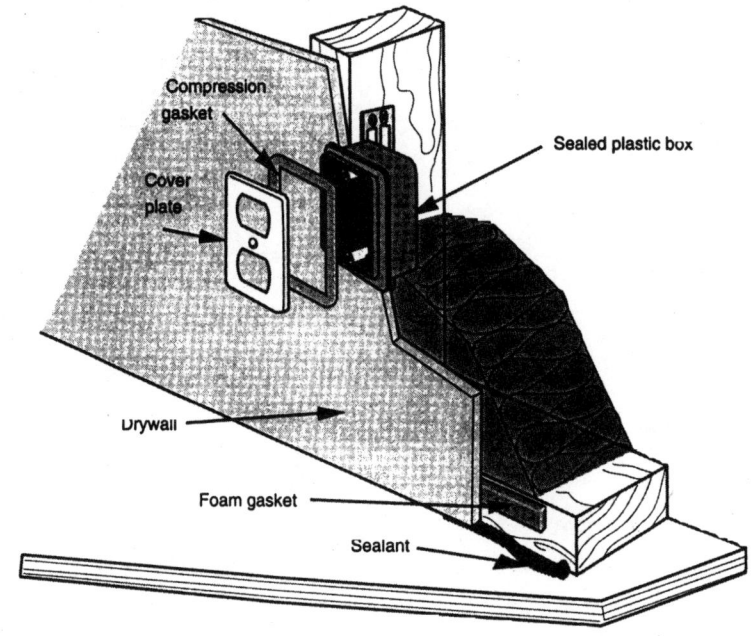

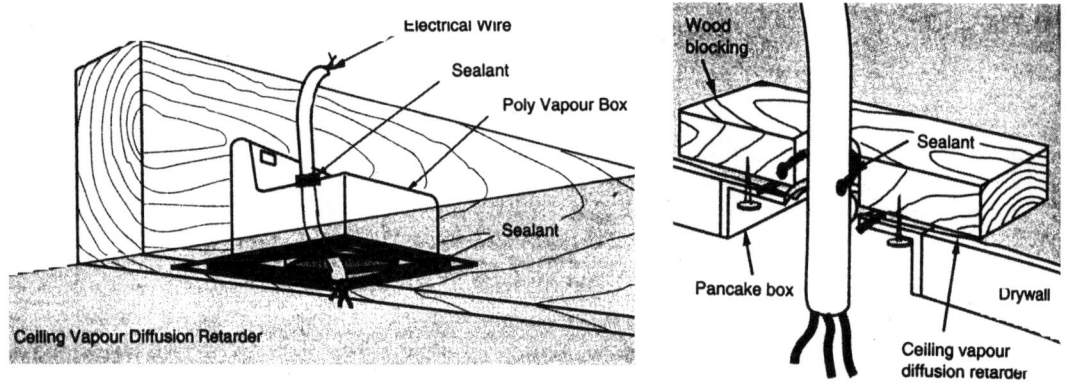

Figure 5.35 Air sealing at electrical penetrations. (*From Canadian Home Builders Association, Builders' Manual.*)

- Joint sealants must be flexible, last the life of the building, and adhere well to wood, drywall, foam insulation, and metals. Single component urethane sealants have performed well. Flexible panel adhesives can be used effectively between framing members.

The air-sealing qualities of this system will control as much as 90% of the total vapor movement, but drywall is not effective in retarding water vapor diffusion. If the 10% or so of vapor movement attributable to diffusion is considered to be a potential problem, or where building codes specifically require

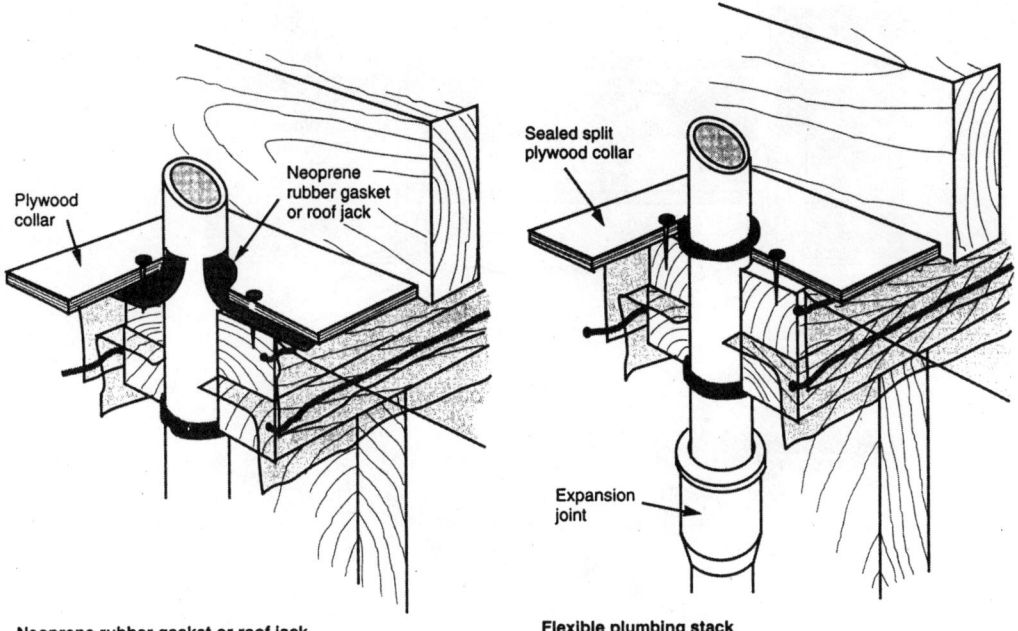
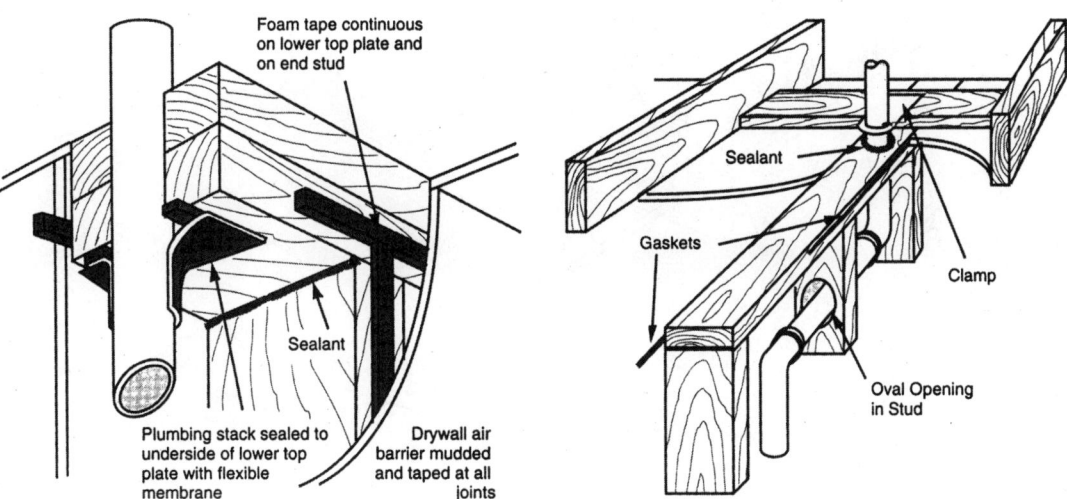

Figure 5.36 Air sealing at plumbing penetrations. (*From Canadian Home Builders Association, Builders' Manual.*)

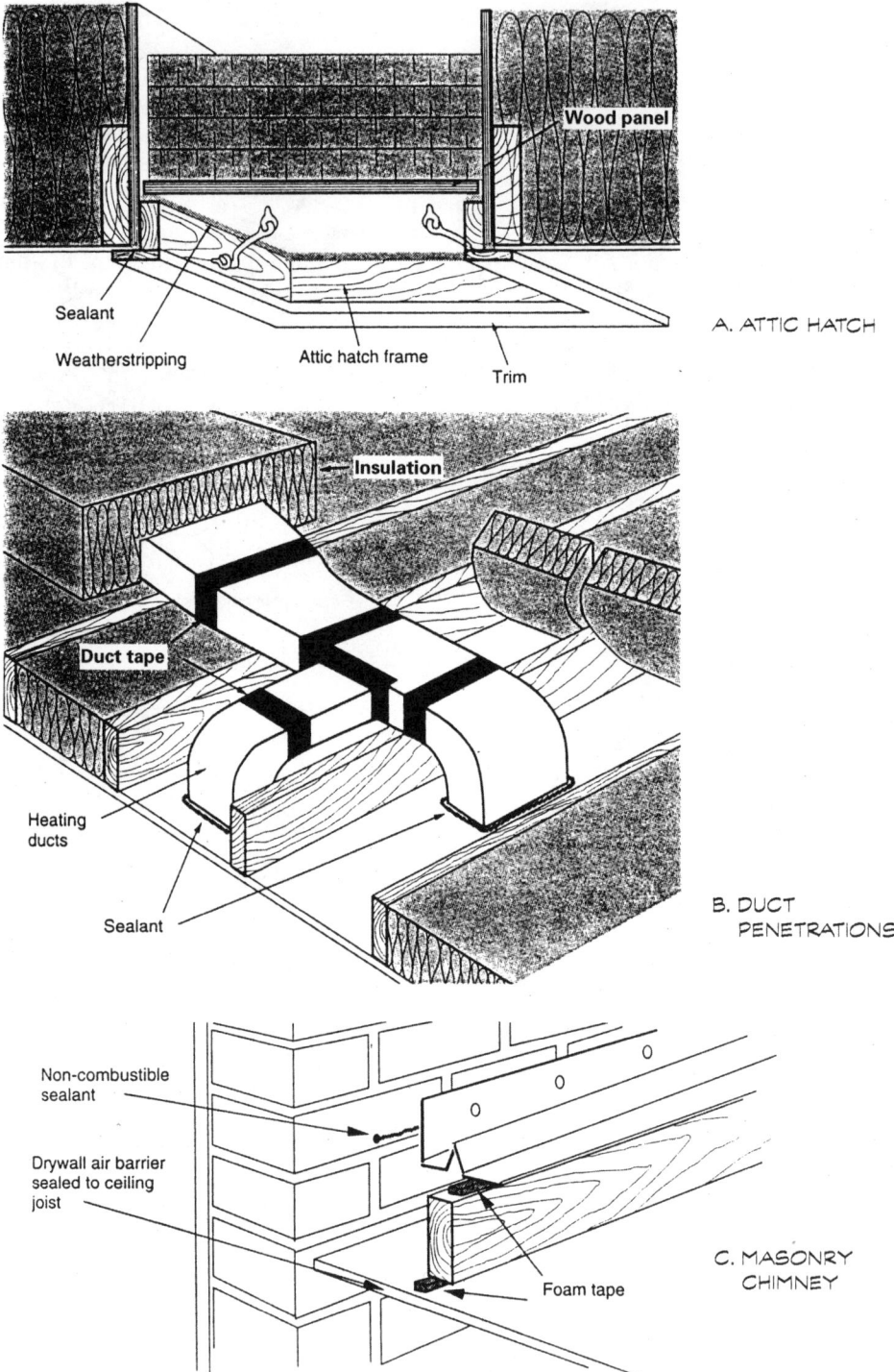

Figure 5.37 Miscellaneous air sealing details. (*From Canadian Home Builders Association, Builders' Manual.*)

a winter vapor diffusion retarder, install a polyethylene membrane as well. Seal penetrations such as those at outlet boxes by either the membrane or drywall method, but seal all other locations as shown for drywall air barriers. Since the polyethylene membrane is not expected to seal against air penetration, neither its seams nor its edges need to be lapped and sealed. If you prefer the belt-and-suspenders approach, or if the building is at high risk of winter condensation because of high indoor humidity levels and air pressure differences, then seal both the membrane and the drywall. Interior vapor diffusion control for cold climates can also be provided by foil-backed gypsum board, vinyl wall coverings, and paints or primers with a low permeance.

5.7.3 Exterior sheathing air barriers

Exterior sheathing board can be used to form an air barrier in frame construction as well as masonry cavity walls. In frame construction, the sheathing is sealed to the framing members in much the same way that gypsum drywall barriers are constructed (*Fig. 5.38*). Like the house wrap membranes, exterior sheathing has the added advantage of fewer electrical and plumbing penetrations which much be sealed, making the installation much simpler. Continuity must be provided at window and door openings with a combination of sheet goods and foam insulation. Where the sheathing is not continuous across floor framing, the air seal must be completed with membrane materials such as self-adhering rubberized asphalt (*Fig. 5.39*).

When foil-faced insulating sheathing is used in exterior walls, it can be air-sealed at the outside face of the insulation only if the predominant seasonal vapor drive is from outside to inside (i.e., hot, humid climates). The foil face of the insulation is vapor-retardant, but a sealed vapor diffusion retarder at this location would not be effective in preventing winter condensation in cold climates. If used in conjunction with an interior vapor diffusion retarder, the sealed foil face would form a trap in which latent construction moisture from masonry, concrete, or lumber would be unable to escape.

5.7.4 Curtain wall air barriers

Pressure-equalized curtain wall systems have internal air seals in the form of gaskets, glazing tape, or wet sealants. Where the curtain wall passes in front of a floor slab, spandrel panels must be of special construction to ensure the continuity of the air barrier (*Fig. 5.40*). Air leakage paths may occur through the interconnected passages in hollow frame sections at intersections and corners. Curtain wall systems can be specified with single-source responsibility for design, fabrication and installation, and by established industry performance standards for air infiltration (ASTM E283 *Standard Method of Test for Rate of Air Leakage Through Exterior Windows, Curtain Walls, and Doors*). Large projects should always include full-scale mockup testing of curtain walls for air and water penetration.

Similar to curtain walls in some ways, pre-engineered metal panel wall systems can also be designed, fabricated, and installed to provide a barrier to air

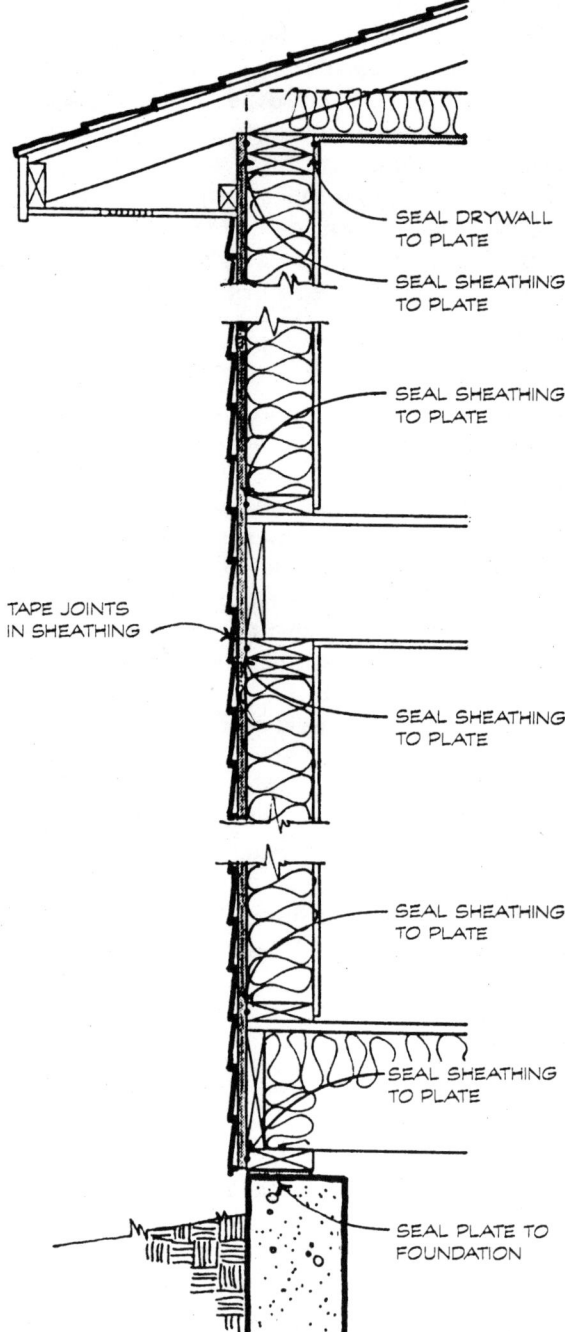

Figure 5.38 Exterior sheathing air barrier. (*From Lstiburek and Carmody, Moisture Control Handbook.*)

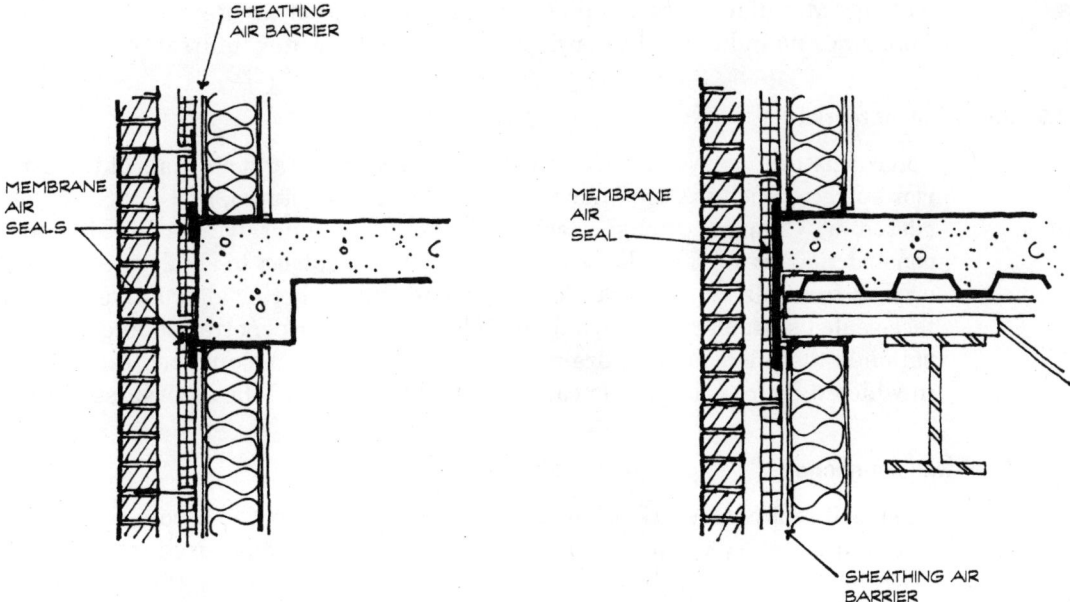

Figure 5.39 Continuity of exterior seal at floor penetrations. (*From Quirouette, The Difference Between a Vapor Barrier and an Air Barrier.*)

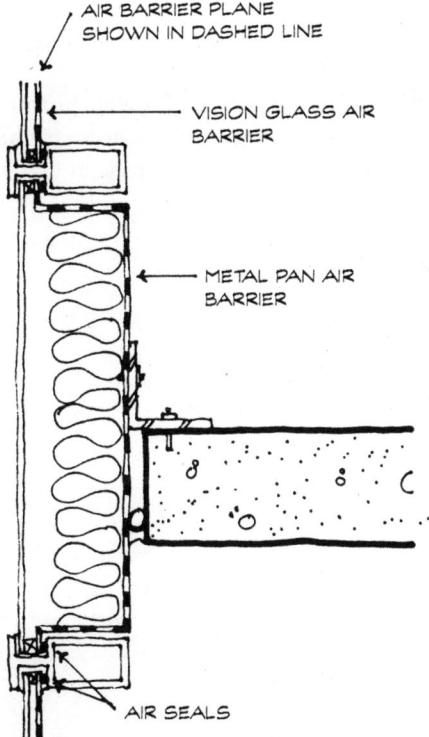

Figure 5.40 Air barrier sealing and continuity at curtain wall spandrel at floor line. (*From Quirouette, The Difference Between a Vapor Barrier and an Air Barrier.*)

leakage. Manufacturers should be consulted for specific performance information, since no industry standards exist for air and water infiltration testing.

5.7.5 Face seal air barriers

Some construction systems lend themselves to weather and air seals at the exterior surface. Such systems include precast concrete, stucco, and EIFS. As discussed in Chap. 4, though, face seals are very unforgiving of design and construction errors, and the sealant materials are subject to accelerated deterioration from ultraviolet radiation, wind and rain. A better alternative to most face-sealed systems from the standpoint of both water penetration and condensation is the provision of a drainage cavity or a pressure-equalized rain screen in which the air seal function can be moved back behind the exterior surface.

5.7.6 Roof-wall connections

In sloped-roof construction, air barriers are usually carried across the ceiling of the upper story by construction methods similar to those used in the walls. With low-slope roofs, the roofing membrane itself is usually the air barrier, although some installations will include a bottom side vapor-retardant air barrier. It is critical that roof and wall membranes be detailed in such a way that the entire building envelope is protected. Figures *5.41* through *5.43* show several methods of accomplishing air barrier continuity at roof/wall intersections.

5.7.7 Weatherstripping

The final component in an air barrier system is the weatherstripping of operable windows and doors. Weatherstripping can be accomplished with compression seals such as rubber or metal gaskets, or friction seals like pile.

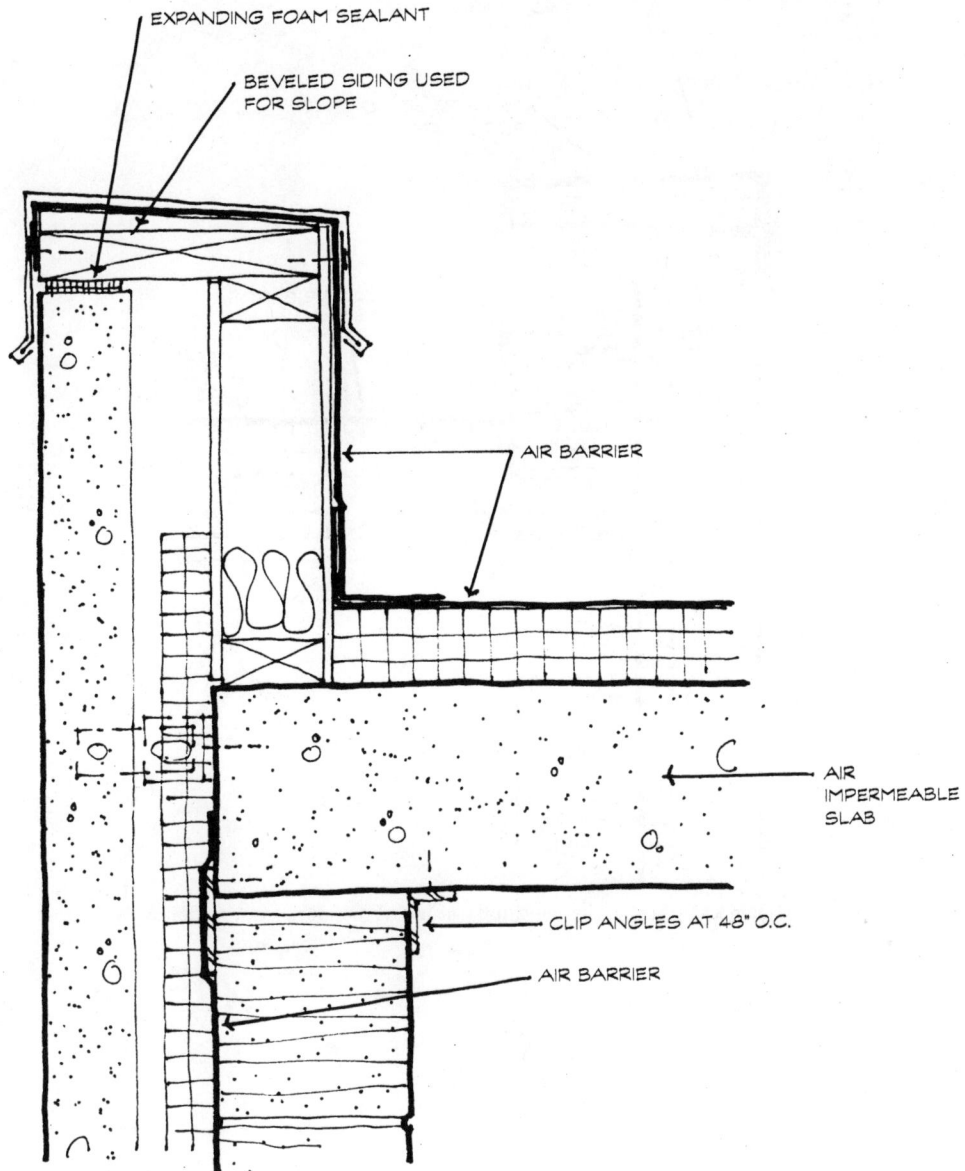

Figure 5.41 Air barrier continuity at concrete roof slab intersection with parapet. (*From National Research Council of Canada, Roofs That Work.*)

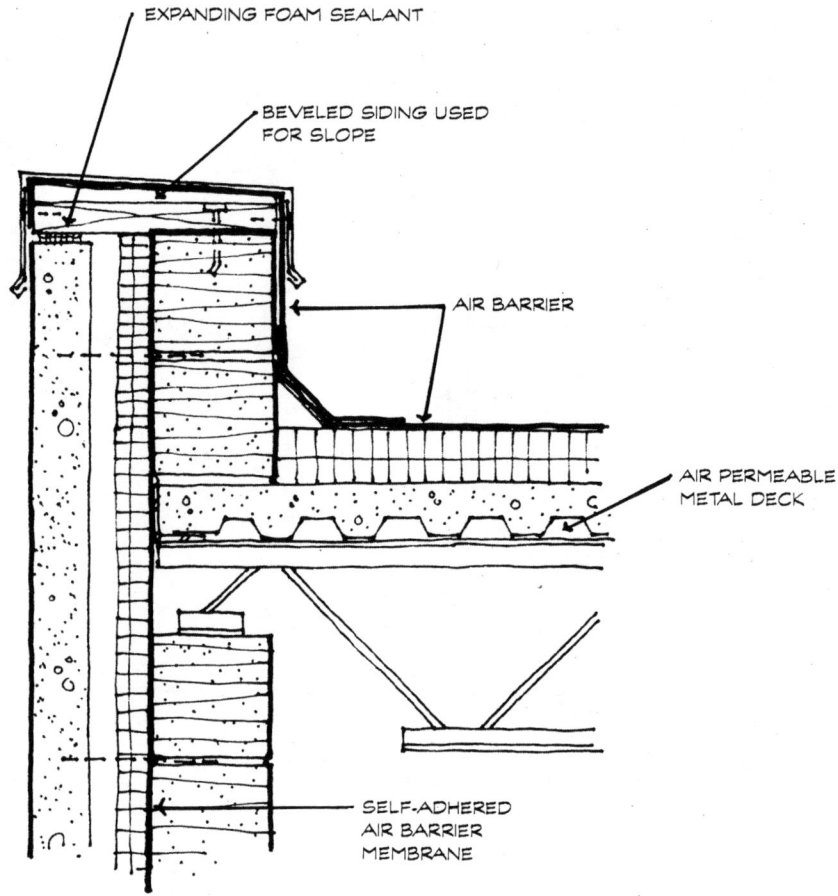

Figure 5.42 Air barrier continuity at metal roof deck intersection with parapet.

Vapor Condensation 211

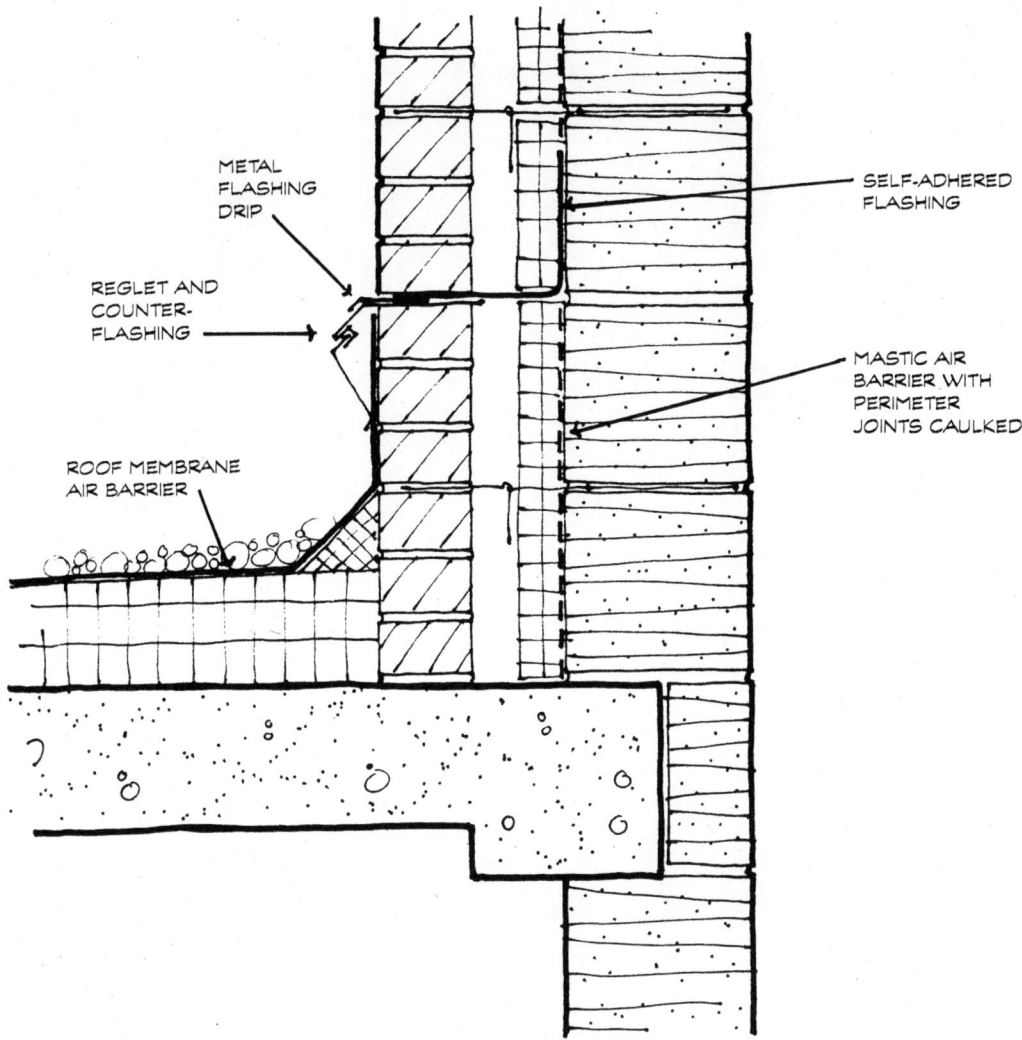

Figure 5.43 Air barrier continuity at intersection of roof with high wall.

Chapter 6

Roofing

The most exposed part of a building is the roof. It is subject to both structural and environmental loads which must be resisted or accommodated to ensure acceptable and durable performance. This chapter covers basic principles of roofing design and detailing. For exhaustive coverage of issues related to roofing, two excellent references are available. The first is the National Roofing Contractors Association (NRCA) *Roofing and Waterproofing Manual*. The fourth edition is greatly expanded and covers both steep and low-slope roofing. The second is C. W. Griffin's *Manual of Low-Slope Roof Systems*, third edition. The text covers built-up roofing and single-ply membranes as well as low-slope metal roofing. Refer to App. C, Bibliography, for more information on these publications.

6.1 Service Conditions

The components which make up a roofing assembly and the roofing assembly as a whole must be capable of withstanding the service conditions to which they will be exposed—rain, snow, ice, wind, sun, and applied loads.

6.1.1 Drainage

Unless they are resistant to or protected from wetting, the roofing components may be damaged by corrosion, rot, expansion, loss of strength, or freeze-thaw deterioration. Drainage of water from the roof surface is of primary importance. The Asphalt Roofing Manufacturers Association (ARMA) ranks roof slope as follows:

- *Level slope*—up to $1/2$ in./ft
- *Low slope*—$1/2$ to $1\,1/2$ in./ft
- *Steep slope*—over $1\,1/2$ in./ft

Both level and low-slope roofs are sometimes referred to as "flat" roofs. Truly flat, or dead level, roofs without slope are not recommended. Slopes of $1/4$ in./ft are the minimum required for adequate drainage, and most people use the term low-slope roof to refer to anything under $1\frac{1}{2}$ in./ft.

If drainage is rapid enough because of steep slopes, the roof covering need not be sealed, but can function effectively if the individual shingles, tiles, or panels are simply overlapped in the direction of water flow. At the edges of steep roofs, water can be collected and drained away from the building with gutters and downspouts. Drainage may be impeded by leaves or by snow and ice, and the edges of steep roofs require protection from water penetration caused by ice dams or other blockages.

On low-slope roofs, drains and overflow scuppers are used to remove water. The roof membrane must form a continuous barrier against water penetration because drainage is much slower than on steep roofs. Water may pond in low areas or at clogged drains, and drainage may sometimes be impeded by leaves or by snow and ice. The National Roofing Contractors Association defines positive drainage as "the drainage condition in which consideration has been made for all loading deflections of the deck, and additional roof slope has been provided to ensure complete drainage of the roof area within 24 hours of rainfall precipitation." Positive slope to the drains is achieved by sloping the structural frame and deck, or with tapered fill or insulation.

Deck deflection should be limited to $1/240$ of the span, and drains should be located at midspan where deflection is greatest. In a 50-ft span, for example, a deflection of $1/240 \times 50$ ft $= 2\frac{1}{2}$ in., which will cause water to flow naturally toward the center (*Fig. 6.1A*). If drains must be located near columns or bearing walls where deflection is least, deck slope should be increased to compensate for restricted drainage (*Fig. 6.1B*). Maintaining a roughly level line between one support and a $2\frac{1}{2}$-in. midspan deflection requires raising the other support 5 in. After deflection slopes have been calculated, additional slope should be designed into the deck or the roof system so that good drainage will be maintained under both minimum and maximum loading conditions. The long-term creep of concrete decks and the camber of some precast concrete decks must also be anticipated in the design (*Fig. 6.1C*). To achieve an effective slope of at least $1/4$ in./ft requires that the structural engineer and the architect collaborate in designing a structural deck and roof system to accommodate the specific span, deflection, and drainage conditions of each individual building design.

6.1.2 Loads

Roof systems must be able to resist applied loads, including thermal expansion and contraction, live loads from foot traffic, snow and rain, and the dead load of ballast and roof-mounted equipment. Deflections of the roof deck, rotation of the structure caused by deflections, and differential movement between the roof and adjoining parts of the building can also occur with the application and removal of live loads and with the long-term creep of concrete.

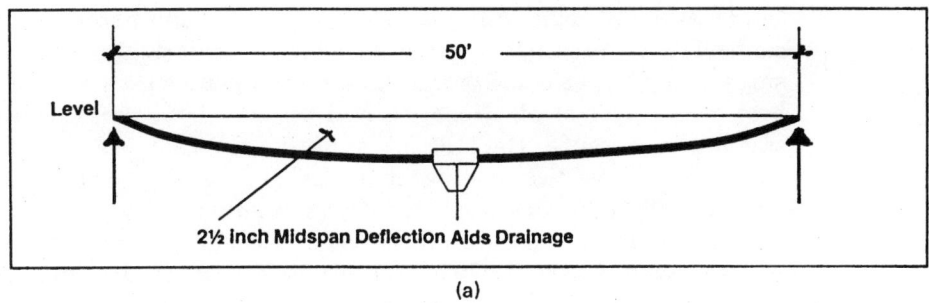

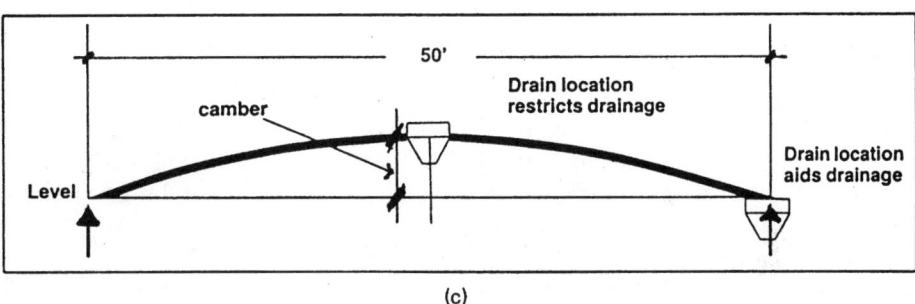

Figure 6.1 Roof deflection, camber, and drain location. (*From NRCA, Roofing and Waterproofing Manual, 4th ed.*)

6.1.3 Wind and air movement

Interruption of wind flow as it blows over and around a building results in areas of turbulence and of positive and negative air pressure. Localized wind forces of 40 psf or more are not uncommon. This is sufficient force to scour gravel ballast, flip concrete pavers, and peel off membranes that are inadequately secured.

Wind uplift is a combination of negative pressure at the upper surface and positive pressure from below. The magnitude of the negative pressure varies with wind speed, wind direction, and perimeter design. Wind speed is affected by geographic location and adjacent terrain as well as building height. The highest localized suction normally occurs when wind strikes a building on a

216   Chapter Six

corner at a 45° angle (*Fig. 6.2*). Roof overhangs and parapets have a significant effect on pressure distribution, either increasing or reducing the pressures. If a parapet is too low, negative uplift can be greater than if there had been no parapet at all. Negative uplift is usually expressed in terms of the stagnation pressure. The stagnation pressure is that which would occur near the center of a vertical wall when a wind strikes it head on. Suction pressure may be more than twice the stagnation pressure at the same wind speed. Most codes prescribe pressure coefficients for various exposures (see Chap. 2). If uplift pressures exceed a roof's ability to withstand the force, wind damage can range from moderate to severe.

Positive pressure uplift occurs when the air pressure inside the building rises due to air infiltration or the opening of doors or windows on the windward side of the building, or when the building is mechanically pressurized. Any transfer of positive pressure to components above the roof deck increases the danger of negative wind uplift damage to the roof. If the roof deck is not airtight because of cracks or joints, and the roofing membrane is not sealed to the deck at edges and penetrations, positive pressure is transferred to the roofing components and combines with negative suction pressure to increase the overall uplift pressure on the roof system (*Fig. 6.3A*). When the edges of the membrane are sealed and the building envelope is airtight, uplift on the roof covering is limited to suction at the upper surface (*Fig. 6.3B*).

6.1.4 Ultraviolet radiation

Roof coverings must be able to resist deterioration from ultraviolet radiation. Asphalt-based materials are sensitive to ultraviolet light, and will deteriorate rapidly if left unprotected. Asphalt shingles and roll roofing are manufactured with a protective granular surface, and asphalt built-up roofs are shielded

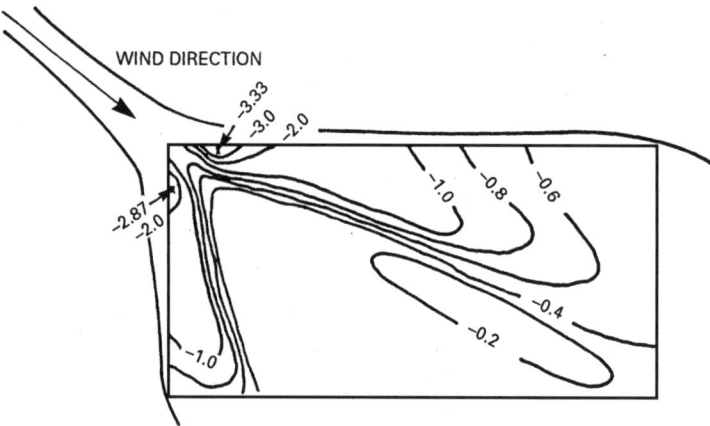

Figure 6.2 Negative pressure uplift on roof. Plan of roof with contours showing negative pressure distribution. Multiply coefficient times the stagnation pressure at same wind speed. (*From Latta, Walls, Windows and Roofs for the Canadian Climate.*)

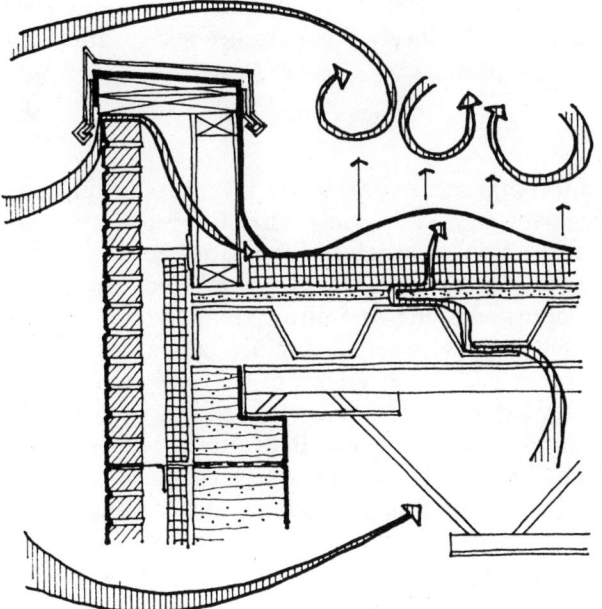

(A) AIR PERMEABLE ASSEMBLY INCREASES UPLIFT PRESSURE.

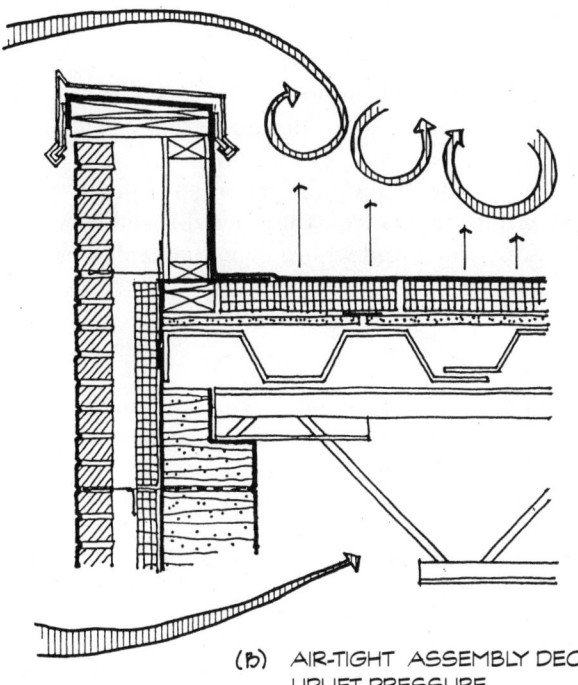

(B) AIR-TIGHT ASSEMBLY DECREASES UPLIFT PRESSURE.

Figure 6.3 Wind uplift on roofs. (*From National Research Council of Canada, Roofs That Work.*)

from exposure by ballast. Other roofing materials, including some single-ply membranes, sealants, and adhesives experience accelerated aging with exposure to ultraviolet radiation.

6.1.5 Temperature cycles

Daily and seasonal temperature variations in the roof components will affect their performance over time. Materials with different coefficients of thermal expansion and contraction experience differential rates of movement and, if the materials are bonded together, thermally induced stresses can cause tearing, bending, loss of attachment, and ultimate failure of the roof.

Thermal expansion and contraction produce lateral forces in a roof system. If not restrained, progressive movement of the roof membrane can occur. In built-up roofing very large forces can develop that are capable of splitting the membrane and tearing flashings loose. To avoid rupture, stresses in the membrane must be transferred to the deck through each of the intervening components by either adhesive bonding or mechanical fastening. In low-slope roofs, expansion joints and area dividers are used to limit the effect of lateral movement and stress buildup.

6.1.6 Water vapor

Air flow through a building envelope can carry with it significant amounts of water vapor. Air escaping from a building through a 1-in. hole can, over a single heating season, transfer 32 oz of moisture into a roofing assembly, and water vapor diffusion through 1 sq yd of 6-in-thick concrete can deposit nearly 10 oz of moisture in a roof. The progressive accumulation of condensed moisture can lead to corrosion, rot, expansion, loss of tensile or compressive strength, and freeze-thaw deterioration. Some buildings may require a vapor retarder, air barrier, or ventilation to control the flow of water vapor into the roof. Recommendations for when and where vapor control is needed are given later in this chapter.

6.1.7 Foot traffic

Most roofs are subject at one time or another to foot traffic, and must have a surface which can withstand the wear and tear or receive a protective cover along the service paths. Where roof-mounted equipment will require periodic maintenance or servicing, permanent protection should be provided against damage to the roof covering or membrane.

6.2 Roof Components

A typical roofing assembly is a combination of several components. NRCA defines the term *roof assembly* as the interacting roofing components (*including* the roof deck) that are designed to weatherproof and, normally, to insulate a building's top surface. The term *roof system* does *not* include the roof deck. The components used in any given roof assembly depend on the style of the building and the service conditions to which they will be subjected:

- Structural decks provide support.
- Vapor retarders prevent water vapor below the roof from diffusing into the roof system.
- Underlayments help shed water and isolate the roofing materials from the substrate.
- Insulation retards heat transfer.
- Steep roof coverings shed water.
- Low-slope roof membranes form a barrier against water entry.
- Flashing and counterflashing protect the edges of the roofing.
- Attachments secure the roof system components to the deck.

The success or failure of a roof is determined not only by the properties of the materials used for these individual components, but also by the interaction of the components in resisting exposure to water, heat and cold, solar radiation, wind and air, and applied loads. The characteristics of each component affect the performance of other components. Many roof failures occur because a deficiency in one component puts an excessive strain on other components. Thermal, mechanical, and moisture property limitations of each component are controlled or compensated for by one or more of the other components in the system.

In addition to the properties that are primary to their function, each roofing component also has other important properties which will influence their selection for a particular application, including thermal expansion/contraction coefficients; tensile, compressive and shear strengths; vapor permeability and moisture absorption characteristics; aging characteristics including embrittlement and loss of thermal resistance; ductility; and fire resistance.

6.2.1 Structural deck

Structural decks may be of wood, metal, or concrete. *Structural concrete decks* can be either cast in place or precast in the form of tees, double tees, channel slabs, flat slabs, or hollow-core planks. Precast concrete decks are problematic because of the numerous joints. Even with a topping slab, precast decks are more prone to movement than cast-in-place concrete decks. The precast panels should be securely tied together to prevent relative movement and cracking of the topping slab. Cast-in-place decks may be of normal weight concrete (density 150 lb/cu ft) or structural lightweight concrete (density 85 to 120 lb/cu ft). The deck must be cured and its surface must be dry before roofing can begin. To test for dryness, tape an 18-in-square clear plastic sheet to the concrete surface, being careful to seal all the edges. If condensation forms on the bottom of the plastic after a minimum of 16 hours, or if the concrete under the sheet looks darker than the concrete that has not been covered, the deck is still too wet to begin roofing application. ASTM D4263 *Test Method for Indicating Moisture in Concrete by the Plastic Sheet Method* can be specified for field quality control.

Lightweight insulating concrete decks are made from a mixture of portland cement and lightweight aggregate such as perlite or vermiculite, or from cellular or "foamed" concrete. Lightweight insulating concrete should not be confused with lightweight structural concrete. The compressive strength and thermal resistance of lightweight insulating concrete depends on both mix design and composition, but density ranges from 20 to 40 lb/cu ft. Lightweight insulating concrete is usually cast as a fill or topping slab over metal decking or form boards. To improve its thermal resistance, lightweight insulating concrete decks sometimes incorporate expanded polystyrene insulation sandwiched between two layers of concrete. The insulation boards are fabricated with a series of holes through which the concrete can flow to form a mechanical key between top and bottom layers. Lightweight insulating concrete is often mixed with a high water-cement ratio and can be slow to dry. The National Roofing Contractors Association recommends that lightweight insulating concrete be used only over venting substrates such as slotted metal deck. The underside venting provided by the holes in the deck allow additional drying of the concrete after the roofing assembly has been installed. If lightweight insulating concrete is used over cast-in-place or precast concrete decks, NRCA recommends a vented base sheet mechanically attached to help dissipate moisture. Except in reroofing where dew point calculations show a need for it, additional insulation should not be installed directly over lightweight insulating concrete decks because it can retard drying and may also absorb moisture in sufficient quantities to cause a loss of thermal resistance and blistering of the membrane.

Metal decks are fabricated from cold-rolled steel sheets. Many different depths, metal gauges, and rib profiles are available. Type A is a narrow rib design, Type B is an intermediate rib, and Type C is a wide rib. Common metal gauges are 16, 18, 20, and 22. Metal finishes include prime coat only, prime coat and finish, or galvanized coating. Decks manufactured with ventilating slots are designed to allow moisture to dry from the bottom of the roofing assembly.

Wood roof decks may consist of tongue-and-groove solid wood planks, or of plywood or oriented-strand board (OSB) sheathing. Only plywood and OSB properly rated for use as a roof deck should be specified for this purpose so that the panels will provide the necessary strength and durability. Some codes prohibit the use of OSB sheathing for roof decks in areas subject to high wind loads. Plywood roof sheathing should meet the requirements of the Department of Commerce Standard PS 1-83, or the American Plywood Association (APA) Performance Standard PRP-108. Oriented-strand board should meet the requirements of APA's PRP-108. All plywood and OSB roof decks should be exterior-rated or an interior type with exterior glue, bear the APA trademark and labeling (*Fig. 6.4*), and be secured to structural supports with annular threaded ring or barbed shank nails, or by approved pneumatically driven fasteners. Unsupported panel edges should be secured with H

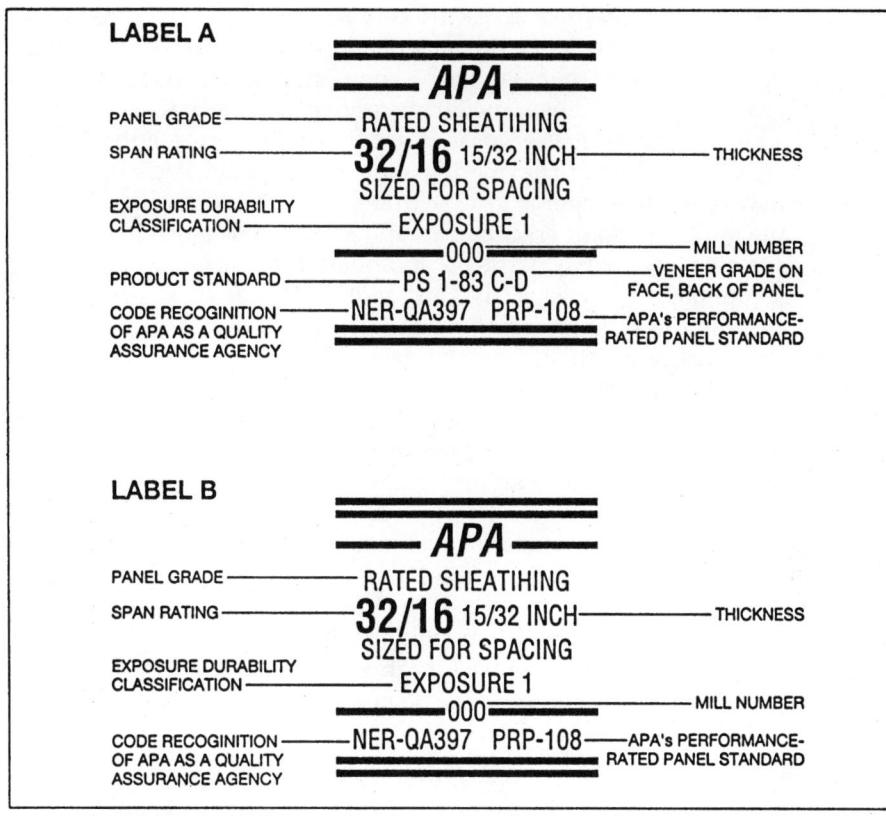

- Label A is found on all-veneer plywood conforming to PS 1-83.
- Label B is found on wood panels conforming to PRP-108, which includes oriented strand board, waferboard, and also all-veneer plywood that does not comply with PS 1-83 in certain respects.

A label similar to one of these should be found on each wood panel used as roof decking.

Note: The Span Rating (32/16 in the sample labels) is of particular significance. The number on the left-hand side of the Span Rating (32 in these labels) indicates the maximum recommended center-to-center spacing of supports (in inches) when the panel is used for roof decking with the long dimension of the panel running across the supports. The right-hand number of the Span Rating (16 in these labels) indicates the maximum recommended spacing of supports (in inches) when the panel is used for subflooring in double-layer construction with the long dimension of the panel running across the supports. Therefore, the Span Rating of 32/16 in these labels means that a particular wood panel may be used either for roof decking over supports that are spaced 32 inches (813mm) on center (maximum), or for subflooring over supports that are spaced 16 inches (406mm) on center (maximum). In all cases, panels are assumed to be continuous over two or more spans.

Note: Though the industry terminology refers to plywood sheathing with exterior glue as C-DX, the "X" (which refers to exterior glue) is not used in Label A. The EXPOSURE-1 classification is typically interior type panel with exterior glue.

Figure 6.4 American Plywood Association label. (*From NRCA, Roofing and Waterproofing Manual, 4th ed.*)

clips and end joints between panels should be staggered and fully supported by framing members. All joints should be gapped about $1/8$ in. to allow for expansion. Fire-retardant treated plywood should not be used as roof deck sheathing because it suffers premature deterioration and loss of strength in roof environments.

Cement–wood fiber decks are composed of treated wood fibers that are bonded together with portland cement or other materials and then compressed or molded into flat panels. These composite materials provide some acoustical properties as well as a moderate amount of thermal resistance.

Roof decks often serve as a substrate for direct adhesive application of a vapor retarder or the roofing membrane itself. Although these membranes may have some elasticity to accommodate minor movement, most do not have the ability to span cracks or joints. Even small movements at cracks or joints can exert excessive stress on an adhered membrane if the movement must be accommodated over a narrow width (*Fig. 6.5*). Concrete decks should be designed and reinforced to minimize shrinkage cracking. Small cracks and construction joints in concrete decks should be patched or filled before roofing work begins. With fully adhered roofing membranes, small joint movements can be accommodated by leaving an unbonded strip 4 in. wide on either side of a joint, a strip of roofing felt or membrane 8 to 12 in. wide can be loose-laid over the joint before adhering the membrane, or an expansion joint can be installed (*Fig. 6.6*).

When a vapor retarder or roof membrane is applied over a fluted metal deck, an underlayment such as insulation board is required to provide a smooth, flat substrate. On built-up roofs, if the insulation is of a type subject to damage from hot bitumen, a composite insulation with a protective overlay of a different type of material may be required, or in roofs without a vapor retarder, the base ply of the roofing membrane may be mechanically attached rather than adhered.

6.2.2 Vapor retarders, air barriers, and underlayments

In low-slope roofs with insulation below the membrane, a vapor retarder is sometimes needed between the insulation and the structural deck to prevent

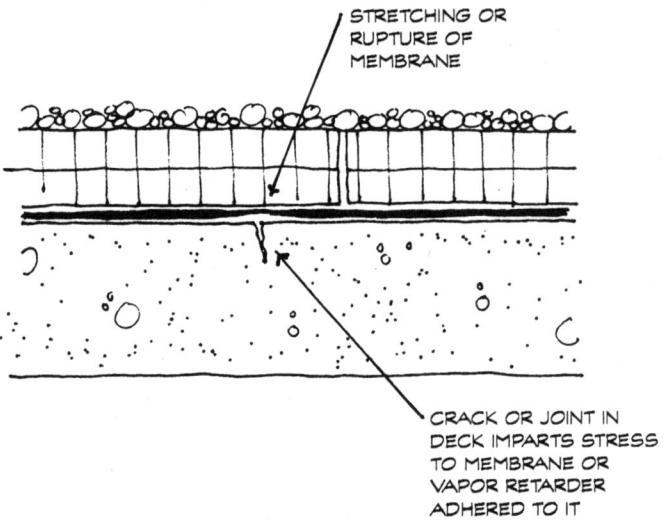

Figure 6.5 Crack or joint in deck stresses membrane. (*From National Research Council of Canada, Roofs That Work.*)

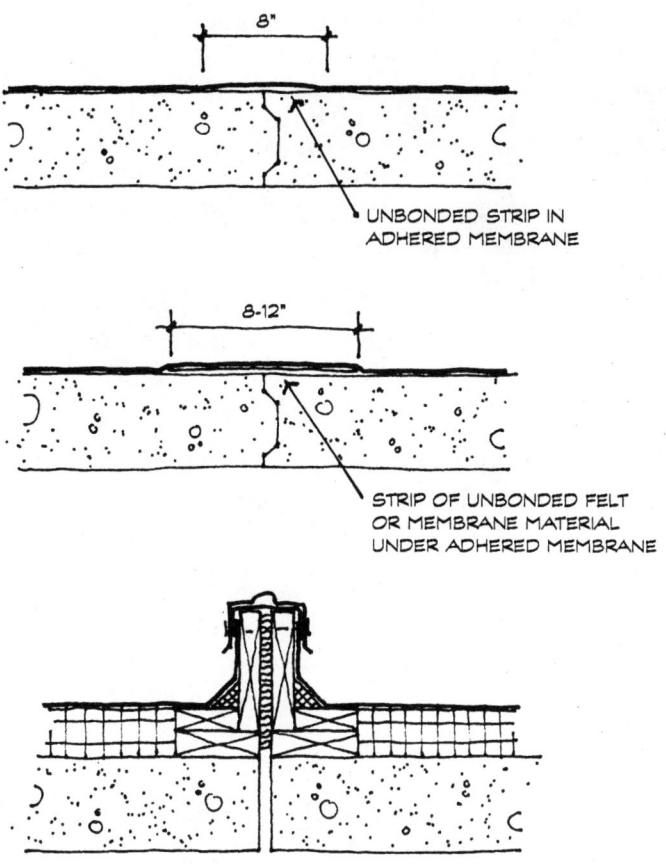

Figure 6.6 Bridging joints in roof deck.

moisture migration into the roof system. Moisture can condense on the bottom of the roofing membrane, wetting the insulation and reducing its thermal resistance or dripping back into the building. This is particularly true in areas with cold winter temperatures and occupancies with high indoor humidity such as swimming pools, museums, computer facilities, hospitals, and some industrial processes. A vapor retarder may also be required in extremely cold climates over spaces with normal humidities. An analysis of the interior and exterior environmental conditions can determine the potential for moisture condensation and accumulation under any proposed service conditions using any combination of roofing components. General recommendations for the use of vapor retarders in low-slope roofs are given below in Sec. 6.4, "Low-Slope Roofing." Common vapor retarder materials used in low-slope roofing applications include built-up membranes consisting of two layers of roofing felt bonded together with hot bitumen, PVC and polyethylene sheets, and foil facings on rigid board insulation. PVC and polyethylene sheets generally are not recommended for use in roofing systems because they are easily damaged by hot bitumen and subject to wind uplift and damage before the adhesives used for attachment and lap seals have fully cured.

Vapor retarders are not required on steep roof decks because the roof covering is unsealed, and allows water vapor to escape. Vapor retarders may sometimes be required, however, to prevent moisture migration into and condensation within the attic spaces below steep roofs. A discussion of attic condensation is included below in Sec. 6.3, "Steep Roofing." On steep roofs surfaced with shingles, tile, or slate, underlayments are frequently placed over the deck or supporting system to serve a variety of purposes depending on the climate and type of roof covering. Some underlayments provide a second layer of defense against water penetration which might occur through small defects in the primary covering or might be blown under the primary covering. Underlayments may also be used to provide an air seal which reduces the penetration of wind-driven rain and snow.

In hot, dry climates, tile roofs are supported on an open wood framework without an underlayment to enhance the cooling effect of increased air flow through the roof. Wood shingles and shakes should also be applied over spaced roof boards without an underlayment because the increased ventilation extends their service life. In cold climates, the underlayment at overhanging eaves should be a sealed waterproof barrier to prevent the entry of water which may back up under the roof covering due to ice dams formed when melting snow refreezes at the uninsulated low edge of the roof.

6.2.3 Insulation

Most roof assemblies must provide resistance to heat transfer. Materials must be selected and assembled to control conduction, convection, and radiation. In steep roofs, insulation may be installed above the ceiling or below the roof deck. Batt- and blanket-type insulations are the most commonly used. In low-slope roofs, rigid insulation boards are more common, including glass fiber, mineral fiber, wood fiberboard, cellular glass, perlite, polystyrene, polyurethane, and polyisocyanurate. Insulation boards are typically installed with end joints staggered and the long edges aligned. They can be applied in a single layer, but it is best to use two or more layers with the joints in each layer offset from those in the layer below. Tapered boards are available in some materials, designed to be placed in patterns which provide positive roof drainage. Lightweight insulating concrete is also used in roof deck construction, instead of or in combination with foam insulation.

Insulation used in roofing applications does more than provide thermal resistance. It also provides a substrate for the application of roof coverings and roofing membranes, and may provide fire resistance in certain low-slope roof assemblies. Because it is subjected to the rigors of roofing operations, foot traffic and equipment loads, and a variety of service conditions, the insulation used in roofing must possess certain physical properties beyond those typically required for other applications. The NRCA *Roofing and Waterproofing*

Manual identifies the following characteristics as desirable in an "ideal" roofing insulation:

- *Compatibility.* Should be able to withstand the effects of adhesives, solvents, and hot bitumen, and be compatible with other components of the roof assembly.
- *Impact resistance.* Should have strength, rigidity, and density to resist impact damage.
- *Fire resistance.* Should be noncombustible and comply with requirements of insurance underwriters and building codes.
- *Moisture resistance.* Should resist the effects of water vapor and free water without deterioration of physical properties, loss of attachment capability, or adverse effects on adjacent materials over the life of the building.
- *Thermal resistance.* Should have low thermal conductivity (k-value) so that the highest possible thermal resistance (R-value) can be achieved in the least possible thickness.
- *Thermal stability.* Thermal conductivity should remain constant with age.
- *Attachment capability.* Should accommodate secure adhesive or mechanical attachment to the underlying substrate and have sufficient peel strength to resist delamination of adhered membranes against wind uplift pressures.
- *Dimensional stability.* Should be dimensionally stable under varying temperature and moisture conditions.

Variable service conditions, design requirements, and type of roof coverings will always make some properties more important than others in a given situation, so the selection of roof insulation should be determined on a project-by-project basis. Always check the roofing system manufacturer's specifications for recommendations or restrictions on specific types of insulation.

6.2.4 Roof coverings and membranes

As long as a roof sheds water quickly, it can be made of almost anything. In steep roofs with a slope greater than 3 in/ft, unsealed lapped roof coverings are used, including shingles, shakes, slate, clay and concrete tile, metal panels, and even thatch. The slope needed for the satisfactory performance of each depends on the individual material, how it is installed, and the underlayment materials used. Climatic factors including rainfall amounts, and freezing temperatures also add certain slope, underlayment, and attachment requirements for the various materials.

In low-slope roofs, a continuous sealed membrane must be formed which is capable of preventing the passage of water. With built-up roofs, watertightness is achieved with multiple layers of felt and bitumen. With single-ply roofs, sheet materials from 3 to 40 ft wide are lapped and sealed with applied adhesives, sealing tapes, heat welding, or solvent welding.

6.2.5 Flashing and counterflashing

In low-slope roofs, base flashing covers the edges of the membrane and cap flashing or counterflashing shields the upper edges of the base flashing. In steep roofs, flashing is used at intersections with adjacent vertical surfaces and at penetrations. ASTM D1079 *Definitions of Terms Relating to Roofing, Waterproofing, and Bituminous Materials* defines *flashing* as a system used to seal membrane edges at walls, expansion joints, drains, gravel stops, and other places where the membrane is interrupted or terminated. *Counterflashing* is defined as formed metal or elastomeric sheeting secured on or into a wall, curb, pipe, rooftop unit, or other surface to cover and protect the upper edge of a base flashing and its associated fasteners. Roof flashing requires more maintenance and is a more frequent cause of problems than the remainder of the roof. Flashing failures can be caused by improper or inadequate details, improper materials, and faulty installation, or may be related to other deficiencies such as inadequate membrane attachment which pulls the flashing apart. Durability and performance of the materials used to form roof flashing are related to the compatibility of the flashing with other roofing system components and to adequate strength and thickness. *Table 6.1* contains recommendations for the minimum weight, gauge, or thickness required for the most common metal roof flashing materials in various types of flashing applications.

Roof flashing metals can be of the watershed type or system component type. Watershed flashings include counterflashing, expansion joint covers, perimeter curb covers, wall closures, weathercaps, storm collars and copings. Watershed flashings do not have to be waterproof, and the joints do not have to be sealed. They must have adequate slope to drain water and should never be sealed or attached to the roofing assembly. Provision should be made for movement of or between metal sections by lapping or interlocking vertical sections or by mechanical joints. System component flashings are attached or sealed to the roofing membrane and must be solidly anchored to wood nailers or to the roof substrate to keep thermal movement of the metal from damaging the membrane.

The greater the coefficient of expansion of a given metal, the more compensation must be made in the design and installation of metal flashings to control or accommodate movement. The effects of metal movement under thermal loading conditions as well as wind loading require that the flashing be either fully restrained or completely separated from the roofing membrane. *Figure 6.7* shows the relative thermal expansion of the metals most commonly used for roof flashing and accessories. Light-gauge metal exerts much less strain on fasteners than heavy metal sections even though the coefficient of expansion is the same. Because of its high coefficient of expansion, aluminum movement is difficult to control by mechanical attachment, so it should be completely

TABLE 6.1A Recommended Sheet Metal Thickness for Roof and Wall Flashing

Item	Copper, oz.	Stainless steel, in.	Galvanized steel, gauge	Aluminum, in.	Terne-coated stainless steel, in.	Terne, gauge
Exposed Flashings						
Cap	16	0.018	26	0.040	0.015	
Through wall	16	0.018	26	0.040	0.015	
Roof projections	16	0.018	26	0.040	0.015	
Roof penetrations	16	0.018	26	0.040	0.015	
Concealed Flashings						
Heads	10	0.010	28		0.015	
Through wall	10	0.010	28		0.015	
Lintels, shelf angles	10	0.010	28		0.015	
Roofing						
Flat seam	16	0.018	26		0.015	28
Standing seam	16	0.018	24	0.040	0.015	28
Batten seam	16	0.018	26	0.032	0.015	28
Copings	16	0.018	24	0.032		
Expansion joints	16	0.018	24	0.040		
Gravel stops	16	0.018	26	0.032	0.015	
Gutters	20, 24	0.018	26	0.040	0.015	
Leaders	16, 20	0.018	26	0.032	0.015	
Reglets	16	0.015	26			
Scuppers	16	0.018	24	0.032		

SOURCE: Rosen and Heineman, *Architectural Materials for Construction.*

divorced from the membrane to avoid damage. When roof flashings are to be restrained, provide treated wood nailers beneath all flanges, and fasteners long enough to penetrate the wood for a depth of at least 1 in. and spaced no more than 4 in. on centers.

Movement in metal flashings and accessories can be accommodated by the use of mechanical joints between sections, interlocking laps or expansion joints, and the use of shorter metal sections (*Fig. 6.8*). Complex metal flashing configurations are difficult to make waterproof, and soldered sections generally break loose after moderate exposure to thermal loading. The use of mechanical joints, such as standing seams in closure details like copings, provides stiffness, minimizes dishing, and creates an effective expansion joint between

TABLE 6.1B Approximate Thickness Equivalents and Gauge Numbers for Sheet Metal

Approx. thickness, mm	Hot- or cold-rolled steel	Stainless steel	Galvanized steel	Non-ferrous material
0.4	28	28	30	26
0.5	26	26	28	24
0.6	24	24	26	22
0.8	22	22	22 or 24	20
1.0	20	20	20	18
1.2	18	18	18	16
1.6	16	16	16	14
2.0	14	14	14	12
2.5	12			10
3.0		12	12	
3.5	10	10		
4.0	8			
4.5		8		6

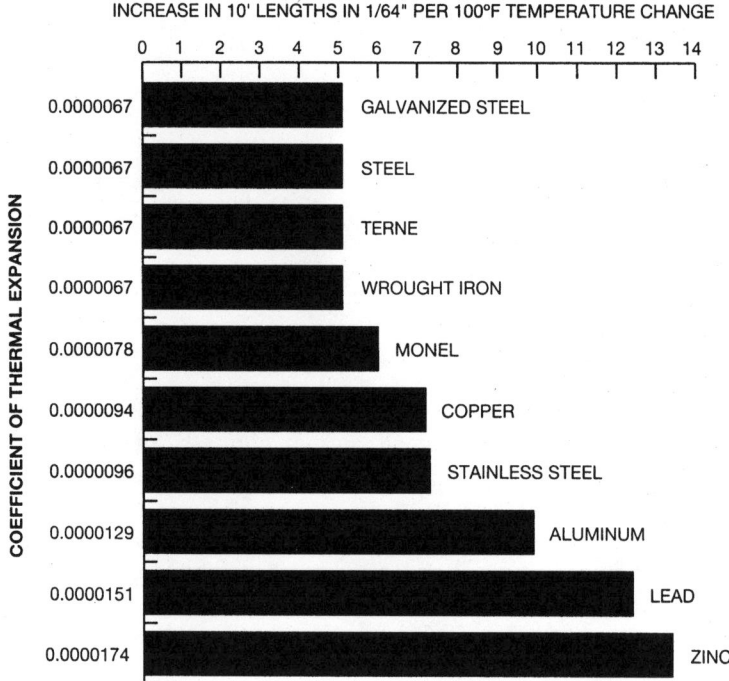

Figure 6.7 Thermal expansion of metals. (*From NRCA, Roofing and Waterproofing Manual, 4th ed.*)

Roofing 229

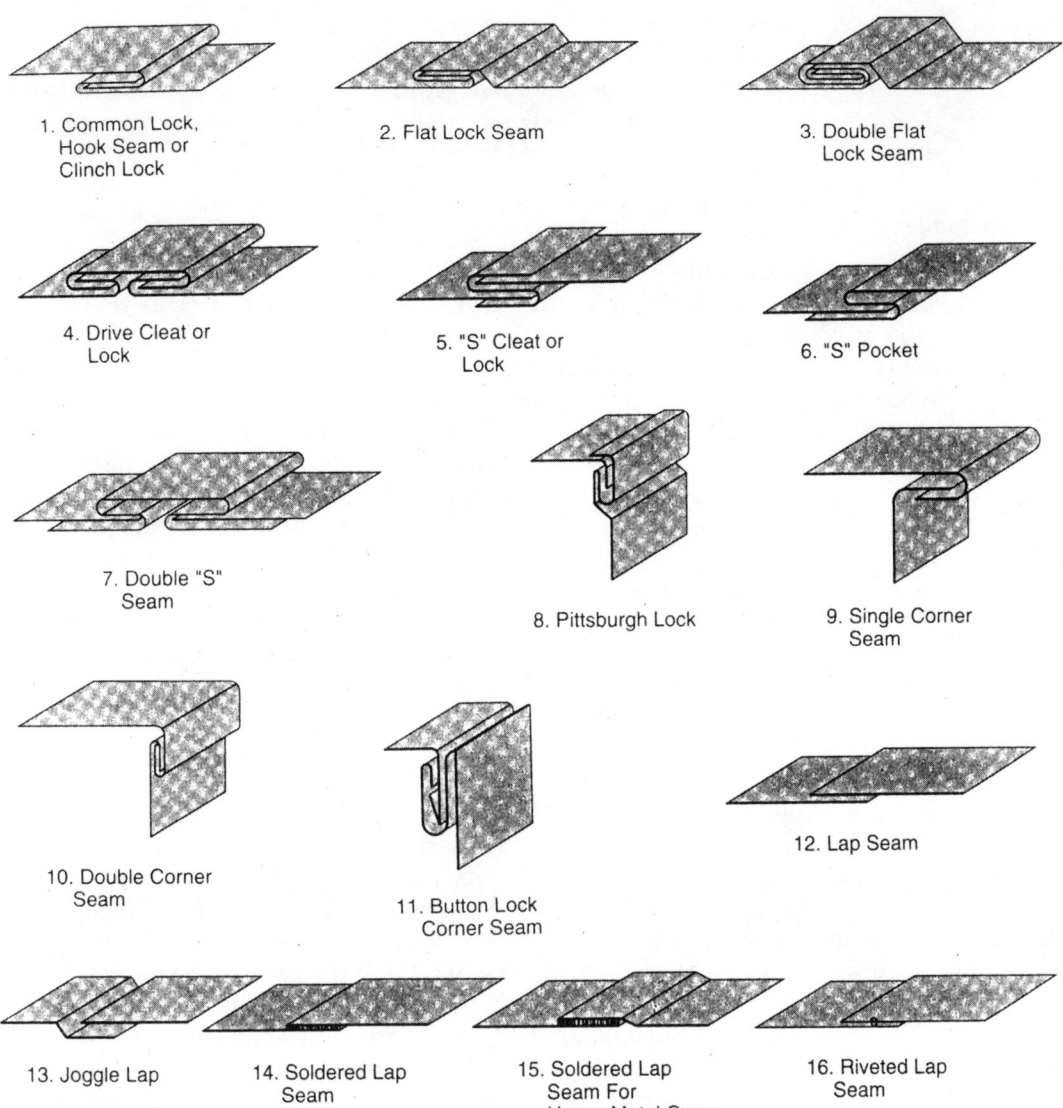

Figure 6.8 Sheet metal locks and seams. (*From SMACNA, Architectural Sheet Metal Manual, 5th ed. Used with permission.*)

sections of metal. Watershed metal such as counterflashings at the edges of the roof covering may be lapped or installed with interlocking joints. Vertical sections are often secured by the use of cleats or hook strips, but fasteners can also sometimes be used on the vertical face of metal sections. Installing fasteners at the overlap of heavy metal sections will result in the fastener backing out or being sheared off by the expansion and contraction movements. Fasteners installed in the middle of a metal section do not require special treatment, but fasteners at other locations should always be installed through elongated holes in the vertical section. Screws used to attach metal flashing should have neoprene washers at least $1/4$ in. larger than the holes in the metal. Corner sections should be installed with legs not exceeding 18 in. in each direction. A lapped or interlocked joint between the corner section and the adjoining metal section will prevent buckling.

6.2.6 Attachments

All of the components in a roof system must be securely attached to prevent wind damage and excessive thermal movement. The type of deck as well as the type of covering or membrane influences the type of attachment needed to secure the roofing. Vapor retarders, insulation, and membranes on low-slope roofs may be either laminated to structural decks and to each other with bitumen or other adhesives, attached with mechanical fasteners, or attached with a combination of mechanical fasteners and adhesive. Single-ply membrane manufacturers stipulate approved methods of attachment which may include adhesives, individual fasteners, or clamping bars attached with fasteners. Although some single-ply membrane systems are "loose laid" and held in place by gravel or paver ballast, mechanical attachments are typically provided at the roof perimeter and at penetrations. Shingles and most other steep roof coverings are attached with nails or special staples, and metal roofing is typically attached with metal clips to subframing or plywood decks.

6.3 Steep Roofing

Steep roofs shed water quickly because of their slope. Shingles, shakes, tile, slate, and metal panels are commonly used as roof coverings for both residential and commercial steep roofs. Because of their rapid drainage, steep roof coverings do not have to form a continuous sealed membrane. Instead the materials are overlapped in successive layers in the direction of water flow. The lower the slope, the slower the drainage, and the greater the requirements for lapping and sealing. Each type of steep roof covering has specific requirements for minimum slope (*Table 6.2*), and the number and spacing of nails or fasteners is also determined by the type of roof covering.

Plywood sheathing for decks on steep roofs should meet the recommended performance criteria of the American Plywood Association and the National Roofing Contractors Association. NRCA recommends that all plywood products used for roof decking be APA-rated sheathing complying with the Department of Commerce Product Standard PS 1-83, be interior type with exterior glue, and be graded C-D or better. For 16-in. rafter or truss spacing, a minimum thickness of $15/32$ in. should be used, and for 24-in. framing, a $5/8$-in. minimum thickness. Plywood panels should be spaced at least $1/8$ in apart to allow for expansion. End joints of the panels should be staggered and supported directly over framing members. Tongue-and-groove panels, wood blocking, or special corrosion-resistant H clips should be used at unsupported edge joints. For spans of 48 in. or more, two clips per panel should be used. Plywood decking should be attached with annular ring or barbed shank nails.

TABLE 6.2 Minimum Slope for Steep Roofing Materials

Roof covering material	Minimum deck slope	Deck type
Asphalt roll roofing	1:12 to 4:12, depending on method of application	Solid wood plank, plywood, or OSB sheathing
Asphalt shingles	$2\frac{1}{2}$:12 to 4:12 depending on type of underlayment and method of its application (i.e., sealed or unsealed)	Solid wood plank, plywood, or OSB sheathing
Wood shingles	4:12	Solid wood plank, plywood or OSB sheathing, or 1×4s spaced on centers equal to the weather exposure of the shingle
Wood shakes	4:12	Solid wood plank, plywood, or OSB sheathing, or 1×6s spaced on centers equal to the weather exposure of the shake
Tile roofing	4:12	Solid wood plank, plywood, or OSB sheathing, or spaced wood plank
Slate roofing	5:12	Solid wood plank, plywood, or spaced wood plank
Metal panel roofing	Varies with type of roofing system and seaming techniques	Solid plywood or OSB sheathing for architectural metal roofs, or open steel framing for structural metal roofs

NOTE: Minimum slope recommendations apply generally throughout the United States. Designers may elect to decrease slopes in warm, arid regions where rainfall is moderate and there is no risk of snow or ice, and increase slopes in coastal and northern regions with significant wind-driven rain or snow and ice.

Lumber for wood plank decks should be air-dried or kiln-dried, nominal 1 in. or thicker, and for solid plank decks should be tongue-and-groove design. For wood plank decks with gaps between the boards to allow air circulation under the roof covering, straight-edge boards may be used, and should be spaced on centers equal to the weather exposure of the shingle, shake or tile. Cracks or knots over $\frac{1}{2}$ in. in diameter should be covered with sheet metal. A separator sheet of rosin-sized sheathing paper is recommended for use over solid wood plank decks, and is necessary over wood plank decks made from lumber that has been treated with an oilborne preservative.

Steep roofs in the northern two-thirds of the United States and nearly all of Canada experience problems with the formation of ice dams, which can result in water penetration at the eaves. When the air temperature is below freezing and a snow-covered roof is warmed by solar radiation and heat loss from the building, the bottom layer of snow melts, runs down to the eave and freezes again on exposure to the air or to an unheated eave overhang. As the melting and refreezing continues, layers of ice build up to form a dam, and water is trapped behind the ice. If there is no positive seal between the overlapping layers of the roof covering, water will begin to seep through the roof (*Fig. 6.9*). To control problems with ice dams, minimize the snow melting effects of heat loss from the interior by adding attic or ceiling insulation. Provide ventilation between the insulation layer and the roof deck to reduce the temperature of the deck, and provide a waterproof underlayment along the eave. Extend this barrier membrane onto the roof a minimum distance beyond the inside wall line as recommended in *Fig. 6.10*. On lower slope roofs, ice damming may be more extensive, and barrier membrane underlayments should also be considered along the valleys and on low roofs below higher eave lines as well. A buildup of leaves at a roof edge or along the valleys between slopes causes similar damming and leakage because the obstruction prevents water from draining properly. At valleys of roofs with a pitch of 4:12 or higher, a single width of ice and water barrier membrane can be centered along the length of the valley, but for slopes less than 4:12, the membrane should extend 36 in. on either

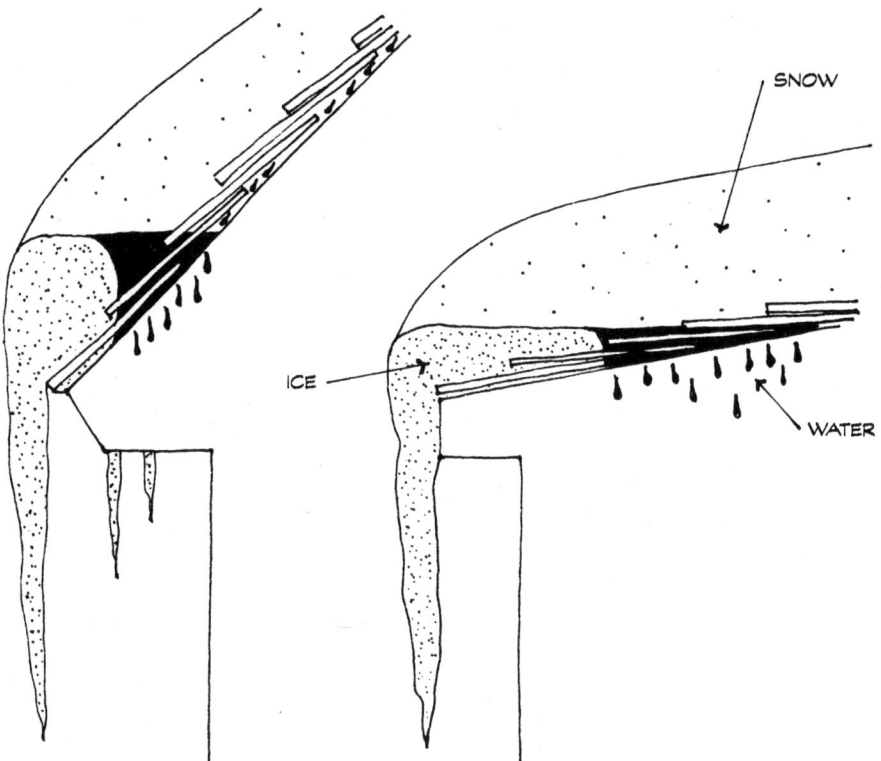

Figure 6.9 Ice dams at roof overhangs. (*From Latta, Walls, Windows and Roofs for the Canadian Climate.*)

Width of Ice and Water Barrier Membrane at Sloped Roof Eaves

Minimum Dimension A, inches	Minimum Roof Pitch
24	over 4:12
36 inch minimum or to roof ridge	4:12 and lower

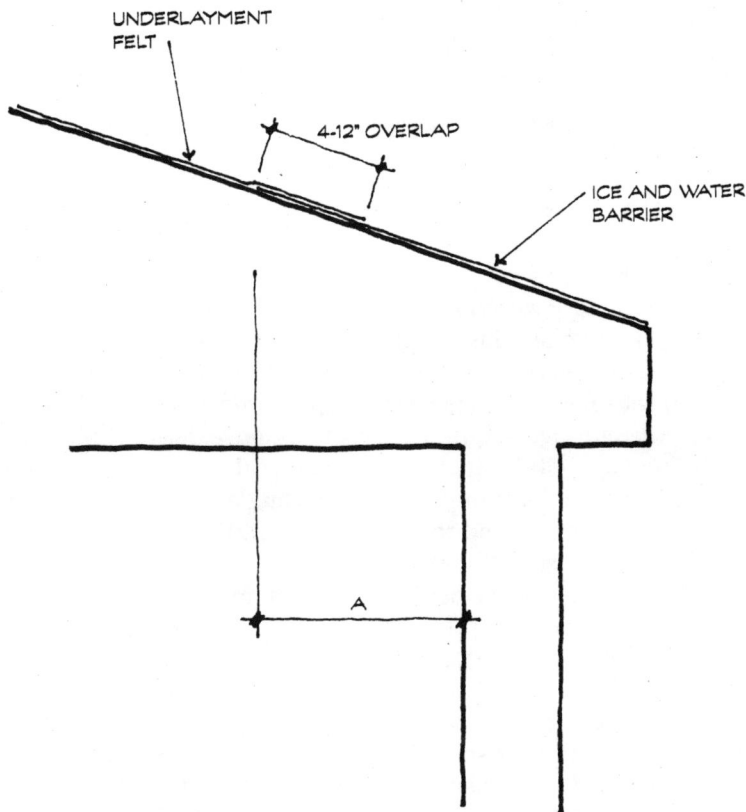

Figure 6.10 Width of ice and water barrier membrane at sloped roof eaves.

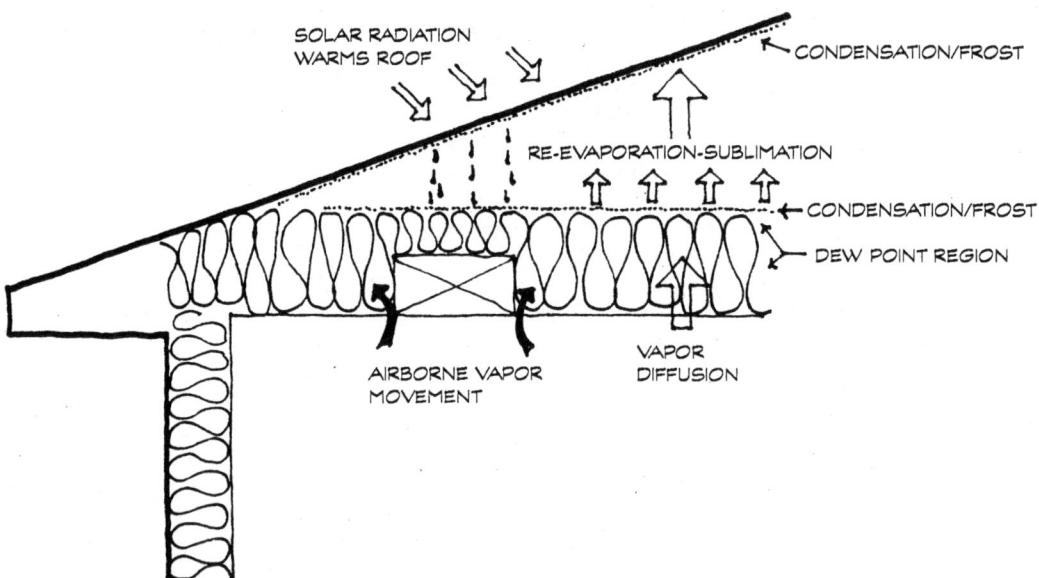

Figure 6.11 Vapor migration and condensation on underside of cold roof deck. (*From Lstiburek and Carmody, Moisture Control Handbook.*)

side of the valley. Underlayment felts should overlap the ice and water barrier membrane 4 to 12 in.

Some architects consider ice and water barrier membranes as inexpensive insurance and cover the entire roof area with the material. The membrane itself should be self-healing because roofing nails are driven through it. Self-adhering rubberized asphalt is a popular material for ice and water barrier membranes because it is easy to apply and reseals around fastener penetrations. Chronic ice damming problems on existing buildings are sometimes addressed by installing electric heating cables along the eaves. The heaters do not prevent ice damming, but by melting small tunnels through the ice, they prevent ponding behind the dams, prevent the dams from becoming very large, and reduce the risk of leaks.

Increased thermal resistance in attics and roofs to minimize ice damming or reduce energy consumption can create a different type of problem. The higher the thermal resistance, the less heat transfer there is to the roof, and the colder the temperature at the under side of the deck. Condensation can occur if vapor retarders and air barriers are not provided to prevent water vapor migration from the interior spaces. Warm, moist air that escapes into the attic or roof space can condense when it comes in contact with the cold fasteners which penetrate the roof deck, or with the roof deck itself (*Fig. 6.11*). Water can accumulate in the insulation and reduce thermal resistance, or drip back to the interior through penetrations, causing what appear to be roof leaks. In addition to using vapor retarders and sealing against air leakage (see Chap. 5), providing adequate ventilation can remove moist air from the attic or roof space and minimize the chances of condensation.

Where steep roofs intersect exterior walls, ceiling insulation can complicate the winter thermal and moisture performance of the envelope. If the insulation thickness is compressed at the perimeter, or if ventilation air through the soffit reduces the effectiveness of the insulation, heat loss is increased and the surface temperature at the interior wall-ceiling intersection drops. As this surface temperature drops, the relative humidity of the air at the interior surface increases. When the surface humidity is above 70%, mold and mildew growth can occur. To provide for adequate ventilation of the attic or roof assembly, a wind baffle should allow unobstructed air flow without creating cold interior surfaces (*Fig. 6.12*). Attics should have a ventilation ratio of 1 sq ft of free vent area for each 300 sq ft of insulated ceiling if a vapor retarder is present, or 1 sq ft of vent for each 150 sq ft of insulated ceiling if no vapor retarder is present (*Fig. 6.13*). Vents should be divided equally between eave and ridge.

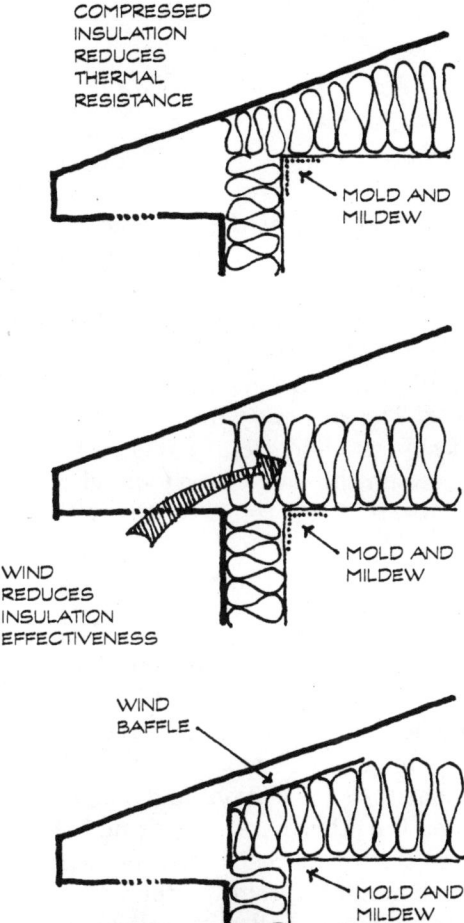

Figure 6.12 Wind baffle at attic perimeter. (*From Lstiburek and Carmody, Moisture Control Handbook.*)

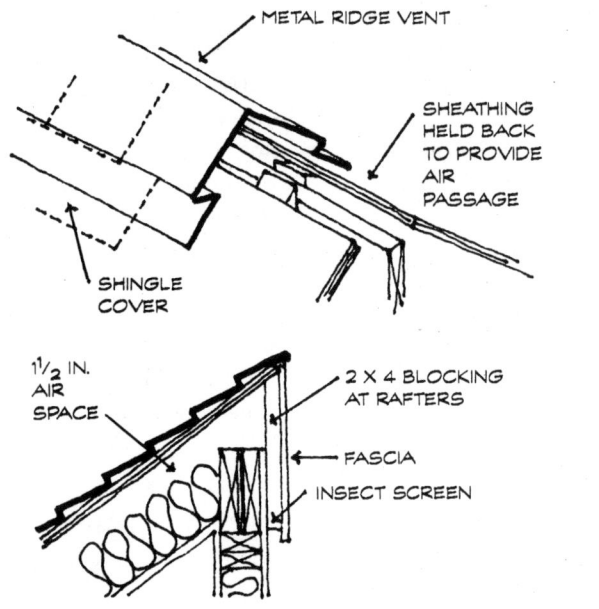 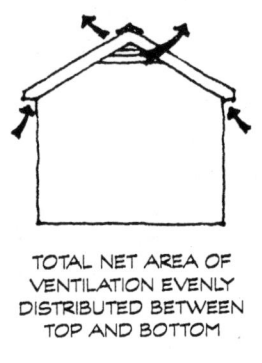

Figure 6.13 Attic ventilation.

Gutters and downspouts can be used at the eaves of steep roofs to direct runoff away from the building wall and foundation line. Gutters do interfere with snow and ice melt, and can aggravate problems with ice dams. If gutters are necessary to control surface water drainage, they should be sized on the basis of expected rainfall (*Figs. 6.14* and *6.15*).

A complete discussion of all of the various types of steep roof coverings is beyond the scope of this book. The sections which follow provide basic guidelines for the use of asphalt shingles and metal panel roofing. For more complete details, and for guidance on the use of wood shingles and shakes, clay and concrete tile, and slate roofing, refer to the NRCA *Roofing and Waterproofing Manual*.

6.3.1 Asphalt shingles

Asphalt shingles are reinforced with either organic or glass fiber felts, coated with a specially formulated asphalt, and covered with mineral granules. Glass fiber reinforced asphalt shingles are sometimes referred to as *fiberglass shingles*. Glass fiber reinforced shingles are more fire-resistant, more mildew-resistant, less affected by moisture absorption, and weigh less than organic shingles. Organic shingles, on the other hand, show underlayment irregularities less noticeably and have better wind resistance and higher tear strengths, so they are less likely to crack or split in cold weather.

Most asphalt shingles are manufactured with self-sealing adhesive strips which supplement mechanical fastening to prevent blow-back of the tabs in high

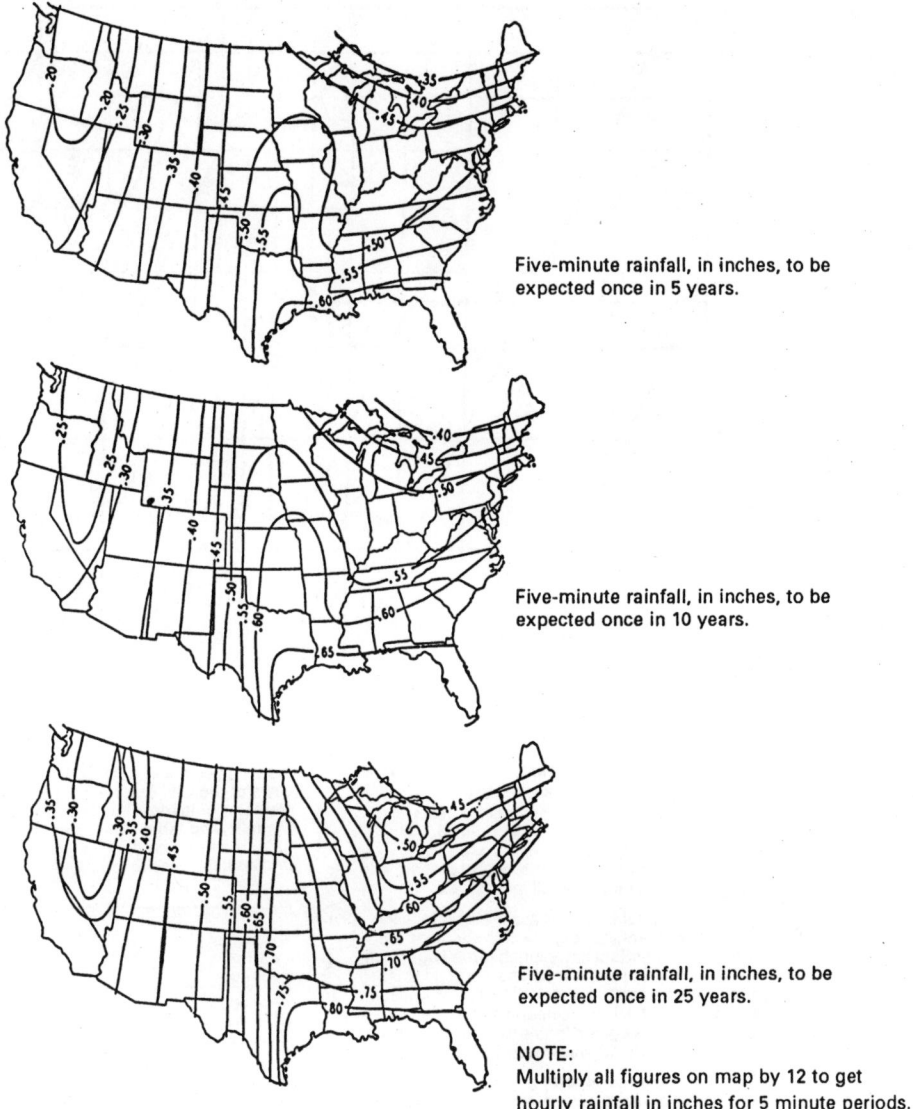

Figure 6.14 Rainfall data. (*From Revere Copper Products, Inc., Copper and Common Sense.*)

winds. Asphalt shingles are available in a variety of weights, colors, and styles, including strip shingles, three-tab strip shingles, and laminated architectural shingles. Laminated shingles have a heavy texture to create a three-dimensional look. Asphalt shingles reinforced with glass fiber felts should meet the minimum requirements of ASTM D3462 *Specification for Shingles, Asphalt, Made from Glass Felt and Surfaced with Mineral Granules,* and those reinforced with organic felts should conform to the requirements of ASTM D225 *Specification for Shingles, Asphalt, Surfaced with Mineral Granules (Organic Felt).*

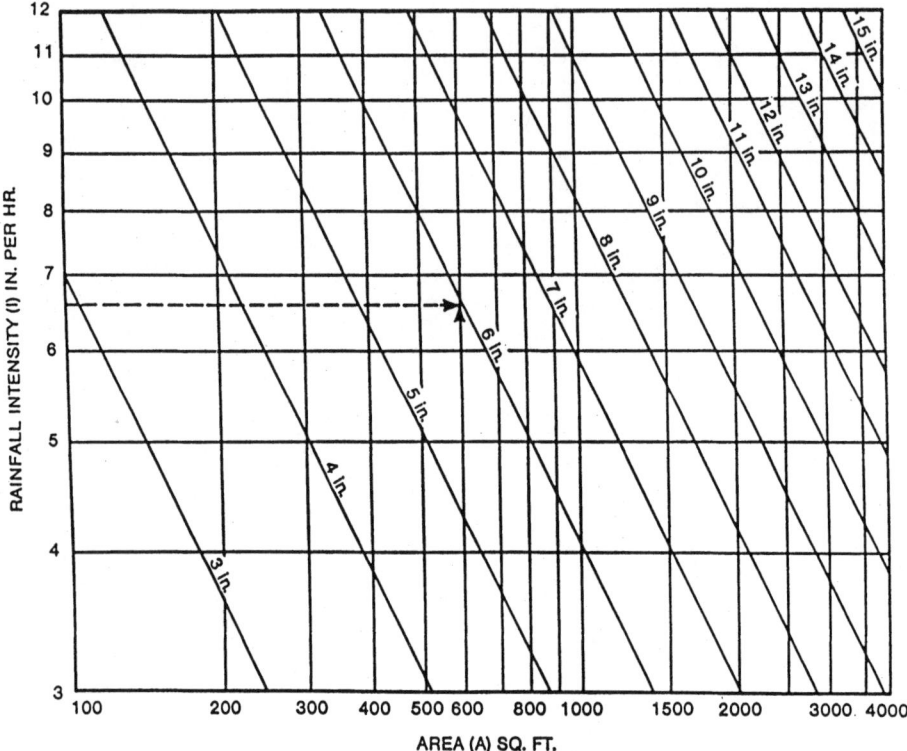

Rectangular box gutters and gutters of irregular cross section

To simplify the solution of empirical formulas deduced from the results of tests on level rectangular gutters, Charts 2 and 3 are used together to determine the width of gutter required for given roof area and rainfall intensities.

The architectural style and scale of the building in many cases determine the size and shape of the gutter. Sloping gutters carry much more water than level gutters of the same size. The graphs for level gutters may be used to check dimensions chosen for sloping gutters. If the selected dimensions are greater than those given by the charts, it is obvious that the gutter will be unnecessarily large.

The required sizes of level gutters of other than semicircular or rectangular shapes can be determined approximately by finding the semi-circle or rectangle of the same area which most closely fits the irregular cross-section.

Leader connections

Gutter outlets very commonly are too small and are the usual cause of overflowing gutters. The Building Laws of the City of New York specify that the diameter of leaders shall not be less than required in the following table. The diameter of the leader shall be determined on the basis of the roof area (horizontal projection) that each leader serves.

Diameter of Leader in Inches	Roof Area in Sq. Ft. that Each Leader Serves
2	500
2½	900
3	1500
4	3000
5	5500
6	9000
8	19000

In other cities where the rainfall is more or less than that of New York City, the diameter of the leader should be in proportion to the expected rainfall as shown in Fig. 6-14.

Figure 6.15 Sizing gutters and leaders from rainfall data. (*From Revere Copper Products, Inc., Copper and Common Sense.*)

Asphalt shingles are fire-retardant, and are labeled by the Underwriters Laboratory (UL) as Class A, B, or C. Class A shingles provide satisfactory protection under severe fire exposures, Class B under moderate fire exposures, and Class C under light fire exposures. Asphalt shingles reinforced with organic felts are generally rated Class C, while fiberglass reinforced shingles are rated Class A. To be labeled wind-resistant, shingles must resist being lifted or torn off by the prevailing or expected maximum winds of the area, as tested in a laboratory wind tunnel.

All asphalt shingle roofs require an underlayment as a redundant or backup watershedding layer. Most frequently used is a 15-lb unperforated asphalt felt nailed to the deck with 2-in. head laps and 4-in. side or end laps. On roofs sloped less than 4:12, a two-ply underlayment should be installed, and in severe climates, the two plies should be fully adhered to the deck and to one another. Remove and replace any underlayment felts that have taken on moisture and become wrinkled prior to shingling. Asphalt shingles should be nailed to the deck with large-headed, hot-dip galvanized, annular ring shank nails. Nails used in nailing guns are less corrosion-resistant because the zinc coatings are thinner to prevent clogging the firing mechanism. Although stapling is still permitted by some codes, shingles attached by stapling are more susceptible to wind damage. On most shingles, the nails should be placed just above the tab slot and below the seal strip to ensure maximum holding, resistance to wind damage, and double nailing of the undercourse. Nails should be of sufficient length to penetrate the deck $^3/_4$ in. Nails should never be exposed, and never penetrate valley flashings. Areas exposed to high winds can be strengthened by using six rather than four nails per shingle, by installing a two-ply underlayment, and using 4-in. rather than 5-in. exposure. *Figure 6.16* shows the basic installation of asphalt shingles. Underlayment and fastening requirements may vary slightly under different building codes.

Steep roofs have special flashing requirements at valleys, eaves, rakes, roof penetrations, and adjacent walls. Valley flashing is especially prone to damage by traffic on a roof. The construction of the underlayment and the water shedding valley liners should complement the type of roof covering to be used. Valleys can be "open" with a separate valley flashing of metal or roll roofing, "closed-cut," where one side of the roof is continued underneath the other, or "woven," where the shingle courses are alternately integrated. The closed-cut valley looks the best, but is impractical with thick laminated shingles. Metal drip edge flashing at rakes and eaves is used to prevent water seepage into the roof deck (*Fig. 6.17*). Where a chimney or other large obstruction occurs in the downslope of a steep roof, a cricket should be constructed on the high side to provide positive drainage (*Fig. 6.18*). At vertical walls, metal flashing should turn up 3 to 5 in. (*Fig. 6.19*).

6.3.2 Metal roofing

The earliest metal roofs were handcrafted from soft, ductile metals such as copper and lead. Panels were attached to a solid roof deck with clips that folded over and locked the panels in place along the seams. Zinc-coated sheet steel roofing was developed in the mid-1800s, and by the early 1900s galvanized, corrugated steel roof panels were becoming widely used because they were strong and inexpensive.

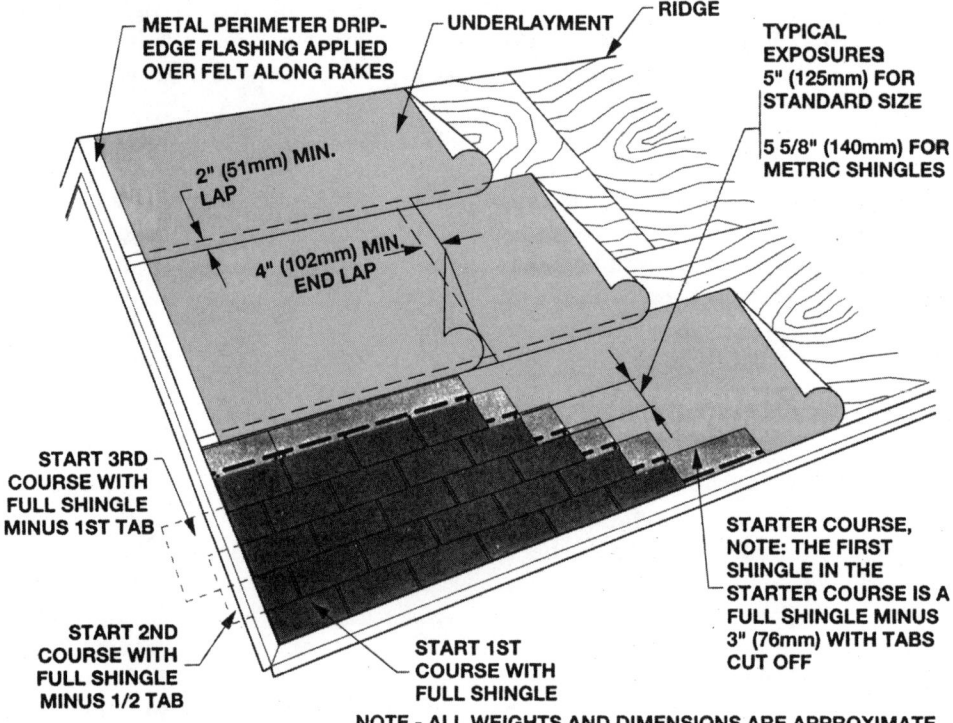

Figure 6.16 Installation of asphalt shingles. Three-tab, square-butt strip shingles—cutouts are centered over the tabs in the course below (the 6-in. method). (*From NRCA, Roofing and Waterproofing Manual, 4th ed.*)

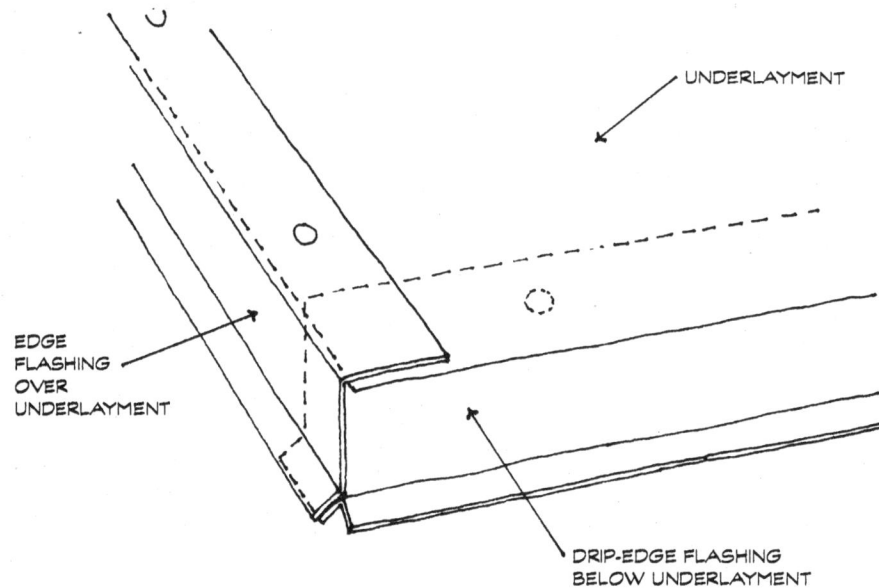

Figure 6.17 Metal drip edge flashing.

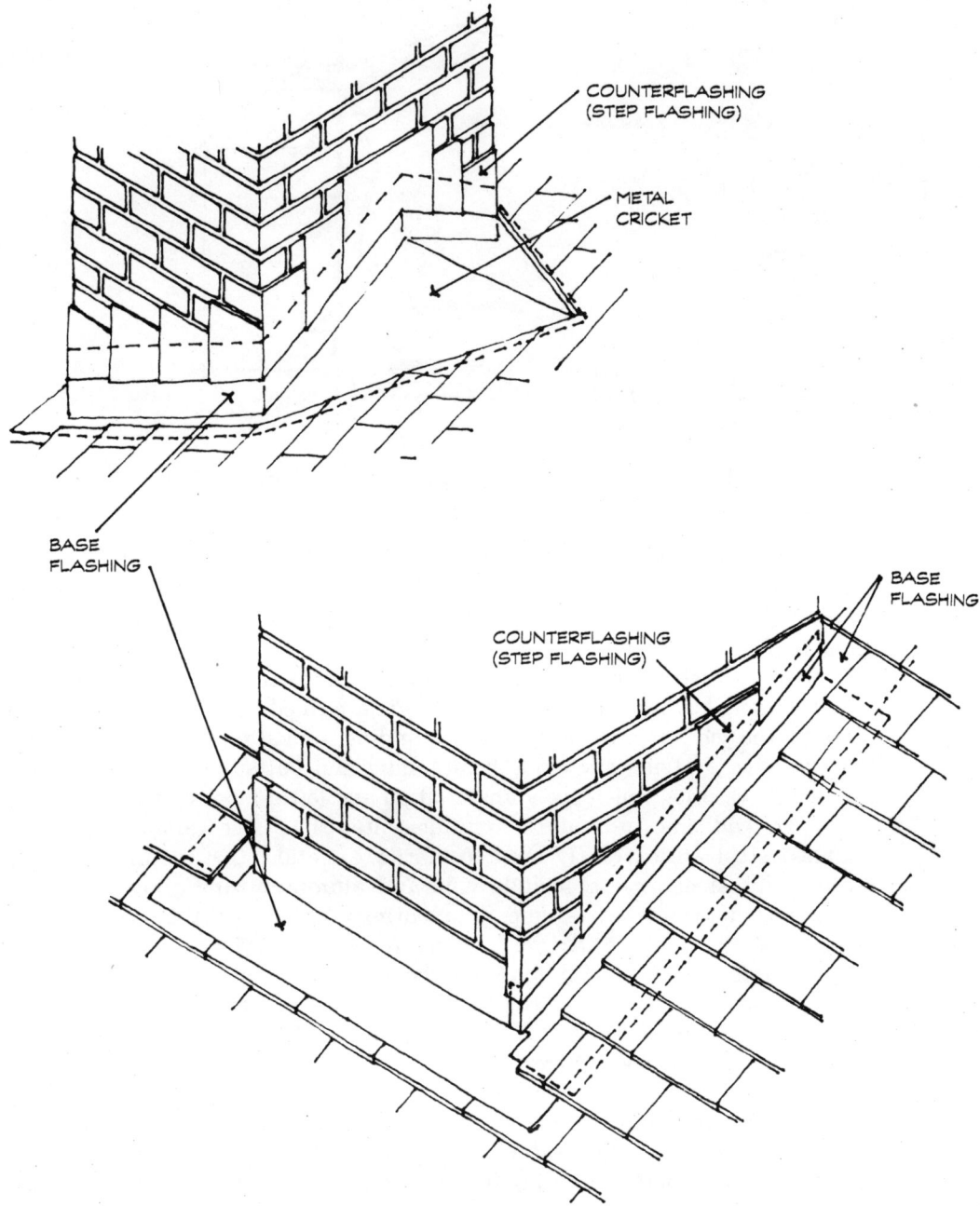

Figure 6.18 Chimney flashing.

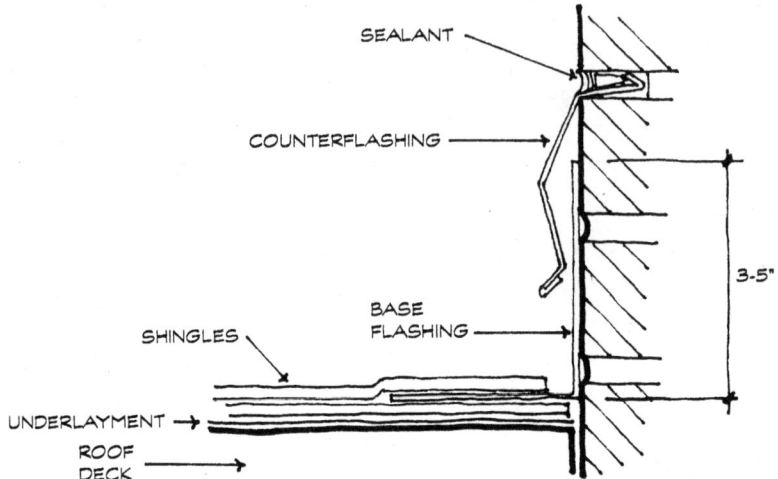

Figure 6.19 Flashing at shingle roofing should turn up 3 to 5 in. at edges. (*From NRCA, Roofing and Waterproofing Manual, 4th ed.*)

Contemporary metal roofing includes shingles designed to simulate the appearance of wood, tile, or slate roofing, but the term *metal roofing* most often refers to long panels joined together by one of several seaming methods. Metal panel roofing is typically attached with concealed clips, and may be either self-supporting structural sheets designed to span between supports, or lighter weight panels for use over solid decking. Field-formed roofing is fabricated in place at the job site with either flat, standing, or batten seams (*Fig. 6.20*). Prefabricated systems are made in standing and batten seam designs with snap-on accessories which speed installation and lower labor costs.

Copper, lead, zinc, aluminum, metallic-coated sheet steel, and terne-coated stainless steel are currently used to fabricate metal panel roofing. Metallic coatings for steel roofing include zinc and aluminum-zinc alloys patented under the trade name Galvalume. Zinc coatings should meet the requirements of ASTM A525 *Specification for General Requirements for Steel Sheet, Zinc-Coated (Galvanized) by the Hot-Dip Process* with a G90 coating weight. An aluminum-zinc alloy coating provides the corrosion resistance and heat reflectivity of aluminum coatings with the formability and galvanic protection at cut edges of zinc coatings. Copper and lead roofing are typically field-formed, while metallic-coated steel is used in the majority of prefabricated metal roofing systems, with either exposed metal, paint finish, or laminate coatings. Paint finishes include polyesters, siliconized polyesters, and fluorocarbons applied over the metallic coating in a coil coating process at the manufacturing plant. Polyester paints are relatively hard and abrasion-resistant. Siliconized polyesters are durable, chalk-resistant, and have good gloss retention. Fluorocarbons are also very durable, heat- and chalk-resistant, and have good color retention. Laminates are thick plastic films adhesively bonded under pressure and heat in a coil coating process. Plastic laminates prevent chalking, fading, and peeling. Both acrylic and fluorocarbon laminates are commonly used in roofing applications.

There are many different ways to describe metal roofing. The most common terms are *structural* and *architectural* roofing. Structural metal roofs are capable of spanning between framing members and carrying imposed wind and snow loads without a supporting deck. Architectural metal roofs are generally of lighter-gauge metals and installed over solid decks. Architectural roofs are typically prominent design features installed at steep slopes with high visibility, while structural metal roofs are often installed at low slopes with little or no visibility to passers by. Other terms used to describe metal roofing include *hydrostatic* and *hydrokinetic* systems. A hydrostatic roof uses waterproof seaming and flashing details which can withstand the hydrostatic pressure of ponded water at slopes as low as $1/_4$ in./ft. A hydrokinetic roof uses water shedding rather than waterproof detailing and requires steeper slopes to drain water quickly from the roof surface. Because structural metal roofs are usually installed at low slopes with hydrostatic details, many in the roofing industry use the terms *structural* and *hydrostatic* interchangeably. Similarly, since architectural metal roofs are usually installed at steep slopes with hydrokinetic details, the terms *architectural* and *hydrokinetic* are often used interchangeably. However, a structural metal roof is not hydrostatic unless it is specifically designed to be so, and an architectural roof can be made hydrostatic with proper detailing. Structural roofs can be and are often installed at steep slopes with hydrokinetic details, and architectural roofs can be and are often installed at lower slopes with hydrostatic details. Structural capability and water resistance are two different performance characteristics that should be identified separately.

Structural metal roofs are typically designed to span about 5 ft between roof purlins or joists, and the panels are attached with concealed clips. Because of their strength and economy, structural metal roofs of metallic-coated sheet steel are frequently used in large industrial and commercial structures and in pre-engineered metal building systems. Corrugations of flat or trapezoidal profile add strength and stiffness to the panels, which are fabricated in widths ranging from 10 to 30 in. Prefabricated lengths are limited to 45 ft because of highway transportation restrictions, but panels can be jobsite-fabricated in lengths up to 200 ft. Structural metal roofing panels are interlocked along the

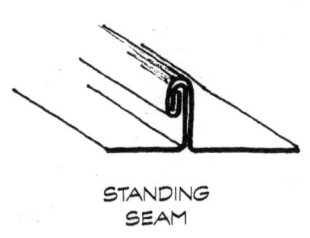
STANDING SEAM

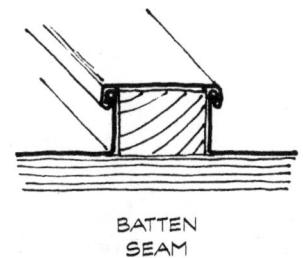
BATTEN SEAM

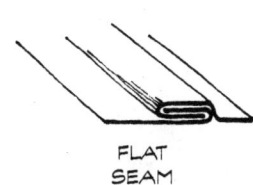
FLAT SEAM

Figure 6.20 Seams for metal roofing.

longitudinal seams either with a portable lock forming machine, with T seams held together by a cap strip crimped to the seam with a portable crimper, or with a snap-together seam. Structural metal roofs can be fabricated to meet UL-90 wind uplift resistance requirements. Wind uplift performance is governed by the yield strength of the panel, its thickness and width, clip design, fastener type and spacing, rib height and configuration, and the yield strength, gauge, and spacing of secondary structural framing. ASTM E1514 *Standard Specification for Structural Standing Seam Steel Roof Panel Systems* covers performance requirements, materials, anchorage, and the sealing of side and end laps.

Architectural metal roofs are typically installed at slopes of 3:12 or greater. Panels are of lighter-gauge steel than structural metal roofs, are fabricated in shorter lengths and in widths of 10 to 24 in., and are flat rather than ribbed. Because the panels are thinner and do not have the added strength of corrugations, architectural metal roofs are applied over a solid plywood deck or metal deck. Plywood decks should be at least $5/8$ in. thick, and screws rather than nails should be used to attach fastener clips because they have greater pullout strength. A 30-lb roofing felt or other underlayment is installed between the roofing and the deck to provide a redundant or backup water shedding layer, and to isolate the metal panels from the corrosive effects of chemically treated plywood. Panels are joined at the side laps with snap-together or batten seams or by mechanical interlocking. Batten seams can be either traditional box battens, cap battens, or integral battens.

All metals expand and contract at relatively high rates when exposed to changes in temperature. A 60-ft steel panel subjected to a temperature change of 100°F will expand approximately $1/2$ in. The length of a 100-ft-long steel roofing panel can change over 1 in. from summer to winter. Aluminum and copper panels will expand even more. The clips and fasteners used to attach metal roofs to the deck or the structural frame must be designed to accommodate this thermal expansion and contraction without inducing stress in either the panels or the attachment assembly itself. Two-piece clips have sliding tabs, and one-piece clips allow the panels to slide over the clip.

Metal roofs can be fixed (i.e., through-fastened to the roof structure) at the ridge and allowed to expand and contract at the eaves, or fixed at the eaves and allowed to expand and contract at the ridge (*Fig. 6.21*). On very long panels, it may be desirable to fix the roof near the midpoint of the slope so that thermal expansion and contraction are divided evenly between the eave and ridge details. When installed at low slopes, the panels are usually fixed at the eave so that the eave trim and fascia can be detailed to resist the hydrostatic pressure of ponded water. When fixed at the eave, the roof "floats" by expanding and contracting at the ridge line. Where through-fastening is required, fasteners should incorporate rubber gaskets, and tape sealants should be placed between the panel and the deck or frame connection to seal the penetrations. When installed at steeper slopes, panels are usually fixed at the ridge with exposed fasteners or concealed fasteners under the ridge trim, and allowed to expand and contract at the eaves. Because of the steeper slope, ponding water

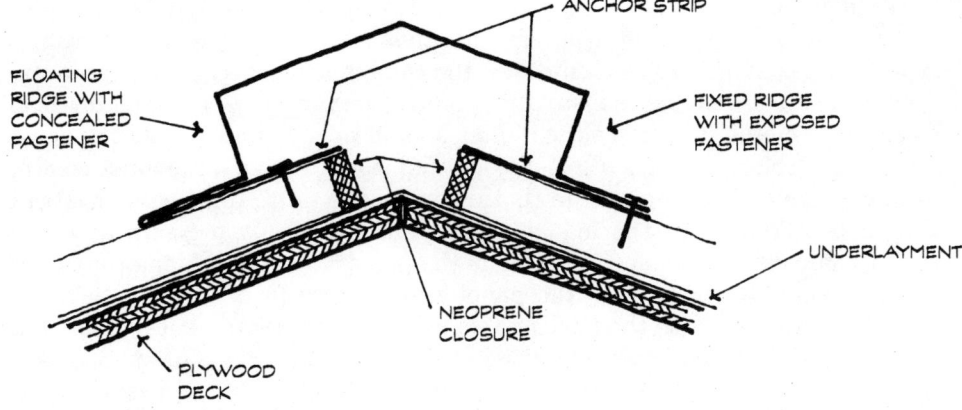

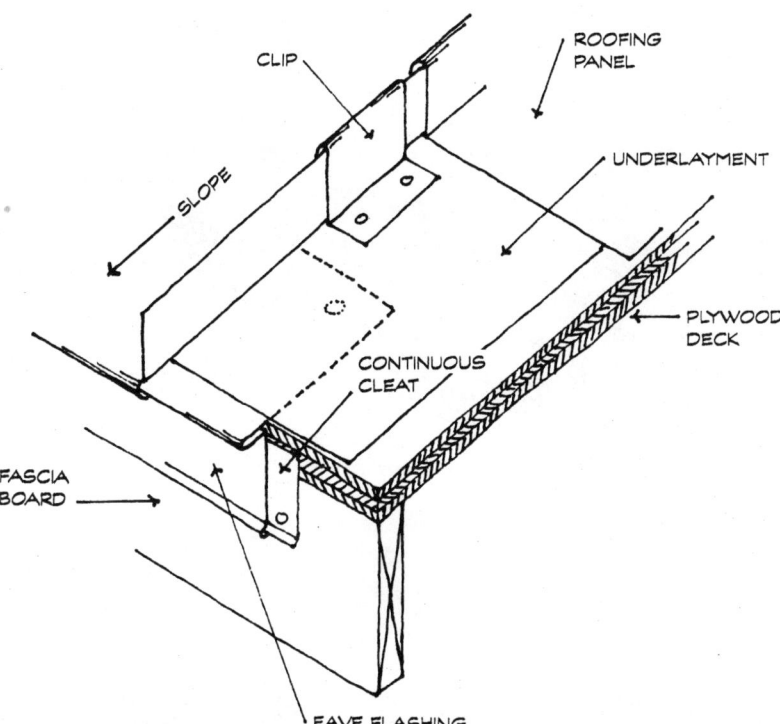

Figure 6.21 Allowance for expansion and contraction in metal roofing attachment. (*From NRCA, Roofing and Waterproofing Manual, 4th ed.*)

is less likely, so hydrostatic detailing is not critical at the eaves. A flexible ridge cap is attached to the metal panels on either side, and flexes as the panels expand and contract. In cold climates, the combination of expansion and contraction at the eaves and the possibility of ice damming presents a high risk of leakage, so metal roofs in cold climates should always be fixed at the eaves rather than the ridge regardless of slope. Gable end rake details must be able to accommodate movement in both fixed-ridge and fixed-eave systems (*Fig. 6.22*). Transverse thermal movements are typically accommodated by trapezoidal ribs or by a slight bowing of the panel (*Fig. 6.23*). Along rakes and parapets that run parallel to the panel seams, fixed flashings and terminations must interact with the floating roof panels whose movement increases as the distance from their point of fixity increases. The moving joint at such locations should be between the counterflashing and base flashing, several inches above the drainage plane, but this may make it difficult to secure the roof against wind uplift at these same locations (*Fig. 6.24*). At curbs and equipment penetrations, detailing should also permit thermal expansion and contraction (*Fig. 6.25*).

An aesthetic problem related to the expansion and contraction of metal roofs is "oilcanning," or the distortion of flat panels. Oilcanning is impossible to eliminate because of the nature of the materials, but it can be minimized by proper attachment of the panels to permit unrestricted expansion and contraction. Tightly crimped seams can sometimes bind at the attaching clips,

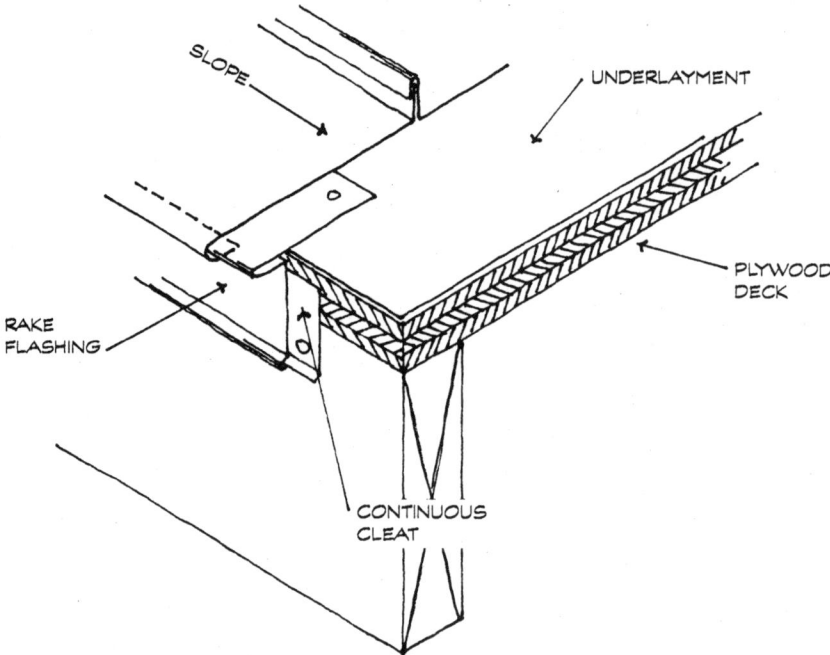

Figure 6.22 Rake flashing to allow expansion and contraction. (*From NRCA, Roofing and Waterproofing Manual, 4th ed.*)

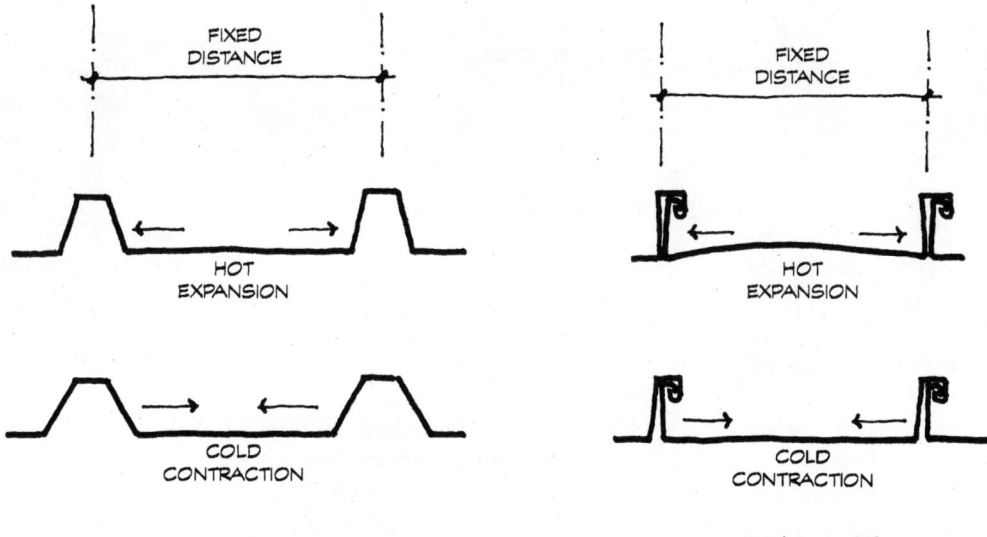

Figure 6.23 Rib shape accommodates movement. (*From Tobiasson and Buska, Snow, Ice and Standing Seam Roofs.*)

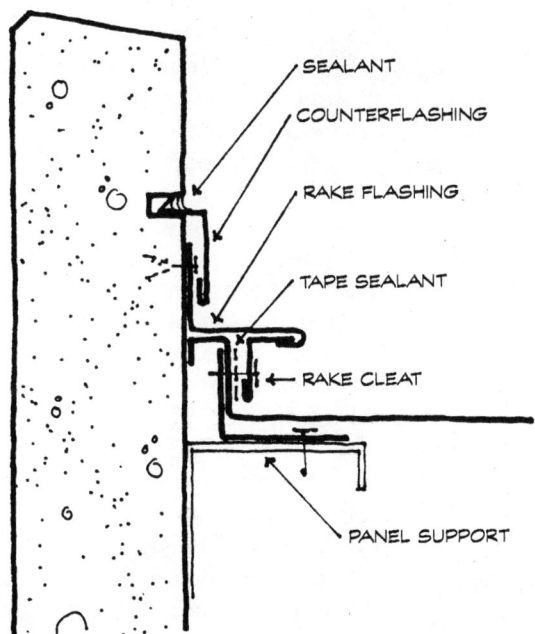

Figure 6.24 Flashing at edge of floating roof panel. (*From Scharff, Roofing Handbook.*)

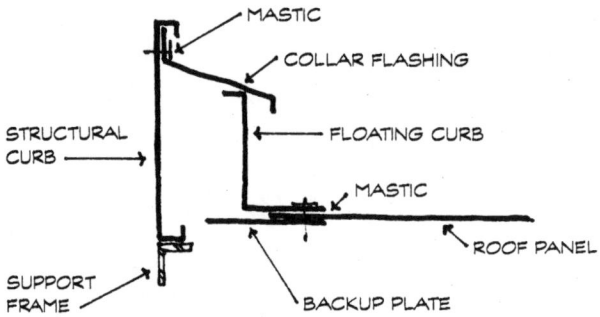

Figure 6.25 Equipment curb flashing at floating roof panel. (*From Scharff, Roofing Handbook.*)

causing some distortion, and tightly nested pans will always distort. Clips and attachments should serve as hold-down devices without restricting movement. To minimize the visual effects of oilcanning, use ribbed or textured panels rather than smooth flat panels, and specify matte rather than glossy finishes.

Low-slope roof panel seams must be sealed to resist hydrostatic pressure when submerged in ponded water. Hydrostatic side lap seams are made waterproof with high-performance continuous gasket seals applied either during the manufacturing process or on site. The most widely used sealants are non-hardening, non-skinning polyisobutylene in either tape form or gun grade. Tape seals are preferred for field installations because of their higher solids content and dimensional stability. Butyl sealants can be used only in concealed locations because they are susceptible to ultraviolet deterioration. Mechanical or snap-together side lap seaming compresses the gasket forming a continuous seal along the rib. End laps are sealed with strips of butyl tape or sealant placed between the two panel surfaces and screwed together to form a compression seal. The metal roofing industry recommends a minimum $1/4$ in./ft slope for hydrostatically detailed roofs. In cold regions, some researchers recommend a minimum slope of 1 in./ft. Hydrostatic roofs should be flood-tested with ponded water by submerging the seams and end laps to a water depth of 3 in. or more above the seam and 6 in. or more above the drainage surface of the panel. End laps, flashing, curbs, and penetrations must also be designed to resist hydrostatic pressure and to maintain their integrity under thermal movement. Hydrokinetic details are sometimes used at the parapets of low-slope metal roofing, but these details must be at least 6 in. above the panel drainage plane. Hydrokinetic details may also be used at the ridge where ponded water is unlikely.

Steep-slope roof panel seams are usually designed as water-shedding, but not waterproof, and do not include continuous gasket seals. Different hydrokinetic seaming details have differing degrees of water resistance, which is measured by ASTM E331 water infiltration tests. The effects of wind-driven rain are simulated by applying a negative pressure below the panel seam or joint, and applying a water spray on top. Where two pieces of sheet metal are overlapped without sealing, there is the possibility of capillary suction. The tighter the lap, the greater the capillary force. Lap joints can be made watertight either by installing a continuous sealant (*Fig. 6.26A*) or by providing a capillary break (*Fig. 6.26B*). Capillary breaks are more common in handcrafted metal roofs, and sealant joints are more common in prefabricated systems. Underlayments used in hydrokinetic roofs should be laid *over* metal eave trim to assure that water drains out of the system (*Fig. 6.21*).

The minimum slope required for acceptable performance with water shedding hydrokinetic details depends on several factors, including roof geometry, extent and type of underlayment, climatic exposure, orientation with respect to solar radiation and snow melting patterns, and the consequence of occasional minor infiltration. Ice damming is possible not only at uninsulated eave overhangs, but also at dormers and valleys. Changing to hydrostatic seams in these high-risk areas will provide some safeguards, as will installing water-

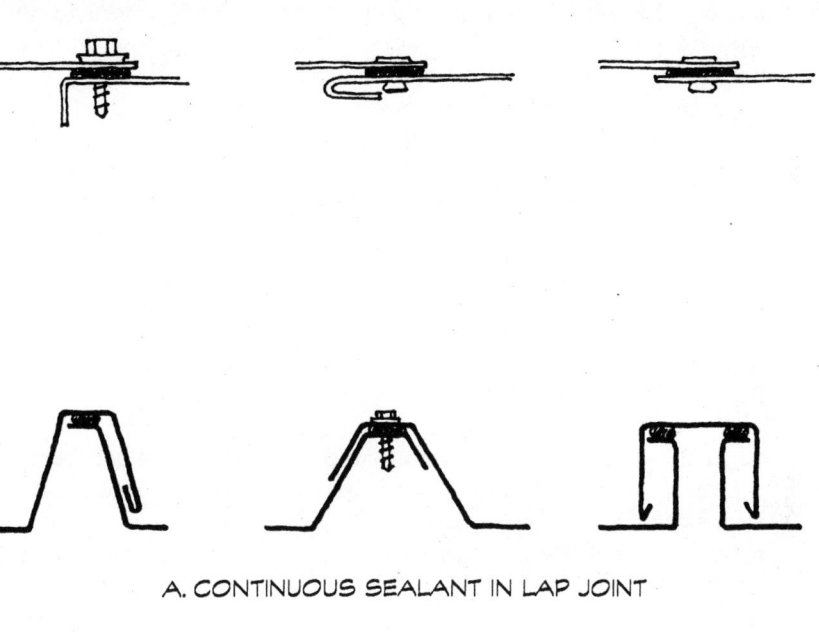

A. CONTINUOUS SEALANT IN LAP JOINT

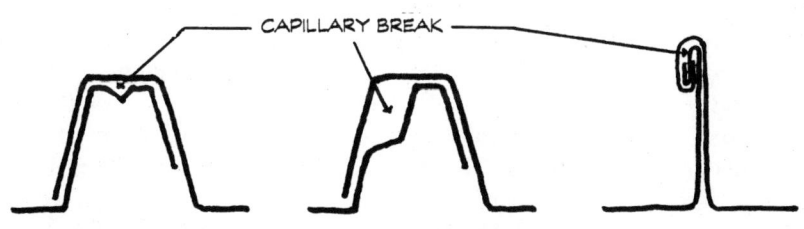

B. CAPILLARY BREAK IN LAP JOINT

Figure 6.26 Watertight joints using sealant or capillary break. (*From CSI Monograph 07M611.*)

proof underlayments such as self-adhering rubberized asphalt membranes. Designs which avoid complicated geometry, cold eaves, and other potential ponding areas pose fewer risks of water penetration. If persistent water penetration keeps underlayment materials saturated for long periods, wood decking can rot and roof panels can corrode from the inside out. Since there is no universal rule of thumb on minimum slope for architectural metal roofs, each project should be evaluated individually, and the manufacturer's recommendations considered regarding infiltration characteristics of the panel systems's seams and flashing details.

It is difficult to create a waterproof detail where gutters that are free to expand and contract are joined to the eave of a metal roof, and often no attempt is made to seal between the gutter and the metal panels. Where water flow is unimpeded, problems may not develop, but if the gutter is blocked by leaves, ice, or other obstructions, leakage is likely. For this reason, gutters are often omitted, especially in cold climates. Eave overhangs should be at least 6 in. and preferably 12 in. to avoid wetting the building walls from roof runoff or icicle formation. Fascias should be fabricated with a drip edge along the bottom.

The SMACNA *Architectural Sheet Metal Manual* shows basic detailing for field-formed batten seam and standing seam metal roofs, as reproduced in *Figs. 6.27* and *6.28*. For more in-depth information, the NRCA *Roofing and Waterproofing Manual* contains a complete chapter on metal roofing and recommended details.

Thermal bridging and condensation are common problems in metal roofs. Where structural metal roofs are installed over batt insulation, thermal bridges occur where the batts are compressed at panel attachment to the purlins. Rigid insulation blocks should be used to increase thermal resistance at these locations (*Fig. 6.29*). The blocks help keep metal purlins warmer in winter and can prevent interior condensation on the cold metal surfaces.

Sliding snow presents a hazard to pedestrians around metal-roofed buildings because the metal surface does not provide any frictional resistance. Proprietary snow guards can be used to hold the snow in place until it is melted by warmer weather. It is best to use several rows of snow guards spaced well apart rather than a single row of guards. Keep the lowest row of snow guards upslope of a vertical line 1 ft from the inside wall surface (*Fig. 6.30*). Mechanical attachment is more reliable than adhesive attachment. Adhesively attached snow guards have failure loads ranging from 400 to 700 lb per unit, while mechanically attached guards have failure loads up to 4000 lb. Snow guards that are attached by fasteners which penetrate the roof restrict the natural expansion and contraction of the metal, and are likely sources of leaks. To minimize the risk of leaks, a combination of silicone bedding sealant and neoprene washers can be used. Snow guards should not be installed on cold overhangs because they can exacerbate the formation of ice dams.

Exposed fasteners can be used to attach metal roofs, but even with neoprene washers, the many penetrations required to attach a roof provide opportunities for leaks. If the fasteners are not tight enough, thermal expansion and contraction of the metal panel relative to its support can enlarge the hole or shear off

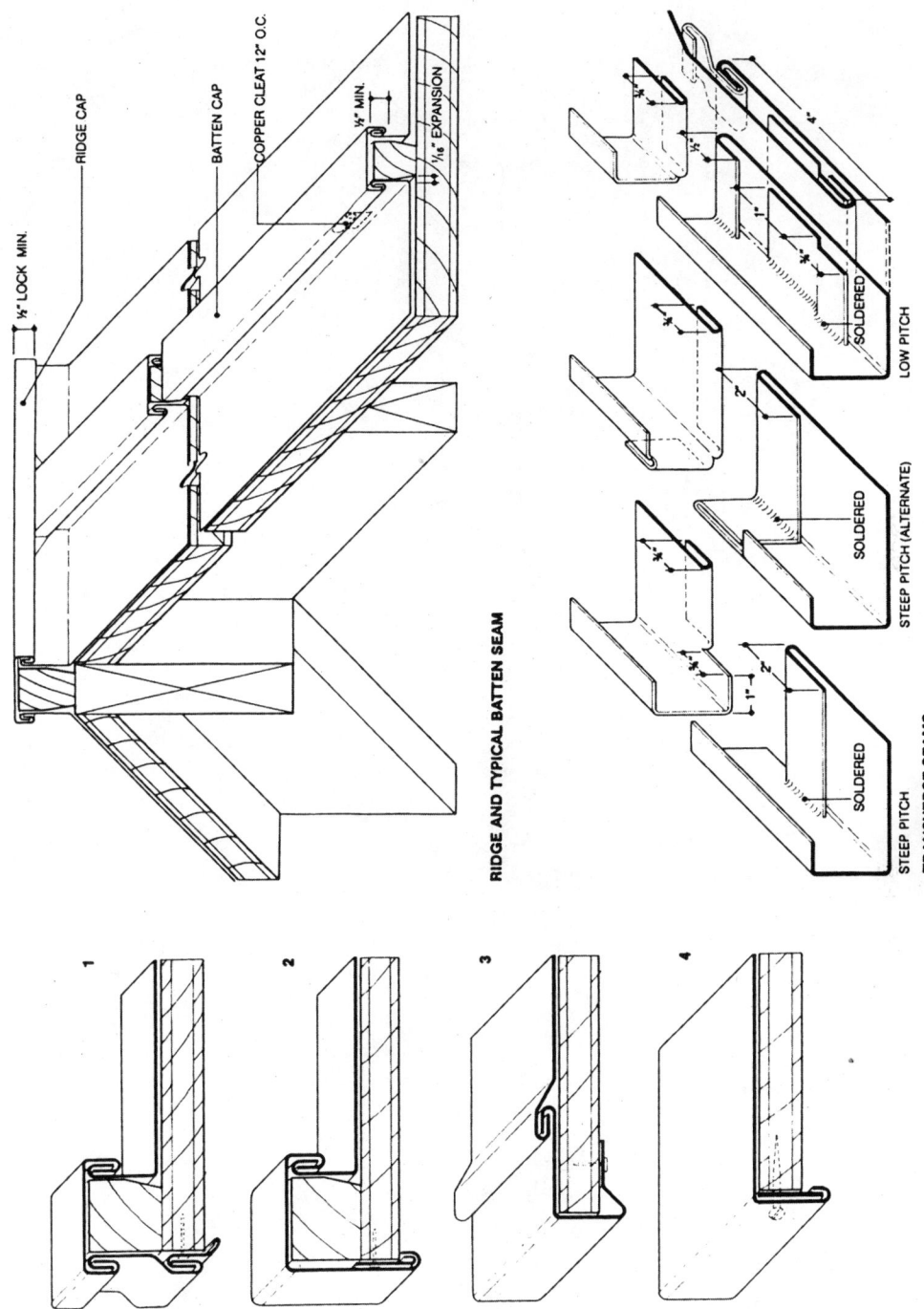

Figure 6.27 Batten seam roofing details. (*From Copper Development Association, Sheet Copper Applications. Used with permission.*)

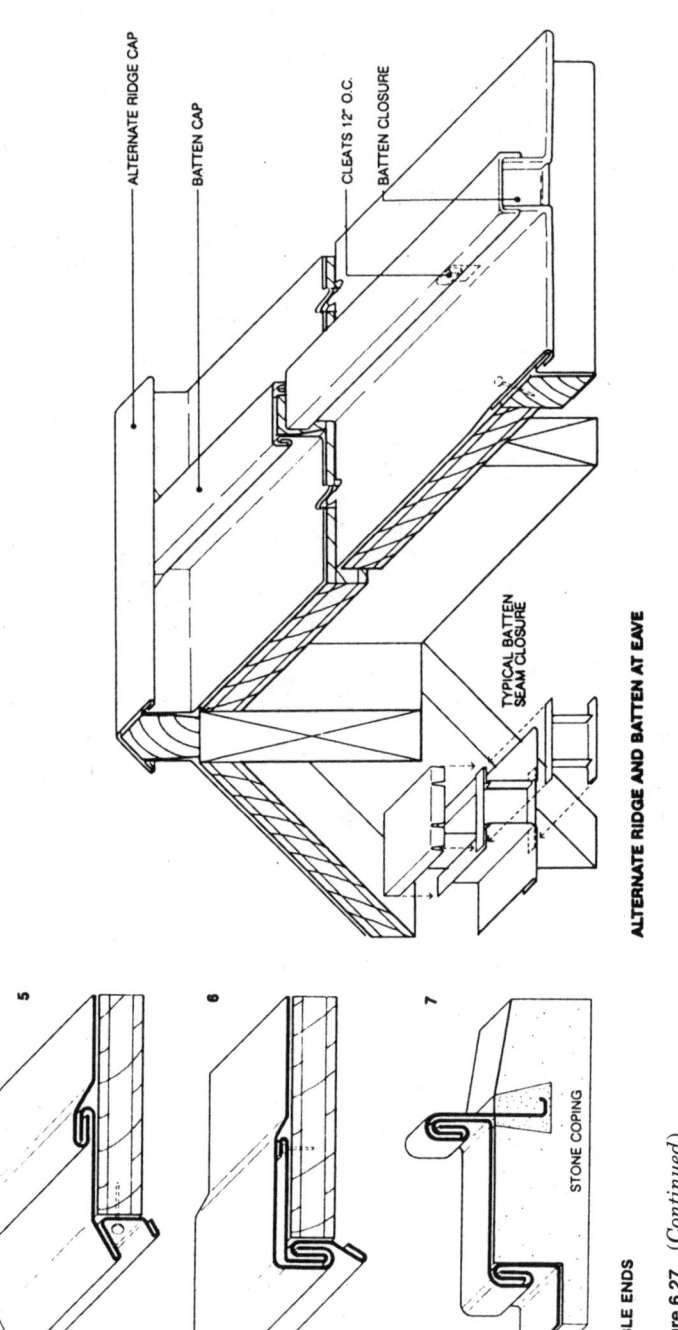

GABLE ENDS

Figure 6.27 (*Continued*)

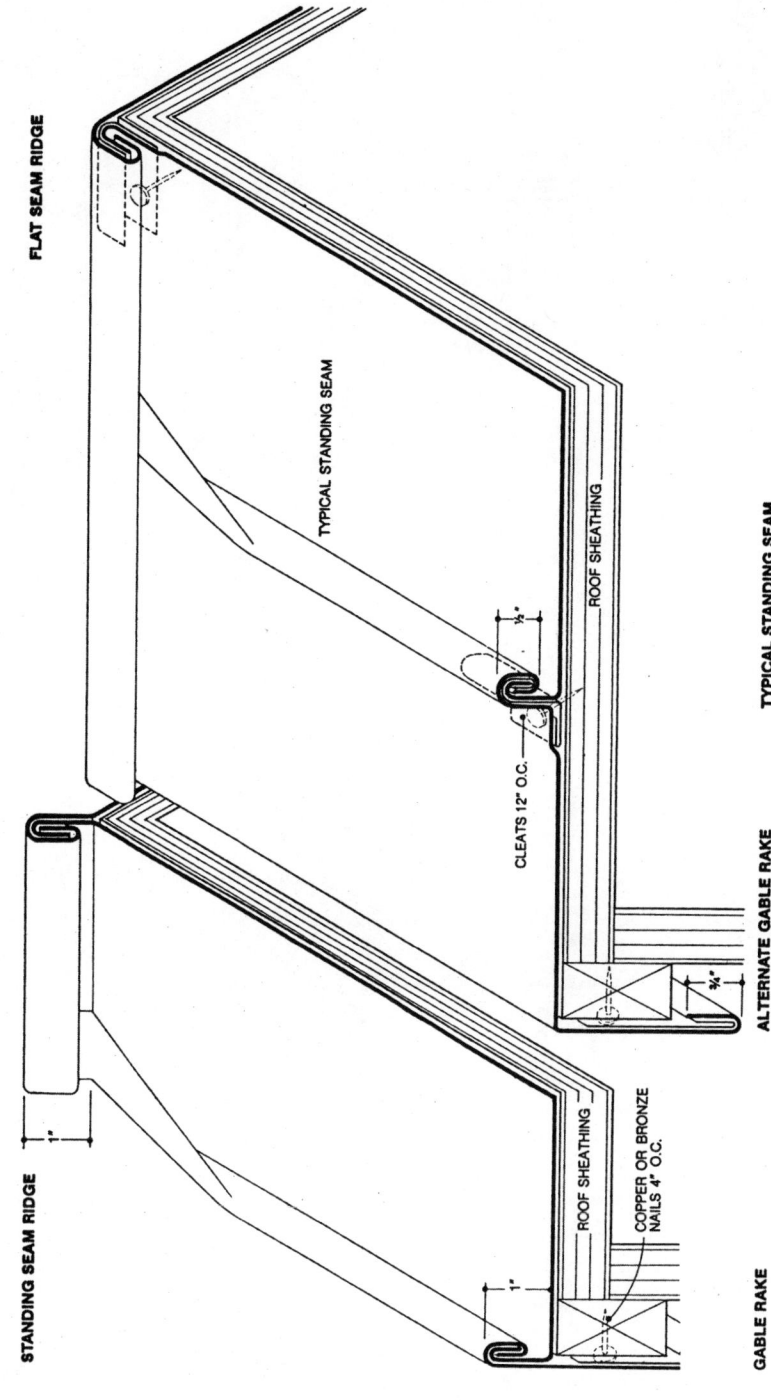

Figure 6.28 Standing seam roofing details. (*From Copper Development Association, Sheet Copper Applications. Used with permission.*)

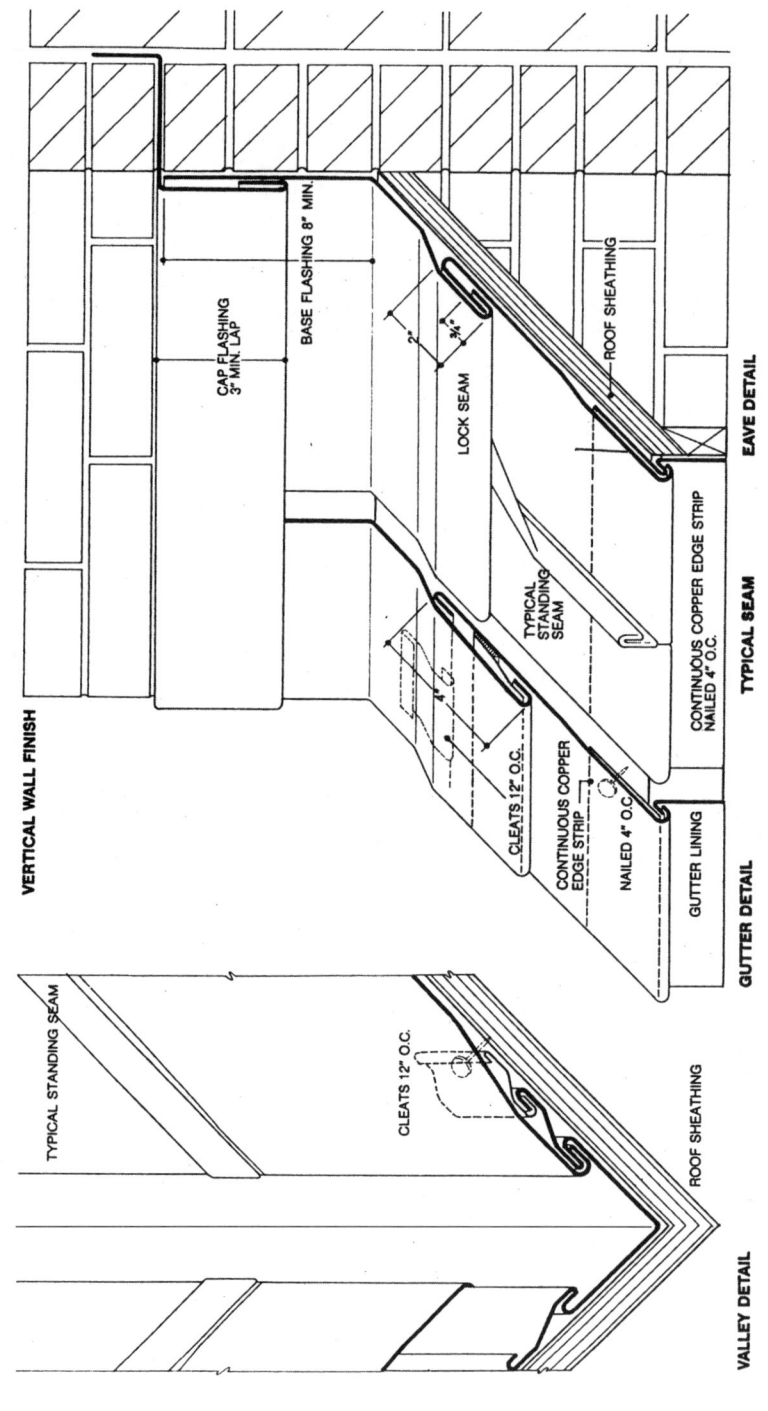

Figure 6.28 (*Continued*)

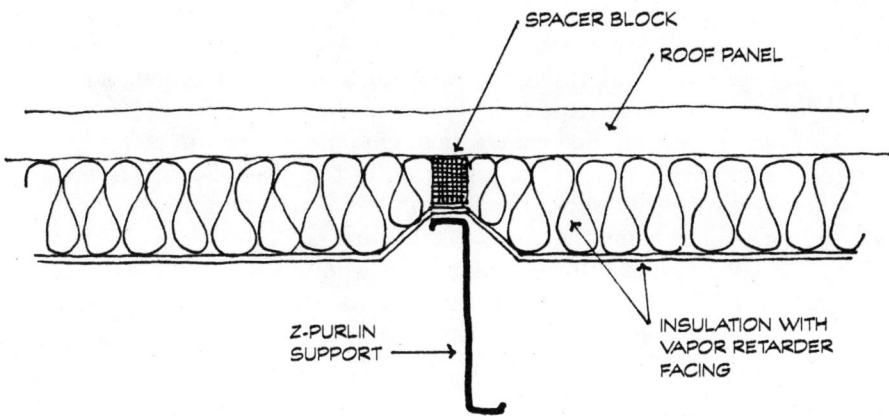

Figure 6.29 Rigid insulation blocks at metal roofing. (*From NRCA, Roofing and Waterproofing Manual, 4th ed.*)

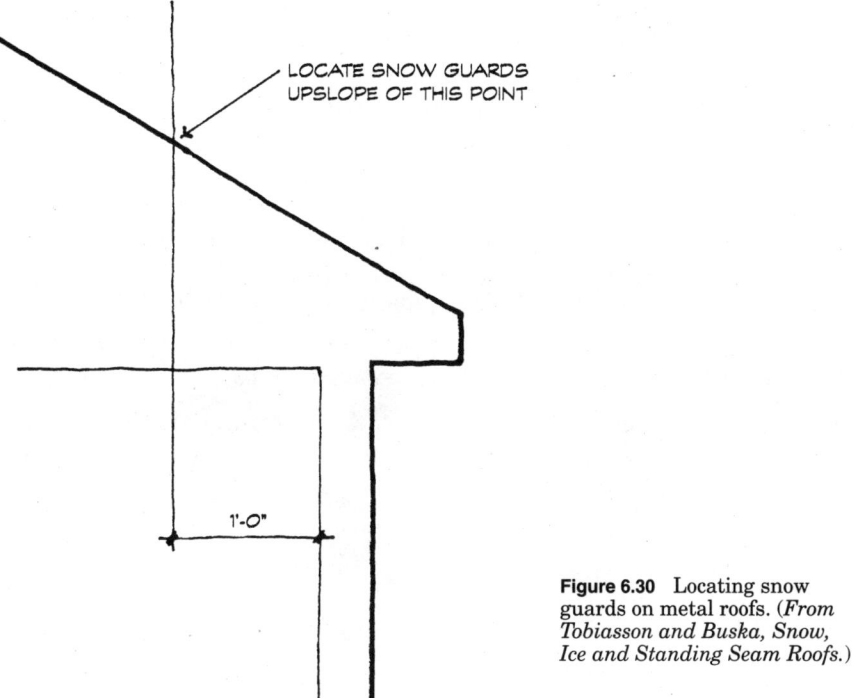

Figure 6.30 Locating snow guards on metal roofs. (*From Tobiasson and Buska, Snow, Ice and Standing Seam Roofs.*)

the fastener. If the fasteners are too tight, the neoprene washer can be squeezed out of the hole. When properly tightened to prevent panel movement relative to the supporting frame, thermal expansion and contraction is accommodated by the flexing of light-gauge metal purlins. If the purlins are inadvertently braced to prevent this flexing, or if the attachment is to a rigid support, repeated expansion and contraction will eventually enlarge the hole or shear off the fastener. Concealed fasteners offer better protection against water infiltration. Two-piece clips are used for long panel sections in which large expansion and contraction movements are expected. One-piece clips are used for shorter panel lengths with less expansion and contraction. One-piece clips can be designed so that one leg attaches rigidly to the panel while the other leg slides along the supporting purlin (purlin slip), or with one leg fastened to the substrate while the other leg slides along the panel (panel slip).

Combining different metals in a metal roofing system can cause galvanic corrosion. To avoid galvanic reactions, provide isolation between *all* dissimilar metals. Fasteners and clips should be of corrosion-resistant materials that are compatible with the metal panels (*Table 6.3*).

6.4 Low-Slope Roofing

A variety of low-slope roofing systems are used in conventional and protected membrane systems, differing both in type of membrane and method of fastening or securing the components to the structural deck and to one another. *Table 6.4* lists various types of roofing membranes. Only built-up roofing and the most commonly used single-ply membranes will be discussed in this chapter. The successful, long-term performance of low-slope roofing systems requires that the membrane provide a waterproof barrier, withstand all weather exposures during its service life, and resist various stresses from internal or external causes during manufacture, installation and in service. *Table 6.5* summarizes the performance characteristics which must be achieved in membrane roofing during manufacture and installation and in service.

6.4.1 Conventional and protected membrane roofing systems

In conventional roofing systems, the waterproofing membrane is outside the insulating layer and exposed to the elements. A conventional roofing assembly consists of five basic components.

- Structural deck
- Vapor retarder (where needed)
- Insulation
- Membrane and flashing
- Surfacing or ballast (where needed)

In addition to the basic components, the system includes adhesives and fasteners to secure the components to the deck and to one another (*Fig. 6.31A*). The first layer of insulation on a conventional roof can be adhered to the structural deck with bitumen or mechanically fastened. Mechanical attachment is less affected by adverse weather conditions and applicator error, and eliminates bitumen dripping through metal decks during application. Subsequent

TABLE 6.3 Fastener Compatibility in Metal Roofing

Metal	Fastener compatibility	Corrosion resistance/durability
Aluminum	Stainless steel or aluminum alloy	Wet materials containing lime, concrete, or other masonry materials will corrode aluminum. Lead and run-off from lead will corrode aluminum.
Copper	Copper, brass, or stainless steel	Limestone, stucco, concrete, and other light-colored porous materials can show staining from moisture runoff from copper. Iron, galvanized metal, and acidic solutions from some trees can stain copper. Acid leaching from cedar roofing can impede patina development and in severe cases cause localized thinning of the copper. Bitumen and fire-treated woods containing salt are corrosive to copper. Copper should not be in contact with steel or galvanized steel.
Lead-coated copper	Copper, brass, or stainless steel	Used to avoid staining associated with copper. However, lead oxide can cause staining (see "Lead" below).
Lead	Copper, brass, or stainless steel	Inert, atmospheric corrosion has little effect. Lead oxide can stain glass, stainless steel, and other materials. Lead is attacked by free lime (found in fresh concrete). Lead can be stained by rust from steel.
Steel:		
Galvanized	Galvanized or stainless steel	Do not use in conjunction with copper. Do not use in harsh, corrosive environments.
Aluminized		Do not use in conjunction with copper. Do not use in harsh, corrosive environments.
Galvalume	Stainless steel, aluminum, or Zn/Al alloy	Do not use in conjunction with copper. Do not use in harsh, corrosive environments.
Stainless steel	Stainless steel	Considered hygienic. Does not stain adjacent surfaces.
Terne metal (terne-coated carbon steel)	Stainless steel or galvanized	Do not nail through metal; use cleats. Avoid contact with aluminum, copper, or acidic materials. Must be painted.
Terne-coated stainless steel (TCS)	Stainless steel	Does not stain.
Zinc	Galvanized or stainless steel	Not compatible with bituminous roofing materials. Copper generates a corrosion compound that attacks zinc. Sulfur dioxide inhibits the development of zinc's carbonate film. Wood preservatives can corrode zinc.

SOURCE: NRCA, *Roofing and Waterproofing Manual,* 4th ed.

TABLE 6.4 Types of Roofing Membranes

Classification	Membrane
Bituminous roofing	Built-up roofing
	Modified bituminous sheets
	SBS (styrene butadiene styrene) modified
	APP (atactic polypropylene) modified
	IPP (isotactic polypropylene)
Single-ply membranes	Thermoplastic sheets
	PVC and PVC blends
	EIP (ethylene interpolymer)
	CPA (copolymer alloys)
	Elastomeric (synthetic rubber) sheets
	Vulcanized
	EPDM (ethylene propylene diene monomer)
	Neoprene (polychloroprene)
	Non-vulcanized
	CSPE (chlorosulfonated polyethylene)
	CPE (chlorinated polyethylene)
	PIB (polyisobutylene)
	NBP (butadiene-acrylonitrile)
Cast in situ	Hot-applied rubberized asphalt
	Cold-applied liquid compounds (various polymeric and bituminous materials)
	Polyurethane foam roof with protective coating

SOURCE: NRCA, *Roofing and Waterproofing Manual,* 4th ed.

layers of insulation are typically adhered with bitumen to avoid continuous thermal bridges through mechanical fasteners. If a vapor retarder is needed to prevent moisture migration into the roof system, adhesive attachment is required to avoid puncturing the vapor retarder.

Conventional roofing systems protect the insulation from wetting and mechanical damage, which allows for a wide choice of insulating materials and, in theory, maintains the insulation material's thermal resistance. Conventional systems allow for easy inspection of the membrane and some do not require ballast, which reduces the overall weight that must be supported by the structural system. The membrane in a conventional roof, however, is exposed to temperature extremes. High temperatures accelerate aging and large temperature changes cause potentially damaging physical stresses in the membrane. In systems which incorporate a vapor retarder, moisture from air or water leaks can become trapped between the vapor retarder and the membrane and can accumulate over time (*Fig. 6.32*).

As an alternative to conventional roofing systems, insulation may be placed above the membrane in what is called a protected membrane roofing (PMR). A protected membrane roof assembly consists of four basic components.

- Structural deck
- Membrane and flashing
- Insulation
- Surfacing or ballast (where needed)

In addition to the basic components, the system includes adhesives and fasteners to secure the components to the deck and to one another (*Fig. 6.31B*).

The inverted location of the insulation and wearing surface in a protected membrane roof prevents mechanical damage to the membrane and provides

TABLE 6.5 **Performance Requirements of Roofing Membranes**

Requirement	During manufacture	During installation	In service
Tensile strength	•	•	•
Elongation		•	•
Crack bridging ability			•
Fatigue resistance			•
Thermal shock resistance	•		•
Tear resistance	•	•	•
Abrasion resistance			•
Lap joint integrity	•	•	•
Static puncture		•	•
Impact resistance		•	•
Low-temperature flexibility		•	•
Weatherability		•	•
Resistance to heat aging	•		
Dimensional stability	•	•	•
Granule imbedment			•
Interply adhesion		•	•
Membrane attachment		•	
Flashing attachment		•	
Materials compatibility	•	•	•
Wind uplift resistance			•

SOURCE: NRCA, *Roofing and Waterproofing Manual*, 4th ed.

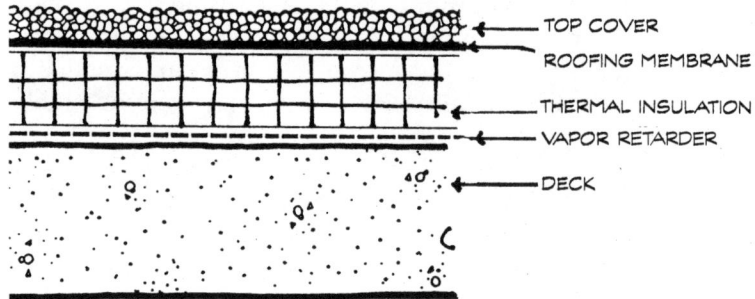

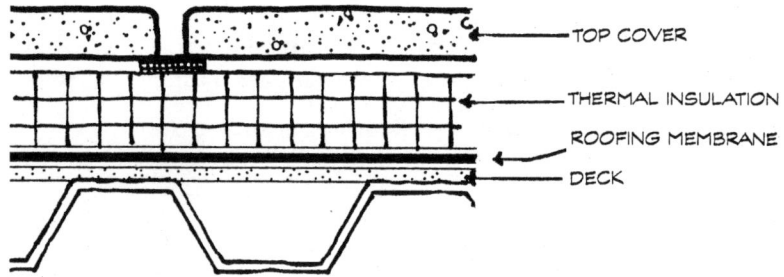

(a) Advantages

- Protects insulation against wetting and physical damage, which allows a broad selection of insulating materials and maintains thermal resistance.
- May not require ballast, depending on type of membrane, which reduces weight of system and structural support requirements.
- Allows easy inspection of membrane.

Disadvantages

- Exposes membrane to extreme temperatures, accelerating aging and increasing physical stress on membrane.
- Can trap moisture between vapor barrier and roof membrane.

(b) Advantages

- Protects membrane against physical damage.
- Maintains membrane at fairly constant temperature, reducing stress.
- Protects Membrane from UV radiation, slowing aging.
- Reduces risk of trapping condensation in insulation.
- Can serve as plaza deck waterproofing.

Disadvantages

- Insulation is exposed to water from rain and melting snow, reducing material options.
- Ballast is required to hold insulation in place increasing weight of system and structural support requirements.
- Difficult to inspect membrane.

Figure 6.31 Conventional versus protected membrane roofing. (*From National Research Council of Canada, Roofs That Work.*)

the option of using the roof for recreational purposes such as an observation deck or roof terrace. The membrane is maintained at fairly constant temperatures and protected from ultraviolet exposure and accelerated aging. Protected membrane roofs allow membrane installation directly on a concrete deck which is a more stable base than insulation. PMR systems never need separate vapor retarders because the topside location of the insulation eliminates the danger of condensation by keeping the membrane temperature above the dew point.

In a protected membrane roof, the insulation must be able to maintain its thermal resistance, compressive strength, and dimensional stability even though it is continually exposed to water. Extruded polystyrene meets these requirements and is the insulation of choice for PMRs. It has a low rate of

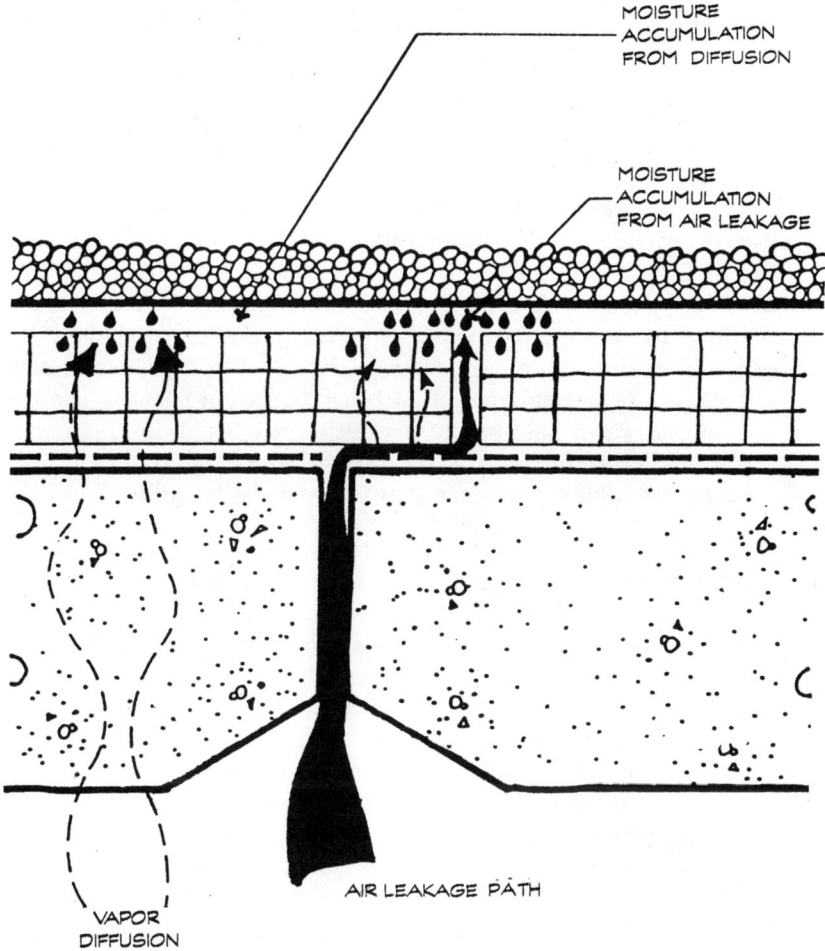

Figure 6.32 Air and vapor movement into roof system. (*From National Research Council of Canada, Roofs That Work.*)

water absorption and does not degrade in the presence of moisture. The insulation may be held in place against wind uplift with a layer of gravel ballast over porous fabric, or with pavers. Some polystyrene is fabricated with interlocking edges and an integral concrete surface to increase its weight and uplift resistance and eliminate the need for separate ballast. When PMRs are installed over metal decks, an additional layer of insulation is placed over the deck flutes to provide a flat, smooth substrate for membrane attachment. The optimum relative insulation thickness above and below the membrane should be analyzed to ensure that the membrane temperature is kept above the dew point (see Chap. 5 for method of condensation analysis). A good rule of thumb is to place no more than one-third of the total roof system insulation R-value below the membrane. Placing a protected membrane between two layers of insulation increases the temperature range to which it is exposed and accelerates aging. Protected membrane roofs should not be used over cold overhangs because ice can form beneath the top layer of insulation and displace it.

6.4.2 Condensation in conventional low-slope roofs

Condensation in conventional low-slope roofs is controlled either by ventilating the interior space or the roof cavity (if one exists) or both, or by the incorporation of vapor retarders. A ventilated low-slope roof, or "cold roof," uses joist cavities to provide air circulation between the insulation layer and the bottom of the roof deck (*Fig. 6.33A*). In a compact roof or "warm roof," the insulation is above the deck, there is no roof cavity to ventilate, and a vapor retarder may be required in cold and temperate climates to prevent winter infiltration of moisture vapor into the roofing system (*Fig. 6.33B*). A protected membrane roof is not subject to the dangers of such condensation because the membrane is on the winter warm side of the insulation. In hot, humid climates, a protected membrane is on the cold side of the insulation in summer, but its low vapor permeance and high water resistance prevent condensation and leakage problems.

Ventilated roofs. Ventilated roofs are sometimes referred to as *framed roofs* or *cold roofs* because the insulation is below the roof deck, between the framing members. Low-slope assemblies are much more difficult to ventilate effectively than steep roofs where natural ventilation is maximized by the stack effect of rising warm air. Without that stack effect, even low-slope roofs with continuous soffit vents can harbor sufficient moisture vapor to pose condensation problems. The addition of power, turbine, or gravity vents is not practical because each of the isolated joist cavities typical in low-slope roofs would require a separate vent. In a design that uses deeper joists or thinner insulation to create larger cavities, or one that uses open truss joists, air movement and moisture vapor removal will be more effective. As an added measure of protection against condensation, ventilated low-slope roof assemblies in cold climates may require an interior vapor retarder or air barrier to minimize the amount of indoor moisture infiltrating the roof cavity.

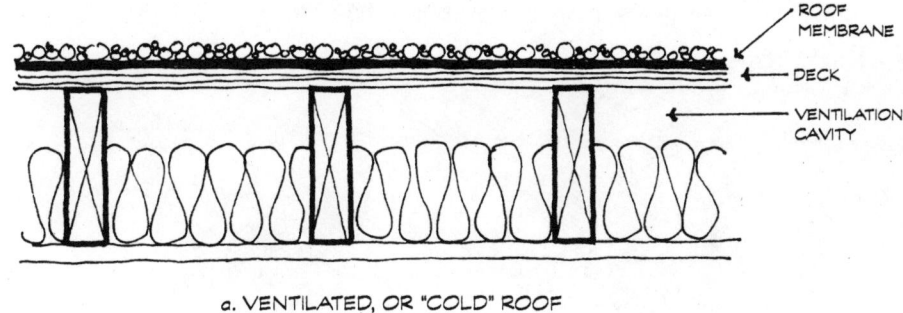

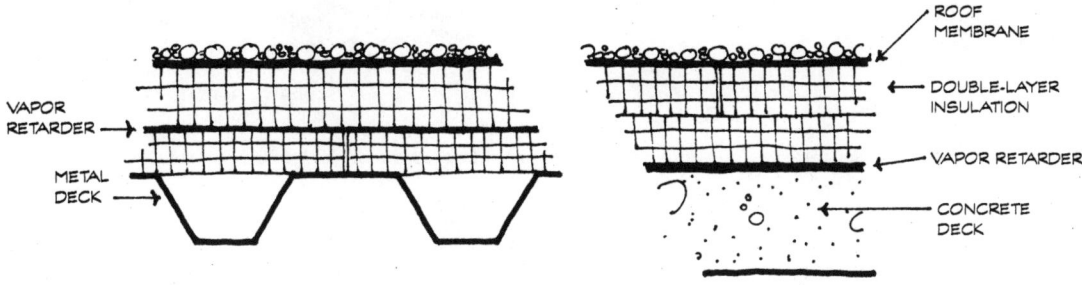

Figure 6.33 Details of "cold" and "warm" roofs. (*From NRCA, Roofing and Waterproofing Manual, 4th ed.*)

Ventilation of low-slope roofs with attic or joist spaces should not be confused with "venting" of compact roofs. One-way relief vents in compact roof assemblies were once thought to aid in preventing condensation, evaporating condensed moisture, and drying wet insulation, but they have been proven ineffective and are no longer considered a part of good roof design.

Vapor retarders. Industry recommendations for the use of vapor retarders in low-slope roofs have changed over the last few years. In the past, vapor retarders were recommended for condensation control in roofs with both of the following conditions:

- Mean average outdoor January temperature below 40°F
- Indoor relative humidity of 45% or higher

This effectively included a wide range of occupancies in a large part of the country (*Fig. 6.34*). More recently, however, research at the U.S. Army Cold Regions Research and Engineering Laboratory (CRREL) has provided revised guidelines based on anticipated indoor temperature and humidity. The

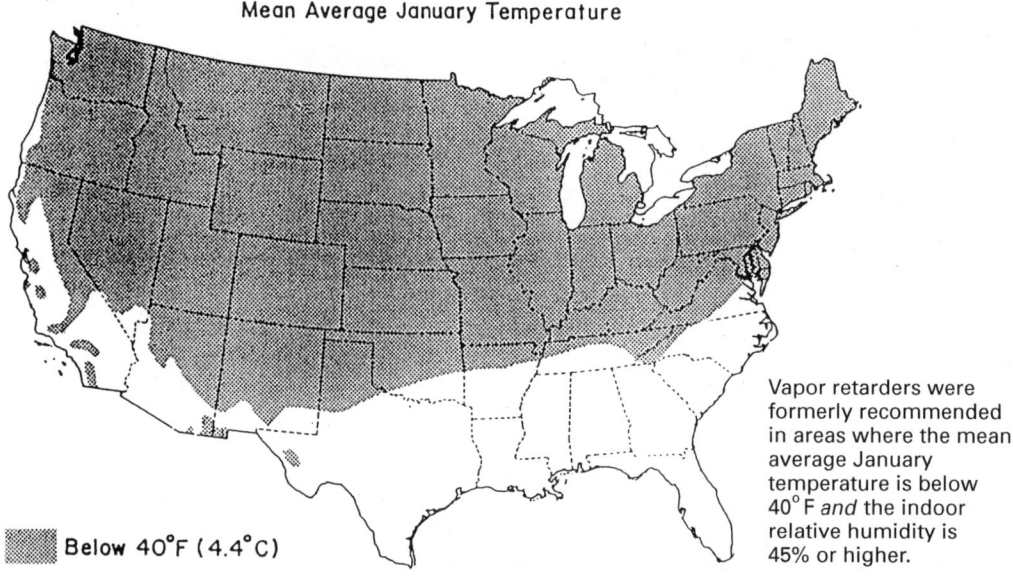

Figure 6.34 Previous vapor retarder recommendations included large parts of the United States. (*From NRCA, Roofing and Waterproofing Manual, 4th ed.*)

CRREL guidelines for various geographic regions call for incorporation of a vapor retarder when the indoor temperature is 68°F and the indoor relative humidity is expected to be above that shown in (*Fig. 6.35*). If the indoor design temperature is something other than 68°F, a correction graph is provided to approximate what the maximum indoor relative humidity can be at a given temperature. In warm climates, vapor retarders are not often used in roofing unless the indoor relative humidity is very high. Vapor retarders used in roof assemblies should provide a complete envelope of protection for the insulation and membrane and, if necessary, be sealed to the wall system vapor retarder or air barrier to prevent infiltration of moist air at the perimeter. Refer to Chap. 5 for more information on vapor condensation, dew points and calculation methods. NRCA recommends that roof vapor retarders be sandwiched between two layers of insulation. The bottom layer of insulation may be mechanically attached, and the vapor retarder and top layer of insulation are then adhesively attached to prevent puncturing the membrane. To assure that the membrane temperature is above the dew point, Tobiasson has developed a graph showing the maximum percent of R-value that can be below the vapor retarder (*Fig. 6-36*).

Since their purpose is to inhibit the flow of moisture vapor into the roofing system, materials used for vapor retarders must have a low permeance. *Table 6.6* lists some common vapor retarder materials used in roofing and their permeance ratings. By definition, a vapor retarder must have a permeance less than 1.0 when tested in accordance with ASTM E96 *Test Methods for Water Vapor Transmission of Materials,* but NRCA recommends that vapor retarders in low-slope roofs have a permeance of 0.5 or less. Vapor retarders in roofs must

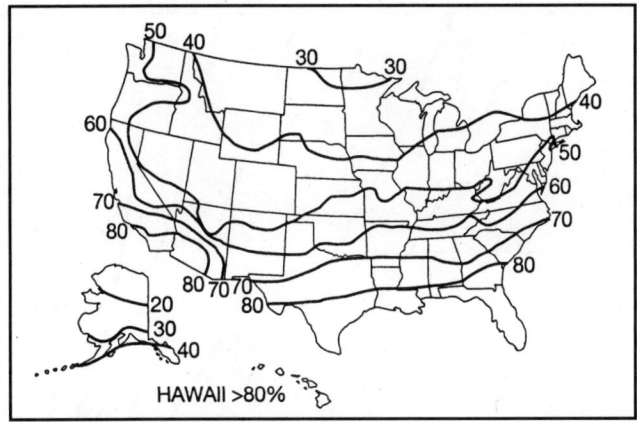
Indoor relative humidity at 68°F (20°C).

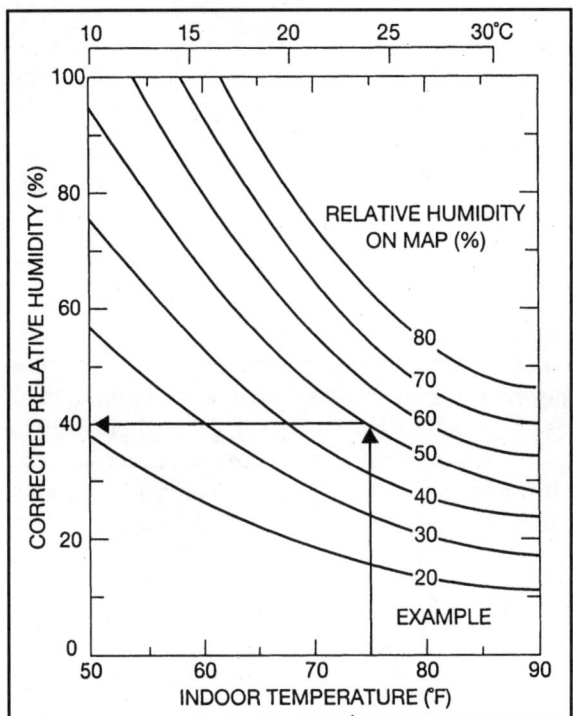
Correction for differing indoor temperature.

Figure 6.35 CRREL recommendations for vapor retarders. (*From Tobiasson and Harrington, Vapor Drive Maps of the U.S.*)

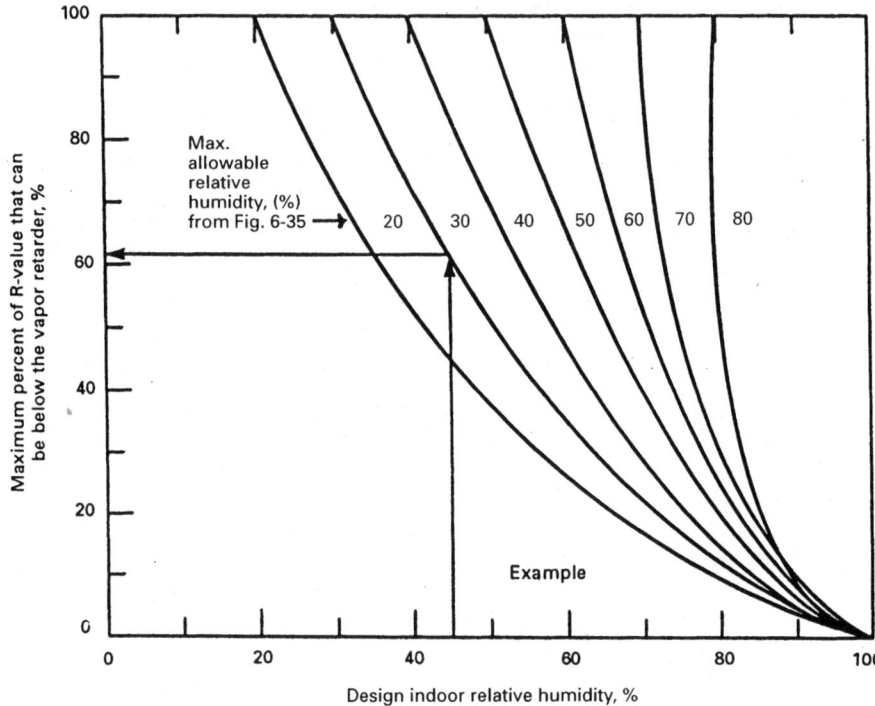

Figure 6.36 Maximum percent of *R*-value which can be below the vapor retarder. (*From Tobiasson, Vapor Retarders Keep Moisture, the Enemy Within, At Bay.*)

be completely sealed to prevent vapor infiltration at laps and roof penetrations and at the roof perimeter, and should be resistant to damage from the hot bitumen or adhesives specified for the project. PVC and polyethylene sheets generally are not recommended for use in roofing systems because they are easily damaged by hot bitumen and subject to wind uplift and damage before the adhesives used for attachment and lap seals has fully cured.

Self-drying roofs. The NRCA *Roofing and Waterproofing Manual* defines self-drying roofs as those designed without vapor retarders. Self-drying roofs can tolerate some moisture vapor accumulation during cold weather when the normal vapor drive is upward into the roof *only* if they are capable of evaporating that same moisture into the interior space during the summer months when the vapor drive is downward. Drying capacity must exceed wetting potential or there will be a progressive accumulation of moisture, subsequent loss of thermal resistance, and eventual damage to the roofing membrane. For a self-drying roof to function effectively, it must have adequate solar exposure and warming during the summer months, a vapor permeable roof deck, and an interior ventilation system capable of dissipating or evacuating the moisture driven downward through the roof deck into the building. Refer to Chap. 5 for a discussion of seasonal wetting and drying cycles.

6.4.3 Built-up membranes

In built-up roofing membranes, multiple plies of reinforcing felts are bonded to one another with hot bitumen, providing redundant layers of protection. In conventional roofing systems, a vapor retarder (if needed) is attached directly to the structural deck or to a thin layer of insulation, additional insulation is attached to the vapor retarder, and the membrane to the top layer of insulation. The waterproof barrier membrane consists of multiple layers or plies of roofing felt and bitumen, with a final flood coat of bitumen in which gravel is embedded to protect against weathering and mechanical damage. A built-up membrane weighs about 2 lb/sq ft. The complete system with gravel ballast weighs about 4 lb/sq/ft. For application over corrugated steel decks, recommended design and installation practices are covered in ASTM E936 *Standard Practice for Roof System Assemblies Employing Steel Deck, Preformed Roof Insulation, and Bituminous Built-Up Roofing.*

The bitumen in built-up membranes provides adhesive attachment and water resistance in a roof but, without reinforcing felts, it lacks tensile strength and durability. Bitumen may be either asphalt or coal tar. ASTM D312 *Specification for Asphalt Used in Roofing* describes four types of roofing asphalt, each of which has a different softening point temperature range for application on roofs with varying slopes (*Table 6.7A*). ASTM D450 *Specification for Coal-Tar Pitch Used in Roofing, Dampproofing, and Waterproofing* describes three types of coal tar (*Table 6.7B*). Type I is often referred to as "old style coal tar pitch." Type II is used for below-grade water-

TABLE 6.6 Permeance of Vapor Retarder Materials

Materials	Thickness, mm	Permeance, perms
Polyethylene sheet	0.15	0.08
Built-up roof membrane	9.5	0.01
Modified bitumen roofing membrane	3.7	0.06
EPDM sheet	1.4	0.06
PVC sheet	1.0	0.09
Concrete	100.0	0.80
Thermal insulation		
Extruded polystyrene	50.0	0.6
Expanded polystyrene	50.0	1.0–3.0
Cellular glass	50.0	0.0
Rigid glass fiber	50.0	56.0

proofing applications, and Type III is less of an irritant to applicators during hot mopping.

Because they have a lower softening-point temperature than asphalts, coal tar systems are limited to lower slope roofs. The low softening point means that when the bitumen warms, it flows and is therefore self-healing of small voids and fractures. Coal tar has excellent weathering properties, and is highly resistant to degradation in corrosive environments and under ponded water conditions. Roofing asphalt is produced by an oxidation process, the duration of which determines what "Type" designation it is given. In general, the greater the degree of oxidation, the higher the softening point and the poorer the weathering characteristics. In general, the steeper the roof slope, the higher the softening point required to prevent slippage. Type I asphalt has the lowest softening point of the four types available, and is most similar to coal tar in its use on very low slope roofs, its self-healing characteristics, and its good weathering properties. Type I asphalt is recommended only for roof slopes less than $\frac{1}{4}$ in./ft. Such low slopes are likely to suffer from ponded water, though, and since asphalt degrades easily under ponding conditions, Type I asphalt is not often used. As the roof slope increases to avoid ponding water, concerns about membrane slippage require a higher softening point. Type II asphalt has the next highest softening point and is "harder" than Type I. Type II asphalt is used on slopes $\frac{1}{4}$ to 1 in./ft and is also used more often in northern climates because the warm temperatures in southern climates can cause roof slippage. Type III is the most commonly used asphalt. Type III asphalt has a higher softening point, but poorer weathering properties than either Type I or Type II. Type III asphalt is usually less expensive than either Type I or Type II, and is an excellent insulation adhesive. Type IV asphalt has the highest softening point and is recommended for use only in very high slope applications. In hot climates, though, Type IV asphalt can be used in all slope applications. Both coal tar and

TABLE 6.7A ASTM D312 Asphalt

Type	Name	Roof Slope
I	Dead-level asphalt	Up to $\frac{1}{4}$ in./ft
II	Flat asphalt	$\frac{1}{4}$ to 1 in./ft
III	Steep asphalt	1 to 3 in./ft
IV	Special steep asphalt	3 to 6 in./ft

TABLE 6.7B ASTM D450 Coal Tar

Type	Name	Roof Slope
I	Coal tar pitch	Up to $\frac{1}{2}$ in./ft with organic felts
II	Waterproofing pitch	NA
III	Coal tar bitumen	Up to $\frac{1}{4}$ in./ft with glass felts

asphalt become harder and more brittle with age, and should be protected from heat and ultraviolet exposure by an opaque surfacing aggregate.

Kettle-modified asphalts are relatively new. Compared to the modified bitumen sheet membranes discussed below, these modified mopping asphalts have better weatherability and ultraviolet resistance and better temperature stability. Compared to a normal Type III roofing asphalt, the modified asphalts have much greater elongation and recovery, higher fatigue resistance, and greater penetration and ductility.

Coal tar is more expensive and more hazardous to workers than asphalt, and is used far less frequently even though it has superior weathering characteristics and suffers only minimal degradation from ponded water. Asphalt and coal tar are not chemically compatible and should not be mixed in the same kettle. Asphalt mixed in a kettle that contains coal tar residue, and vice versa, can produce a bitumen with seriously impaired adhesive properties. The chemical incompatibility between asphalt and coal tar are greatest when the two materials have similar softening points. Type I and Type II asphalt therefore should generally not be used with coal tar roof systems. For the same reason, a coal tar system should not be used to re-roof over an old asphalt system, and vice versa, unless an isolation board is provided between old and new roof to separate the two types of bitumen. Coal tar systems must use asphalt flashings because higher softening points are required for these steeper applications, but the flashings should be Type III or Type IV asphalt. Coal tar systems can also be used with asphalt coated or saturated felts, and were commonly installed this way for many years because glass fiber felts were available only in asphalt. Coal tar–impregnated glass felts have been introduced more recently, however, and probably provide better performance with coal tar membranes.

Roofing felts may be of organic fibers or glass fibers. Glass fiber felts (or mats) were once more commonly used in asphalt built-up roofs and organic felts in coal tar built-up roofs. More recently, however, glass fiber felts are the preferred reinforcing for both asphalt and coal tar systems. The felts are coated or saturated with a bitumen that is compatible with the intended interply bitumen. That is, for asphalt built-up roofs, asphalt coated or saturated felts are best, and for coal tar built-up roofs, coal tar–coated or –saturated felts are best. *Ply sheets* may be used for both the first and successive plies of built-up roofs. Polyester felts are the most recent development, and are used primarily with kettle-modified asphalts. *Base sheets* are heavier felts that are often used for the first layer of the built-up membrane. If the first ply of roofing is to be nailed rather than mopped to the deck with bitumen, a base sheet should be used instead of a ply sheet because of its greater strength. *Venting base sheets* are designed to allow for the lateral venting of water vapor, but are capable of moving only small amounts of moisture. Venting base sheets are typically used over concrete decks to allow some residual drying of construction moisture.

Built-up roofing membranes are surfaced with aggregate ballast, pavers, liquid-applied coatings, or a cap sheet. Aggregate ballast materials used on both coal tar and asphalt built-up roofs include gravel, crushed rock, and slag. Typical coatings used on asphalt built-up roofs include cutback asphalts and asphalt emulsions. Cutback asphalt is a solvent-thinned asphalt which may include aluminum pigments for increased reflectivity. Asphalt emulsions consist of clay and asphalt particles dispersed in water. Some emulsions may also include aluminum pigments or titanium dioxide for reflectivity. Both cutback asphalt and asphalt emulsions are available in fibrated and nonfibrated grades—that is, with or without reinforcing glass fibers. Cap sheets for asphalt built-up roofs are heavy coated felts that are fabricated with a mineral granule surface.

Metal flashing. Metal flashings at penetrations and at the perimeter of built-up roofs are a common source of leaks. In low-slope roofs, base flashing is sealed to the roof membrane and extended onto adjacent walls, parapets and penetrations to a height that will prevent entry of ponded or windblown water. Eight inches is the recommended minimum base flashing height, and twelve inches is the maximum height. Wind-driven rain and water running down from above are prevented from entering the top of the base flashing by a metal counterflashing attached above and lapping over the base flashing a minimum of 4 in. The lapping allows limited movement to occur between the roofing and the adjacent element without rupturing the membrane, flashing or attachments. *Tall masonry parapet walls should not be sealed on the back side* with a roofing membrane or base flashing. Masonry parapets may be treated with water repellents or covered with metal panels if necessary, but they must be able to breathe so that internal moisture can evaporate. Applying a dark roofing membrane also increases differential movement between the backing and facing of multi-wythe masonry parapets because it creates differential heat absorption. Differential movement that cannot be absorbed or accommodated can cause cracking in the masonry and subsequent leakage.

Continuous metal cleats can be used to secure the front edge of metal wall copings against high winds. At roof edges where there are no parapets, metal gravel stops or edge strips are placed over the membrane and securely attached to the deck. Strips of roofing felt or other membrane material are sealed over the connection to make it watertight. The membrane-to-metal connection should be raised above the rest of the roof membrane an inch or more to decrease its vulnerability to ponded water. At drains, flashing is secured to the drain housing with a clamping ring. The roofing membrane itself or a sheet of lead or elastomeric material is clamped to the drain assembly and sealed to the adjacent membrane. Metal pitch pans at pipe penetrations and equipment supports should be avoided because they are a sure source of leaks as the bitumen used to fill the pan dries and shrinks away from the edges. Small penetrations such as pipes and vent stacks should be designed with a hooded flashing that eliminates the need for pitch pans.

The NRCA *Roofing and Waterproofing Manual* is an excellent source for dozens of recommended flashing details. Some of the most common conditions are shown in *Figs. 6.37* to *6.42*. Where counterflashing is used at parapets and adjacent walls, it should be a two-piece system which will allow easy installation of the base flashing and easy replacement of the roofing (*Fig. 6.43*).

Expansion joints and area dividers. Expansion joints are used to relieve tensile stresses in roofing membranes. Expansion joints usually consist of raised loops of flexible materials, or capped curbs (*Fig. 6.44*). Flat edges on either side of the loops are sealed to the membrane and stripped in with flashing. Generally, expansion joints should be installed where

- Expansion or contraction joints occur in the structure
- Steel framing, structural steel, or steel decking changes direction
- Separate junctures or wings of L, U, T, or similar shaped configurations intersect
- The type of decking changes
- Additions are connected to existing buildings
- Interior heating or cooling conditions change
- Movement between vertical walls or parapets and the roof deck may occur

Where expansion joints are not provided, area dividers can be used to relieve stress in the membrane on large roofs. Area dividers are similar in construction to expansion joints except that the physical separation does not penetrate through the deck (*Fig. 6.45*). Area dividers should be installed every 150 to 200 ft, forming smaller areas that are roughly square in shape.

The use of raised curb expansion joints and area dividers is preferred over elastic, preformed joints that are designed to be installed in the flat plane of the roofing system. Raised expansion joints and area dividers should not impede the flow of water after a rain and should be located in conjunction with drainage patterns to assure proper membrane performance as well as rapid water drainage.

Modified bitumen membranes. These are prefabricated sheets of polymer-modified asphalt with polyester or glass fiber reinforcement. Sheet thickness may range from $1/8$ to $1/4$ in. Proper modification produces membranes whose performance characteristics are far superior to those of typical unmodified bitumens. Some modified bitumen membranes are frequently applied by torching the underside of the sheet as it is being unrolled, using an open flame device. Others have a self-adhesive backing or can be adhered with a mopped-on adhesive. Since open-flame torching is a fire hazard, some manufacturers have introduced an electric heat welding process. Overheating modified bitumen degrades the mastic and leads to poor adhesion and weak lap joints.

The two types of polymers most commonly used to modify the asphalt in modified bitumen sheets are atactic polypropylene (APP) and styrene butadiene

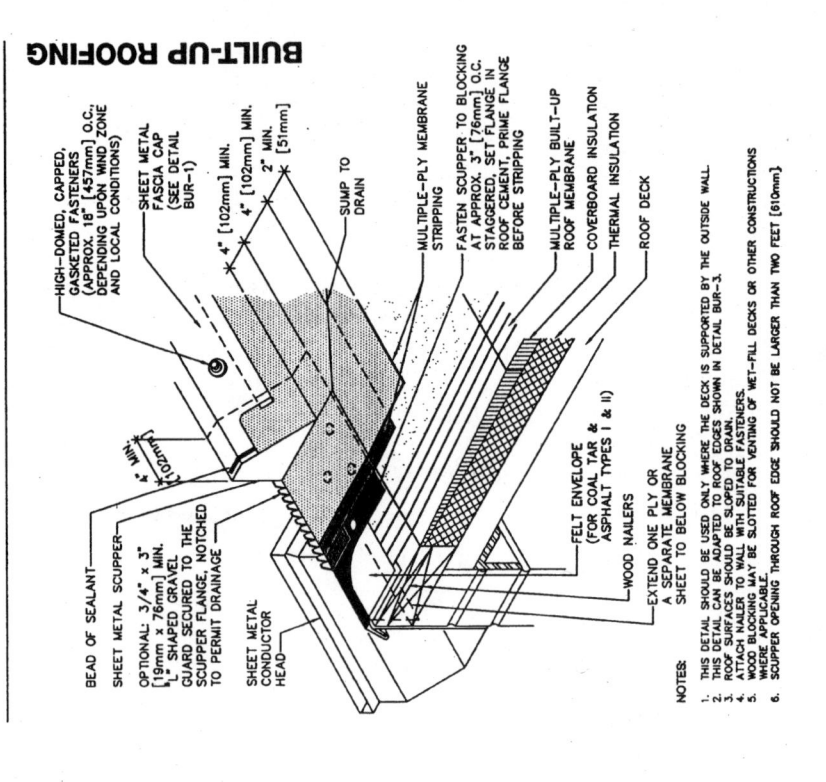

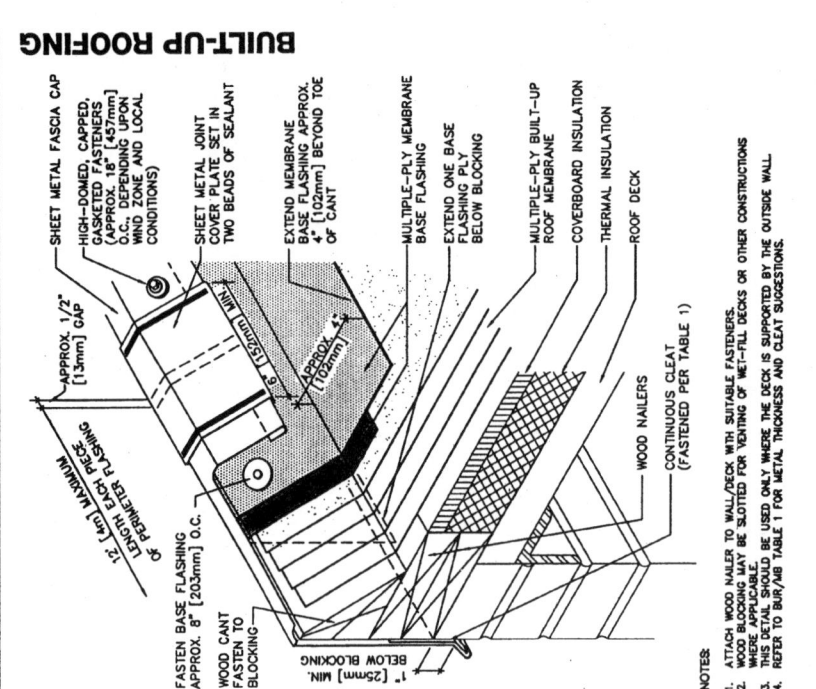

Figure 6.37 NRCA fascia cap and scupper details. *(From NRCA, Roofing and Waterproofing Manual, 4th ed.)*

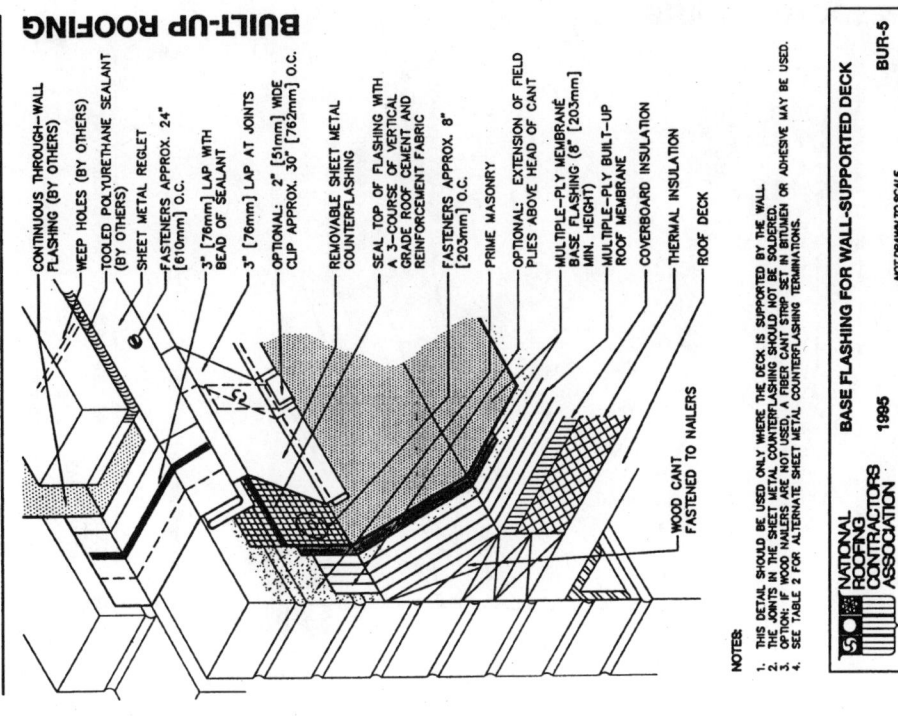

Figure 6.38 NRCA base flashing details. *(From NRCA, Roofing and Waterproofing Manual, 4th ed.)*

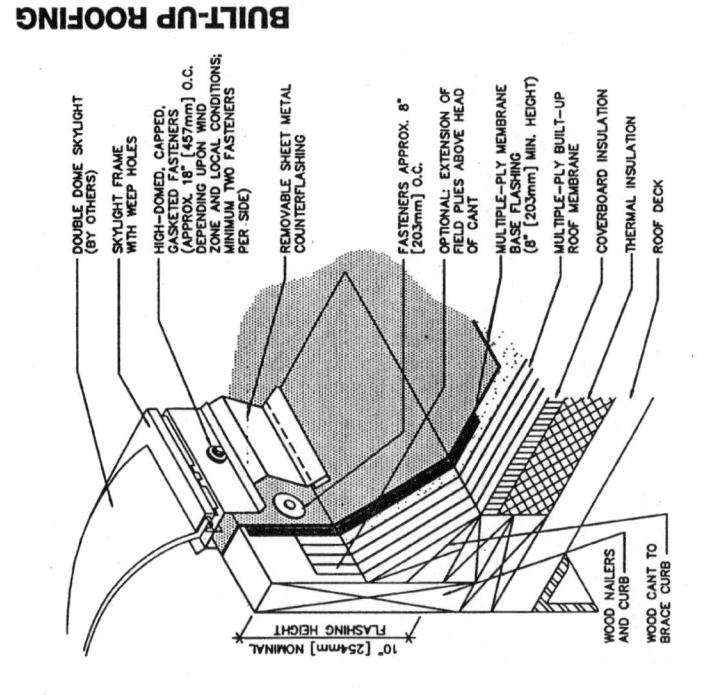
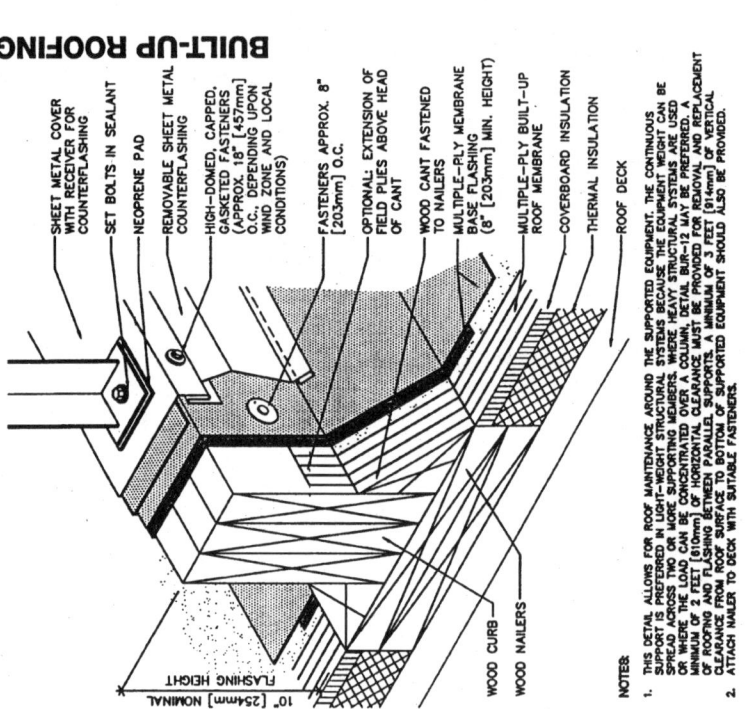

Figure 6.39 NRCA curb details. *(From NRCA, Roofing and Waterproofing Manual, 4th ed.)*

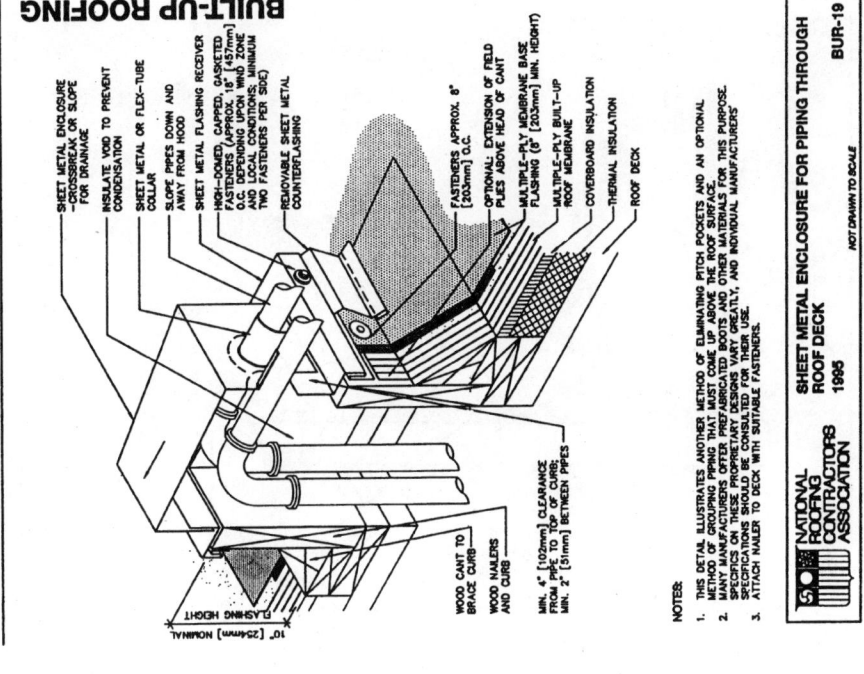

Figure 6.40 NRCA piping penetration details. (*From NRCA, Roofing and Waterproofing Manual, 4th ed.*)

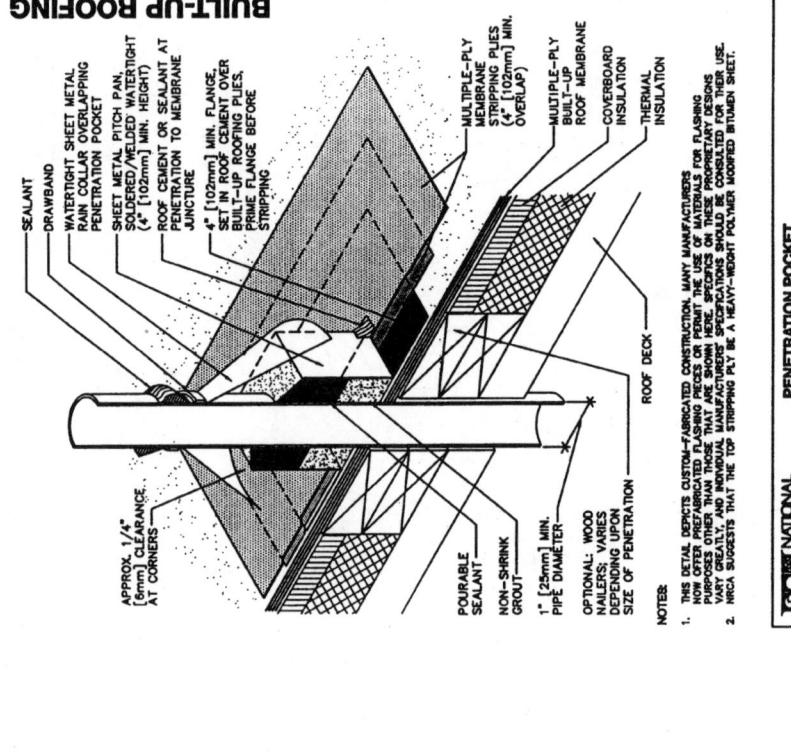

Figure 6.41 NRCA pipe support and hooded pitch pan details. (*From NRCA, Roofing and Waterproofing Manual, 4th ed.*)

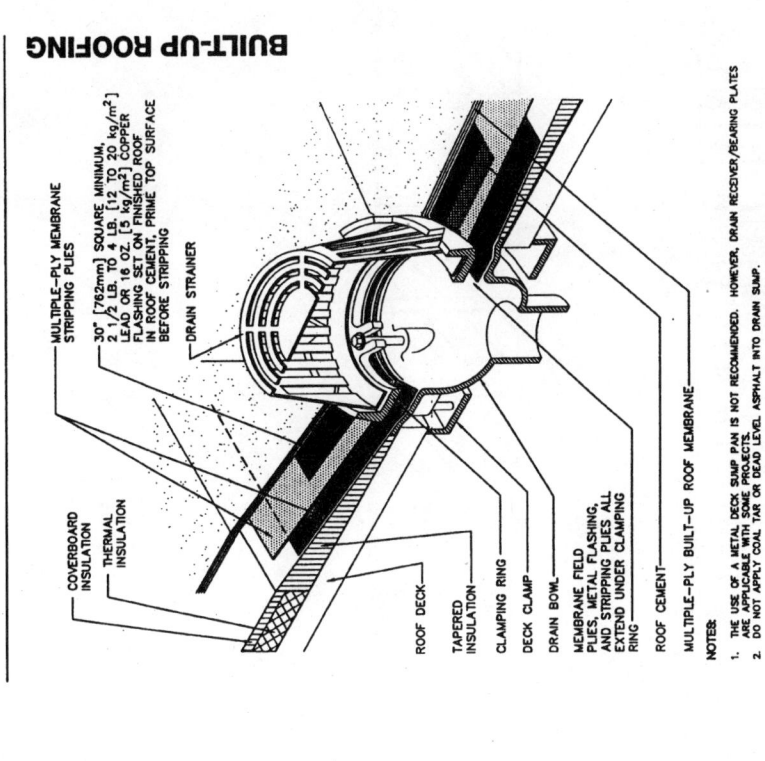

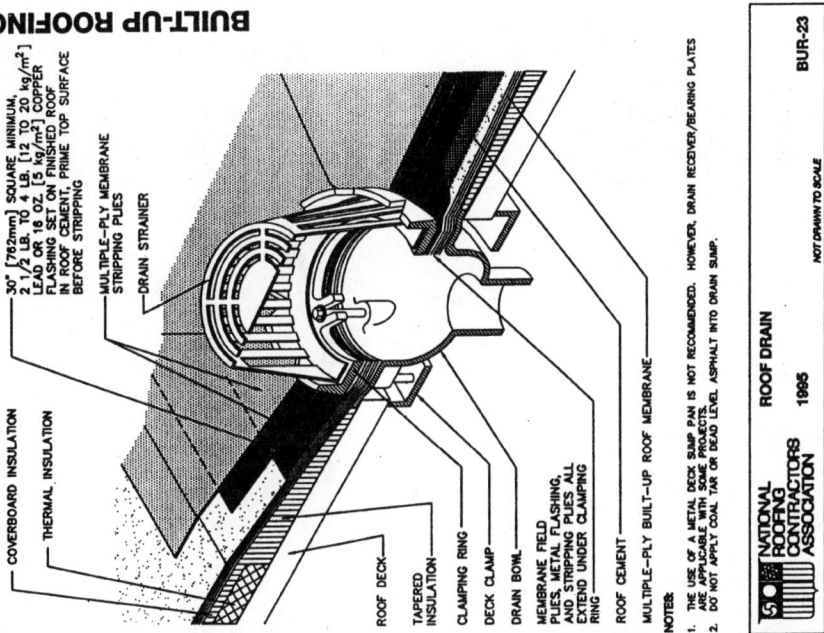

Figure 6.42 NRCA roof drain details. *(From NRCA, Roofing and Waterproofing Manual, 4th ed.)*

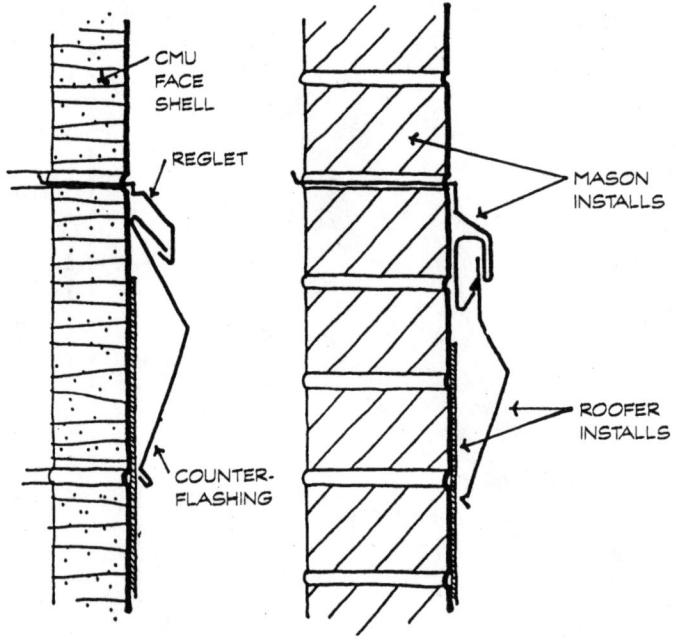

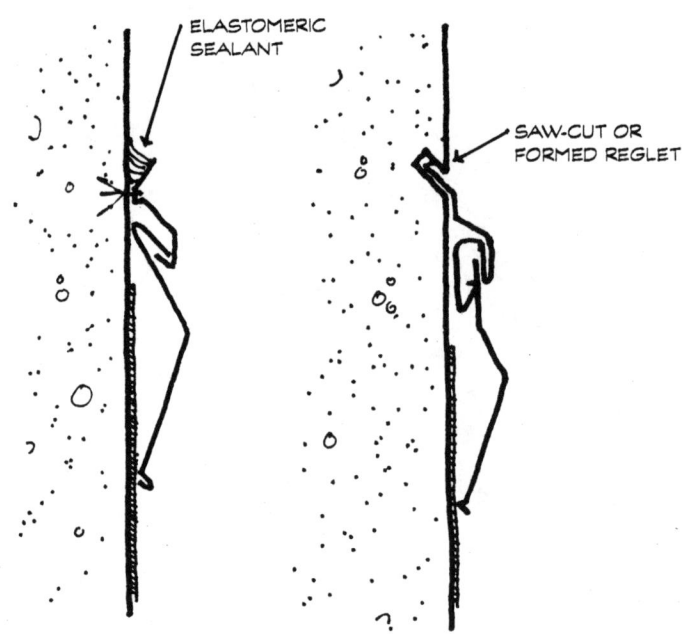

Figure 6.43 Counterflashing details.

NOTES:

1. THIS DETAIL ALLOWS FOR BUILDING MOVEMENT IN BOTH DIRECTIONS.
2. FLASHING REQUIREMENTS TYPICAL FOR BOTH SIDES OF EXPANSION JOINT.
3. ATTACH NAILER TO DECK WITH SUITABLE FASTENERS.

NATIONAL ROOFING CONTRACTORS ASSOCIATION | EXPANSION JOINT WITH PREMANUFACTURED COVER | 1995 | NOT DRAWN TO SCALE | BUR-9A

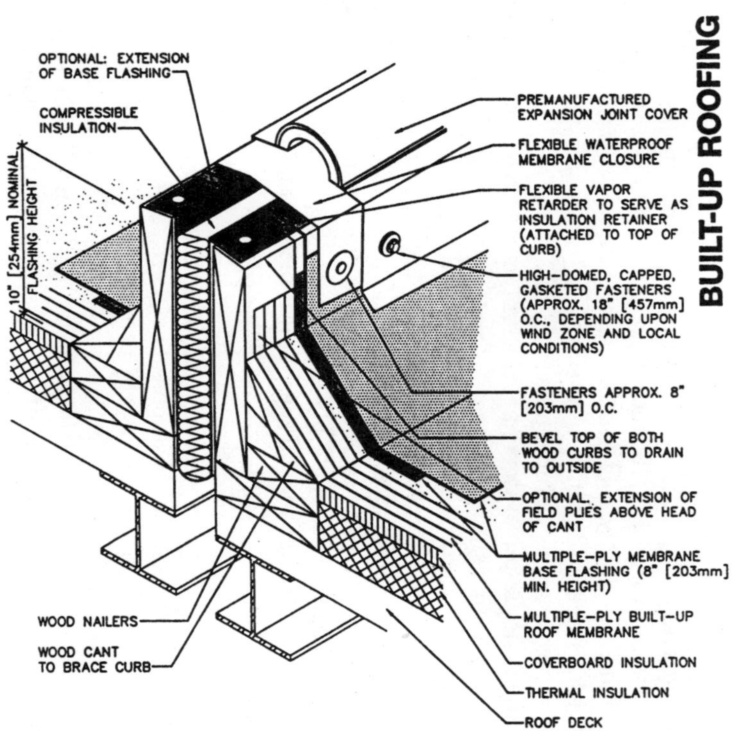

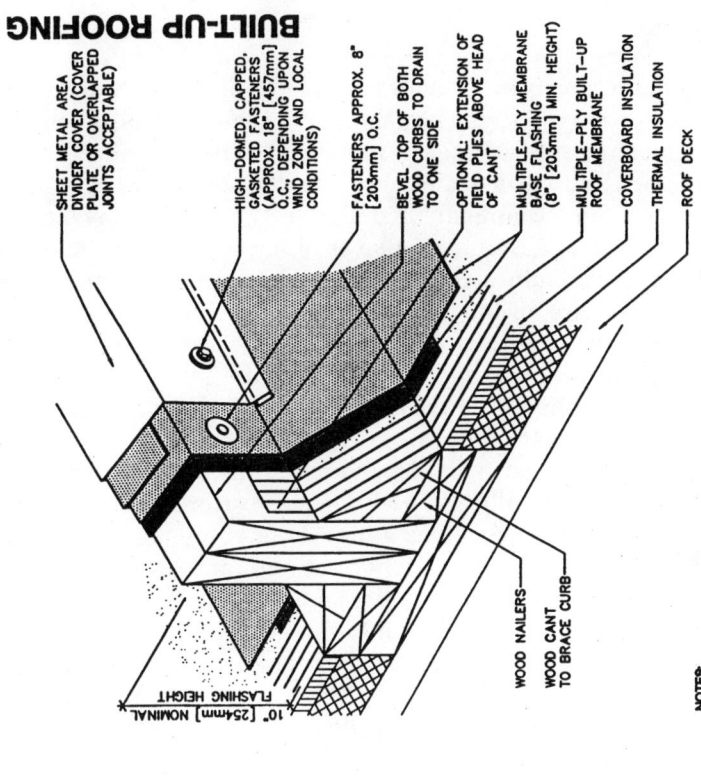

Figure 6.44 NRCA expansion joint detail. (*From NRCA, Roofing and Waterproofing Manual, 4th ed.*)

NOTES:

1. AN AREA DIVIDER IS DESIGNED SIMPLY AS A RAISED DOUBLE WOOD MEMBER ATTACHED TO A PROPERLY FASTENED WOOD BASE PLATE THAT IS ANCHORED TO THE ROOF DECK. AREA DIVIDERS SHOULD BE CONSIDERED TO BE LOCATED BETWEEN THE ROOF'S EXPANSION JOINTS AT 150 TO 200 FOOT [46m TO 61m] INTERVALS, DEPENDING UPON CLIMATIC CONDITIONS AND AREA PRACTICES. AREA DIVIDERS SHOULD NEVER RESTRICT THE FLOW OF WATER.
2. FLASHING REQUIREMENTS TYPICAL FOR BOTH SIDES OF AREA DIVIDER.
3. ATTACH NAILER TO DECK WITH SUITABLE FASTENERS.

NATIONAL ROOFING CONTRACTORS ASSOCIATION | AREA DIVIDER IN ROOF SYSTEM | 1995 | NOT DRAWN TO SCALE | BUR-10

Figure 6.45 NRCA area divider detail. (*From NRCA, Roofing and Waterproofing Manual, 4th ed.*)

styrene (SBS). Modified bitumen roof membranes are typically installed over a base sheet which may or may not also be composed of modified bitumen, and the system may also sometimes include a ply sheet. When originally introduced, modified bitumen roofing systems were usually installed in a single layer, but today they are more commonly installed in two or more layers. SBS membranes are usually set in a continuous layer of hot asphalt, but may also be torch-applied or set in cold adhesive. Self-adhering styrene ethylene propylene styrene (SEPS) sheets are also available. SBS and SEPS membrane must be protected from ultraviolet light, so the sheets are typically manufactured with applied mineral granules, coatings, or reflective foil facings. APP membranes are usually torch-applied, and most are coated with cutback asphalt, asphalt emulsion, latex, or mineral granules.

SBS membrane performance can vary depending on the amount of polymer content in the formulation. Generally, more SBS means better low-temperature flexibility and fatigue resistance as well as a higher softening point and wider temperature use. APP membranes have higher strength and lower elongation than SBS membranes. A small quantity of filler adds rigidity, but large quantities reduce flexibility and adhesion. Modified bitumen perimeter and penetration flashings are often used in conjunction with built-up roofing. Flashing principles and details recommended for modified bitumen roofing are similar to those for built-up roofing, and are extensively illustrated in the NRCA *Roofing and Waterproofing Manual*.

6.4.4 Single-ply membranes

Single-ply roof systems are available in a variety of membrane materials, and may be installed loose-laid and ballasted, fully or partially adhered, or mechanically attached. Single-ply roof systems eliminate the logistical problems of raising hot bitumen and gravel to the roofs of multi-story buildings, and unballasted membranes also reduce the weight of the roof system. Unlike built-up roofs, though, single-ply systems provide only a single line of defense against leakage rather than redundant layers of protection. Seams and flashings are most vulnerable to defects in workmanship and are, in fact, the weak point of these systems. Most membranes, including any required adhesives and mechanical fasteners, weigh between 0.5 and 1.5 lb/sq ft. Loose-laid and ballasted systems, however, depending on the type of membrane and ballast used, can weigh 12 to 20 lb/sq ft. Single-ply membranes experience problems with seam failures, shrinkage, leakage at mechanical fasteners, holes in membranes as delivered or due to damage in place, embrittlement, checking, and cracking.

Loose-laid and ballasted single-ply membranes are generally attached only at the roof perimeter and around penetrations. Joints are sealed by heat welding or adhesives, and wind uplift is resisted by gravel ballast. If crushed rock is used as the ballast, a protective sheet of non-woven polyester fabric or other porous material must be placed between the ballast and the membrane to prevent punctures. *Fully or partially adhered* single-ply membranes use adhesives or bitumen for attachment to the substrate, and they are sometimes also mechanically fastened at the roof perimeter and penetrations. *Mechanically attached* single-ply membranes use individual or bar fasteners. Where subject to significant wind uplift loads, mechanically attached membranes should be reinforced to resist elongation. In conventional roofs, insulation for single-ply systems may be loose-laid or secured to the substrate with adhesives or mechanical fasteners. Mechanical fasteners in single-ply roofs form continuous thermal bridges through the membrane, insulation, and vapor retarder which increase heat transfer and the likelihood of condensation. In protected membrane roofs, the insulation is loose-laid on top of the membrane and held in place by ballast.

Billowing of a loose-laid membrane from wind uplift pressures usually will not occur if there is adequate ballast to resist the applied forces. However, any air that is trapped under a loose-laid or mechanically fastened membrane may collect at the region of greatest uplift pressure (i.e., the windward corner) and cause billowing. If wind has scoured the gravel away from the corners of a loose-laid membrane, billowing will be greater because there is less ballast to hold the membrane in place.

Although modified bitumen sheets are sometimes considered single-ply membranes, the group of products most commonly considered as single-ply systems includes thermoset and thermoplastic sheets such as chlorosulfonated polyethylene (CSPE), polyisobutylene (PIB), chlorinated polyethylene (CPE), ethylene propylene diene monomer (EPDM), polychloroprene (neoprene), polyvinyl chloride (PVC), and PVC blends. Thermoset materials cure during manufacture or in service, and once cross-linked, can only be bonded at seams with adhesive. Thermoplastic materials are uncured and therefore capable of being hot-air-welded or solvent-welded at lap seams.

Polyvinyl chloride (PVC) membranes are *thermoplastic* sheets available in a variety of colors, and should conform to the requirements of ASTM D4434 *Specification for Polyvinyl Chloride Sheet Roofing*. Plasticizers are used to impart flexibility to PVC sheets and to improve processing. Fillers and extenders such as calcium carbonate are used primarily to lower the cost of the compound, but also improve processing and other mechanical properties such as the hardness and dimensional stability of the finished sheet. Stabilizers protect PVC against heat during processing and against ultraviolet radiation during service. Pigments are added to color the material. Loss of plasticizer and subsequent embrittlement was once a problem with PVC roofing membranes, but performance has been improved through the use of high-molecular-weight plasticizers that have less tendency to volatilize or migrate out of the PVC resin. PVC sheets may be unreinforced, lightly reinforced with fibers or fabrics that act as carriers for the resin and add dimensional stability to the sheets, or reinforced with glass and/or polyester fibers or fabrics which increase tensile strength. PVC membranes have good resistance to industrial pollutants, bacterial growth, and extreme weather conditions. Minor damage to the membrane during installation or in service can be easily repaired by the same techniques used for field seaming, by patching the hole using heat or solvent welding to form new molecular linkage and a tight seal.

Elastomeric membranes such as CSPE, PIB, and EPDM are *thermoset,* synthetic rubber materials which may be either vulcanized or non-vulcanized. Vulcanization is an irreversible curing process during which an elastomeric compound is chemically cross-linked and changes from a soft, tacky thermoplastic to a partially thermoset material with improved elastic properties. Non-vulcanized, or uncured, sheets typically cure slowly in service from exposure to solar heat. Once cured, their behavior is similar to that of a cured elastomer. Non-vulcanized sheets that are not self-curing have properties similar to thermoplastic membranes. In general, elastomeric sheets have good tensile and other mechanical properties, excellent resistance to ultraviolet radiation, ozone, many oils, and solvents.

Chlorosulfonated polyethylene (CSPE) is commonly known by the trade name Hypalon. It is a non-vulcanized elastomeric membrane which cures after application on the roof, and is usually white in color. CSPE roofing should conform to the requirements of ASTM D5019 *Specification for Reinforced Nonvulcanized Polymeric Sheet Used in Roofing Membranes,* Type I. *Polyisobutylene* (PIB) membranes are non-vulcanized elastomeric membranes available in both black and white. PIB roofing should meet the requirements of ASTM D5019, Type III. *Ethylene propylene diene monomer* (EPDM) is the most commonly used elastomeric membrane. Although EPDM is available in white, black sheets are more durable and much more common. EPDM membranes are manufactured by laminating two plies, with or without reinforcing. EPDM membranes should conform to the requirements of ASTM D4637 *Specification for Vulcanized Rubber Sheet Used in Single-Ply Roof Membranes,* Type I.

Since vulcanized elastomeric materials such as EPDM can be bonded only with adhesives, lap seam integrity is critical to performance. Single-ply membranes do not have the redundant protection of a multi-ply built-up membrane, so even small gaps or openings at the seams can cause leaks. Successful field seaming requires use of proper adhesives, care in surface preparation, skillful application, and adequate curing time before loads are applied. Sheets are coated with a powder to prevent the rolled product from sticking to itself during shipping and handling, and surface preparation must include thorough cleaning and priming of the seaming area. Factory seaming to create the largest practical membrane panels reduces the amount of field seaming that is required and minimizes the possibility of problems.

Since single-ply roof membranes are typically sold as proprietary systems, it is important to follow the manufacturer's recommended details and specifications. General recommendations for good roofing practice still apply, modified only as needed to accommodate the specifics of a particular membrane. The NRCA *Roofing and Waterproofing Manual* contains dozens of details for both thermoplastic and thermoset single-ply membranes.

Chapter 7

Waterproofing

Chapter 4 contained a discussion of the three simultaneous conditions which can cause water penetration through a building enclosure:

- The presence of water
- An opening through which water can enter
- A physical force to move the water

Moisture in the soil provides a source of water in below-grade construction, and rain and melting snow provide a source of water at plaza decks. Openings through which water can penetrate these elements may occur at capillary pores in the substrate or at cracks, joints and penetrations through the enclosure. The physical forces which push water through openings in plaza deck and below-grade waterproofing membranes are

- Gravity
- Capillary suction
- Hydrostatic pressure

Barrier-type waterproofing membranes require complete and continuous coverage and must eliminate all holes in the substrate if they are to provide effective protection against water penetration under these forces. Barrier membranes are very unforgiving of design and application errors, and on most waterproofing applications, there is only one chance to get it right. Once the backfill is in place or a plaza deck has been laid, the membrane is not easily accessible for repairs. A more practical approach to waterproofing combines surface and subsurface drainage, capillary breaks, vapor retarders, and waterproofing or dampproofing membranes to control the forces which move water into the structure and to provide a factor of safety in the performance of the systems.

7.1 Moisture Movement in Soils

Water moves through the soil by gravity flow and capillary suction, and exerts hydrostatic pressure against below-grade walls and slabs. Water vapor diffuses through soils because of vapor pressure differentials between areas of different temperature. To prevent moisture penetration, both water and vapor movement must be accommodated in the design of below-grade structures, foundations, and slabs on grade.

7.1.1 Water movement

At some elevation below every building site, there is water in the ground because of rain penetration and seepage into the soil, and because of the natural water content of the soil. This groundwater may be close to the surface, or well below grade. The top elevation of groundwater is called the groundwater level or *water table*. Water table varies with climate, amount of rainfall, season and, to some extent, with type of soil. The water table follows the general contours of the land, but is closer to the surface in valleys and farther from the surface on hills, ridges and plateaus. Water moves laterally through the soil by gravity flow to lower elevations. The direction of groundwater flow is always in the direction of lower elevations until the water emerges in a spring, stream, or other open body of water (*Fig. 7.1*).

A geotechnical survey can identify the soil types which will be encountered below a building site, as well as the elevation of the water table. Since water table can vary with climate and amount of rainfall, it is important to understand that the water table listed in a geotechnical report should not be taken as an absolute. If the soil tests were performed during the rainy season, the elevation of the water table may be at its highest expected level, but if the tests were done during a period of drought, the water table may be unusually low and misrepresentative of the normal conditions which would be encountered. The geotechnical survey should include comprehensive and reliable information on subsurface water table levels and the hydrology of the soil strata, his-

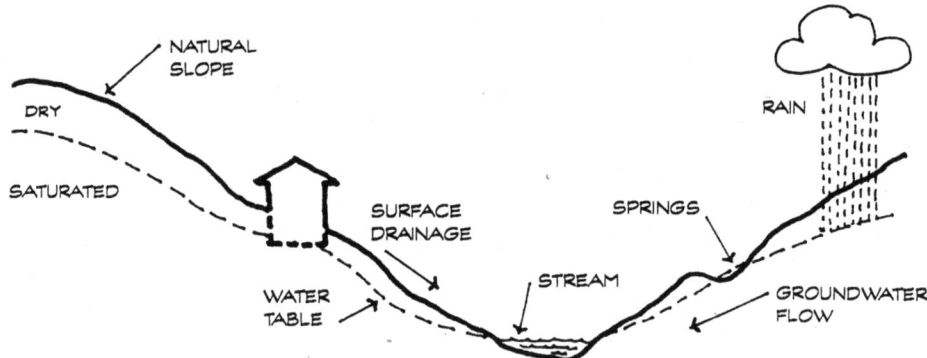

Figure 7.1 Groundwater flow. (*From Callendar, Timesaver Standards for Architectural Design Data.*)

torical data on surface flooding and hydrology, projected changes resulting from ongoing or anticipated development in the area, and the effects of landscape irrigation systems.

Hydrostatic pressure is defined as the pressure at a point in a fluid at rest due to the weight of the fluid above it. The hydrostatic pressure exerted by ground water at any point against a below-grade wall is equal to the depth of that point below the water table times the unit weight of water (62.4 pcf). If the bottom of the wall is 8 ft below the water table, the hydrostatic pressure at that point is $8 \times 62.4 = 499.2$ pounds per square foot (psf) of wall area. The lateral pressure of the soil itself is somewhat reduced because of the buoyancy of the water it contains, but the added hydrostatic pressure increases the overall structural load on the wall substantially (*Fig. 7.2*). The hydrostatic pressure on the bottom of a slab is calculated in the same way. Multiply the depth of the bottom of the slab below the water table by the unit weight of water. Both the structure and the waterproofing membrane must be able to withstand the lateral and uplift loads created by hydrostatic pressure. If a membrane has the ability to span cracks in a substrate, then it must also have the strength to withstand hydrostatic pressure across that span where the membrane is unsupported. Groundwater can be diverted away from a below-grade structure by installing subsurface drains to lower the water table below the structure. Draining water away from the structure reduces both structural loads on the floor, walls, and foundation and hydrostatic pressure on the waterproofing membrane.

In addition to lateral gravity flow, water can move upward through soil from the water table by *capillary action*. The rate at which this capillary rise occurs depends on the particle size and distribution and the resulting pore size of the soil. Clay soils have the finest pore structure, and can draw capillary moisture upward from a water table many feet below. Coarse, sandy soils generally have a pore structure so large that capillary suction is minimal. The capillary moisture content of soil varies in direct proportion to the fineness of the soil (*Fig. 7.3*). Capillary moisture cannot be drained out of soil because the surface tension within the pore structure holds the water tightly. Moist soil can be dried and the water evaporated by ventilation and exposure to dry air.

Soil particles less than $\frac{1}{480}$ in. in size are called *fines*. Laboratory tests of soils that contained 56% fines showed moisture constantly rising to the surface and evaporating at an average rate of about 12 gal per 1000 sq ft per 24 hours with a water table as much as 30 in. below the surface. Field tests have also shown that substantial amounts of moisture migrate upward through fine soil even when the water table is as much as 20 ft below the surface. *Figure 7.4* indicates the height of capillary moisture rise which can be expected with various soil types. Any crawl space, basement, or slab on grade built without protection on moist soil would be exposed to a continuous capillary migration of moisture into the structure.

To prevent the capillary rise of water into a slab on grade or below-grade slab, an intervening layer of material must be added which is either impervious to moisture penetration or has a pore structure large enough to preclude

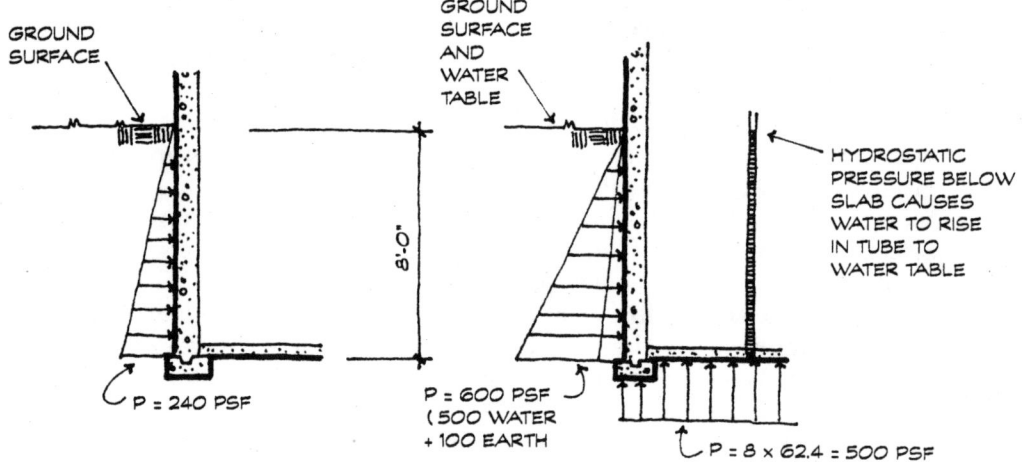

Figure 7.2 Hydrostatic pressure. (*From Callendar, Timesaver Standards for Architectural Design Data.*)

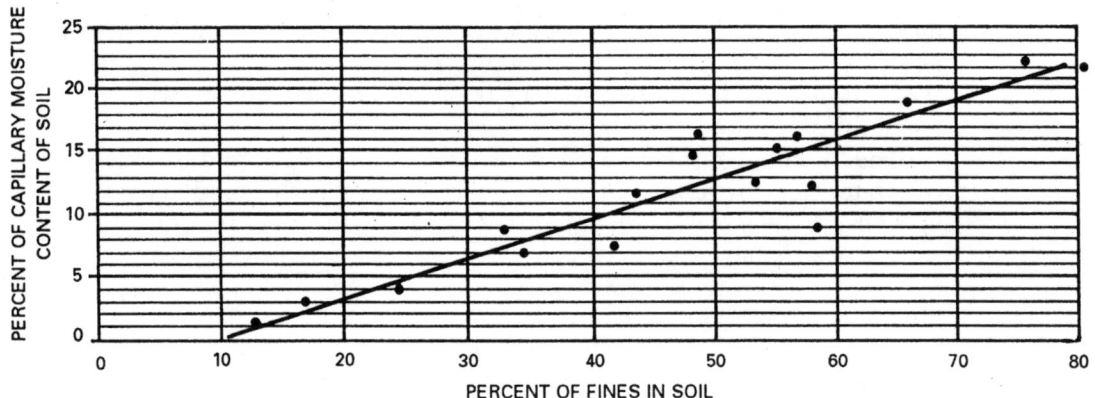

Figure 7.3 Capillary moisture content of soil varies in direct proportion to fineness of soil particles. (*From W. R. Meadows, Inc., The Hydrologic Cycle and Moisture Migration.*)

capillary suction. Gravel and crushed rock are the materials most commonly used to provide a capillary break under a slab on grade or below-grade slab. The aggregate should be mostly single-graded and of $3/4$-in. maximum size. The bottom of an undrained granular base course must not be below the adjacent finished grade, or it will become a reservoir for water. The drainage characteristics of various soils are shown in *Table 7.1*. Capillary water penetration can also be prevented by installing dampproofing or membrane waterproofing as a barrier against capillary movement (*Fig. 7.5A*). Historic masonry structures incorporated a damp course of dense, impervious masonry such as slate or hard-fired clay brick at or just above grade to prevent "rising damp" upward through the porous brick. Contemporary structures supported on masonry

TABLE 7.1 Soil Types and Properties

Division	Designation	Description	Frost action	Drainage
Coarse-grained soils:				
Gravels (more than half of coarse fraction larger than $1/4$" in size):				
Clean gravels	GW	Well-graded gravels, gravel-sand mixtures, little or no fines	None	Excellent
	GP	Poorly graded gravels or gravel-sand mixtures, little or no fines	None	Excellent
Gravels with fines	GM	Silty gravels, gravel-sand-silt mixtures	Slight	Poor
	GC	Clayey gravels, gravel-sand-clay mixtures	Slight	Poor
Sands (more than half of coarse fraction less than $1/4$" in size)				
Clean sands	SW	Well-graded sands, gravelly sands, little or no fines	None	Excellent
Sands with fines	SP	Poorly graded sands or gravelly sands, little or no fines	None	Excellent
	SM	Silty sands, sand-silt mixtures	Slight	Fair
	SC	Clayey sands, sand-clay mixtures	Medium	Poor
Fine-grained soils:				
Silts and clays (liquid limit less than 50)	ML	Inorganic and very fine sands, rock flour, silty or clayey fine sands or clayey silts with slight plasticity	Very high	Poor
	CL	Inorganic clays of low to medium plasticity, gravely clays, sandy clays, silty clays, lean clays	Medium	Impervious
	OL	Organic silts and organic silty soils, elastic silts	High	Impervious
Silts and clays (liquid limit more than 50)	MH	Inorganic silts, micaceous or diatomaceous fine sandy or silty soils, elastic silts	Very high	Poor
	CH	Inorganic clays of high plasticity, fat clays	Medium	Impervious
	OH	Organic clays of medium high plasticity, organic silts	Medium	Impervious
Highly organic soils	PT	Peat and other highly organic soils	Slight	Poor

foundations should incorporate a "damp check" of through-wall flashing just above grade (*Fig. 7.5B*).

7.1.2 Vapor movement

Temperature and humidity, by creating differences in vapor pressure, cause water vapor migration within the soil, from soil to air, and from air to soil. Cold winter air is generally very dry, and large temperature differences between the soil and the air create substantial vapor movement from the soil to the air (*Fig. 7.6A*). In the spring and fall, vapor movement is also generally from the soil to the air, but since the air and soil temperatures are closer together, vapor movement is relatively slow (*Fig. 7.6B*). Warm moist summer air that is at higher vapor pressures than cooler, drier soil causes vapor to move from the air to the soil (*Fig. 7.6C*). When the air and surface soil temperatures are below 32°F, vapor in the soil migrates toward the surface, and condenses as frost. The depth of the frost line depends on the length and severity of the cold season (see Chap. 3).

Below-grade vapor pressures within the soil, particularly if capillary moisture is present, are usually higher than vapor pressures within buildings. This

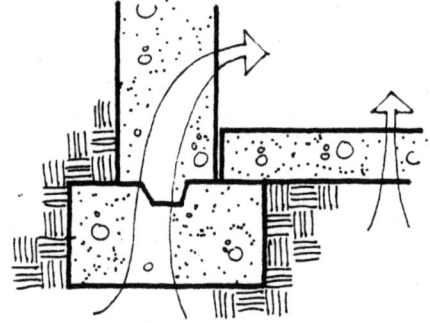

CAPILLARY RISE OF MOISTURE

Soil Type	Saturation Zone, ft	Capillary Rise, ft
Clay	5 +	8 +
Silt	5 +	8 +
Fine Sand	1 - 5	3 - 8
Coarse Sand	0 - 1	1 - 3
Gravel	0	0

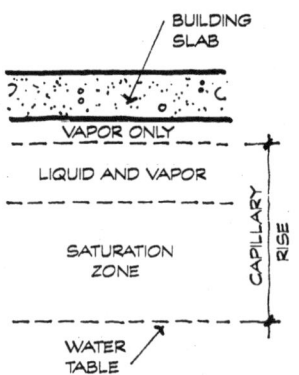

Figure 7.4 Height of capillary moisture rise above water table for various soils. (*From Olin, Construction Principles, Materials and Methods, 5th ed.*)

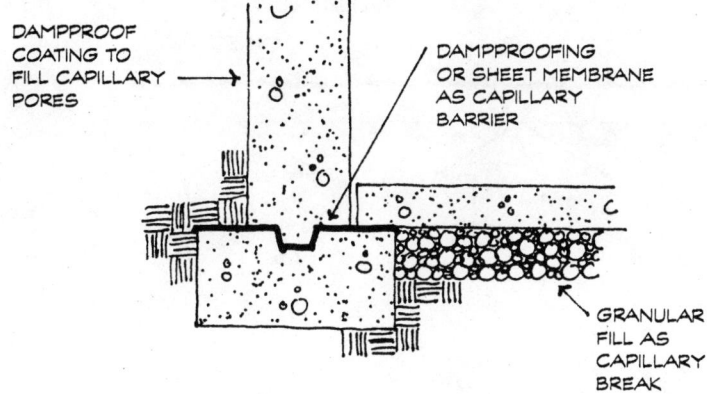

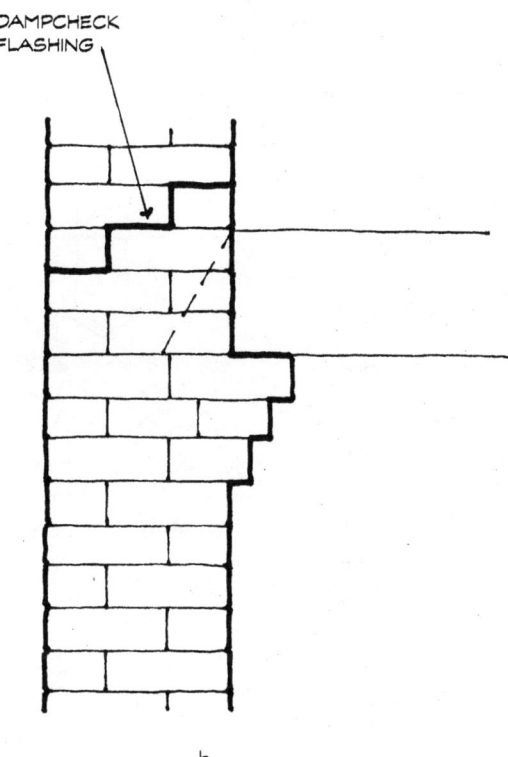

Figure 7.5 Membrane barrier to capillary moisture rise.

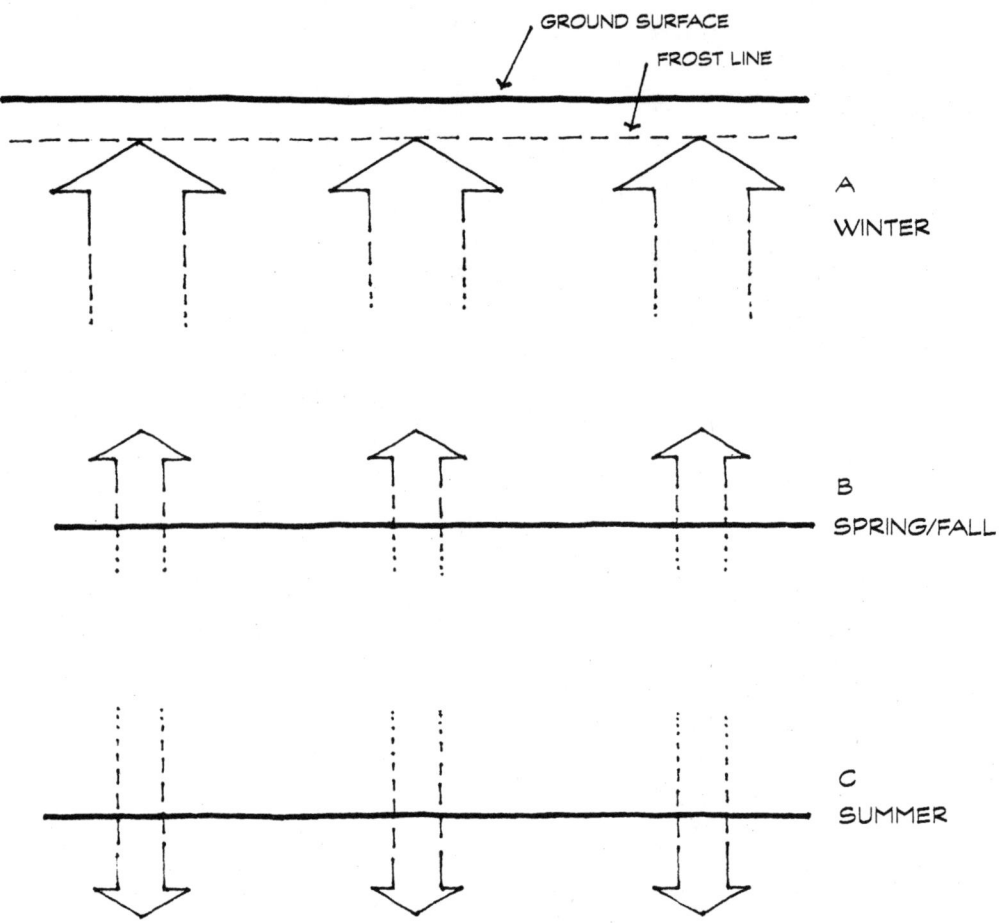

Figure 7.6 Seasonal vapor exchange between soil and air. (*From W. R. Meadows, Inc., The Hydrologic Cycle and Moisture Migration.*)

pressure differential creates a flow of vapor from the soil to the structure regardless of season or interior heating or cooling cycles (*Fig. 7.7*), and vapor can then migrate through a concrete slab or framed floor structure into the building. Vapor migration from the soil, if unimpeded, can provide a continuous supply of below-grade moisture flowing into the structure and then migrating outward through the walls and roof. If cooled below its dewpoint, this continuous supply of moist air will condense to liquid on interior surfaces, or condense as liquid or frost within the walls or roof of the building envelope.

7.2 Basic Waterproofing Principles

Waterproofing did not become a real issue in building design and construction in the United States until the twentieth century. Large commercial structures began to locate mechanical and electrical equipment in below-grade space, residential

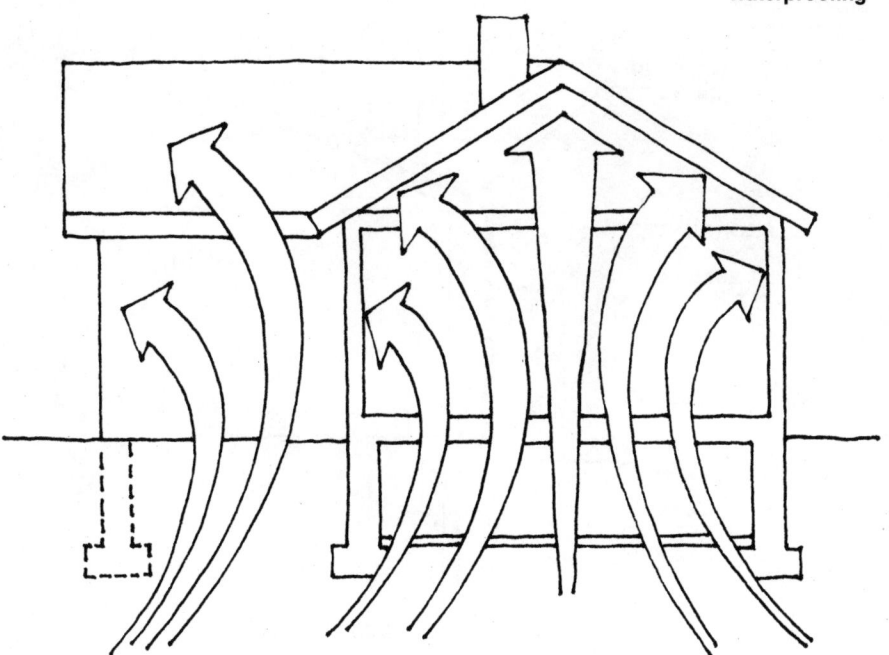

Figure 7.7 Vapor migration into buildings from soil. (*From W. R. Meadows, Inc., The Hydrologic Cycle and Moisture Migration.*)

basements became usable living areas, and less desirable land with poor drainage was pressed into development as populations grew. Today's materials and technology allow the construction of complete underground structures that are capable of comfortably housing computers and other moisture-sensitive equipment as well as human occupants.

7.2.1 Surface drainage

Surface drainage should be an integral part of every waterproofing system. Groundwater can be controlled to a great extent by reducing the rate at which rainwater and surface runoff enter the soil adjacent to a building enclosure. Roofs typically concentrate collected rainwater at a building's perimeter, where it can cause serious groundwater problems (*Fig. 7.8*). Water that is drained quickly away from a building at the ground surface cannot enter the soil and contribute to below-grade moisture problems. Roof overhangs, gutters, and downspouts provide effective control for sloped roofs by diverting the runoff away from the building (*Fig. 7.9*). Drains for low-slope roofs are typically connected to a storm sewer or remote outfall. For optimum surface drainage, overflow scuppers on low-slope roofs should be connected to downspouts that are directed away from the foundation. Site selection, building orientation, and grading should provide slopes away from the building, and ground swales and troughs can also be used to redirect surface runoff.

292 Chapter Seven

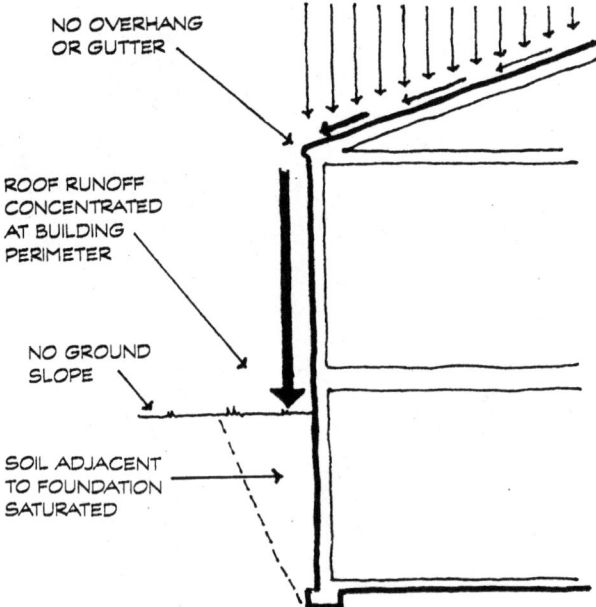

Figure 7.8 Roof runoff saturates foundation perimeter. (*From Lstiburek and Carmody, Moisture Control Handbook.*)

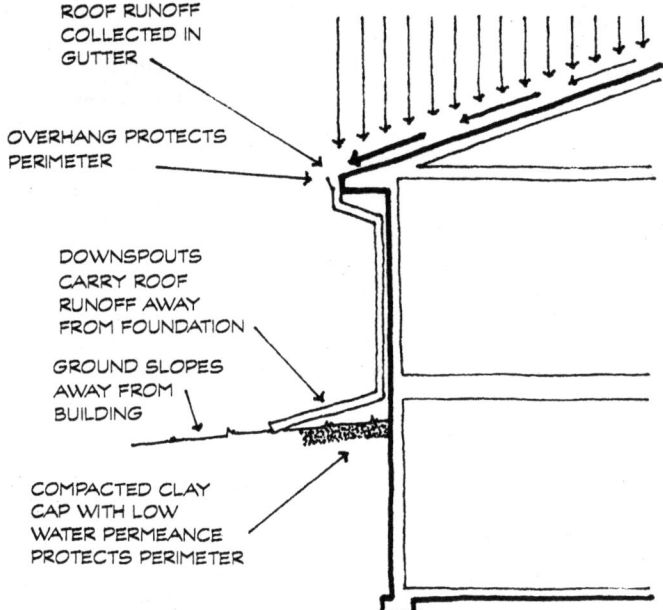

Figure 7.9 Protecting foundation perimeter. (*From Lstiburek and Carmody, Moisture Control Handbook.*)

Backfill adjacent to a building should be compacted sufficiently to prevent settlement and the possibility of ponding water which might drain toward the foundation wall. Backfill materials that contain a high percentage of fines may absorb and hold surface water and rainwater, concentrating the moisture immediately adjacent to the building. A low-permeance cap of compacted clay soil can be installed under grassy areas. Planting beds located next to the building walls should always be well drained to avoid concentrating moisture along the foundation line. Sidewalks located adjacent to a building can prevent groundwater absorption, but may cause backsplash and soiling on the walls. Sidewalks should always be sloped away from the building a minimum of $1/2$ in./ft. The joint between the sidewalk and the building should be sealed with a traffic-grade elastomeric sealant if substantial rainfall, accumulated snow drifts, or exposure to roof or site runoff is expected.

7.2.2 Subsurface drainage

Subsurface drainage systems can collect and divert groundwater away from the walls and floor of a below-grade structure, and relieve hydrostatic pressure. The most common method of keeping groundwater away from basement structures is to provide a *perimeter drain* or *footing drain* in the form of perforated, porous, or open-jointed pipe at the level of the footings. Perforated drains are generally preferable to the porous pipe and open-jointed systems. When perforated drains are used, they should be installed with the perforations on the bottom so that water rises into the pipe. The drainage pipe must be protected from soil infiltration and clogging with a filter of geotextile fabric. Perimeter drains artificially lower the water table below the elevation of the floor and relieve hydrostatic pressure against the walls and the bottom of the slab. The drains must be placed below the floor level, but above the bottom of the footing to prevent washout. As a rule of thumb, the bottom of the footing should be at least 4 in. below the bottom of the drain to prevent undermining the footing stability. Crushed stone or gravel is always placed above and below perimeter drains to facilitate water flow.

Perimeter drains can be laid level without slope. To create water flow into a level drain requires that the pipe be placed far enough below the floor slab to build up sufficient hydraulic head. The example shown in *Fig. 7.10* illustrates the method used to calculate the required depth of the drain. The outfall pipe into which the perimeter drains flow is then sloped to a natural gravity outflow or to a sump. If perimeter drains are to be sloped, the distance between the floor slab and the bottom of the footing must be sufficient to accommodate the required slope between the two elevations so that the bottom of the pipe is always at least 4 in. above the bottom of the footing (*Fig. 7.11*).

For clay soils, which have poor drainage and only limited amounts of groundwater flow, a nominal 4-in. drain is usually adequate. For sandy soils with better drainage and more groundwater flow, a nominal 6-in. drain is needed. For gravely soils with good drainage and large groundwater flow,

Example

Given: Inflow of 100 gpm, established by a pumping test during excavation
Length of drain—350 ft
Friction factor for pipes—n = 0.015

Find: Underdrainage design required to keep cellar dry

Solution: Assume a 6-in. pipe. From nomograph, required slope for a 6-in. pipe and 100 gpm discharge is 0.0024. Hydraulic drop is 350 x 0.0024 = 0.84 ft. Bottom of drain should be, therefore, 0.84 + 0.5 (thickness of floor slab) = 1.34 ft below surface of floor.

To adjust for n = 0.019, multiply inflow by 0.019/0.015 and then use nomograph.
To adjust for n = 0.013, multiply inflow by 0.013/0.015 and then use nomograph.

Example

Inflow of 100 gpm
Friction factor—n = 0.019

Solution

100 × 0.019/0.015 = 126
Taking line from 126 on Discharge bar through 6-in. Drain Diam. gives Slope of 0.0040

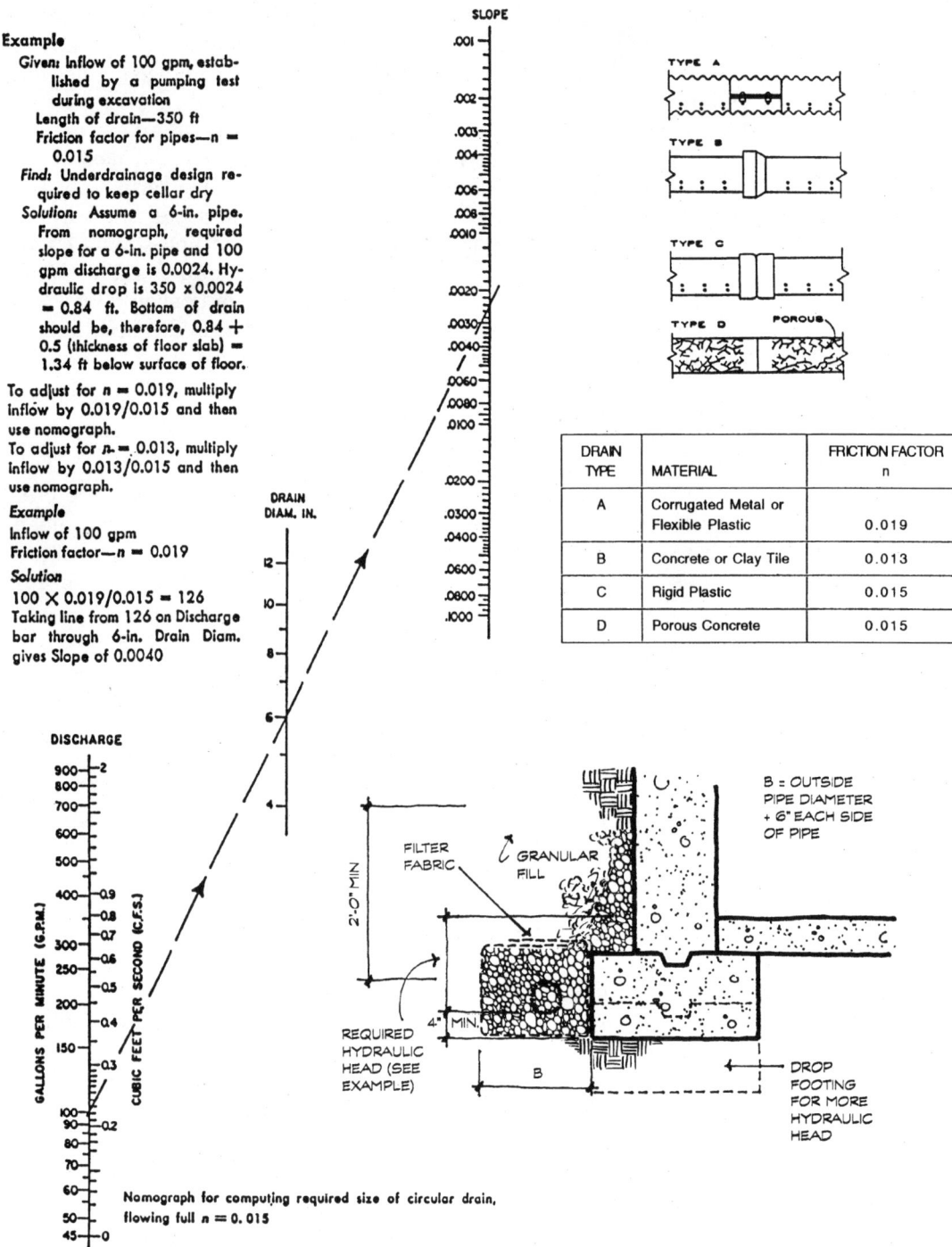

Figure 7.10 Sizing and locating subsurface drains. (*From Callendar, Timesaver Standards for Architectural Design Data.*)

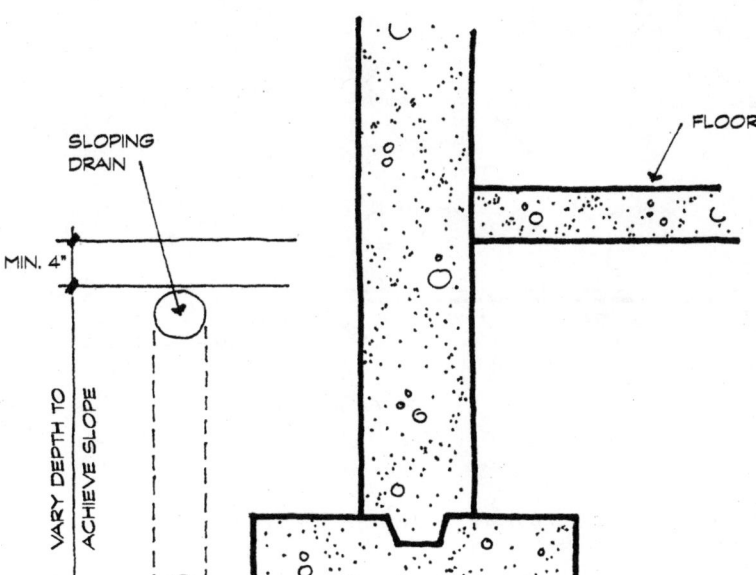

Figure 7.11 Dropped footing for sloping drain.

pumping tests should be made during excavation to determine flow quantity and size of pipe. *Table 7.1* summarizes various soil types and their drainage characteristics.

At the center of a building with perimeter drains, the water table will be slightly higher than at the edges (*Fig. 7.12*). If the floor area is large, underfloor drains may have to be used in addition to the perimeter drains to keep this raised water level away from the bottom of the floor. The depth and spacing of subsurface drains for underfloor and perimeter combinations for various soil types are shown in *Table 7.2*.

Subsurface drainage can also be used to relieve hydrostatic pressure against the full height of a below-grade wall. A free-draining, porous backfill that extends the height of the wall allows groundwater to flow by gravity down to the level of the drain (*Fig. 7.13A*). This free-draining material should be carried up the wall to within a few inches of the ground surface with only a covering of top soil for landscaping purposes. Proprietary drainage mats and insulation boards with vertical drainage channels can be used in lieu of the porous backfill as long as their capacity is sufficient to handle the expected flow of water (*Fig. 7.13B*). These mats and drainage boards are generally easier to install than gravel backfill, particularly on multistory below-grade applications. This method of subsurface drainage is sometimes referred to as a

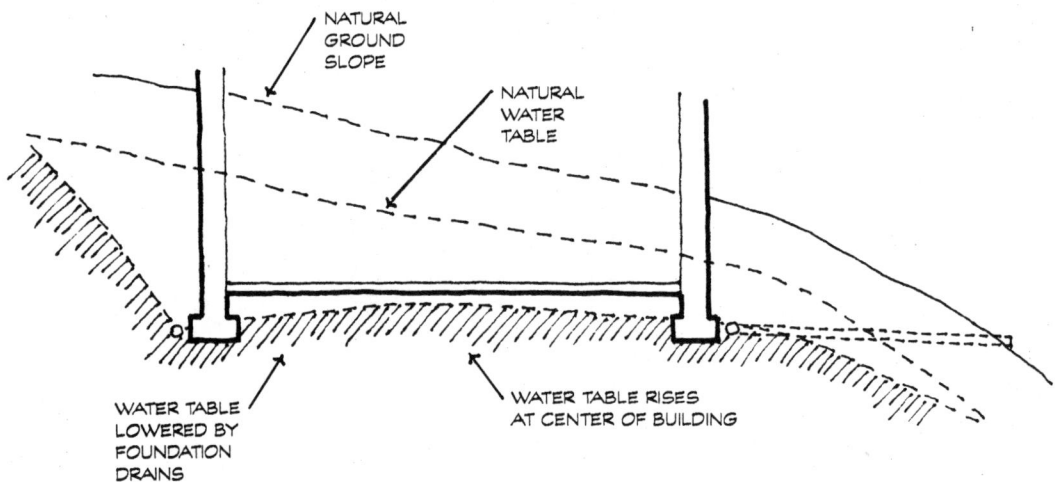

Figure 7.12 With perimeter drains, water table rises slightly in center of building. (*From Callendar, Timesaver Standards for Architectural Design Data.*)

drain screen since it functions in a manner similar to a rain screen on an above-grade wall—by neutralizing the forces which would typically push water through the enclosure.

In arid climates with little rainfall and deep water tables, or on well-drained sites with no history of groundwater problems, and no possibility of a rising water table, a drain screen type of subsurface drainage system may be all that is necessary to cope with water which may periodically enter the ground at the perimeter of a building. In moderate climates, a dampproof coating may be added to the basement wall to inhibit the absorption of any groundwater which might reach the wall surface. In wet climates, or on sites with high water tables, fluctuating water tables, or poor drainage, a waterproofing membrane will be necessary in addition to subsurface drains and drainage media. This redundant protection provides a second line of defense as a backup system, and affords a factor of safety under extreme conditions when drainage alone is not adequate to prevent water penetration. By draining groundwater away from the building and relieving hydrostatic pressure, though, a drain screen reduces the performance requirements on the waterproofing membrane, and makes the system more forgiving of minor application errors.

Drainage of the water collected by the perimeter drains should be by gravity outflow to an exposed lower elevation, a dry well that is above the water table, or to an approved storm sewer system. When these disposal methods are not feasible or practical, it will be necessary to collect the water in a sump and pump it out mechanically. Perimeter drains are often located just inside rather than outside the footings, particularly with a sump. Weeps should be located at the base of the foundation wall or at the top of the footing to allow any water which builds up on the outside of the wall to flow into the gravel bed inside the footing and then into the drain (*Fig. 7.14*).

7.2.3 Waterproofing membranes and dampproof coatings

The difference between waterproofing and dampproofing is one of degree. ASTM defines *waterproofing* as "treatment of a surface or structure to prevent the passage of liquid water under hydrostatic, dynamic, or static pressure" and *dampproofing* as "treatment of a surface or structure to resist the passage of water in the absence of hydrostatic pressure." Where waterproofing is defined in absolute terms as *preventing* water infiltration under any circumstances, dampproofing is defined in relative terms as *resisting*, but not necessarily preventing, water infiltration under limited circumstances.

Since a waterproofing membrane must withstand hydrostatic pressure, it is imperative that *all* holes, cracks, and openings in the substrate be eliminated. This is easier to do below grade than it is in above-grade walls because of the

TABLE 7.2 Depth and Spacing of Subdrains Recommended for Various Soil Classes

Soil classes	Percentage of soil separates			Depth of bottom of drain, ft	Distance between subdrains, ft
	Sand	Silt	Clay		
Sand	80–100	0–20	0–20	3–4	150–300
				2–3	100–150
Sandy loam	50–80	0–50	0–20	3–4	100–150
				2–3	85–100
Loam	30–50	30–50	0–20	3–4	85–100
				2–3	75–85
Silt loam	0–50	50–100	0–20	3–4	75–85
				2–3	65–75
Sandy clay loam	50–80	0–30	20–30	3–4	65–75
				2–3	55–65
Clay loam	20–50	20–50	20–30	3–4	55–65
				2–3	45–55
Silty clay loam	0–30	50–80	20–30	3–4	45–55
				2–3	40–45
Sandy clay	50–70	0–20	30–50	3–4	40–45
				2–3	35–40
Silty clay	0–20	50–70	30–50	3–4	35–40
				2–3	30–35
Clay	0–50	0–50	30–100	3–4	30–35
				2–3	25–30

SOURCE: Callendar, *Timesaver Standards for Architectural Design Data.*

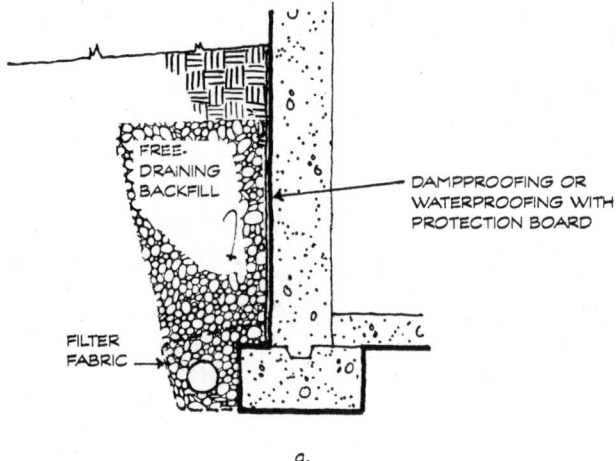

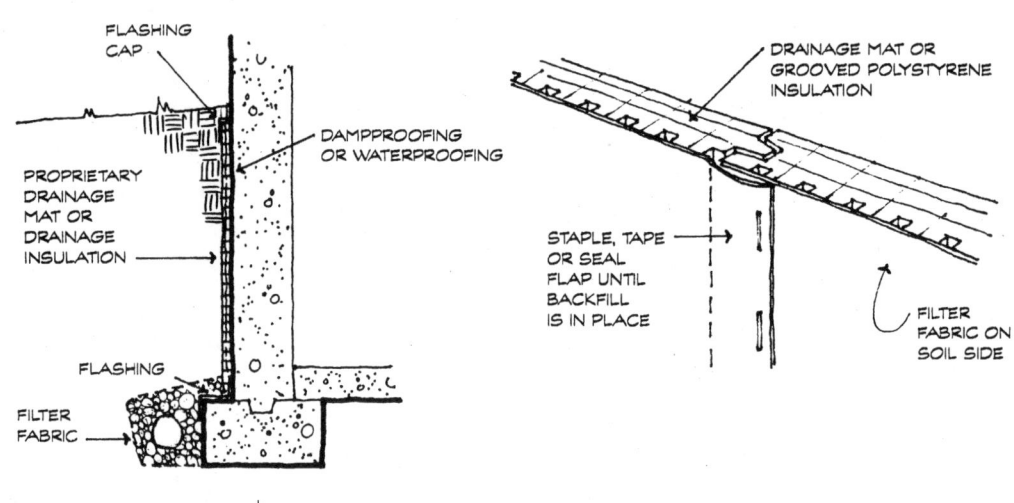

Figure 7.13 Subgrade drains can be used to relieve hydrostatic pressure. (*From Lstiburek and Carmody, Moisture Control Handbook.*)

absence of doors and windows, because there are fewer joints, because thermal expansion and contraction is less with smaller temperature variations, and because there is no ultraviolet deterioration of materials. Perfect barriers, however, are still difficult to achieve, and the barrier concept is very unforgiving of application errors. When combined with effective subsurface drainage, however, a waterproofing membrane can provide good performance even though human error will inevitably introduce minor flaws into the system.

Waterproofing membranes must be fully adhered to the substrate so that water cannot flow behind the membrane, and so that any leaks which occur will be easier to trace to the source. Membranes applied to concrete or concrete block must have sufficient flexibility to span cracks that will inevitably appear as a result of curing shrinkage, and enough elasticity to expand and contract with temperature changes. Steel reinforcing or control joints can be used to limit the amount of shrinkage cracking which will occur and to regulate the location of such cracks. If control joints are used, they must be sealed against water intrusion with an elastomeric sealant that will not deteriorate when submersed in water, that is chemically compatible with any membrane waterproofing or dampproofing which will be applied, and is resistant to any contaminants which may be present in the soil.

There are two general methods of waterproofing. In *positive side waterproofing,* the waterproofing is applied to the same side of the wall or floor on which the water source occurs (*Fig. 7.15*). In *negative side waterproofing,* the

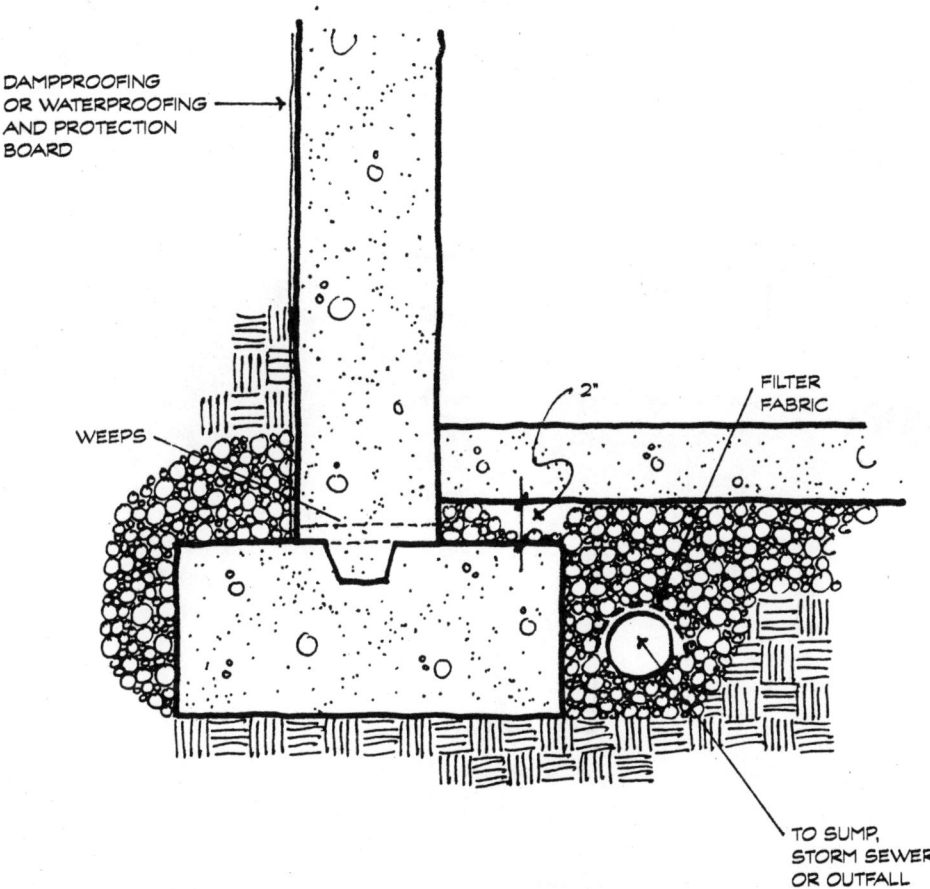

Figure 7.14 Drains inside footing with weeps.

waterproofing is applied on the opposite side of the structure as the water source (*Fig. 7.16*). Positive side waterproofing is always preferable because the structure itself is protected from moisture penetration, as well as the interior spaces. This is particularly important when reinforcing steel may be corroded by prolonged moisture exposure or chloride contamination from the soil. Negative side waterproofing is generally used only as a remedial measure in existing buildings where outside excavation and repair are impossible or prohibitively expensive. Where there is inadequate access to apply conventional waterproofing systems on the outside of below-grade walls, *blind side waterproofing* is possible with some specially developed systems. In these applications, the waterproofing is attached to a soil retention system such as timber lagging, sheet piling or shotcrete, and the concrete walls are then cast against it (*Fig. 7.17*).

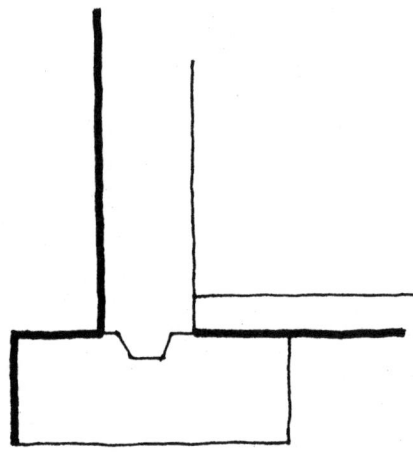

Figure 7.15 Positive side waterproofing.

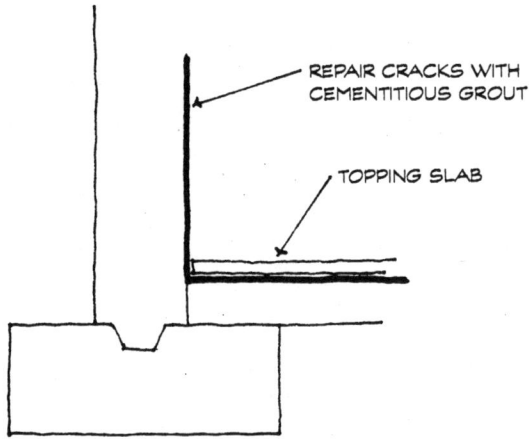

Figure 7.16 Negative side waterproofing.

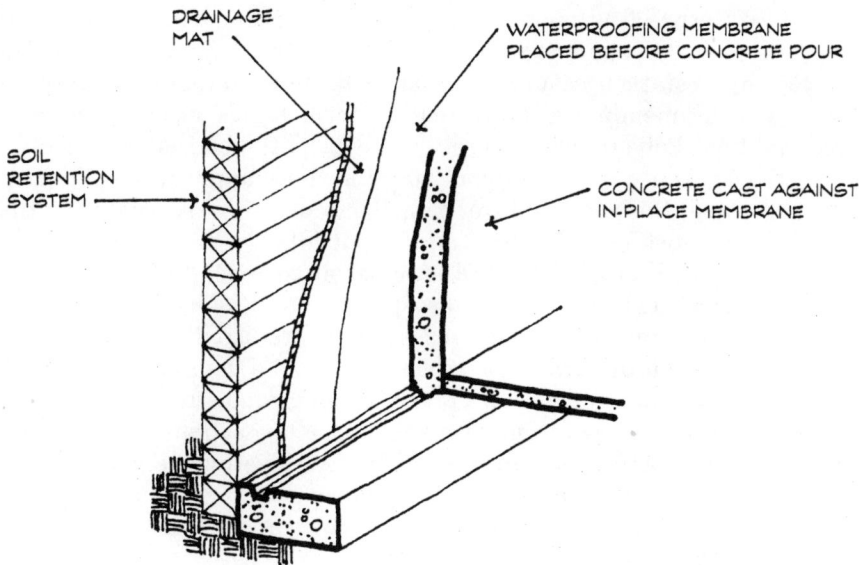

Figure 7.17 Blind side waterproofing.

Horizontal membranes for below-grade slabs are often cast on a thin "mud slab" and the structural slab is then cast on top, or the membrane is installed on the structural slab and a topping slab added as a wearing surface. This provides a stable subbase to support the waterproofing and a protective wearing surface above it. Some types of waterproofing can be placed on compacted subgrade fill and a single structural slab cast on top of it.

Dampproof coatings provide resistance to moisture penetration by closing the capillary pores in concrete and masonry substrates. Dampproofing will not resist moisture penetration under hydrostatic pressure, and the cementitious and mastic materials typically used for these coatings do not have the ability to bridge cracks. Where hydrostatic pressure against a below-grade wall is relieved by a subsurface drainage system, dampproofing may provide sufficient protection for the wall as long as there is no possibility of periodic hydrostatic buildup.

7.2.4 Vapor retarders

Where vapor migration from the soil is a potential problem, vapor retarders are necessary to protect the structure from a continuous flow of moisture. Where vapor-impermeable or moisture-sensitive floor finishing materials are to be used, vapor retarders are particularly important in preventing loss of adhesion, peeling, warping, bubbling, or blistering of resilient flooring or seamed continuous flooring. Vapor retarders can also prevent damage to flat electrical cables, buckling of carpet, and fungal growth and the offensive odors and indoor air quality problems that accompany it.

Waterproofing membranes with vapor permeability ratings below 1.0 also function as vapor retarders. In slabs on grade where there are no frost heave or hydrostatic pressure conditions which would require waterproofing, a vapor retarder membrane can be installed which does not have waterproofing capabilities. Polyethylene or reinforced polyethylene sheets of 6-, 8-, or 10-mil thickness are most commonly used in these applications. For maximum effectiveness, the vapor retarder must lap over and be sealed to the foundation; seams must be lapped 6 in. and sealed with the membrane manufacturer's recommended adhesive or pressure-sensitive tape; and penetrations for plumbing, electrical or mechanical systems must be sealed.

Vapor retarders under slabs on grade are usually installed over a base layer of free-draining gravel or crushed rock as a capillary break. Although vapor retarders themselves will prevent capillary moisture movement, they are usually used in conjunction with a drainage layer to provide a margin of safety in case of punctures or lap seam failures. Polyethylene vapor retarders are generally not reliable enough to function as part of a radon mitigation system, but waterproofing membranes with very low vapor permeability are acceptable for this use in most code jurisdictions.

ASTM E1643 *Standard Practice for Installation of Water Vapor Retarders Used in Contact with Earth or Granular Fill Under Concrete Slabs* provides guidance on the proper installation of these materials (*Fig. 7.18* through *7.20*). The granular base should be a minimum of 3 in. thick, and of compacted, mostly single-graded, coarse aggregate no larger than $^3/_4$ in. in size. To protect the vapor retarder from puncture, a $^1/_2$-in. layer of fine, compactable sand fill may be rolled over the base. To keep the sand from settling, a geotextile fabric can be placed over the coarse base material. Traditionally, a 2- to 4-in. layer of sand fill is added on top of the vapor retarder. There are two schools of thought on whether this is necessary.

In addition to providing a protection course on top of the vapor retarder, a layer of sand is thought by some to provide a cushion for the concrete and to act as a blotter to absorb excess moisture from the bottom of the slab. This supposedly promotes more even curing of the concrete, prevents excessive shrinkage cracking and slab curling, and permits earlier finishing of high-slump concrete with a high water-cement ratio. Others feel that the vapor retarder can be better protected by a geotextile fabric rather than sand, that low-slump concrete mixes should be used with high-range water reducers or superplasticizers added to provide workability, and that the blotter effect of the sand is not necessary to proper curing and finishing of the slab. In the 1980 edition of the American Concrete Institute's *Guide for Concrete Floor and Slab Construction* (ACI 302.1R), the fill material is described as "sand." In the 1989 edition, the fill is described as "approved granular, self-draining compactable fill." In the current edition, the composition and installation of the fill is more specifically described:

> If a vapor barrier [sic] or vapor retarder is specified due to local conditions, these waterproof membranes should be placed under a minimum of 4" of trimmable, compactable, self-draining granular fill (not sand). A so-called "crusher run" material (usually graded from 1-$^1/_2$" to 2" down to rock dust) is suitable. Following com-

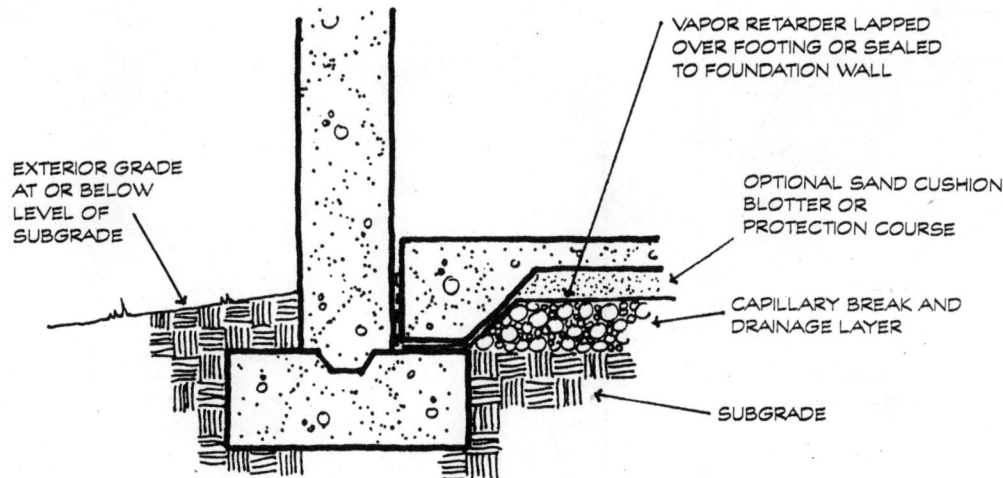

Figure 7.18 Optimum relationship of vapor retarder and waterproofing system components. (*Adapted from ASTM E1643, Standard Practice for Installation of Water Vapor Retarders Used in Contact with Earth or Granular Fill Under Concrete Slabs. Copyright ASTM. Reprinted with permission.*)

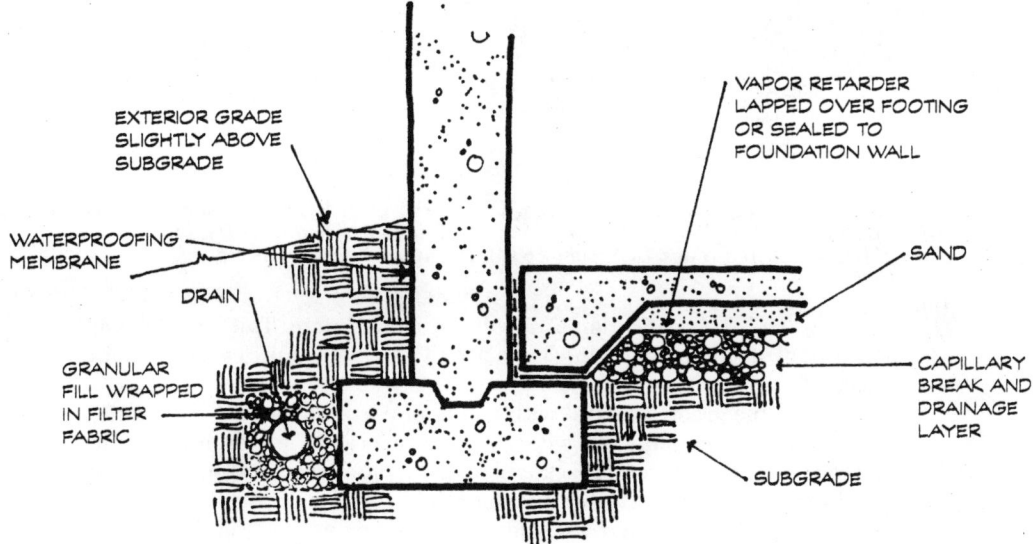

Figure 7.19 Vapor retarder and waterproofing components when exterior grade is slightly above subgrade. (*Adapted from ASTM E1643, Standard Practice for Installation of Water Vapor Retarders Used in Contact with Earth or Granular Fill Under Concrete Slabs. Copyright ASTM. Reprinted with permission.*)

paction, the surface should be choked off with a fine grade material to reduce subgrade friction.

The material should be compactable, trimmable granular fill which will remain stable and support construction traffic. The tire of a loaded ready-mix truck should not penetrate the surface more than $\frac{1}{2}$" when driven across it. The use of "cushion sand" or clean sand with uniform particle size such as concrete sand

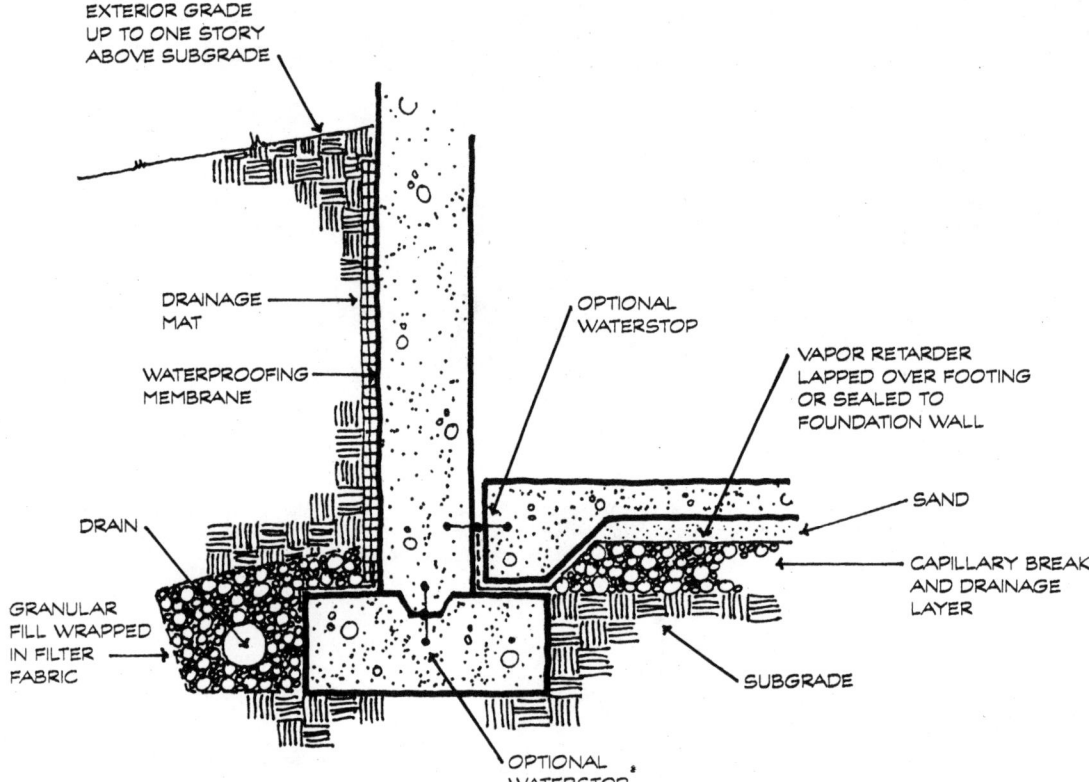

Figure 7.20 Vapor retarder and waterproofing components when exterior grade is up to one story above subgrade. (*Adapted from ASTM E1643, Standard Practice for Installation of Water Vapor Retarders Used in Contact with Earth or Granular Fill Under Concrete Slabs. Copyright ASTM. Reprinted with permission.*)

meeting ASTM Specification C33 will not be adequate. This type of sand will be difficult if not impossible to compact and maintain until concrete placement is complete.

A clean fine grade material with plenty of particles passing a #100 sieve (10 to 30%) but not contaminated with clay, silt or organic material is recommended. Manufactured sand from a rock crushing operation works well; the jagged slivers tend to interlock and stabilize the material when compacted. It is important that the material have a uniform distribution of particle sizes from #4 through #200.

The granular fill and the fine grade material should be sufficiently dampened to be compactable, but still be able to act as a "blotter" when the concrete is placed.

Reinforced polyethylene vapor retarders are more resistant to damage than unreinforced polyethylene, and are manufactured in multiple plies for greater strength. Some products are available with an integral geotextile fabric intended to eliminate the need for a protective sand fill material on top of the vapor retarder. If a sand cushion is not used, concrete mix designs should take into consideration the effect of a low-permeance vapor retarder on concrete curing, shrinkage, and drying time. Depending on the type of finish floor materials specified and the ambient conditions, concrete drying to acceptable moisture levels can take anywhere from 3 to 6 months. If scheduling is a potential problem, consider using a low-slump concrete so that there is a minimum amount of residual mixing water to evaporate after cement hydration has taken place.

7.3 Materials and Properties

Waterproofing membranes were traditionally constructed of multiple layers of roofing felt or fabric cemented together with hot applications of coal tar pitch or asphalt. There are many more options today, including cold-applied asphalt mastics with felt or fabric reinforcement, fluid-applied membranes, sheet membranes, preformed rubberized asphalt membranes, and earthen systems of bentonite clay.

Leaks most often occur at substrate cracks that a waterproofing membrane is not able to bridge. Over repeated cycles of temperature fluctuation, a good waterproofing membrane should be able to span a $1/_8$-in. crack without damage, and, while stretched across the opening, still resist hydrostatic pressure without rupture or leakage. Waterproofing membranes must also have the longevity to last the full service life of the building, since maintenance and replacement are not practical options. Some types of membranes have sufficient elasticity to span structural expansion joints, while others require special details and accessory items or materials. Whenever possible, waterproofing should be specified as a complete system from a single manufacturer to assure adequate detailing and compatibility of components. The manufacturer should always be consulted for recommended flashing, penetration, and termination details that are appropriate to the system. The following are the most commonly used commercial waterproofing materials and accessories.

7.3.1 Built-up membranes

Built-up membranes include both hot-applied and cold-applied bituminous materials of either coal tar pitch or asphalt. Since bituminous materials have very little innate tensile strength, they are reinforced with roofing felts or fabric and applied in multiple layers to achieve the required moisture resistance. The reinforcing adds the ability to withstand the strains of expansion, contraction, vibration, and building movement.

The felts used in built-up waterproofing membranes are the same as those used in roofing applications. Cotton fabric reinforcing should weigh a minimum of 10 oz/sq yd and have a successful history of use in similar applications. Glass fabric is stronger than both cotton fabric and roofing felt, and should have a minimum tensile strength of 75 lb/in. Wood nailers, pressure-treated with preservatives, are required at membrane terminations on vertical surfaces and at 6-ft vertical intervals throughout the height of the wall to permit installation of the membrane in sections (*Fig. 7.21*). Protection boards are used to prevent damage to the membrane from construction traffic or backfilling operations. Prefabricated asphalt protection boards vary in thickness from $1/8$ to $1/2$ in. and may incorporate glass fabric or organic felts for reinforcement; facings of polyethylene are added by some manufacturers for additional moisture resistance. In vertical applications, extruded polystyrene insulation boards are often used to protect the membrane when additional thermal resistance is also needed. Some insulation boards have drainage channels covered with filter fabric which facilitate the gravity flow of groundwater down to subsurface drains.

Hot-applied bitumens are heated in a kettle just as in roofing applications, and temperatures must be controlled to prevent overheating and thermal degradation of the material. Cold-applied bitumens are asphalt emulsions which must be fully cured before coming in contact with water or they will re-emulsify. For this reason asphalt emulsions are generally recommended for the less-stringent performance requirements of dampproofing rather than waterproofing applications.

Hot-applied asphalt waterproofing systems are inexpensive, but asphalt bitumens cannot tolerate standing water. Hot-applied coal tar systems are extremely durable, but unpleasant to work with. Cold-applied asphalt systems are inexpensive, can be sprayed or brushed on, and result in a seamless membrane. Built-up bituminous membranes have added puncture resistance, are more forgiving of rough substrates, and have an added margin of safety because of the multiple plies. Their full adhesion to the substrate prevents lat-

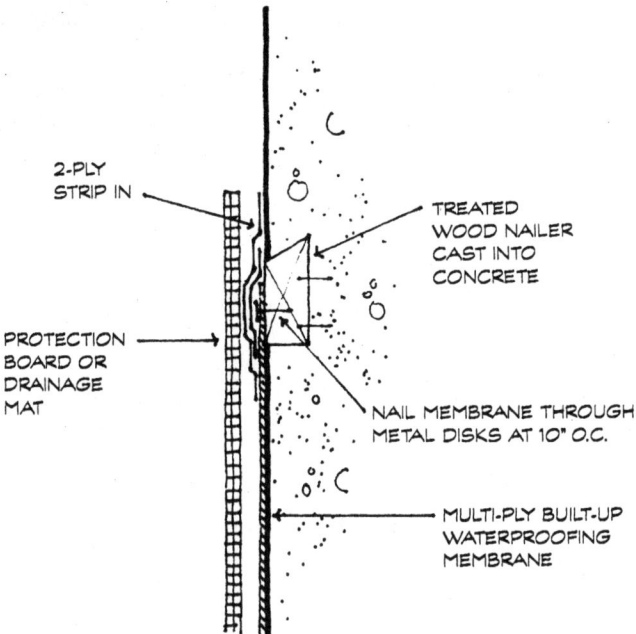

Figure 7.21 Vertical termination of built-up membrane. (*From ACI 515.1R, A Guide to the Use of Waterproofing, Dampproofing, Protective and Decorative Barrier Systems for Concrete.*)

eral migration of water. Both coal tar and asphalt have little elongation and poor crack spanning ability, and asphalt embrittles with age. The hot-applied systems are both labor- and equipment-intensive.

The number of plies required in built-up waterproofing depends on the amount of hydrostatic pressure to which the membrane will be subjected. The National Roofing Contractors Association *Roofing and Waterproofing Manual* recommendations are shown in *Table 7.3*. *Tables 7.4* and *7.5* indicate the appropriate ASTM standards for the various materials. The asphalt bitumen used in waterproofing is slightly different than that used in roofing. It has a lower viscosity and softening point which is more suitable for waterproofing applications. Coal tar pitch and asphalt bitumen are not chemically compatible, so asphalt-saturated felts should be used with asphalt systems, and coal–tar saturated felts with coal tar systems. Refer to App. B for a complete list of ASTM standards with numbers and titles.

Built-up waterproofing membranes cannot bridge expansion or control joints, or large cracks in a substrate. These must be reinforced with a loose laid rubber strip that is mopped into the membrane on either side of the opening so that expansion and contraction movements do not cause rupture or splitting (*Fig. 7.22*).

TABLE 7.3 Required Plies for Built-Up Waterproofing Membranes

Hydrostatic pressure head, ft H_2O	1–10	11–25	26–50
Number of plies	3	4	5

SOURCE: NRCA, *Roofing and Waterproofing Manual*, 4th ed.

TABLE 7.4 ASTM Standards for Hot-Applied Built-Up Waterproofing Membranes

Material	ASTM standard
Coal tar pitch systems:	
Coal tar bitumen	ASTM D450, Type II
Creosote primer	ASTM D43
Coal tar saturated organic felt	ASTM D227
Asphalt bitumen systems:	
Asphalt	ASTM D449, Type I, II or III
Asphalt primer	ASTM D41
Asphalt saturated organic felt	ASTM D226
Asphalt impregnated glass mat	ASTM D2178, Type IV
Fabrics:	
Bitumen-saturated cotton fabric	ASTM D173
Glass fabric	ASTM D1668

TABLE 7.5 ASTM Standards for Cold-Applied Asphalt Emulsion Membranes

Material	ASTM standard
Asphalt emulsion	ASTM D1227
Glass fabric	ASTM D1668, Type III

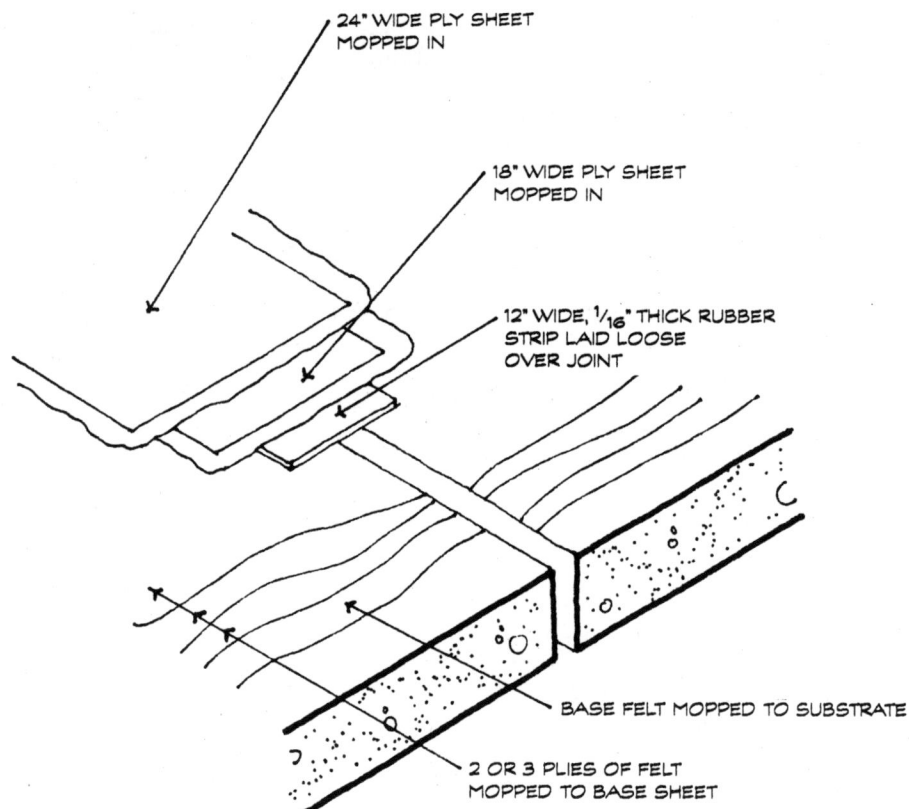

Figure 7.22 Built-up membrane spanning small joint or crack. (*From NRCA, Roofing and Waterproofing Manual, 2d ed.*)

Bituminous mastics and emulsions are also used for dampproof coatings, but are not built up with layers of felt or fabric reinforcing. They may be applied by spray, roller, or trowel. Trowel mastics are applied in a single $\frac{1}{8}$-in.-thick coat. Spray and brush applications should consist of two $\frac{1}{16}$-in. coats. *Table 7.6* lists ASTM standards for several types of bituminous dampproofing products.

7.3.2 Fluid-applied membranes

Fluid-applied membranes include both hot- and cold-applied single- and multi-component elastomeric products such as polychloroprene (neoprene),

neoprene-bituminous blends, polyurethane, modified polyurethane, polyurethane-bituminous blends, epoxy-bituminous blends, and hot rubberized asphalt. Fluid membranes are applied by spray, roller, or trowel, and cure to a rubbery coating. Each type of system has special application procedures that are peculiar to the system, some requiring glass fabric reinforcement, at least over joints or cracks more than 60 mils wide. Most systems require a dry film thickness of at least 50 mils where traffic is expected, and all require application in multiple coats to minimize the effects of pinholes, which are a characteristic of fluid-applied membranes. Even multiple coats, however, will not cover, hide, or level surface irregularities. Fluid-applied membranes should not be used over lightweight aggregate concrete or concrete block, or over thin veneers containing polyvinyl acetate (PVA) or latex bonding agents.

Fluid-applied membranes are sensitive to ambient weather, temperature, precipitation, and humidity. Manufacturers generally set limitations on temperatures for application because below about 40°F, many of the materials are too viscous to form a continuous film of the correct thickness, and the adhesive bond to the substrate is weak. Rain during or immediately after application can also adversely affect the development of required mechanical properties. The seamless membranes require few joints and are fully bonded to the substrate, so lateral water migration is prevented. Fluid-applied membranes are quick to apply, have relatively low in-place costs, and provide good elongation, but there may easily be inconsistencies in membrane thickness because of application errors. Fluid-applied membranes are not recommended over lightweight insulating concrete because of problems with membrane blistering.

Hot rubberized asphalt is perhaps the most popular of the fluid-applied membranes. It is normally installed at 150 to 210 mils thick. Although puncture resistance is low, hot rubberized asphalt is self-healing of minor punctures, and more forgiving on rough substrates than some other membrane types. Rubberized asphalts can be applied at temperatures as low as 0°F, they have good crack-bridging ability, tolerate bad weather almost immediately after application, and can receive application of a protection board almost immediately after application. The materials are packaged and shipped in solid form, and heated at the job site to proper temperature. Material that is not used in one day can be reheated and used the next day, which eliminates waste. Hot

TABLE 7.6 ASTM Standards for Bituminous Dampproofing Materials

Material	ASTM standard
Solvent-based trowel mastic	ASTM D2822, Type I
Solvent-based semi-mastics	ASTM D2823
Non-fibrated bituminous emulsion	ASTM D1227, Type III ASTM D1187, Type I or Type II
Fibered bituminous emulsion	ASTM D1227, Type I or Type IV ASTM D1187, Type I or Type II

rubberized asphalt, however, is better suited to horizontal than to vertical applications.

Modified polyurethane is also a popular fluid-applied waterproofing membrane, typically applied in a 60-mil thickness. It is a versatile material, is easy to apply and economical, and can be used on both vertical and horizontal surfaces. The crack-spanning ability of modified polyurethane is less than that of rubberized asphalt and other fluid-applied membranes, and the material is sensitive to moisture during application and to concrete curing compounds. The two-component elastomers cure faster in cooler weather and under higher humidity than one-component formulations. The substrate must be in excellent condition for a successful application, and the membrane must cure for 24 to 48 hours before a protection board can be put in place. Modified polyurethane also has limited pot life and fairly low puncture resistance.

Fluid-applied waterproofing membranes cannot span expansion and control joints. A rubber strip is used at moving joints to permit the membrane to expand and contract without rupture or splitting (*Fig. 7.23*). There are three applicable standards for cold-applied systems:

ASTM C836 *Standard Specification for High Solids Content, Cold Liquid-Applied Elastomeric Waterproofing Membrane for Use With Separate Wearing Course*

ASTM C957 *Standard Specification for High-Solids Content, Cold Liquid-Applied Elastomeric Waterproofing Membrane with Integral Wearing Surface*

ASTM C1127 *Standard Guide for Use of High Solids Content, Cold Liquid-Applied Elastomeric Waterproofing Membrane with an Integral Wearing Surface.*

7.3.3 Sheet membranes

Sheet membrane waterproofing includes thermoset and thermoplastic materials. Thermoset membranes may be vulcanized or non-vulcanized materials as well as preformed rubberized asphalt sheets. Vulcanized rubber sheets include EPDM (ethylene propylene diene monomer), neoprene (polychloroprene), and butyl (polymerized polyisobutylene) rubber sheets, generally of 45- to 60-mil thickness, and reinforced. Vulcanized sheets should conform to the requirements of ASTM D4637 *Standard Specification for Vulcanized Rubber Sheet Used in Single-Ply Roof Membranes.* Non-reinforced EPDM rubber sheets should meet the requirements of ASTM D4637, Type I, Class U. Reinforced EPDM and neoprene rubber sheets should meet the requirements of ASTM D4637, Type I, Class SR. Non-vulcanized materials include chlorosulfonated polyethylene (Hypalon or CSPE) and chlorinated polyethylene (CPE), generally reinforced with polyester and seamed by heat welding. Polyvinyl chloride (PVC) is a thermoplastic membrane. Sheet membranes have the advantage of allowing protection board placement and backfilling operations to begin immediately after application. Most sheet membranes can be fully adhered to the

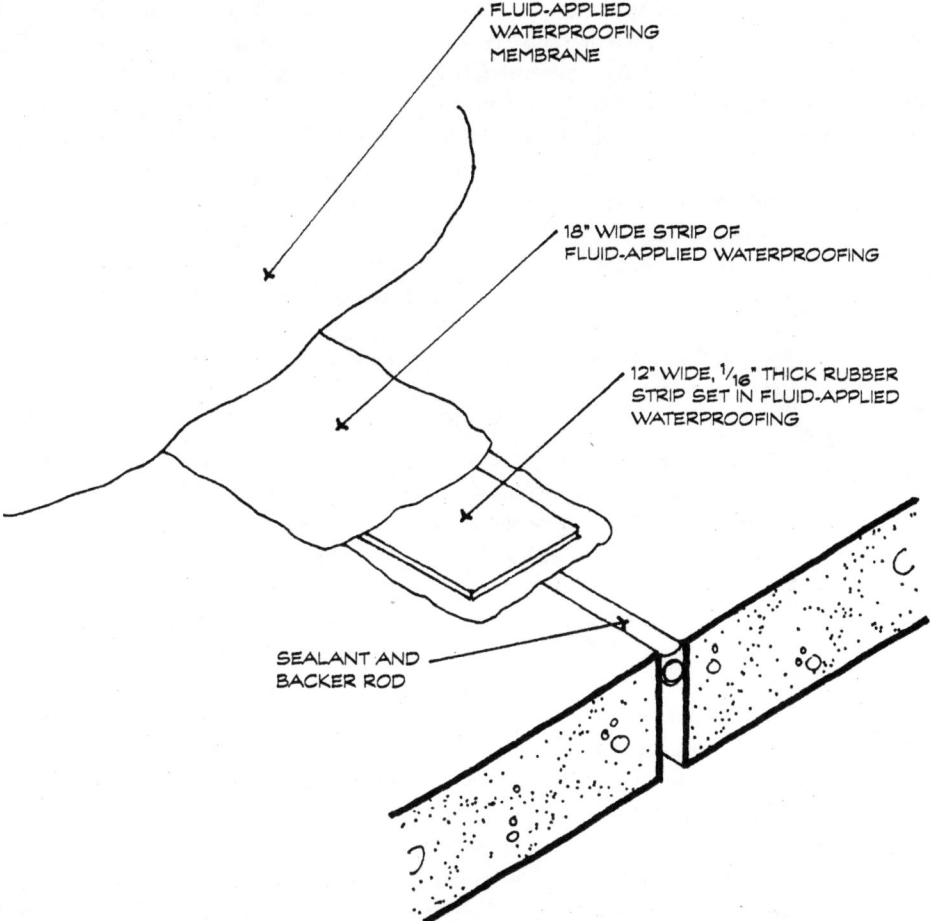

Figure 7.23 Fluid-applied membrane spanning small joint or crack. (*From NRCA, Roofing and Waterproofing Manual, 2d ed.*)

substrate, or loose laid. Fully adhered systems prevent lateral migration of water under the membrane and are not as vulnerable to leaks caused by seam failures as loose-laid systems. Cutting and splicing at penetrations is time-consuming and expensive. Unlike built-up and fluid-applied membranes, sheet membrane waterproofing materials generally have sufficient flexibility to span across typical expansion and control joints without the need for additional stripping and flashing. In fully adhered systems, the adhesive is omitted over substrate joints. Follow the manufacturer's recommended detailing and maximum joint size and spacing recommendations for each type of sheet membrane material.

Vulcanized rubber membranes have excellent crack-bridging ability and high elongation properties to withstand thermal expansion and contraction movements without rupture. Field seams for neoprene and EPDM membranes are spliced with an adhesive. Field seams for butyl membranes are chemically welded with a solvent cement. Butyl membranes can be used on vertical or horizontal surfaces, but EPDM and neoprene are usually limited to horizontal applications. Standing water on adhesive field seams can result in leaks if even minor workmanship defects occur. Vulcanized membranes puncture relatively easily, and applying flashing at penetrations is very time-consuming. Adhesives and cements should be supplied or recommended by the membrane manufacturer.

Non-vulcanized rubber membranes are more resistant to punctures and have hot-air- or solvent-welded rather than adhered or cement-welded seams for greater seam integrity. They are generally reinforced with a polyester mat. Some CPE formulations are subject to plasticizer loss and subsequent embrittlement. Non-vulcanized rubber membranes can be used on vertical or horizontal surfaces, and can withstand ponded water.

PVC membranes have good puncture resistance and crack-spanning ability. Seams are hot-air-welded and provide greater integrity than adhered seams. PVC membranes are generally available in 80- to 120-mil thicknesses. Some formulations are subject to plasticizer loss and subsequent embrittlement. Request manufacturer's data on accelerated weathering and other laboratory tests to assure long-term durability. PVC membranes should conform to the requirements of ASTM D4434 *Standard Specification for Polyvinyl Chloride Sheet Roofing.*

Rubberized asphalt is sometimes referred to as polymer-modified asphalt or modified bitumen. Rubberized asphalts are very popular waterproofing membranes because they are self-adhering, easy to install, and self-healing at small punctures. Seaming at lap splices does not require solvents or adhesives because the membrane sticks to itself to form a tight seal. The asphalt membrane is laminated to a polyethylene film on one side to facilitate handling, and to a siliconized release sheet on the adhesive side. The release sheet is removed immediately before application, and the membrane is rolled to assure full contact with the substrate. Primers may be required to assure good adhesion to concrete surfaces, especially when form-release compounds have been used. Penetrations are easy to seal because of the self-adhesive properties. Expansion and control joints must be reinforced with a separate strip of membrane (*Fig. 7.24*).

7.3.4 Bentonite clay waterproofing

Bentonite is a natural material consisting largely of a mineral called *montmorillonite,* named for the city of Montmorillon, France, where it was originally discovered. In the United States, bentonite is mined primarily in Wyoming and Montana. Bentonite is a soft, waxy clay with extremely fine particles of flat, platelike shape. Its natural affinity for water creates a cohesiveness and plasticity from the surface tension of very thin layers of water between each of these minute particles. As the bentonite absorbs moisture from the soil, it swells to more than 15 times its original volume and, when confined under

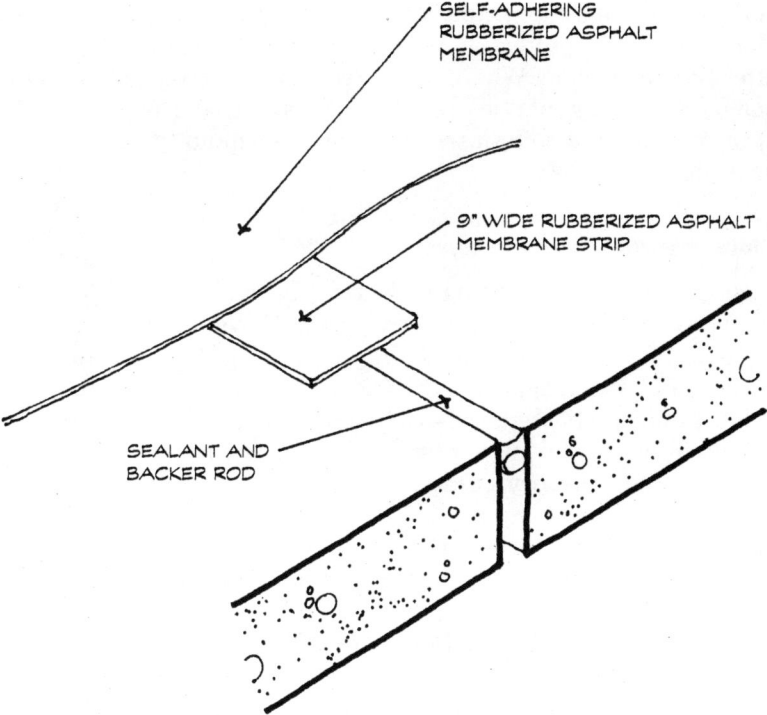

Figure 7.24 Rubberized asphalt membrane spanning small joint or crack.

pressure, it forms an impermeable barrier against subsequent water penetration. A minimum of 30 to 60 lb/sq ft pressure is required to seal the membrane to the substrate. The higher the pressure, the better the seal. Backfill materials and concrete slabs generally provide sufficient pressure to assure good performance. Bentonite can be fabricated into panels between layers of corrugated cardboard for easy handling, and is used for both vertical and horizontal applications. Bentonite can also be sprayed in place for horizontal applications usually at a coverage rate of $1\frac{1}{2}$ to 2 lb/sq ft.

Bentonite waterproofing is self-healing and can be applied to fresh concrete walls which still contain latent moisture. Panels are mechanically attached to walls until the backfill secures it permanently in place. In some horizontal applications, the bentonite can be laid directly on the subgrade, eliminating the need for a mud slab. Bentonite panels can also be used in blind side waterproofing of property line walls where the concrete is cast after the waterproofing is applied to timber lagging or other soil retention system. The moisture in the wet concrete activates the clay, and the swelling membrane forms a tight seal against the wall. Bentonite must be kept dry before it is backfilled or covered by a structural slab, and polymer-modified bentonite must be used in areas where the soil or groundwater is contaminated by chlorides or salt water. There are no temperature restrictions for application, the membrane is very forgiving of rough or uneven substrates, the intimate bond with the substrate surface prevents lateral migration of water, and no protection board is generally needed. If a protection board is required, it can be installed immediately after the membrane. Where vapor resistance is required, a separate polyethylene membrane should be used. At expansion and control joints, a second layer of membrane is usually used to provide adequate protection for the expected movement (*Fig. 7.25*).

7.3.5 Cementitious, metallic oxide, and crystalline coatings

Cementitious waterproofing is inexpensive and easy to apply, but has no elasticity and cannot tolerate joint or crack movement. With acrylic latex additives, bond to the substrate is improved as well as durability, cohesion, and tensile and flexural strength. Cementitious waterproofing can be used in below-grade applications where thermal expansion and contraction are accommodated by movement joints. Cementitious coatings can be applied to both concrete and masonry surfaces and are often used in residential basement applications. Accessory materials are available to seal active leaks, cracks, penetrations, cants, coves and fillets. Coating application is usually $1/8$ to $1/4$ in. thick. Cementitious waterproofing can be used in positive side or negative side applications.

Metallic oxide waterproofing materials include mixes of fine metallic materials which chemically react with, oxidize, penetrate, or expand into the capillaries of porous concrete to reduce its permeability. As many as three to five coats may be required for effectiveness, and moist curing is essential to promote oxidation. A sand/cement coat is usually applied as a finish. These systems require high-quality workmanship and close field supervision to assure that proper mixing and application procedures are followed. Metallic oxide

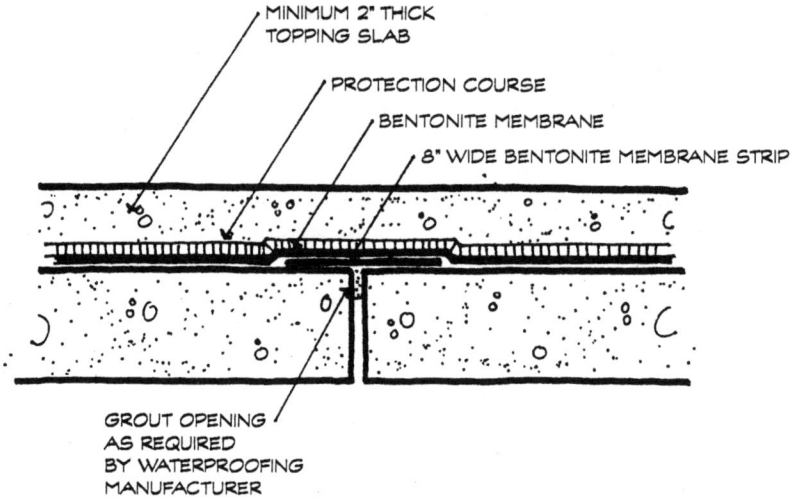

Figure 7.25 Bentonite clay waterproofing spanning small joint or crack. (*Adapted from American Colloid Co.*)

waterproofing is generally used on negative side applications for remedial projects where positive side access is not possible. Metallic oxide waterproofing is rigid, cannot bridge cracks, and cannot tolerate thermal expansion and contraction of the substrate.

Crystalline waterproofing is a compound of cement, quartz or silica sand, and other active chemicals. It may be applied to damp or uncured concrete and, in fact, requires moisture to activate the material. Like metallic oxides, crystalline waterproofing is generally used on negative side applications for remedial projects where positive side access is not possible. Crystalline waterproofing is rigid, cannot bridge cracks, and cannot tolerate thermal expansion and contraction of the substrate.

7.3.6 Waterstops and compression seals

One of the most common sources of water leakage in below-grade structures is through construction, control, and expansion joints. At joints in concrete, waterstops of stainless steel, PVC, rubber, polyethylene, or bentonite clay can be used to form watertight seals. Different shapes and sizes are available to accommodate various expected joint movements (*Fig. 7.26*). Flat ribbed waterstops, dumbbell waterstops, and preformed strips of bentonite clay are designed for use only in joints where no movement is expected. Waterstops with a center bulb will accommodate lateral, transverse, and shear movements. Where high service temperatures are expected, stainless steel waterstops are generally required.

There is some controversy regarding the effectiveness of waterstops. These products are installed at the time of the concrete pour and can easily be damaged during concreting operations. NRCA does not recommend the use of waterstops.

Compression seals are designed to accommodate large movements in horizontal joints. Plaza deck joints that are subject to vehicular traffic are usually formed of a combination of neoprene extrusions and steel armor plates that interlock with the concrete (*Fig. 7.27*). In order to resist water penetration, the seal must be under compression. The neoprene webbing must be flexible through a wide range of temperatures, have good recovery even at low temperatures, and be mechanically interlocked with the steel frame so that the seal is held securely in place.

7.3.7 Flashing

Flashing is required to form terminations and transitions and to seal penetrations in waterproofing systems. No matter how good a membrane is, if the details are not properly designed and installed, the system will fail. As one architect put it, "It's always the details that leak."

Flashing may be formed of strips of the same material as the waterproofing membrane, of a compatible material, or of metal, depending on the type of system and the particular detail. Flashing is perhaps the most important element

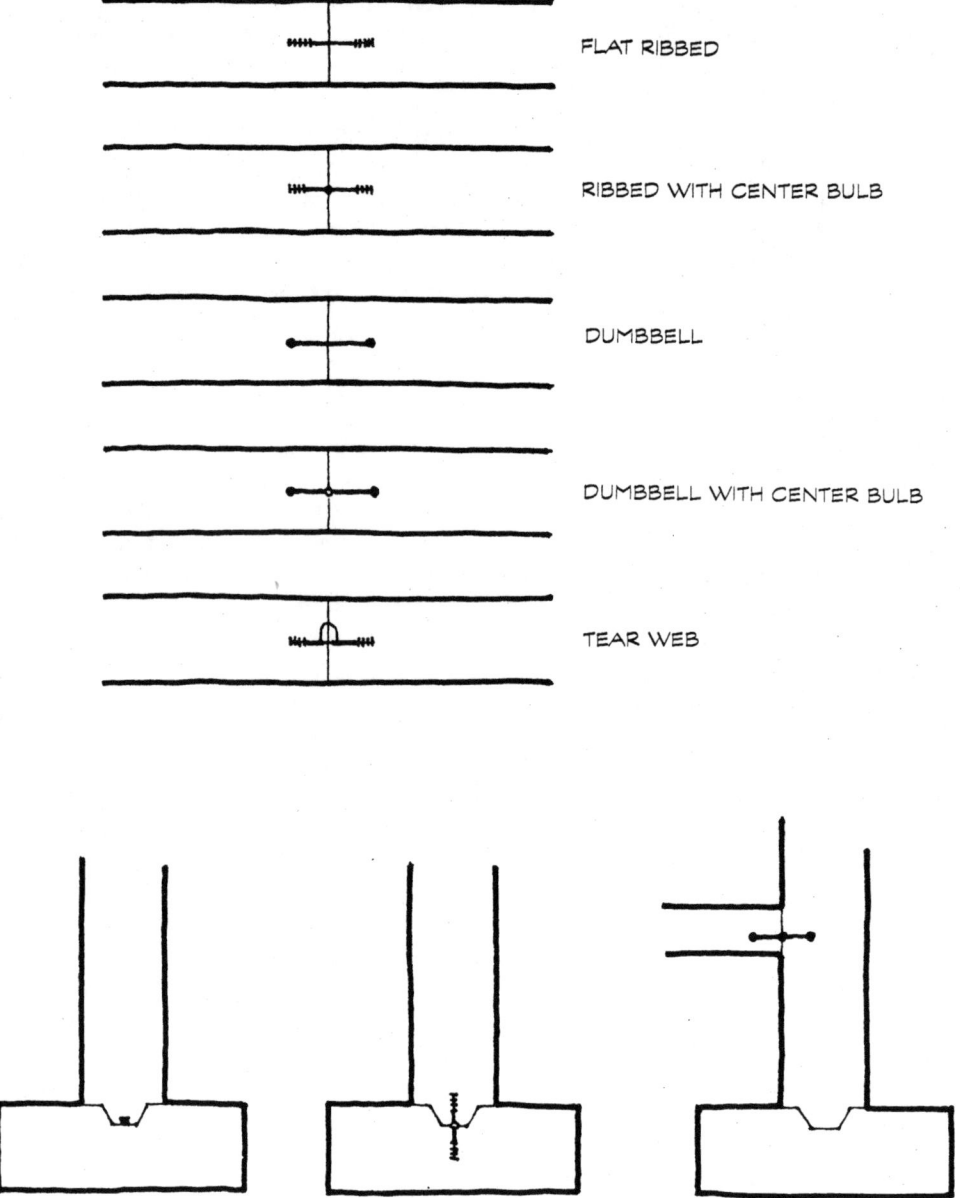

Figure 7.26 Waterstops come in several types.

of a waterproofing system, and also the component most likely to fail. It is critical that flashing details be reviewed with the system manufacturer and amply described and illustrated in the construction documents. Isometric drawings can help in working out suitable solutions to difficult or complex transitions, and in providing the applicator with a clear understanding of what the installation entails. Penetrations should be kept to a minimum, and flashing should be designed to provide simple but effective protection for all exposed edges of the waterproofing membrane.

7.4 Slabs on Grade and Crawl Spaces

Except on very poorly drained sites, most slabs and foundations on grade are not subject to hydrostatic pressure from groundwater. Moisture can enter the structure, however, by capillary action and by vapor migration. Winter heat loss from the slab can also affect frost heave as the moisture in soils freezes and expands. In Chap. 3, the use of perimeter insulation was described as a means of controlling heat flow to permit the use of shallow footings even in areas where the winter frost depth normally requires deep footings (refer to *Figs. 3.26* through *3.30* in Chap. 3). Vapor retarders and capillary breaks can be used in conjunction with perimeter and underslab insulation to control both heat and moisture flow in concrete and masonry foundations (*Figs. 7.28* and *7.29*).

In structures with elevated slabs or framed construction over an open crawl space, adequate ventilation must be provided to dissipate soil moisture vapor and prevent its being drawn into the interior. A minimum of four openings should be provided (one at each corner), placed as high in the foundation wall as possible. The required net area of the openings can be calculated by the formula

$$a = \left(\frac{2L}{100}\right) + \left(\frac{A}{300}\right) \tag{7.1}$$

where a = required net area of all vents, sq ft
L = perimeter of crawl space, lin ft
A = area of crawl space, sq ft

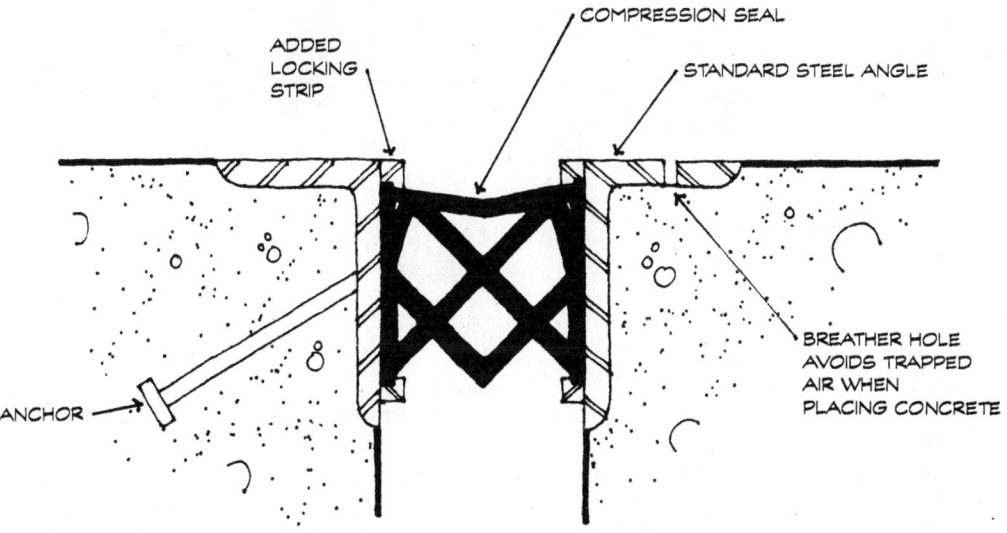

Figure 7.27 Compression seals are used for large horizontal joints. (*From ACI 504R, Guide to Joint Sealants for Concrete Construction.*)

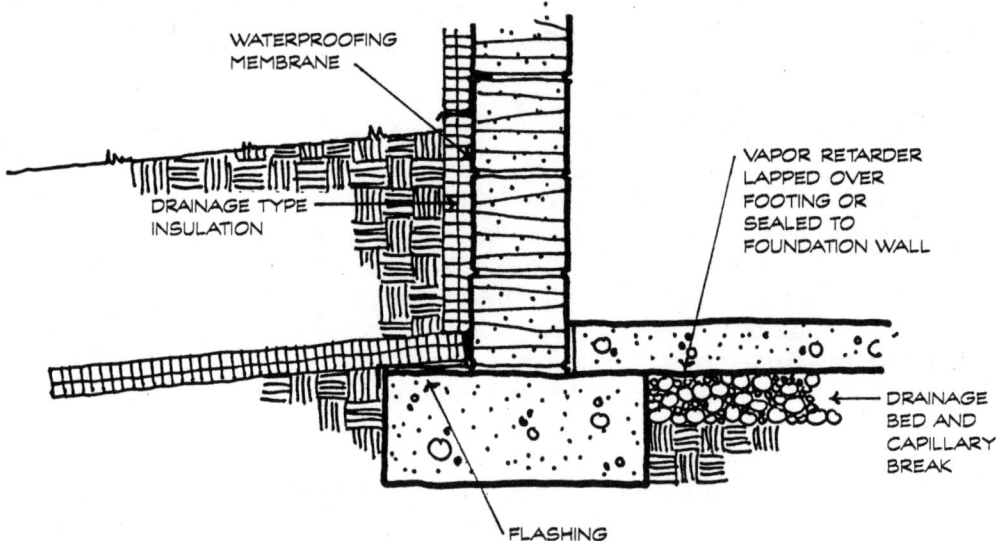

Figure 7.28 Vapor retarder and waterproofing for shallow footing.

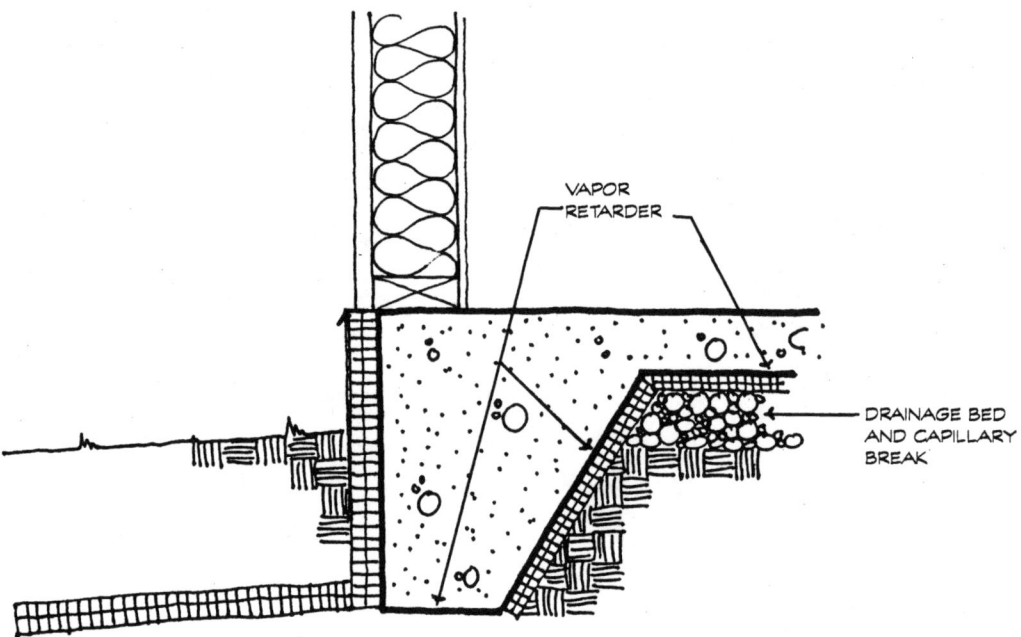

Figure 7.29 Vapor retarder for slab on grade.

Net and gross ventilator areas for different types of screens and louvers are given in *Table 7.7*. After calculating the required net area, multiply by the coefficient shown to determine the overall size or gross area of ventilators needed. Ventilation will dissipate soil moisture vapor, but it also will cool the underside of the floor sufficiently to require insulation to control winter heat loss. An alternative control measure is to provide a covering for the exposed soil. With a vapor retarder of polyethylene film, heavy roll roofing (55 lb), or a proprietary membrane, ventilation could be reduced to as little as 10% of that calculated as necessary for the uncovered soil. Where a crawl space is provided below wood-framed construction, the wood should be separated from the exposed soil by the minimum distances shown in *Fig. 7.30*. In addition to separating the wood framing from the vapor source and allowing for ventilation, these clearances assure adequate access for visual inspection. Particularly in warm, moist climates that are subject to subterranean termite infestations (refer to *Fig. 1.5* in Chap. 1), periodic visual inspection is an important factor in long-term maintenance.

TABLE 7.7 Net and Gross Ventilator Areas*

Ventilator covering	Coefficient
$1/4$" mesh hardware cloth	1
Screening, 8 mesh/in.	1.25
Insect screen, 16 mesh/in.	2
Louvers plus $1/4$" mesh hardware cloth	2
Louvers plus screening, 8 mesh/in.	2.25
Louvers plus insect screen, 16 mesh/in.	3

*Gross ventilator area = required net area × coefficient

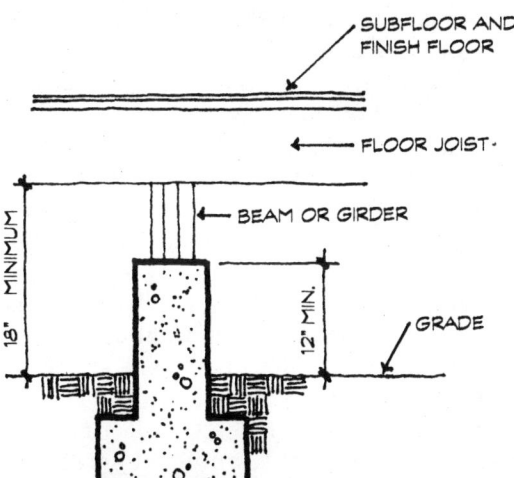

Figure 7.30 Minimum height of wood framing above exposed subgrade.

7.5 Below-Grade Walls and Floors

Some building codes dictate the use of either dampproofing or waterproofing on below-grade structures, depending on the nature of the enclosed space and the ground water conditions. Appropriately installed *dampproof coatings* can resist capillary moisture penetration, but *waterproof membranes* are required to resist hydrostatic pressure. Where no specific code mandates exist, the decision to provide footing drains or a drain screen, dampproofing, waterproofing, or vapor retarders on below-grade walls and floors should be based on the amount of moisture present in the soil, the level of the water table, and the sensitivity of the occupied space for which protection must be provided. Where a below-grade parking structure might tolerate a small amount of moisture seepage under extreme weather conditions, spaces designed for human occupancy or which house mechanical or electrical equipment require higher levels of protection. If there is any chance that a utilitarian space might subsequently be used as habitable space, even by a second or third building owner, the prudent course of action should provide the maximum protection appropriate to the soil-moisture conditions. If the water table may fluctuate under different seasonal or weather conditions, protective measures should include a waterproof membrane capable of withstanding hydrostatic pressure, regardless of whether footing drains or a drain screen are provided. If steel reinforcing is used in concrete or masonry basement walls (including joint reinforcement in concrete masonry), sufficient protection must be provided on the positive (water source) side to prevent moisture from entering the wall and corroding the metal.

For dry or well-drained soils with low water tables, *Fig. 7.31* illustrates appropriate drainage and dampproofing measures. Basement and below-grade walls of concrete masonry are very porous and must be protected against capillary penetration of moisture from the soil. For units with a coarse texture, the surface should first be parged with cement mortar to fill large voids and smooth the surface. A bituminous dampproofing can then be sprayed, brushed, or troweled over the cured parging. If a parge coat is to be applied, mortar joints should be struck flush. If a bituminous dampproofing is to be applied directly to the masonry without a parge coat, the joints should be tooled concave.

Parging consists of a $3/8$- to $1/2$-in.-thick coating of either Type M masonry mortar or a portland cement and sand mortar mix applied in two layers of approximately equal thickness. Type M mortar should conform to the requirements of ASTM C270 for masonry mortar. A portland cement–sand mix should be proportioned 1 part cement to $2\frac{1}{2}$ parts sand by volume using ASTM C150 portland cement and ASTM C144 masonry sand. The masonry should be dampened before parging. The first coat, called a *scratch coat,* should be roughened or scratched to form a mechanical bond with the finish coat. The scratch coat should be allowed to cure for at least 24 hours, then dampened immediately before application of the second coat. This finish coat should be troweled to form a dense surface, and a cove should be formed at the base of the foundation wall to prevent water from accumulating at the wall-footing juncture. The finish coat should be moist-cured for 48 hours to minimize shrinkage cracking and assure complete cement hydration.

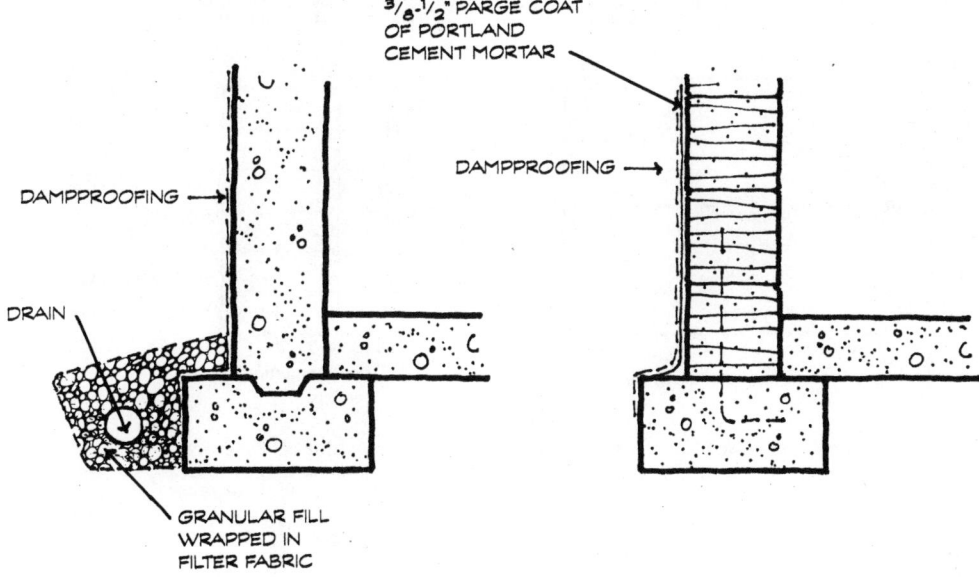

Figure 7.31 Dampproofing for dry or well-drained soils with low water tables.

Bituminous dampproofing can be applied directly to the surface of concrete. Cracks and voids such as form tie holes should be patched or filled in accordance with the membrane manufacturer's recommendations. Mastic dampproofing cannot bridge cracks in the substrate. The permeability of the concrete itself is affected by the water-cement ratio and the length of curing time. The lower the water-cement ratio, the lower the permeability and porosity of the concrete (*Fig. 7.32*). Workability can be maintained with low water-cement ratio mixes by adding a water-reducing (superplasticizing) admixture. Moist curing conditions should be maintained for at least 7 days to obtain a dense, compact concrete. In hot, dry, or windy weather, extra precautions should be taken to prevent premature evaporation of curing water and to assure sufficient moisture for complete cement hydration. The vapor permeability of the interior coating of a dampproofed concrete or masonry wall should be higher than the vapor permeability of the exterior coating so that construction moisture and any soil moisture vapor which permeates the wall can dry to the inside (*Fig. 7.33*).

Other dampproofing materials such as acrylic latex coatings and acrylic modified cement coatings are acceptable under some building codes. For below-grade applications, consult the manufacturer to verify the suitability of the products. Residential basements are often dampproofed by wrapping the walls with 6-mil polyethylene sheets, but these are easily damaged during backfilling.

For wet soils with a high water table or a water table which may fluctuate seasonally or under severe weather conditions, and for deep foundations in multi-story below-grade structures, *Fig. 7.34* illustrates appropriate drainage and waterproofing techniques. Slabs can be waterproofed in different ways,

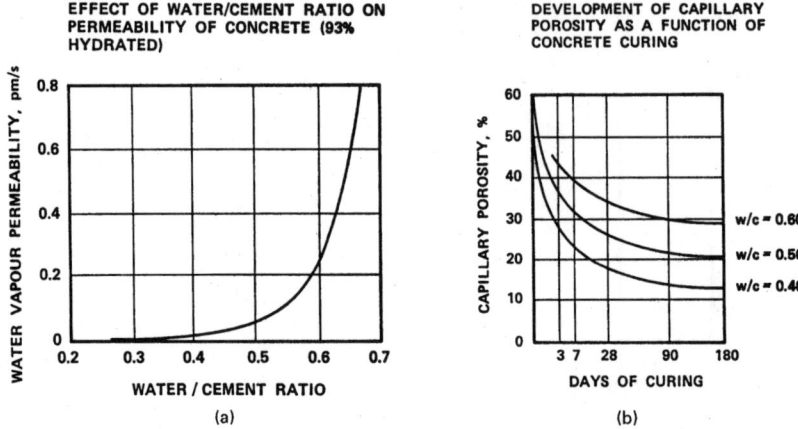

Figure 7.32 Water/cement ratio and curing time affect permeability of concrete. (*From National Research Council of Canada, Performance of Materials in Use.*)

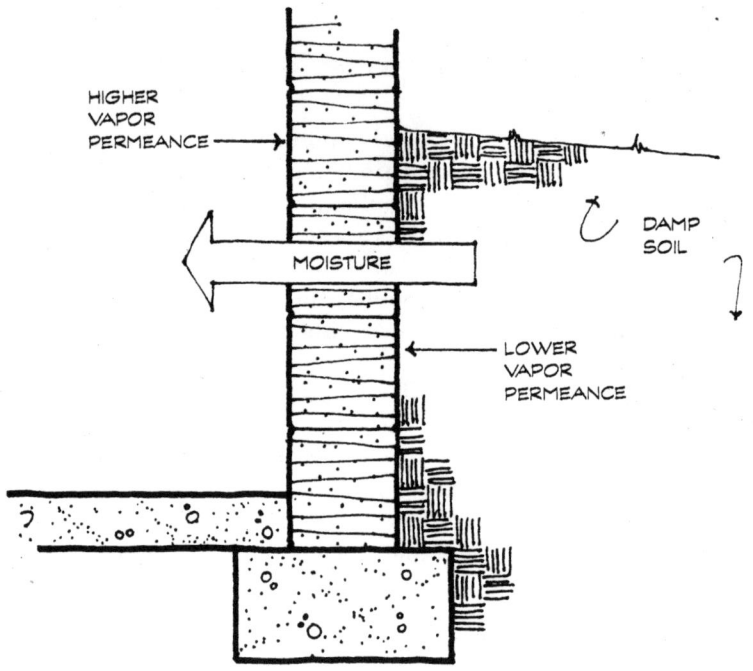

Figure 7.33 Vapor movement through basement walls.

depending on the type of membrane being used. Bentonite waterproofing and some rubberized asphalt systems can be applied directly on a prepared base and the structural slab cast on top, but most membranes require placement of a 3- to 4-in. thick mud slab to act as a substrate for membrane application. After the waterproofing is installed, the structural slab is cast in place. If there will be a delay in placement of the structural slab, a 2-in. protection slab can be poured over the waterproofing to prevent damage to the membrane from construction traffic (*Fig. 7.35*).

Joints in the waterproofing membrane should be lapped and sealed as recommended by the membrane manufacturer. Flashing details for vertical and horizontal pipe penetrations, wall-to-footing intersections, and membrane transitions or terminations at grade should be as shown in *Figs. 7.36* through *7.41* from the National Roofing Contractors Association *Roofing and Waterproofing Manual*, or as recommended by the membrane manufacturer. Built-up membranes always require a cant strip to eliminate 90° bends in the stiff base felts. Without a cant strip, a void is formed at the sharp bend, and pressure from the topping or backfill can break the felts along the bend line.

7.6 Plaza Decks

Plaza decks include promenade decks which accommodate pedestrian or vehicular traffic, terrace decks with landscape coverings, and parking garage decks. Parking garage decks are typically treated with only a deck coating material since they are not over habitable space. For vehicular traffic, the coatings include a first coat of one-or two-part self-leveling polyurethane rubber, and a top coat of one- or two-part modified elastomer (ASTM C957 *Standard Specification for High-Solids Content, Cold Liquid-Applied Elastomeric Waterproofing Membrane with Integral Wearing Surface*. For pedestrian traffic only, deck coatings usually consist of a first coat of a polymeric material

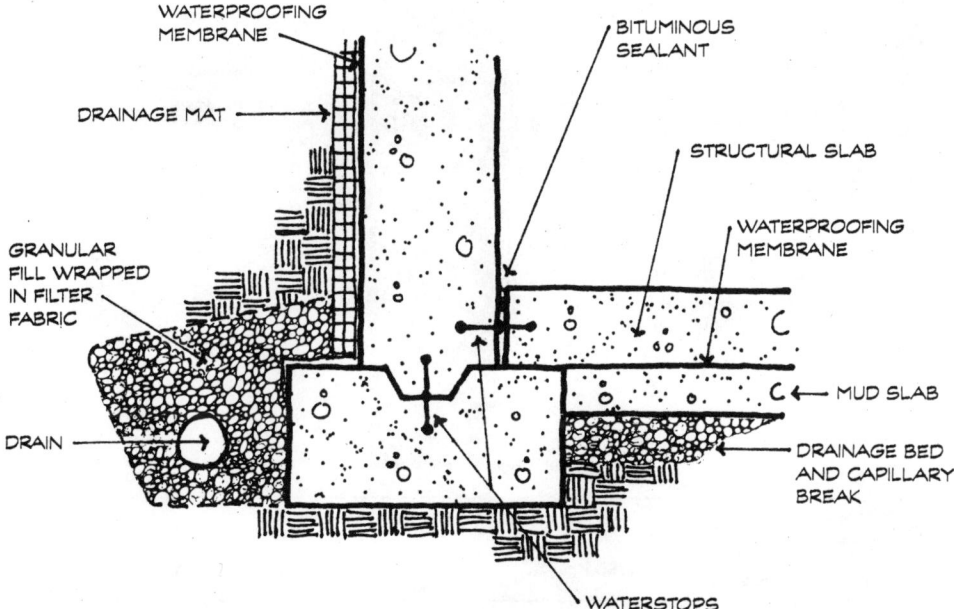

Figure 7.34 Waterproofing for wet soils with high or fluctuating water table.

324   Chapter Seven

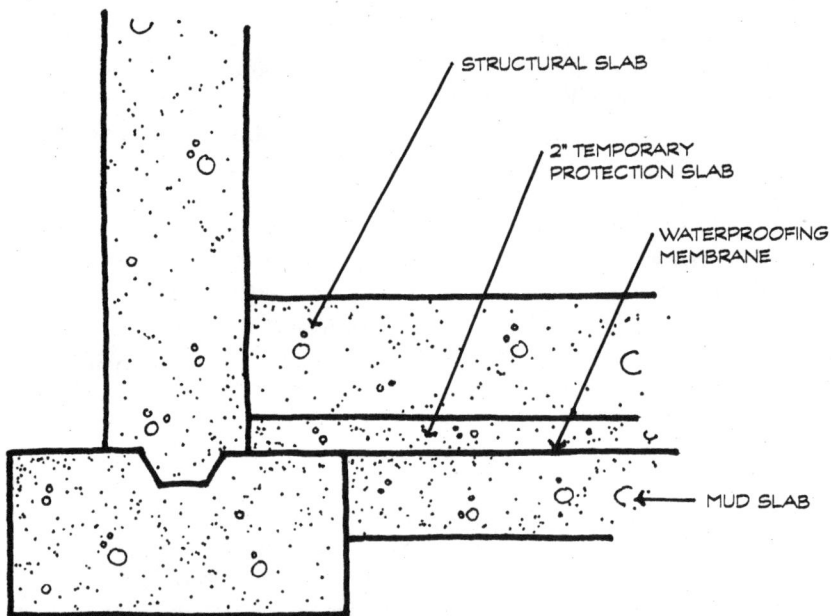

Figure 7.35 Use temporary protection slab if placement of structural slab will be delayed. (*From NRCA, Roofing and Waterproofing Manual, 2d ed.*)

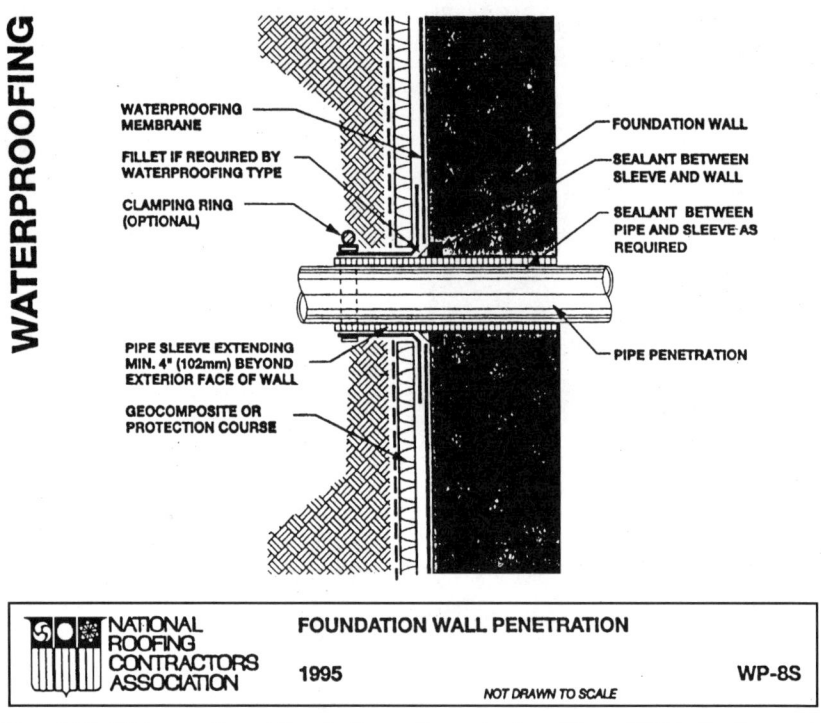

Figure 7.36 Waterproofing at foundation wall penetration. (*From NRCA, Roofing and Waterproofing Manual, 4th ed.*)

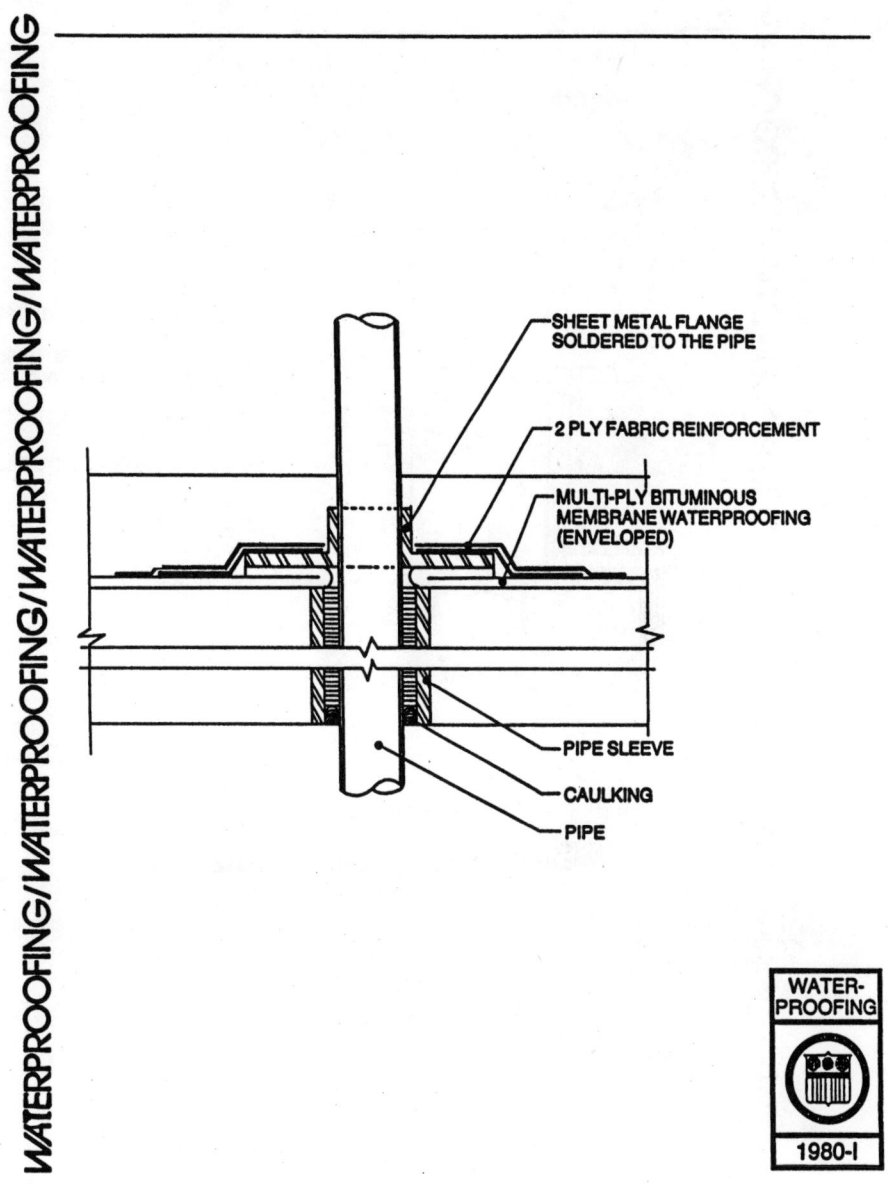

Figure 7.37 Waterproofing at floor slab penetration. (*From NRCA, Roofing and Waterproofing Manual, 2d ed.*)

that reacts with atmospheric moisture for an overnight cure. The second coat is a one- or two-part polyurethane mixture. For both vehicular and pedestrian deck coatings, mineral granules can be broadcast into the wet surface of the top coat to provide skid resistance.

Plaza decks over occupied space essentially function as low-slope roofs and are just as unforgiving of design and installation errors as low-slope roofs. Plaza deck

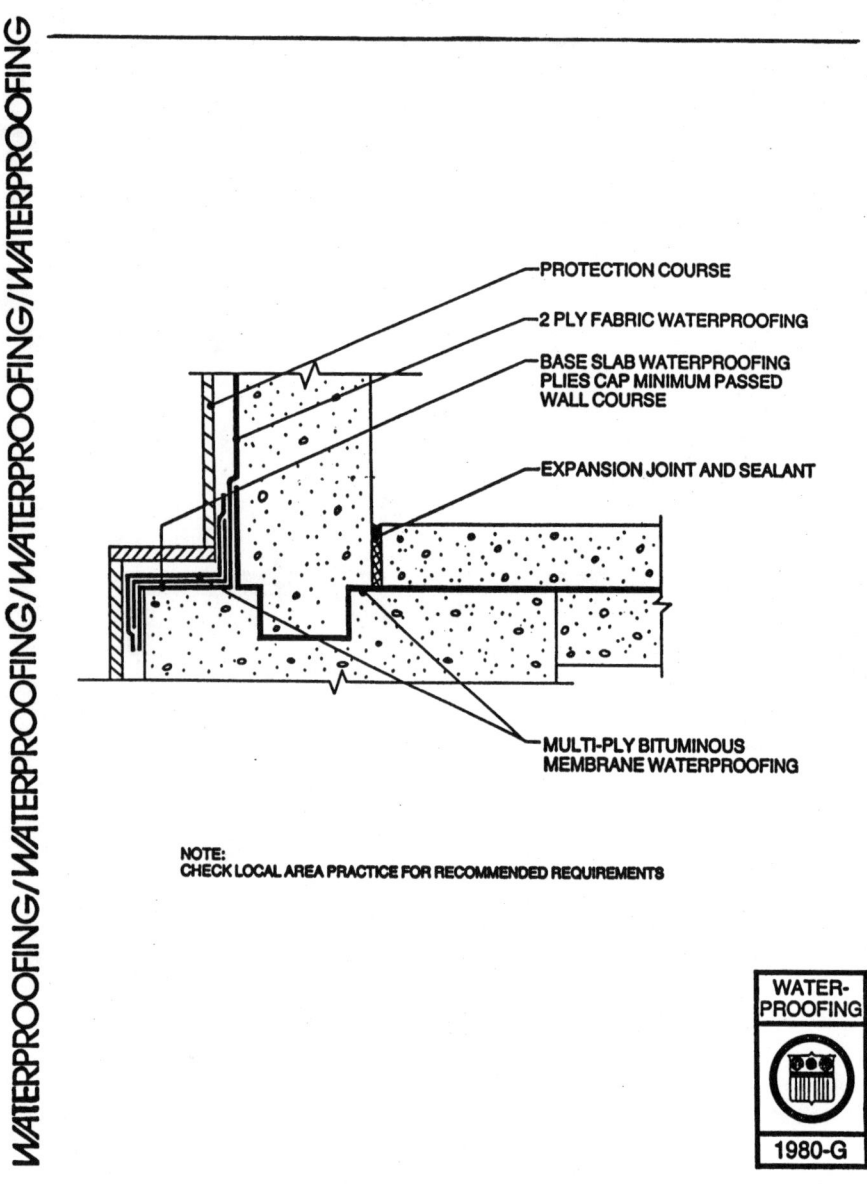

Figure 7.38 Waterproofing at wall/slab/footing intersection. (*From NRCA, Roofing and Waterproofing Manual, 2d ed.*)

waterproofing should include a combination of surface drainage and a waterproofing membrane with appropriate flashing, penetration, and termination details. The basic components of a plaza deck waterproofing system are the structural deck, waterproofing membrane, protection course, percolation layer, insulation, protection or working slab, and wearing course/traffic surface (*Fig. 7.42*).

Although almost any type of waterproofing membrane can be used if the flashing details and drainage are properly designed and installed, fully adhered membranes simplify the detection of leaks after installation. ASTM publishes several standards for waterproofing building decks. The recommendations and details that follow are based primarily on ASTM C898 *Guide for Use of High Solids Content, Cold Liquid-Applied Elastomeric Waterproofing Membrane with Separate Wearing Course* and ASTM C981 *Guide for Design of Built-Up Bituminous Membrane Waterproofing Systems*

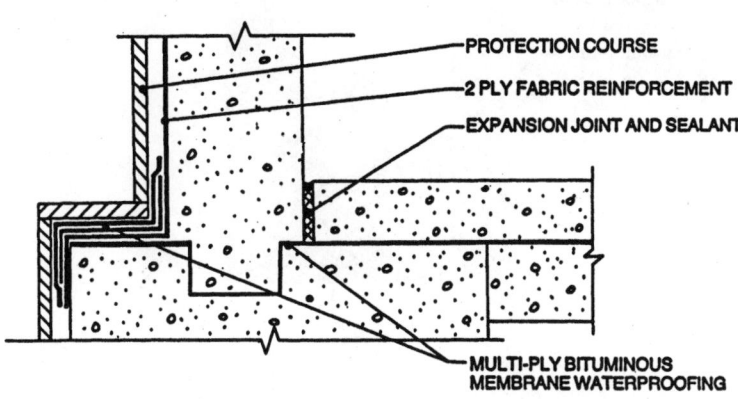

Figure 7.39 Alternative waterproofing at wall/slab/footing intersection. (*From NRCA, Roofing and Waterproofing Manual, 2d ed.*)

328 Chapter Seven

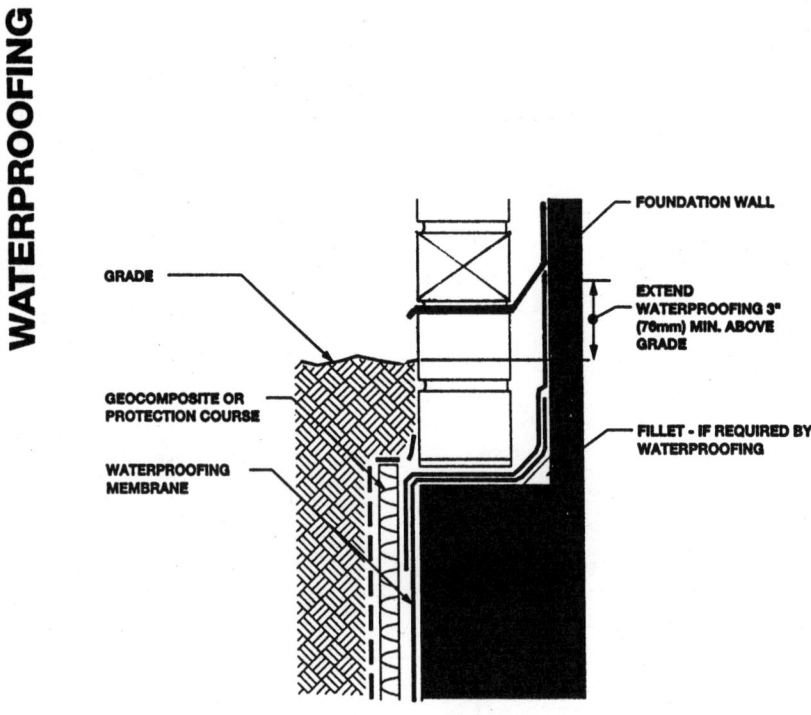

Figure 7.40 Waterproofing at foundation wall brick ledge. (*From NRCA, Roofing and Waterproofing Manual, 4th ed.*)

for Building Decks. Details for other types of membranes, including hot-fluid-applied systems and sheet membranes are similar to those indicated, and the performance concepts are the same. Always consult the membrane system manufacturer for recommended details that are specific to the materials involved.

Waterproofing 329

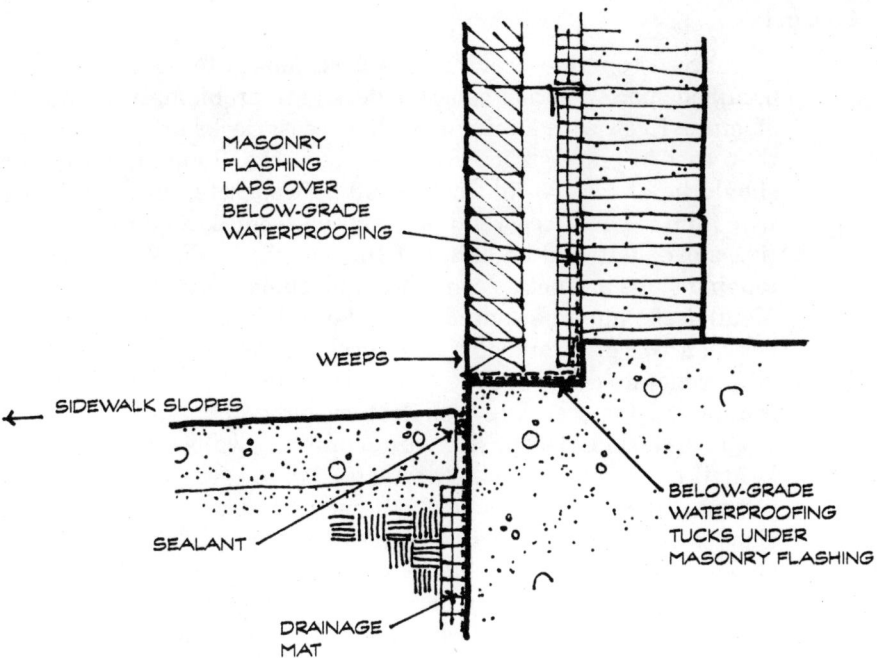

Figure 7.41 Above-grade masonry flashing laps over below-grade waterproofing.

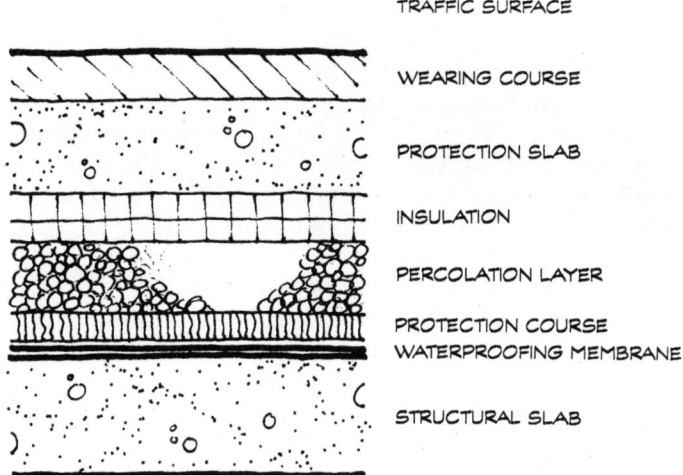

Figure 7.42 Basic components of plaza deck waterproofing system. (*From ASTM C981, Guide for Design of Built-Up Bituminous Membrane Waterproofing Systems for Building Decks.*)

7.6.1 Substrate

Reinforced concrete provides the best substrate for waterproofed decks over occupied space. Precast concrete decks are problematic because of the number of joints. Even with a topping slab, precast decks are more prone to movement than cast-in-place structural concrete decks. Topping slabs for precast decks should be at least 3-in.-thick, 3000-psi concrete, and reinforced with welded wire fabric. The precast panels must be securely tied together to prevent relative movement and cracking of the topping slab. Metal decks with concrete topping slabs are not as common in plaza deck construction as concrete decks. Metal decks must be vented on the bottom so that moisture cannot be trapped between the deck and the waterproofing membrane. Post-tensioned cast-in-place concrete decks are ideal substrates for plaza waterproofing because they are the least susceptible to deflection and cracking.

The deck should slope away from building walls and expansion joints at $1/8$ to $1/4$ in./ft to provide adequate drainage and to prevent deck deflection and water ponding. It is best to slope the structural deck itself rather than to achieve slope with a separate topping slab. Unreinforced topping slabs are prone to shrinkage cracking, and the cleavage plane between the topping and the structural deck can inhibit leak detection in the finished system.

Concrete for building decks should have a minimum density of 110 lb/cu ft and a maximum moisture content of 8% when cured. Lightweight concrete and lightweight insulating concrete cannot meet these criteria and are not appropriate substrates for waterproofing membranes. Polymeric, latex, and other organic chemical–based admixtures and additives should not be used because they can impair membrane bond to the concrete. If any admixtures, additives, or modifiers will be used in the concrete mix, the membrane manufacturer should be consulted about compatibility. The underside of the concrete deck should not have an impermeable barrier such as a metal liner or coating which would retard vapor diffusion. Such a barrier would trap moisture in the concrete and prevent or destroy the adhesive bond of the membrane to the concrete. The concrete should receive a float finish or a troweled finish, with the final troweling omitted. The finished concrete should be moist cured for a minimum of 7 days, and aged 28 days (including curing time) before application of the membrane. The membrane manufacturer should be consulted for maximum allowable moisture content at the time of application. Since the deck will be exposed to the weather, drying time should be allowed after any type of precipitation.

Joints in the substrate must be treated according to size and type. *Reinforced joints* include hairline cracks, cold joints, construction joints, isolation joints and control joints held together with steel reinforcing bars or wire fabric. These are considered static joints with little or no anticipated movement. *Non-reinforced joints* include butted construction joints, and isolation joints not held together with steel reinforcing or wire fabric. Although these are generally considered static joints from a structural standpoint, they are capable of some movement, the magnitude of which is difficult to predict. *Expansion joints* are designed to accommodate movement, and should be of a size and shape appropriate to the anticipated amount of movement and the selected joint sealant materials. Expansion joints should always be located at the high points of a contoured or sloping deck.

7.6.2 Membrane

The waterproofing membrane should be applied in accordance with the manufacturer's instructions to a deck that is clean, dry, and frost-free throughout the depth of the slab. Some types of membranes may require a surface primer to enhance adhesion. Built-up membranes should have a minimum of three plies, and can be installed by the "ply-on-ply," or phased, method, or a combination of the phased method and the "shingle" method. Under no circumstances should all plies of a built-up membrane be installed in shingle fashion at one time on walls or decks. *Figures 7.43* and *7.44* show appropriate membrane treatments at reinforced and non-reinforced joints respectively. Expansion joints in the membrane can be sealed by either the *positive seal* method or the *water shed* method, and for an additional safeguard, a drainage gutter can be added under the joint (*Fig. 7.45*). Positive seal joints are installed at the membrane level, where they are very vulnerable to water penetration. The materials used and their joinery must be engineered by the membrane manufacturer, and field installation requires precision workmanship with no margin for error. Since this type of expansion joint seal has a high risk of failure, it is not recommended. Water shed joints more closely resemble expansion joints in roofing membranes where the joint seal is raised above the membrane level for greater safety. Drainage must be provided at the membrane level to prevent water buildup. The details shown in *Figs. 7.46* and *7.47* are designed for a maximum movement of $\pm 3/8$ to $\pm 3 1/2$ in. Larger joint openings for greater anticipated movements can be accommodated using larger gaskets.

Termination details at membrane edges and transitions are critical components of the waterproofing system. Details for liquid-applied membranes are shown in *Fig. 7.48* and for built-up membranes in *Fig. 7.49*. Built-up membranes must be mechanically anchored on vertical surfaces to prevent slippage. Pipe penetrations can be sealed around a metal pipe sleeve to isolate the membrane from pipe movement (*Figs. 7.50* and *7.51*), but some manufacturers prefer the use of a prefabricated flashing boot (*Fig. 7.52*). If the wearing surface is an open-jointed paver system which allows water to pass through the joints, drains are required only at the membrane level (*Fig. 7.53A*). *Multi-level drains* which remove water at both the traffic surface and the membrane level *are imperative* with monolithic slab surfaces and with pavers that have grouted joints (*Fig. 7.53B*). The inevitable cracks or joint separations at the surface will allow copious amounts of water to penetrate to the membrane level, and without positive drainage at the membrane, leaks or damage to the deck system are almost inevitable. Drains should have an integral base flange at least 2 in. wide to facilitate flashing (*Figs. 7.54* and *7.55*).

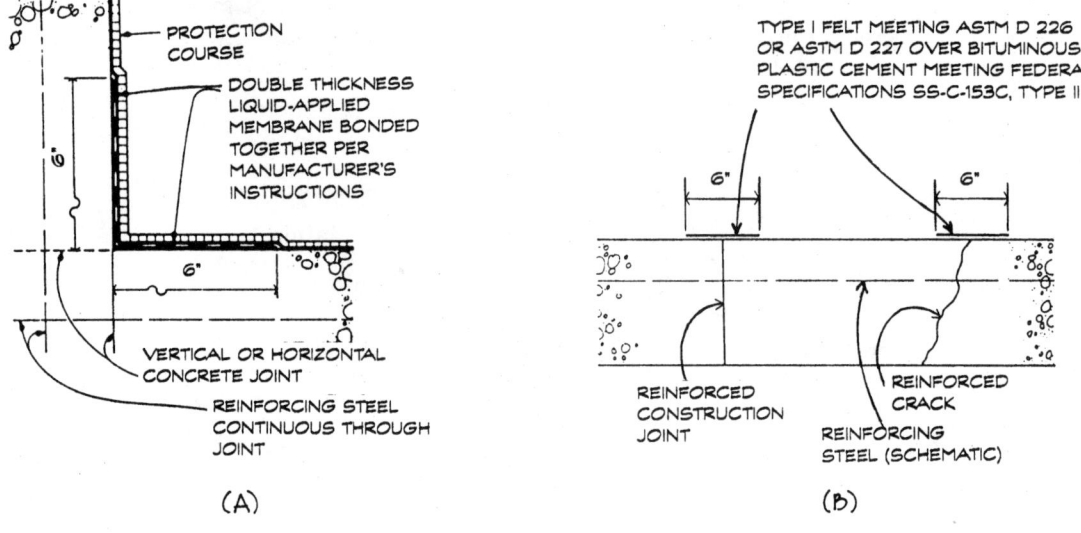
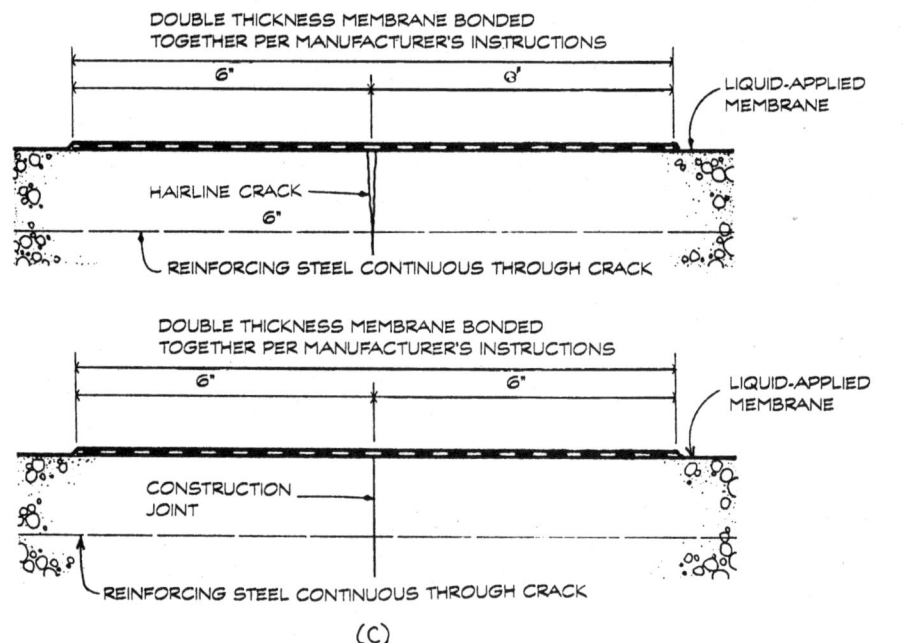

Figure 7.43 Plaza deck waterproofing at reinforced cracks and joints. (*From ASTM C898, Guide for Use of High Solids Content, Cold Liquid Applied Elastomeric Waterproofing Membrane with Separate Wearing Course, and ASTM C981, Guide for Design of Built-Up Bituminous Membrane Waterproofing Systems for Building Decks. Copyright ASTM. Reprinted with permission.*)

Before any additional work is done, the drains should be plugged and the membrane should be flood-tested with 2 in. of water for 24 to 48 hours. Depending on deck slope and drain spacing, intermittent water dams may be required to keep the water from getting too deep and overloading the structure. After the flood test is drained, check for any ponding. Walk every lap seam, watching for expulsion of water under the pressure of your feet, and repair any areas where water has penetrated the seam. The waterproofing system should be tested again after the deck system is completed to assure that the percolation layer and drains are functional and that the membrane has not been damaged during construction.

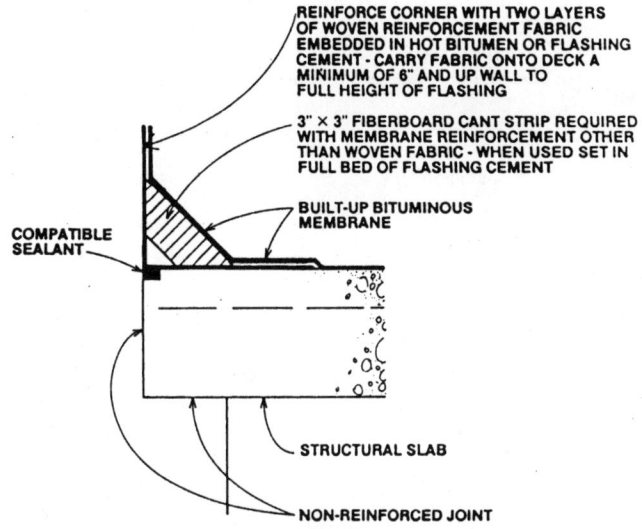

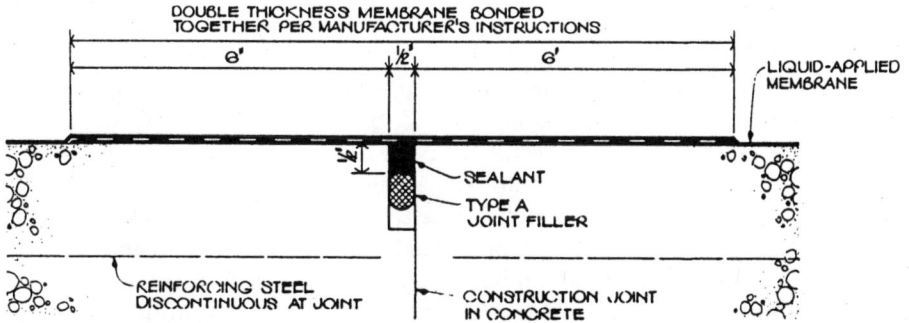

Figure 7.44 Plaza deck waterproofing at unreinforced cracks or joints. (*From ASTM C898, Guide for Use of High Solids Content, Cold Liquid Applied Elastomeric Waterproofing Membrane with Separate Wearing Course, and ASTM C981, Guide for Design of Built-Up Bituminous Membrane Waterproofing Systems for Building Decks. Copyright ASTM. Reprinted with permission.*)

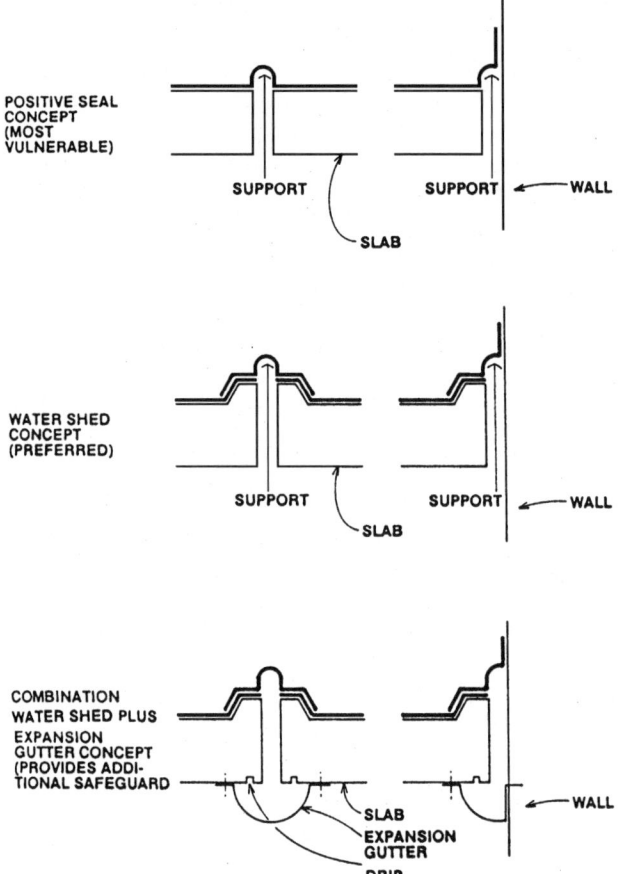

Figure 7.45 Plaza deck waterproofing at expansion joints. (*From ASTM C981, Guide for Design of Built-Up Bituminous Membrane Waterproofing Systems for Building Decks. Copyright ASTM. Reprinted with permission.*)

7.6.3 Protection course

The waterproofing membrane must be protected from damage from construction traffic, completion of deck construction, and subsequent movement and penetration of materials after the deck is completed. The membrane manufacturer should provide or recommend a type of protection board which is compatible with the membrane. With sheet materials and bentonite waterproofing, protection boards can be placed immediately after the membrane is applied, but liquid-applied and built-up membranes must cool to ambient temperatures and tack-free condition before protection board placement.

7.6.4 Percolation layer

It must be assumed that all plaza decks, even those with monolithic surfaces, will experience water penetration to the membrane level, and therefore *all plaza decks must be drained at the membrane level.* The structural deck should be sloped to direct water to the drains, and a percolation layer or drainage course should be installed to increase the rate of water flow to the drains. It is important to drain water from the membrane level as *quickly* as possible to avoid building up a pressure head at vulnerable joints and splices, to avoid freeze-thaw cycling of trapped water which could heave and disrupt the wearing course, and to maximize thermal efficiency. The percolation layer may be an aggregate drainage course or a proprietary drainage medium designed to allow free water flow to the drains. Aggregate used for a drainage course should be washed round river gravel. Crushed rock and other sharp, angular materials should not be used. If no insulation layer will be used in the deck assembly and a concrete working slab or wearing surface will be installed, a perforated polyethylene membrane or other protective layer should be installed over the drainage medium to prevent the fresh concrete from filling the drainage voids. Most proprietary drainage systems have an integral protective surface. On plaza decks which use an aggregate drainage course and on which there will be vehicular traffic, the aggregate should be stabilized with an epoxy binder so that vibrations and lateral thrust do not cause the gravel to shift along the sloped deck.

7.6.5 Insulation

Insulation should not be located directly over the membrane or protection course because it will inhibit free water flow to the drains. Placing the insulation on

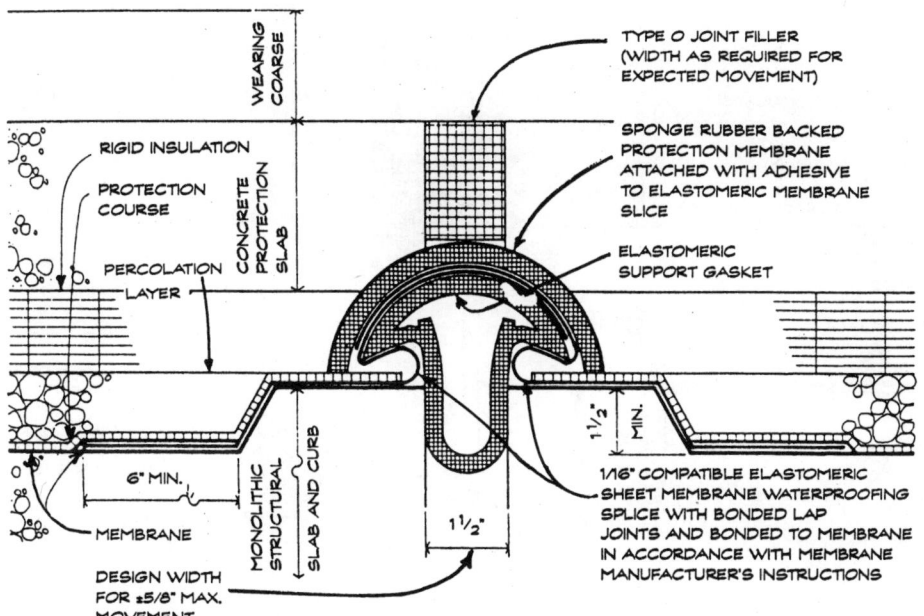

Figure 7.46 Plaza deck horizontal expansion joint detail. (*From ASTM C898, Guide for Use of High Solids Content, Cold Liquid Applied Elastomeric Waterproofing Membrane with Separate Wearing Course. Copyright ASTM. Reprinted with permission.*)

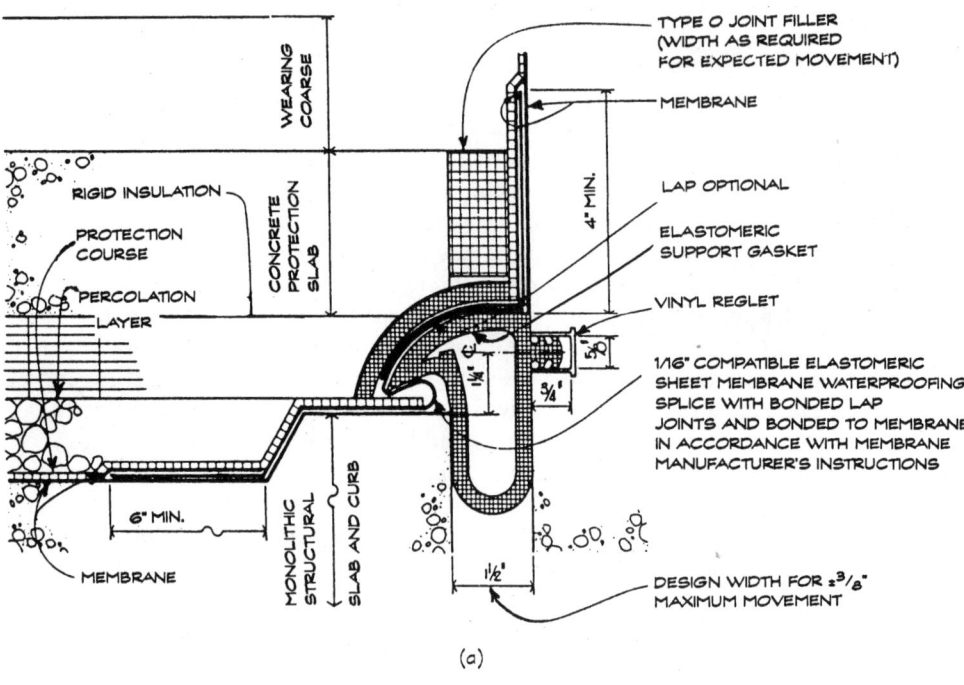

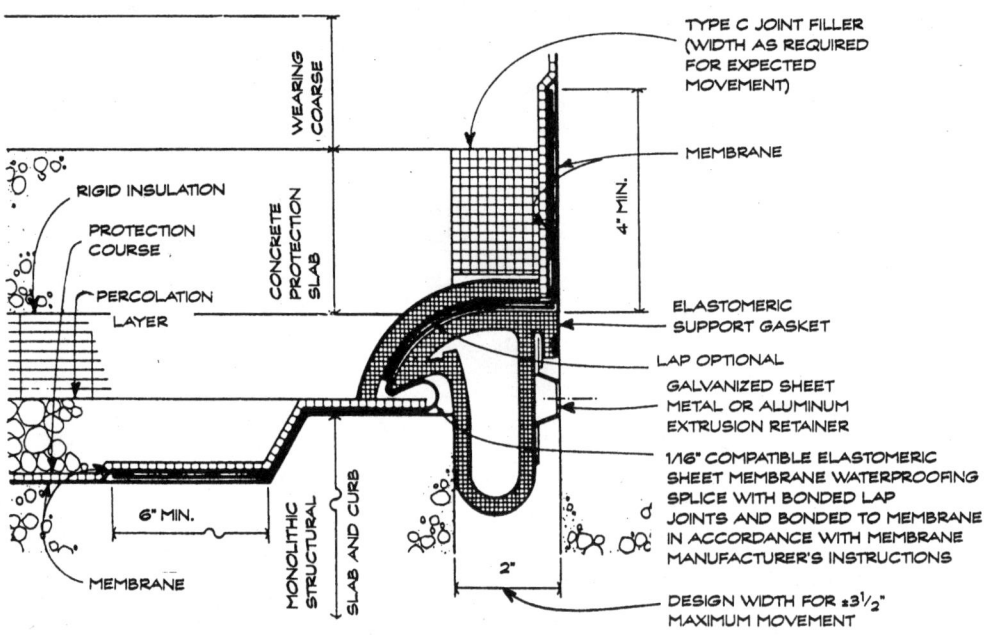

Figure 7.47 Plaza deck expansion joint detail at intersection with vertical wall or curb. (*From ASTM C898, Guide for Use of High Solids Content, Cold Liquid Applied Elastomeric Waterproofing Membrane with Separate Wearing Course. Copyright ASTM. Reprinted with permission.*)

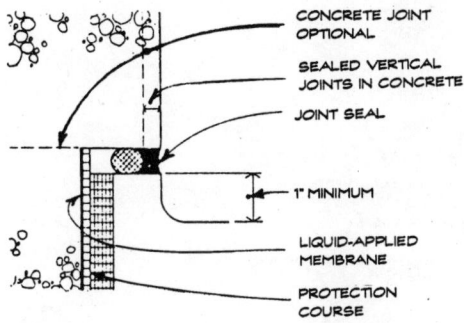

TERMINAL CONDITION ABOVE FINISH GRADE ON CONCRETE WALL

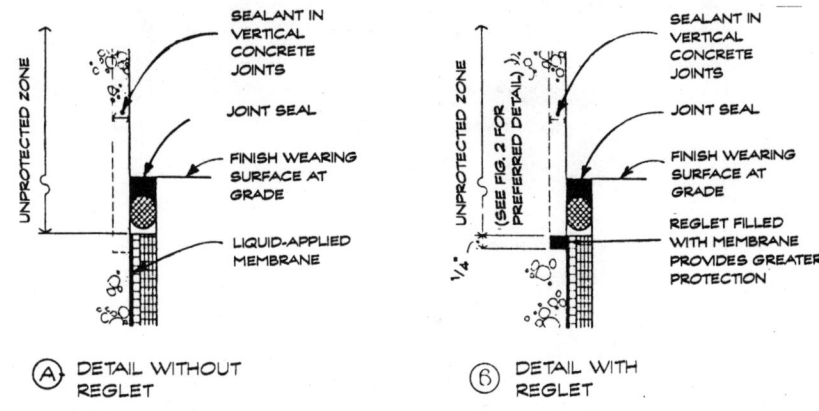

TERMINAL CONDITIONS ON CONCRETE WALL BELOW FINISH WEARING SURFACE AT GRADE

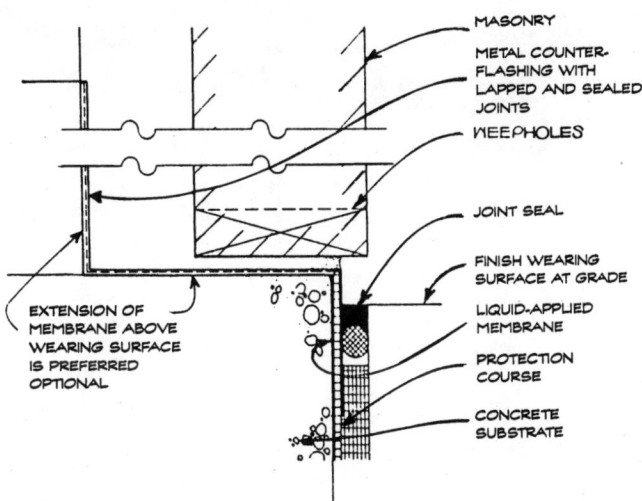

TERMINAL CONDITION WITH MASONRY ABOVE FINISH WEARING SURFACE AT GRADE

Figure 7.48 Termination details for fluid-applied plaza deck waterproofing. (*From ASTM C898, Guide for Use of High Solids Content, Cold Liquid Applied Elastomeric Waterproofing Membrane with Separate Wearing Course. Copyright ASTM. Reprinted with permission.*)

338   Chapter Seven

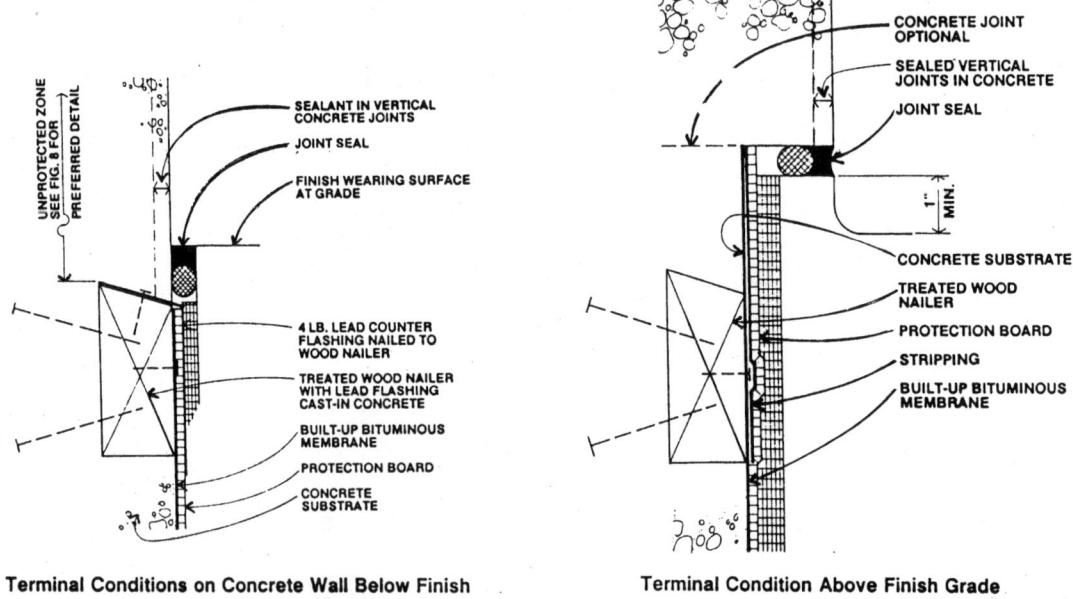

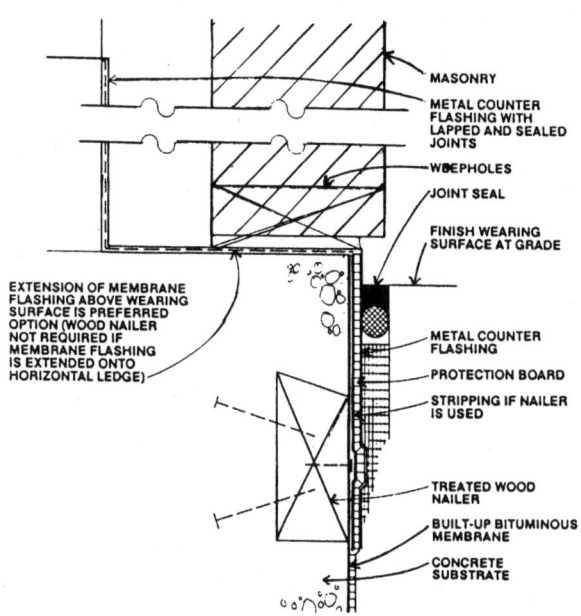

Figure 7.49 Termination details for built-up plaza deck waterproofing. (*From ASTM C981, Guide for Design of Built-Up Bituminous Membrane Waterproofing Systems for Building Decks. Copyright ASTM. Reprinted with permission.*)

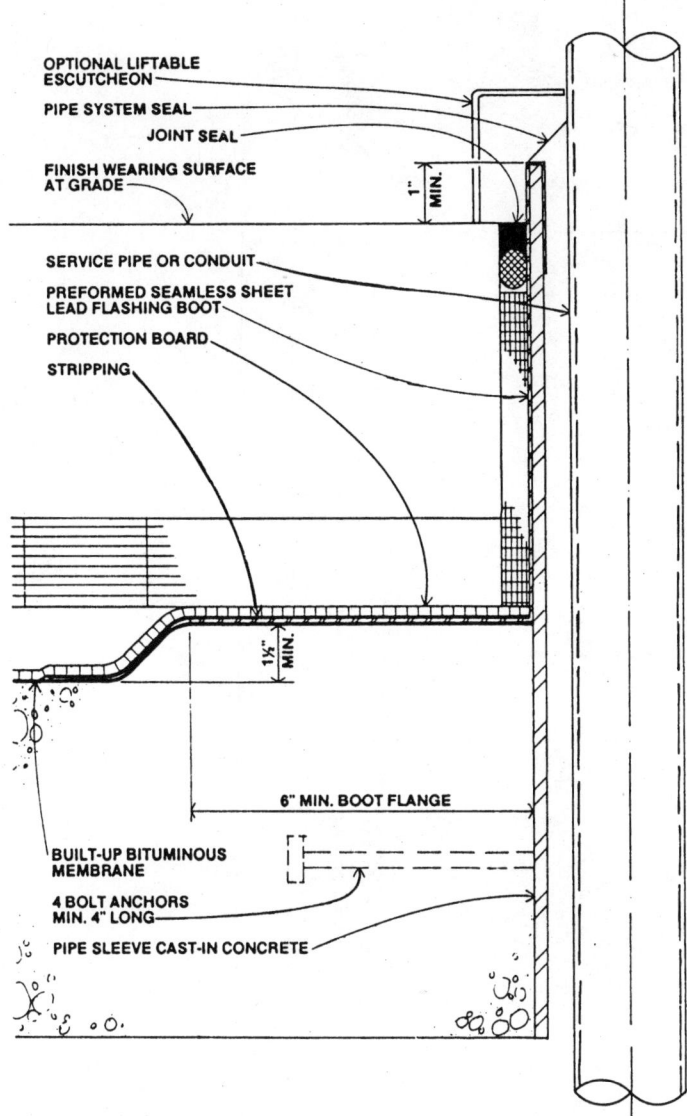

Figure 7.50 Built-up plaza deck waterproofing at pipe penetration. (*From ASTM C981, Guide for Design of Built-Up Bituminous Membrane Waterproofing Systems for Building Decks. Copyright ASTM. Reprinted with permission.*)

the underside of the structural deck exposes the membrane to a wide range of temperature extremes and excessive thermal movement. The best location for insulation on waterproofed decks is just above the percolation layer. This allows water to collect below the insulation without being in direct contact with it. Insulation used in these applications must have a high resistance to moisture

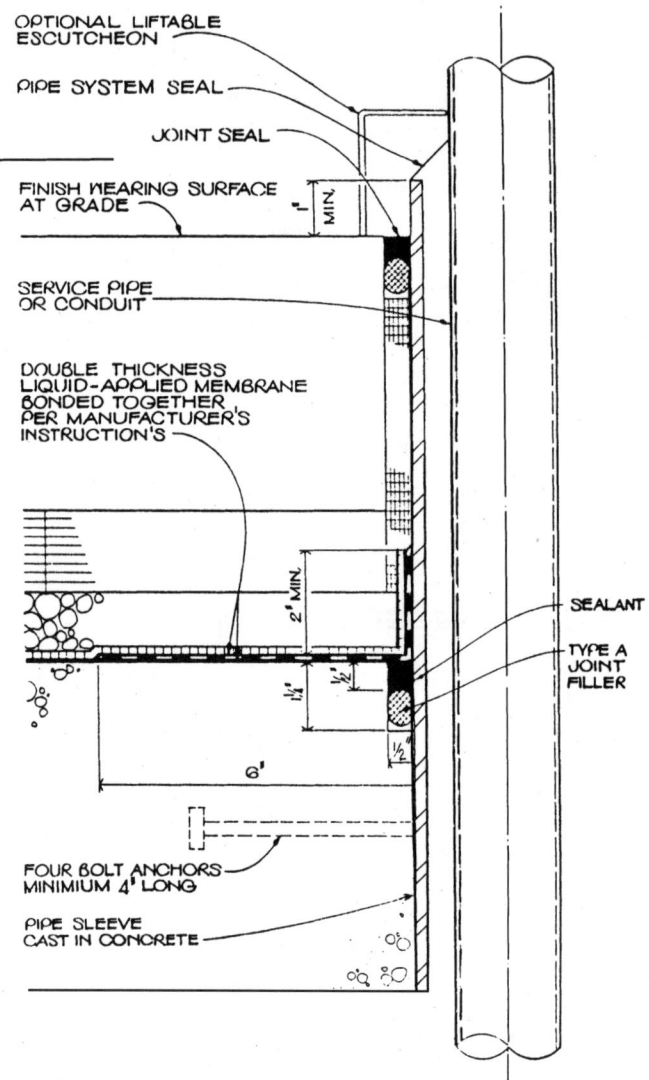

Figure 7.51 Fluid-applied plaza deck waterproofing at pipe penetration. (*From ASTM C898, Guide for Use of High Solids Content, Cold Liquid Applied Elastomeric Waterproofing Membrane with Separate Wearing Course. Copyright ASTM. Reprinted with permission.*)

absorption because plaza decks are wet environments no matter how well drained they are. The insulation must also have sufficient compressive strength to withstand imposed loads and load concentrations. Extruded polystyrene and cellular glass insulation are the only two materials which meet these requirements, and cellular glass is susceptible to progressive deterioration from freeze-thaw cycling, so it should not be used unless the climate is mild enough to preclude freeze-thaw damage.

Waterproofing 341

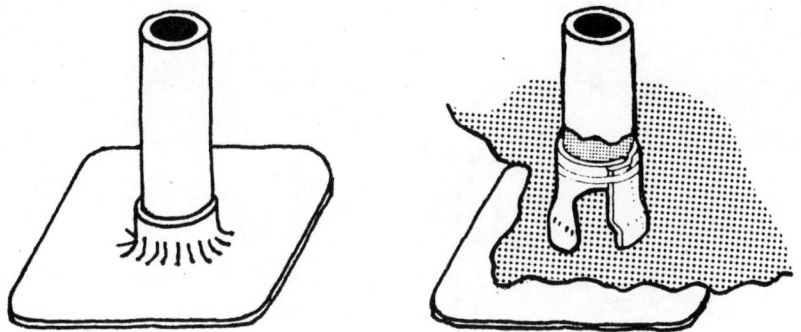

Figure 7.52 Prefabricated flashing boot at pipe penetration. (*From American Hydrotech Co.*)

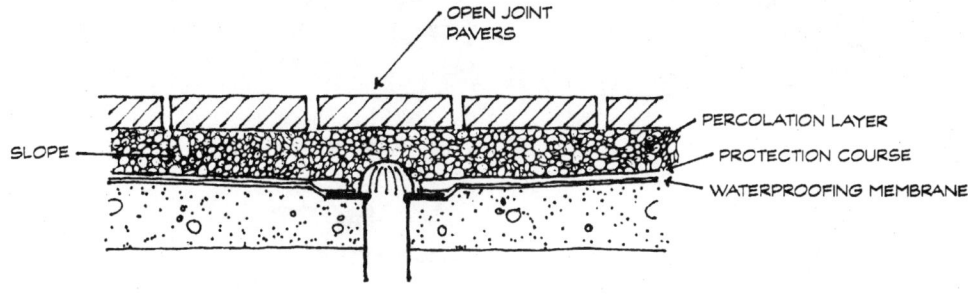

a. SINGLE-LEVEL DRAIN

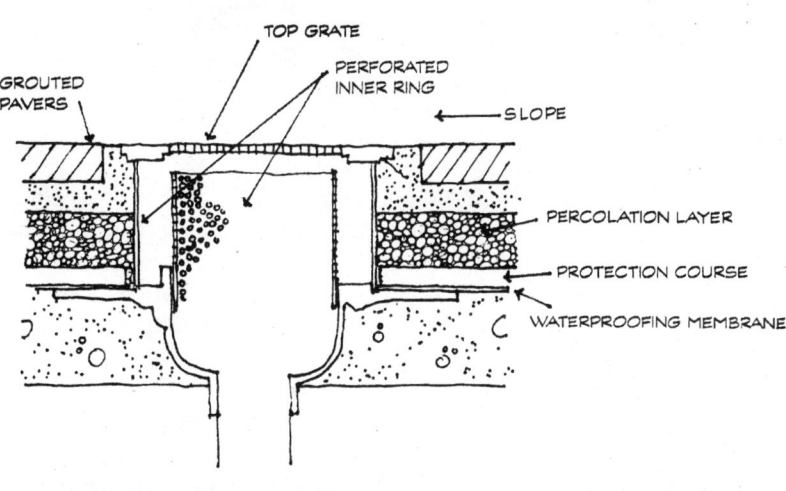

b. TWO-LEVEL DRAIN

Figure 7.53 Single-level and multi-level plaza deck drains. (*From Griffin, Manual of Low-Slope Roof Systems, 3d ed.*)

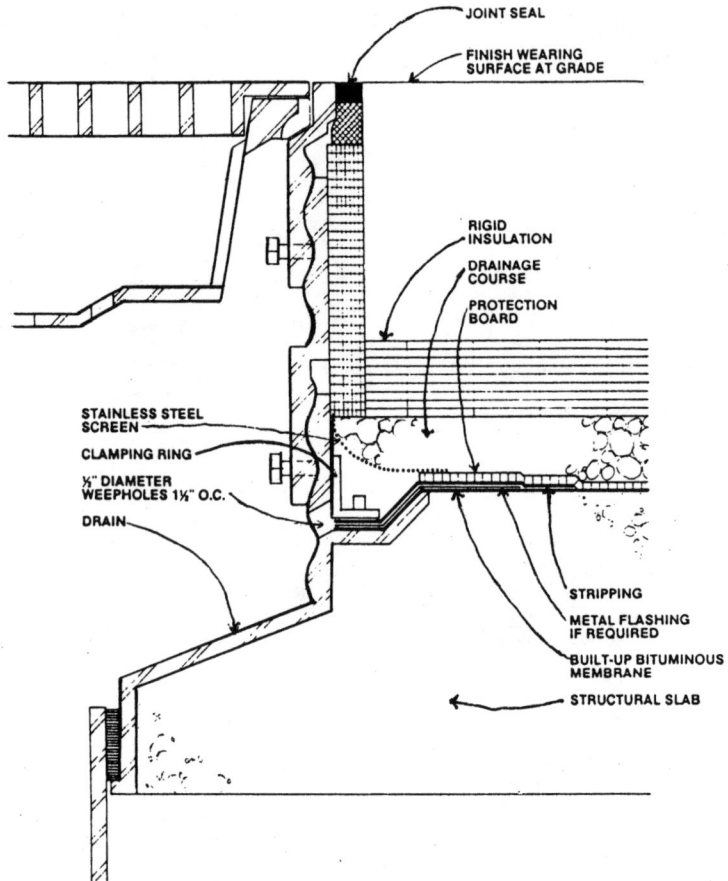

Figure 7.54 Drain detail for built-up plaza deck waterproofing. (*From ASTM C981, Guide for Design of Built-Up Bituminous Membrane Waterproofing Systems for Building Decks. Copyright ASTM. Reprinted with permission.*)

7.6.6 Protection or working slab

Plaza decks are often waterproofed early in the construction sequence so that work can proceed on the enclosed spaces below. If construction traffic, materials storage, or other activities will take place on the plaza, a protection board will not provide adequate safeguards against membrane damage. A final wearing surface slab could be poured, but it too could easily sustain damage from construction operations. In these instances, a protection slab or working slab can be used to protect the membrane, provide a working platform, and serve as a substrate for the finish wearing course materials. The slab should be reinforced to carry the imposed loads and should be at least 3 in. thick. Control joints should be located at about 20 ft on center to minimize cracking and to keep joint size minimal. If the wearing course will be poured concrete or other rigid paving materials, the control joints should align with the joints which will be in the wearing course. Joints in the protection slab do not have to be sealed, but should be fitted with premolded, resilient joint fillers. Water should be able to filter through these joints and into the percolation layer.

7.6.7 Wearing course/traffic surface

The choice of wearing course materials is primarily an aesthetic one. Monolithic concrete slabs with aggregate surfacing, color stains, patterned surfaces,

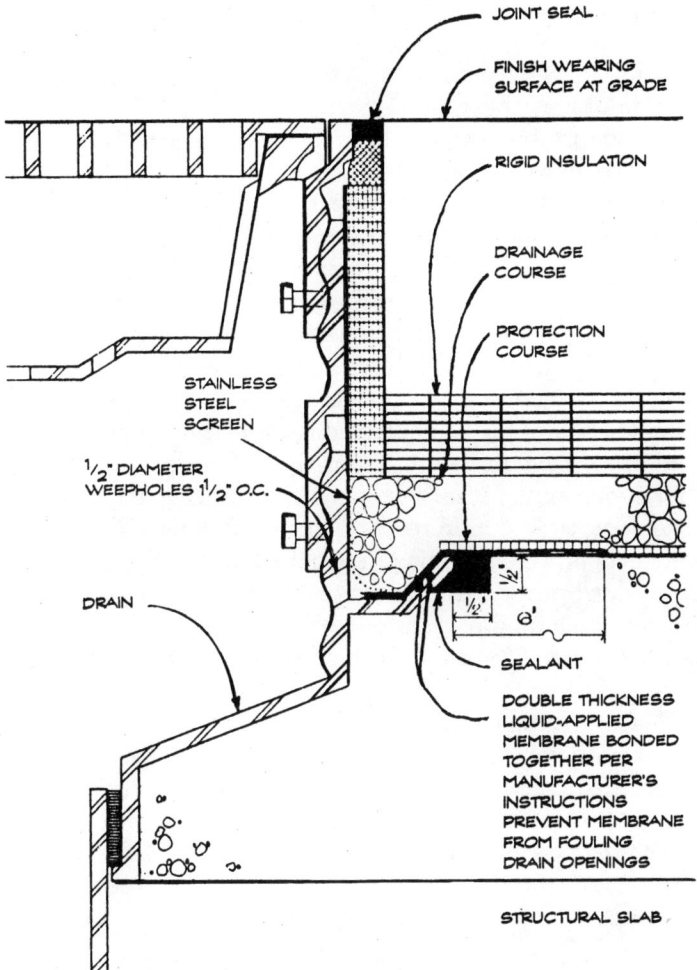

Figure 7.55 Drain detail for fluid-applied plaza deck waterproofing. *(From ASTM C898, Guide for Use of High Solids Content, Cold Liquid Applied Elastomeric Waterproofing Membrane with Separate Wearing Course. Copyright ASTM. Reprinted with permission.)*

mortared brick pavers, sand bed pavers, and large precast concrete pavers each produce a distinct character that can be matched to the surrounding architectural features. From a functional standpoint, wearing courses can be divided into two types:

- *Closed-joint systems* designed to remove most of the rainwater and snow melt rapidly by multi-level drains and allowing only a small amount of water to infiltrate to the membrane.
- *Open-joint systems* in which the vertical joints between pavers are left unsealed so that most of the rainwater and snow melt filters through the joints between insulation boards and down to the percolation layer and membrane level drains.

Closed-joint systems include mortared pavers and cast-in-place concrete. Expansion joints in these systems are typically filled with an elastomeric sealant or a compression seal. The wearing surface should be sloped to multi-level drains to expedite the removal of water. Expansion joints should be sealed with an elastomeric sealant (*Fig. 7.56*). Joint spacing should be coordinated with the location of joints in the structural deck. If wide joints are unacceptable aesthetically, then joint spacing in both the structural deck and wearing surface should be reduced so that there are more joints of narrower width, able to accommodate the same anticipated movement.

Open-joint systems include sand bedded pavers and pavers supported on pedestal systems. The wearing surface can be level instead of sloped, but open-joint systems should not be subjected to vehicular traffic. The joints between individual pavers should be no larger than $\frac{1}{4}$ in. so that they do not create a hazard for pedestrians. If large expansion joints are required to accommodate deck movement, they should receive a compression seal or sliding plate so that the small heels of women's shoes are not caught in the joint.

If sand-bedded pavers are laid directly over the percolation layer (i.e., without a protection slab), a perforated polyethylene sheet or landscape filter mat must be used to keep the sand from clogging the drainage voids. The depth of a gravel drainage course or the sand bed itself can be varied to create a level surface. Pedestal support systems (*Fig. 7.57*) for larger pavers accommodate a sloping deck with variable heights. The space below pedestal-supported pavers is left open. Pedestals should never be placed directly on the membrane. If placed on the protection board, the waterproofing system manufacturer should be consulted to determine if the imposed loads will have any damaging effects on the board and the membrane under the anticipated service conditions. If pedestal supports are placed directly on the insulation layer, the insulation must have sufficient compressive strength to resist the concentrated loading. A pedestal system installed over a protection slab will provide greater long-term durability. Although left open for drainage, the joints between pedestal supported pavers should have resilient spacers to avoid creeping or shifting panels. Access should be provided for drain maintenance to assure that drains do not become clogged with debris which might infiltrate the open joints.

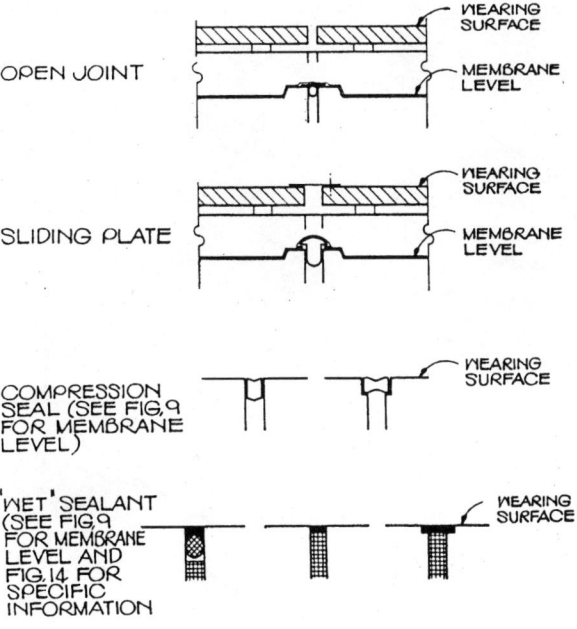

Schematic Expansion Joint Concepts at Wearing Surface Level

(A)

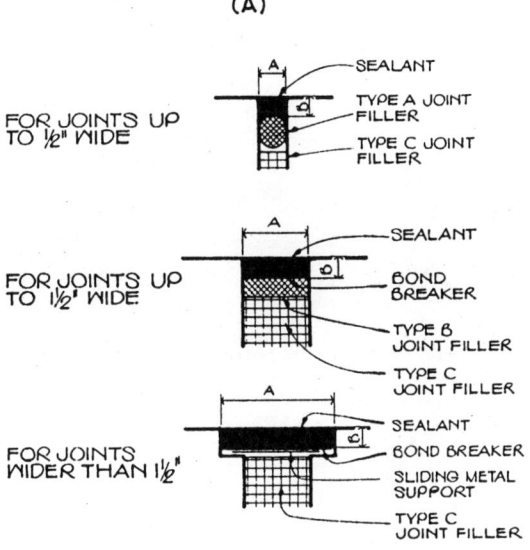

Wet Sealant Details at Wearing Surface

(b)

Figure 7.56 Elastomeric sealant at plaza deck expansion joints. (*From ASTM C898, Guide for Use of High Solids Content, Cold Liquid Applied Elastomeric Waterproofing Membrane with Separate Wearing Course. Copyright ASTM. Reprinted with permission.*)

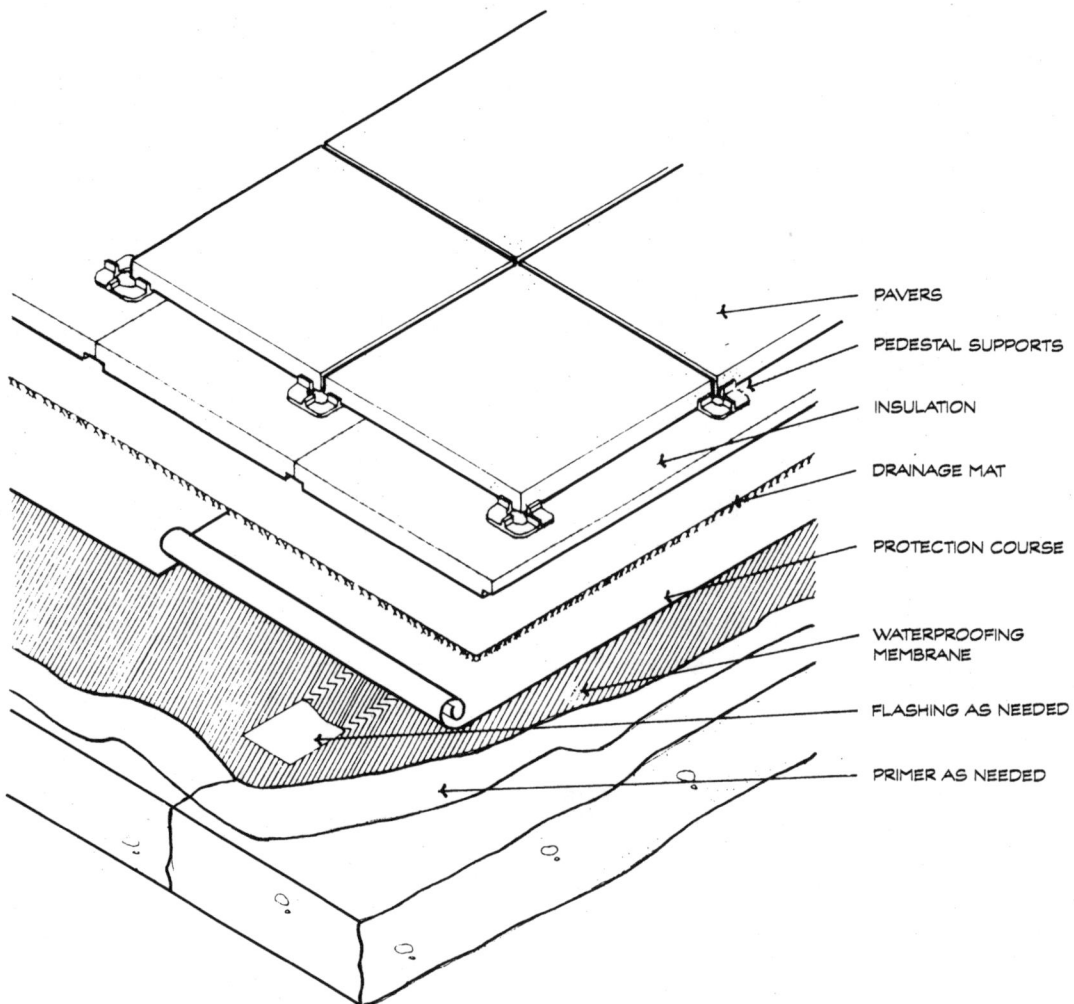

Figure 7.57 Plaza deck waterproofing with pedestal supported paving system. (*From American Hydrotech Co.*)

7.6.8 Earth-covered plazas and planters

In earth-covered plazas where lawns or planting areas are the finished surface, the waterproofing concepts are similar to pedestrian and vehicular plazas, but there are additional criteria to be considered such as depth of soil required for growth of proposed plant materials and soil temperature conducive to plant growth. A percolation layer is essential in earth-covered plazas, since most plants cannot tolerate saturated soil. To prevent the soil from clogging the drainage voids, a permeable filter such as landscape fabric should be placed over the percolation layer (*Fig. 7.58*). If landscape irrigation systems will be used, perforated drainage pipe may be needed in the percolation layer to direct excess soil moisture to a storm sewer system.

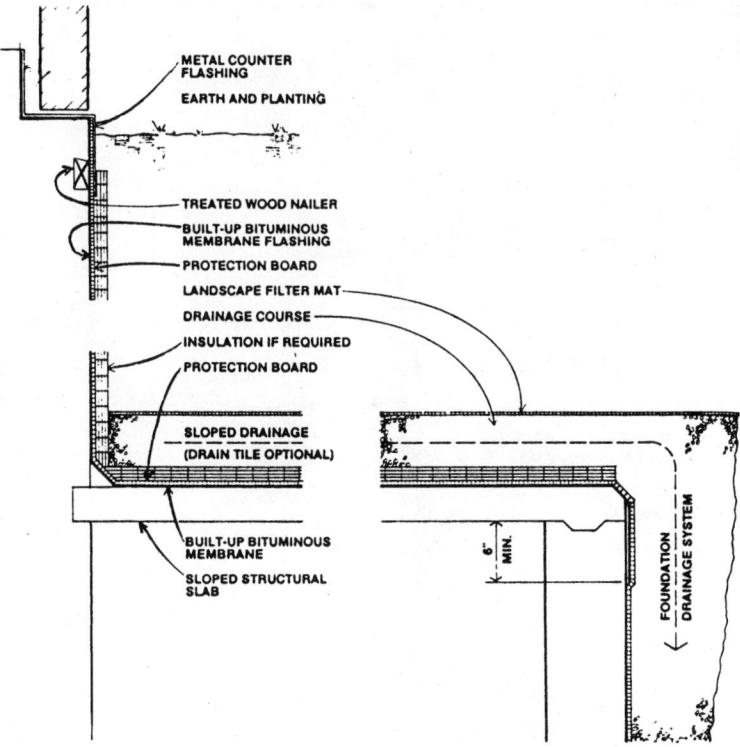

Figure 7.58 Waterproofing at earth fill and planter areas. (*From C981, Guide for Design of Built-Up Bituminous Membrane Waterproofing Systems for Building Decks. Copyright ASTM. Reprinted with permission.*)

Planters are waterproofed or dampproofed in the same way as below-grade walls and plaza decks, and with the same materials. If the planter is over occupied space, waterproofing is critical to prevent leaks, but in other locations, dampproofing should be applied to protect the planter structure itself from saturation and the freeze-thaw damage or unsightly efflorescence it can cause. Planters should have a means of positive drainage, provided either by interior drains (*Fig. 7.59*) or weeps (*Fig. 7.60*). The soil level should be kept at least 3 in. below the top of the waterproofing.

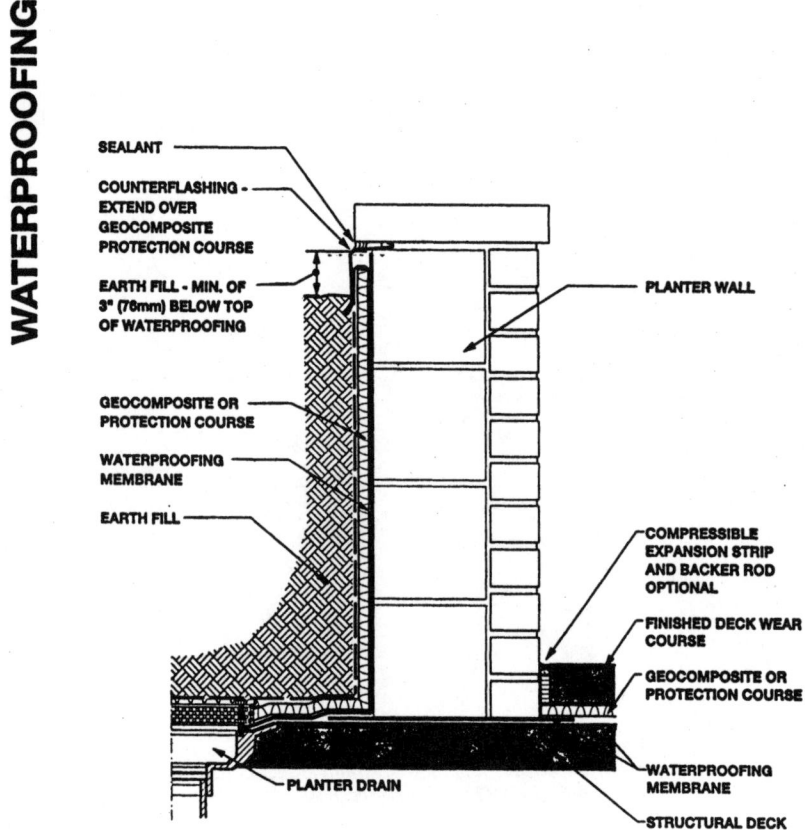

Figure 7.59 Waterproofing at planter with interior drainage. (*From NRCA, Roofing and Waterproofing Manual, 4th ed.*)

Waterproofing 349

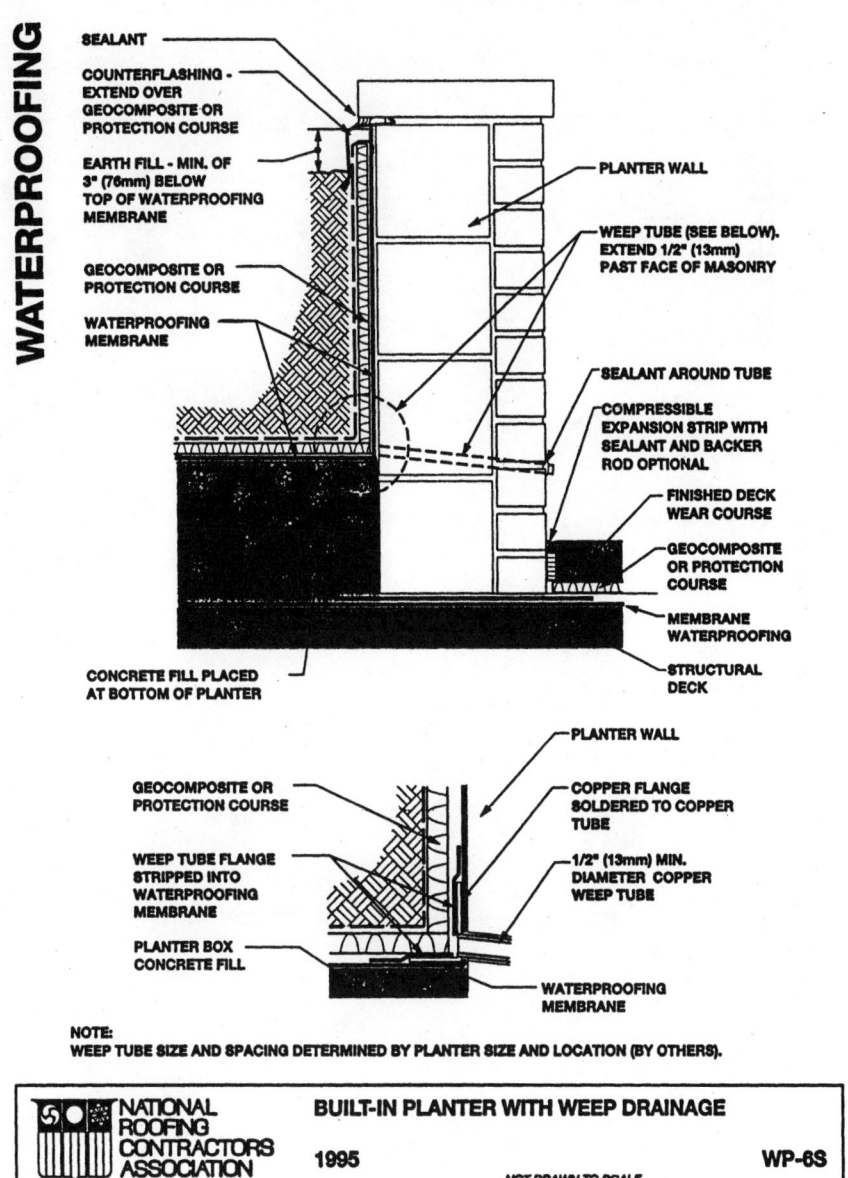

Figure 7.60 Waterproofing at planter with weep drainage. (*From NRCA, Roofing and Waterproofing Manual, 4th ed.*)

Chapter 8

Cladding

A cladding system is defined as a material assembly applied to a building as a non-loadbearing exterior wall, or attached to an exterior wall surface as a protective and ornamental covering. The successful performance of cladding systems is based on maintaining thermal resistance and controlling moisture leakage and condensation.

8.1 Basic Design Principles

It is not possible to keep the walls of a building from getting wet. They can be shielded to some degree by roof overhangs and projections, but these elements will not protect against wind-driven rain. Macro- and microclimatic exposure will determine the relative risk of water infiltration, and a cladding system can be selected or designed for a building on the basis of the level of protection it affords. Water infiltration can be caused by basic design flaws, construction errors, or material failures. Design flaws in cladding systems can be eliminated by careful study of the physical forces acting to move water through a building envelope and by the application of moisture control strategies in the conceptual design and detailing of the system. Chapter 4 outlines three basic strategies used to minimize the risk of moisture damage:

- Limiting water penetration by providing
  - Barriers such as membranes and joint sealants
  - Diversions such as sloping surfaces and gutters
  - Screens such as projections and baffles
- Preventing moisture accumulation by providing
  - Drainage to remove water
  - Drying/evaporation of water
  - Ventilation to remove water vapor

- Neutralizing the physical forces that transport water with
  - Capillary breaks
  - Drips
  - Protected openings
  - Rain screens

Although cladding systems may combine several of these moisture protection strategies, they can generally be divided into three basic wall types, as described in Chap. 4: barrier walls, drainage walls, and rain screens.

8.1.1 Rain barrier systems

Rain barrier concepts are used in low-slope roofing where the membrane must completely exclude water penetration even in the presence of hydrostatic pressure. Barrier concepts can also be used in below-grade waterproofing systems where the exterior surface of the wall is completely covered with a protective membrane. While a cladding system could be designed with the same type of continuous barrier membrane that is used in roofing and waterproofing, it probably would not be aesthetically acceptable to most people. The exterior walls of buildings are usually made up of multiple components and materials that are more vulnerable to water infiltration, especially where different materials or components come together. Although some cladding systems are intended to act as rain barriers, the concept itself requires unrealistic perfection in the construction and assembly of the components. The integrity of any so-called "barrier" system is only as good as the integrity of the joints and the continuity of protection. Any breach in the barrier represents a failure of the system, the consequences of which may be significant or insignificant, depending on the location. Since there is no place for penetrating water to go, moisture accumulates and is trapped inside the wall where it can deteriorate the materials in the substrate and delaminate the finish itself.

Exterior insulation and finish systems (EIFS) have traditionally been marketed in the United States as rain barriers, but their performance has been compromised by both design flaws and construction deficiencies. EIFS failures related to flashing problems, sealant joint problems, and general installation deficiencies have recently led the EIFS industry to develop alternatives to the barrier concept that are more realistic in application and performance expectations.

8.1.2 Drainage wall systems

Unlike rain barrier systems, drainage walls do not attempt to exclude water penetration altogether. The concept is based on an acceptance of some degree of water penetration and the subsequent collection and discharge of that water back to the exterior. The exterior wall surface deflects most of the wind-driven rain that strikes it, but the drainage system provides backup or redundant protection against damaging water infiltration. In some types of cladding, the drainage system can be used to expel condensed moisture as well as any rain which penetrates the exterior.

Drainage walls provide a higher level of protection with less risk of failure than barrier systems because, in light of the physical forces at work, they are based on more realistic performance expectations. Drainage walls are not foolproof, though, and require attention to design and detailing as well as reasonable care in construction. Masonry cavity walls and veneers are typically designed as drainage walls.

8.1.3 Rain screen systems

When the wind blows against a building wall, it creates a differential between the inside and outside air pressures. On the windward side of a building, this pressure differential forces water through even microscopic openings or defects that may exist in the envelope. This is one of the primary reasons that successful performance of barrier walls is so difficult to achieve. The greater the differential pressure, the greater the water infiltration. Although the pressure created by the wind cannot be eliminated, the pressure differential can be equalized by applying an identical opposing force acting outward. This is accomplished by creating a compartmented chamber or cavity immediately behind the facade. Air is allowed to pass freely into this chamber so that the external and cavity air masses are balanced and the inward-acting pressure across the cladding is neutralized (see Chap. 4). This is the basic principle of the pressure-equalized rain screen. However, air leakage through the inner wall of the enclosure will prevent the balancing of pressures acting across the outer wall. A sealed, continuous air barrier between the chamber and the interior building space is required for optimum performance. The air barrier need not be perfect, and it is unrealistic to expect it to be so, but every effort should be made during design and construction to make the inner wall as airtight as possible. It is possible to obtain an adequate degree of airtightness in the air barrier under normal construction conditions. The air chamber must also be compartmentalized in both the vertical and horizontal directions to assure rapid and effective pressure equalization. The pressure-equalized rain screen was first developed in the curtain wall industry and has been adapted with varying degrees of success to other cladding systems.

8.1.4 Second lines of defense

Both drainage walls and rain screens depend on collecting and discharging moisture as a second line of defense against water infiltration. Collection and drainage is usually achieved through a system of flashing and weepholes.

Flashing. Some elements of the building envelope, like windows and curtain walls, often have internal moisture control systems to collect and drain water which penetrates the outer glazing seal or condenses within the framing system.

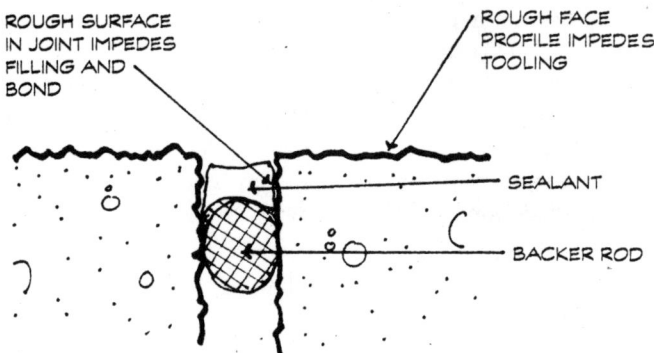

Figure 8.1 Defective sealant joint in rough-surfaced precast panel. (*From Nicastro, Difficult Sealant Joints.*)

Most elements, however, require field fabrication of such secondary components, and field-fabricated or installed connections to other elements. Flashing above doors, windows, and other penetrations must stop the downward flow of water, collect and direct it to drainage holes or weeps. The omission or inadvertent obstruction of weepholes can cause moisture to accumulate within the wall and prolong the natural drying process.

Flashing materials must be selected for long-term durability rather than short-term economy. The cost of replacing failed flashing and other components damaged by water far exceeds any savings realized by skimping on materials, detailing, or workmanship. And since flashing usually forms the last line of defense against moisture infiltration, its performance is critical. Stainless steel and copper flashing are tough and durable materials, but can be difficult to handle in the field, especially for a person not trained in sheet-metal work. Newer products such as self-adhering rubberized asphalt sheet flashing are easier to work with and more forgiving of minor installation defects. Full adhesion keeps water from flowing under rubberized asphalt flashing, its pliability makes it easy to conform to complicated shapes, lap joints are self-sealing, and it is self-healing of small punctures.

For effective performance, flashing must be sealed to the backing wall so that water cannot get behind it, and joints must be lapped and sealed so that water cannot get underneath it. At lateral terminations, the flashing must be turned up to form a pan so that water cannot run off the ends, and the leading edge of the flashing should extend beyond the face of the wall to make sure the collected moisture is not dumped inside rather than outside the wall. Flashing should *never* be terminated short of the exterior wall face, but if it is fully adhered to the substrate, it can be stopped *at* the exterior wall plane. Materials like rubberized asphalt that are pliable and sensitive to ultraviolet deterioration can be used in conjunction with a metal drip at the outer wall face.

Weepholes. Water which penetrates the exterior surface of a cladding and is collected within the drainage cavity must be quickly and effectively expelled.

Weepholes must provide for an unobstructed flow of water, and must be present in sufficient quantity to assure rapid drainage.

8.2 Cladding System Details

There are a wide variety of cladding systems commonly used in the United States. The three basic moisture control strategies can be applied to each of these in different ways. Although this chapter will not provide exhaustive detailing of each system, it will illustrate the application of basic design principles in precast, masonry, and EIFS cladding systems and in windows and curtain walls.

8.2.1 Precast concrete cladding

Precast concrete cladding is a very simple system because it involves so few components. Most precast and tilt-up concrete cladding systems are intended to resist water infiltration by forming a barrier against moisture penetration. Joints between adjacent panels are typically sealed at the exterior face with a backer rod and elastomeric sealant. The problem that occurs most often with face-sealed barrier systems is the impracticality of achieving and maintaining a *100% perfect* seal. Since there is no accommodation of any penetrated moisture, every sealant joint defect represents a system failure. Even with the most conscientious installation, the vulnerability to weather exposure of a face-sealed system requires rigorous inspection and maintenance of the joint sealants to assure long-term performance. A good sealant bond is very difficult to achieve between panels with a rough surface profile like exposed aggregate (*Fig. 8.1*). Because of the irregularities in the surface, the sealant cannot be properly tooled to make good contact with the substrate surfaces on the sides of the joint. Sealant bond is also adversely affected by contamination of the substrate by surface-applied water repellents. To achieve weather resistance with exposed aggregate finishes, precast panel edges should be fabricated smooth and protected from surface contaminants. Alternatively, the joint surfaces should be ground smooth or filled to provide a good substrate for sealant application, or the jointing system should be designed as a rain screen where face sealing does not have to be perfect.

Two-stage joints provide redundant protection for precast cladding systems, and two-stage rain screen joints provide an even higher level of protection. A two-stage rain screen joint is one in which an outer sealant bead repels or diverts most of the water hitting the outside surface, but is not required to act as an absolute barrier. Weep tubes through the sealant allow penetrated moisture to drain back to the outside, and help equalize pressure in the cavity or air space *between* the inner and outer sealant beads so that moisture is not sucked into the building by pressure differentials (*Fig. 8.2*). The inner sealant bead must form an effective air barrier in order to achieve pressure equalization between the two seals. Field installation of two-stage sealant joints can be difficult because of accessibility problems and joint size variations. Where installing the inner seal with a caulking gun is difficult because

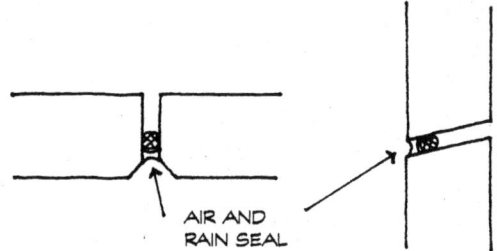

a. ONE-STAGE JOINT

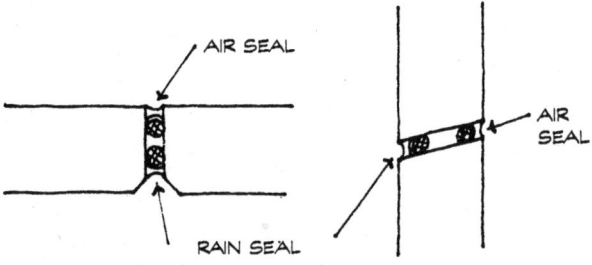

b. TWO-STAGE JOINT

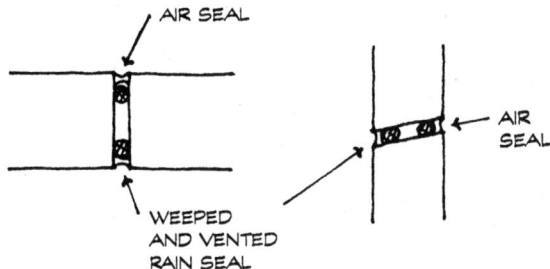

c. TWO-STAGE, PRESSURE-EQUALIZED JOINT

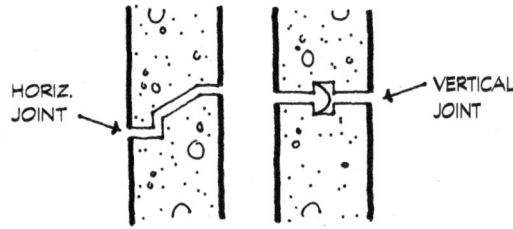

d. BAFFLE JOINT

Figure 8.2 Two-stage sealant joints. (*From Persily, Envelope Design Guidelines for Federal Office Buildings.*)

of poor accessibility, a preformed expanding foam sealant may be more suitable. Foam seals, however, must be adequately compressed to resist the full force of both positive and negative pressure differentials without displacement. *If* the air seal can be maintained at the back of the panel joint, then the face of the joint can simply be fitted with a baffle that will provide adequate weather protection even though it may allow some moisture penetration.

Some precast systems are designed with internal or field fabricated gutters and weeps to collect and expel penetrated and condensed moisture (*Fig. 8.3*). To be effective, water penetrating both vertical and horizontal joints must be diverted into the gutter. If the joints between panels are not designed as two-stage pressure-equalized rain screen joints, or if the envelope has not been properly designed to limit condensation, the gutter may have to handle substantial quantities of water. In cold climates this can be a serious problem if freezing temperatures cause ice to block drainage paths at the gutters and weeps. Depending on its macro- and microclimatic exposure, a building may need the added safety factor and redundant protection provided by both rain screen joints and drainage flashing. If the cavity behind the precast panels can be compartmentalized and vented to create a pressure equalization chamber, then the panels themselves become a rain screen and water penetration is decreased even if the sealant joints are ordinary, less than perfect one-stage joints (*Fig. 8.4*).

Anchorage of precast and tilt-up concrete cladding panels should allow for thermal expansion and contraction so that cracking will not allow water infiltration. A condensation analysis should be performed to help determine the need for and optimum location of insulation, vapor retarder, and air barrier. In some climates and with some occupancies, winter condensation on the back side of uninsulated concrete panels can be substantial if the envelope has air and vapor leaks. A humidified and pressurized building can produce winter condensation sufficient to be mistaken for rain leaks.

Barrier systems such as precast concrete do not typically incorporate flashing because they are based on the optimistic theory that no water will ever penetrate the exterior seal. Because precast concrete is relatively dense, water penetration directly through the panels is not usually a problem. However, flashing should be included at the base of walls and at window and door penetrations to collect and expel water which will likely enter at imperfectly installed or maintained sealant joints. Water repellent treatments can be helpful in keeping precast panels clean and reducing surface absorption, but they are not effective in sealing cracks and should not be relied upon as an integral part of the basic moisture protection system.

8.2.2 Masonry cladding

Historic masonry buildings were constructed with solid walls that acted as a barrier to water infiltration. The mass of porous materials acted as a sponge that absorbed and held rainwater until it could later evaporate. The walls of these buildings were much thicker than those commonly built today, and they were not subjected to the pressure differentials created by heating, ventilating,

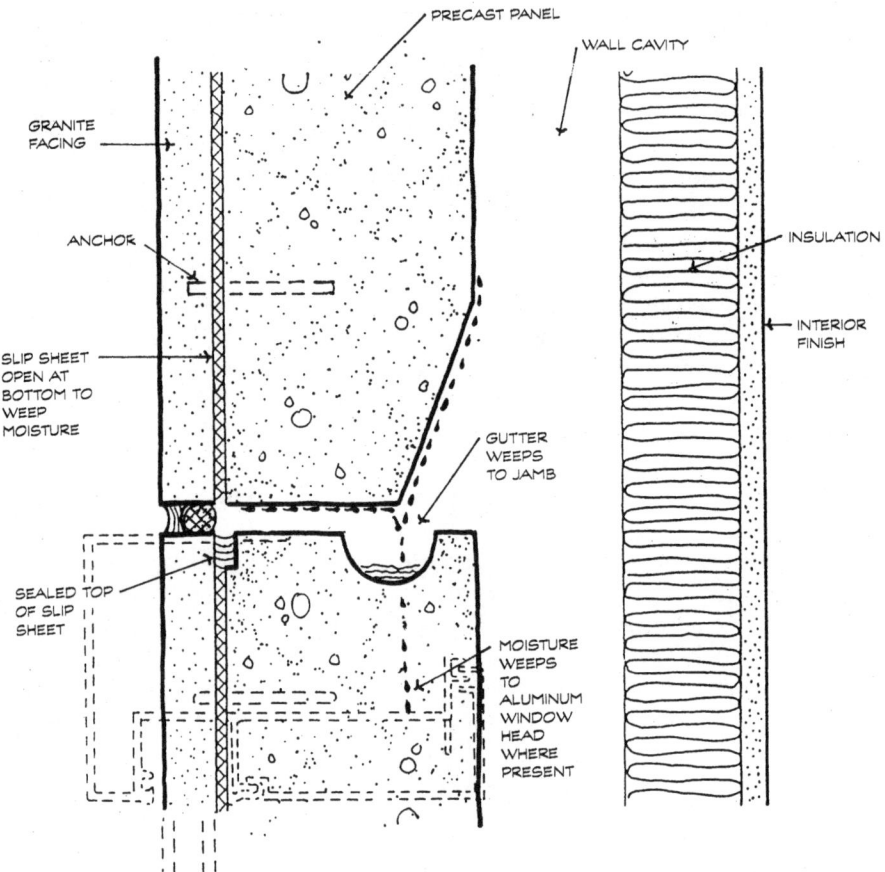

Figure 8.3 Granite-faced precast concrete cavity wall—125 Summer Street Building, Kohn Pederson Fox Architects. (*From Wright, Wall Systems Strive to Foil Moisture Intrusion.*)

and air-conditioning systems and "tight" construction. Solid masonry walls today are usually no more than two units in thickness, and air pressure differentials increase the likelihood of water infiltration to the building interior through their thinner section.

All masonry is permeable to moisture. The materials are porous, the joints are numerous, and the construction is handcrafted under diverse conditions. Although the porosity of the materials permits water absorption at the exterior surface, most water entering through the face of a masonry wall travels through defects at the mortar-to-unit bond line. Even well designed, well built masonry allows some moisture penetration. The key to successful weather resistance is limiting the amount of moisture which enters the wall and expediting the removal of moisture to prevent damage. The primary means of limiting moisture penetration are complete and intimate bond between units and mortar, full head and bed joints, adequate allowance for expansion and contraction to prevent cracking, and prudent details. The primary means of removing moisture from the wall is a well designed and installed system of sealed and continuous flashing, and unobstructed weepholes.

The term *barrier wall,* when applied to masonry construction, can give a false sense of security. Single-wythe walls of hollow units are most vulnerable to water infiltration because the face shells provide only a thin separation from the exterior. Grouted multi-wythe walls of solid units and solidly grouted hollow unit walls are somewhat more resistant to water infiltration because of their increased mass and the ability to absorb and retain water until it evaporates. A grouted collar joint between two wythes of masonry, however, is not an effective moisture barrier unless it is continuous, well consolidated, and without voids or cracks. Cavity walls and anchored veneers which have a complete separation between backing and facing provide the best protection against water infiltration. This *drainage wall* concept permits moisture which enters the wall or condenses within the cavity to be collected on flashing membranes and expelled through weepholes (*Fig. 8.5*). At the base of the wall, and at any point where the cavity is interrupted such as shelf angles, window heads and sills, and door heads, a layer of flashing must be installed, and with it, a row of weepholes. The open cavity in a drainage wall should be at least 2 in. wide (exclusive of insulation or sheathing), because narrower cavities are difficult to keep clean of mortar droppings during construction. In single-wythe hollow masonry walls, the ungrouted cores of the units should be treated as drainage cavities, and flashing and weepholes provided wherever the downward flow of water is interrupted (*Fig. 8.6*). Either proprietary flashing units or membrane flashing can be used. Where through-wall membrane flashing breaks the mortar bond, vertical continuity is restored by the periodic placement of reinforcing

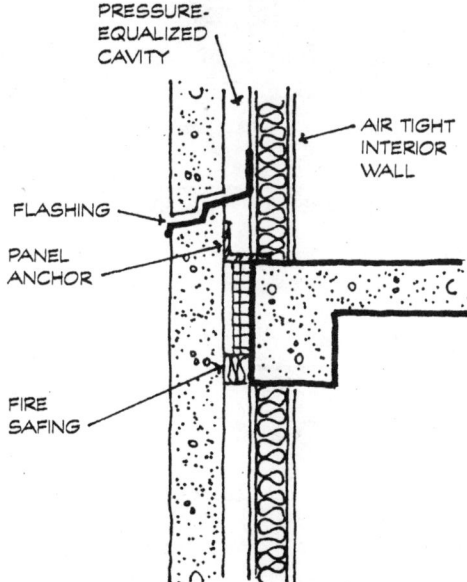

Figure 8.4 Pressure-equalized precast cladding system. (*From Persily, Envelope Design Guidelines for Federal Office Buildings.*)

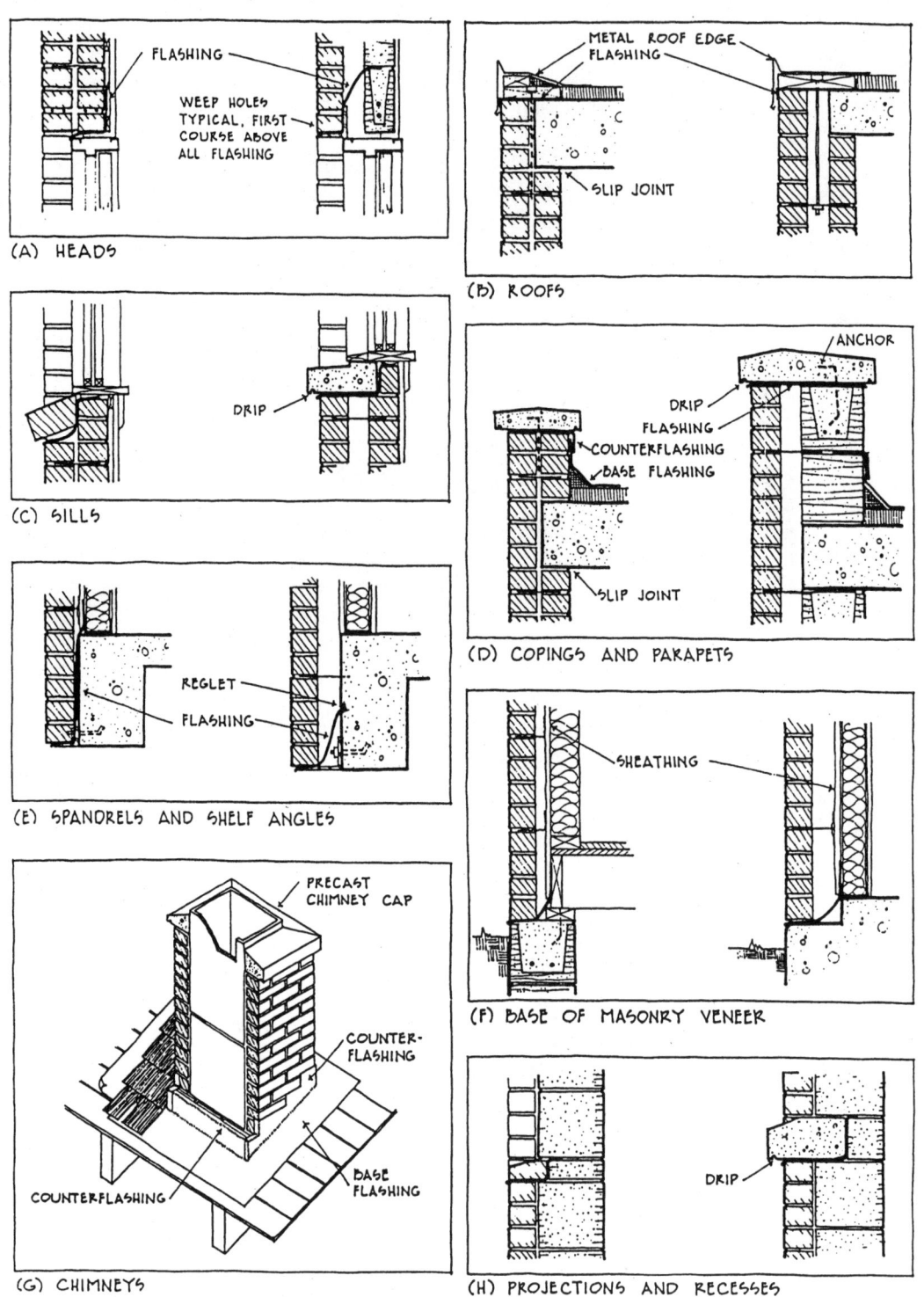

Figure 8.5 Masonry cavity wall and veneer flashing and drainage details. (*From Beall, Masonry Design and Detailing, 4th ed.*)

Cladding 361

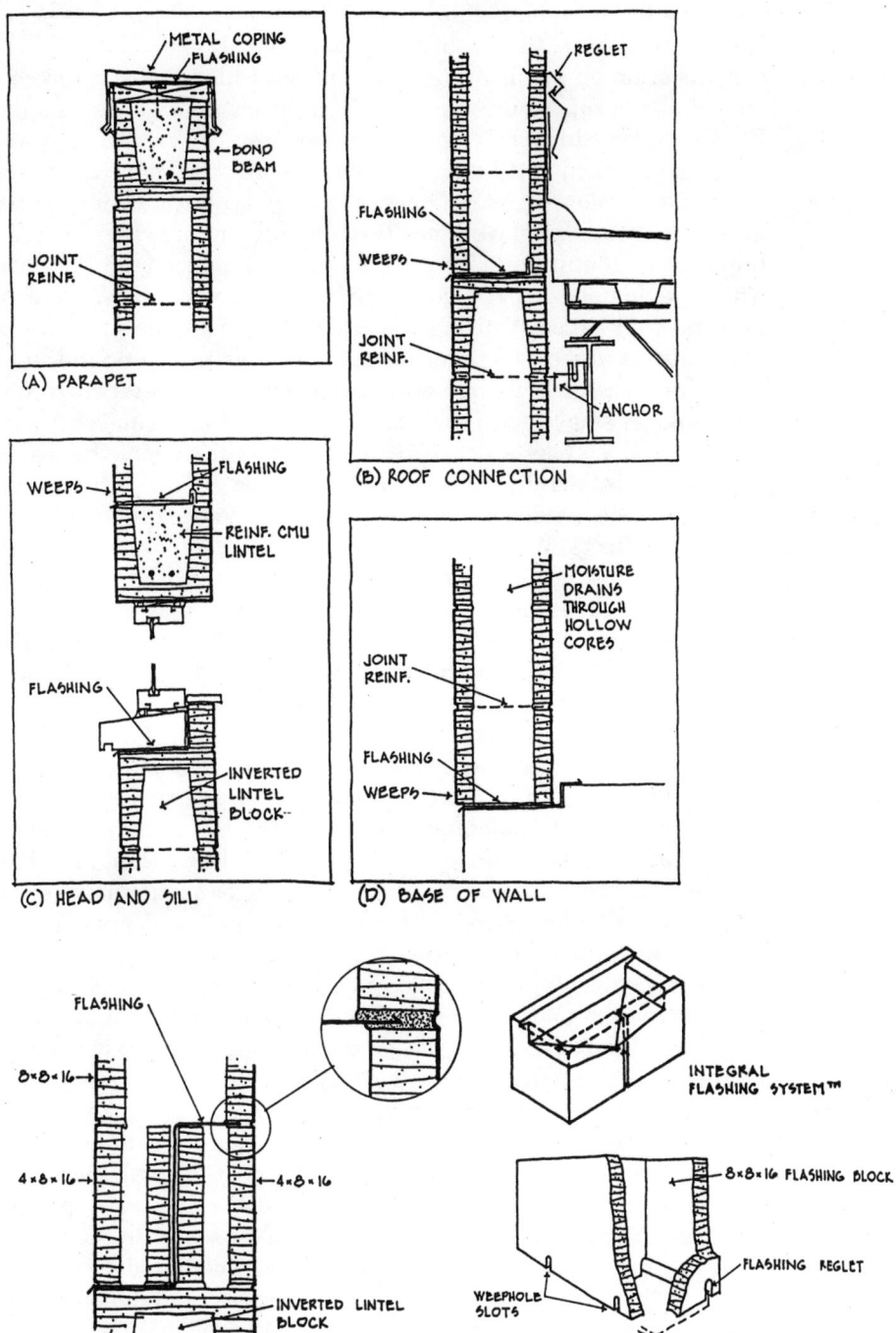

Figure 8.6 Single-wythe masonry flashing and drainage details. (*From Beall, Masonry Design and Detailing, 4th ed.*)

steel grouted into the hollow cores. Wherever the flashing is penetrated by the vertical steel bars, the flashing must be sealed.

A variation on the drainage wall concept called the *rain screen wall* is based on equalizing the air pressure between the cavity and the outside atmosphere. Blowing winds during a rain cause a low-pressure condition in wall cavities. In seeking a natural state of equilibrium, air moves from the high-pressure zone outside to the low-pressure zone in the cavity. With air infiltration, rainwater is carried through the wall face through any minute cracks which may exist at the mortar-to-unit interface. Under such a pressure differential, rainwater which would normally run down the face of the wall is literally driven or sucked into the wall cavity. Venting the cavity equalizes the pressure differential to eliminate the force which pushes or pulls moisture through the wall.

The rain screen principle was developed in the metal curtain wall industry and requires some special detailing for adaptation to masonry construction. To function properly as a pressure-equalized rain screen, the wall section must include an air barrier, and the cavity must be divided into compartments. The cavity must be divided both horizontally and vertically to prevent wind tunnel and stack effects. Without an air barrier and compartmenting, the horizontal flow of air around building corners and through the backing wall prevents pressure equalization in the wall cavity. Shelf angles in conventional masonry cavity wall and veneer construction provide compartmental barriers to the vertical flow of air, but corners require special detailing. *Figure 8.7* shows a sheet-metal fin inserted into expansion joints near the building corners to form a barrier to horizontal air flow. Preformed expanding foam sealants might also be used, if they can be adequately secured against displacement by the full force of both positive and negative pressure differentials. Each "compartment" must be adequately vented so that pressure change occurs as rapidly as possible. Rain screen vents should be located near the top of the wall or panel section, and constructed in the same manner as open head joint weepholes. Metal or plastic inserts can be used to disguise the appearance of both the weeps and the vents.

Wherever flashing is installed in masonry walls, sealed laps, continuity at building corners and the placement of end dams is critical to effective protection (*Fig. 8.8*). Weepholes must provide unobstructed flow, so it is important to keep mortar droppings out of the wall cavity. Open head joint weeps provide the best drainage and should be spaced at a maximum of 24 in. on center. Cotton wick weeps are less effective, and should be spaced at a maximum of 16 in. on center. Narrow weep tubes are too easily blocked by even small mortar droppings and construction debris, and are generally not reliable in providing effective drainage. Insulation placed in the cavity of masonry drainage walls minimizes thermal bridging and reduces the risk of concealed condensation, but it must be of a type that is resistant to moisture absorption and dimensionally stable.

Masonry walls are relatively brittle and are characterized by thousands of linear feet of joints along which cracks can form causing increased water penetration. Thermal and moisture movements and differential movements

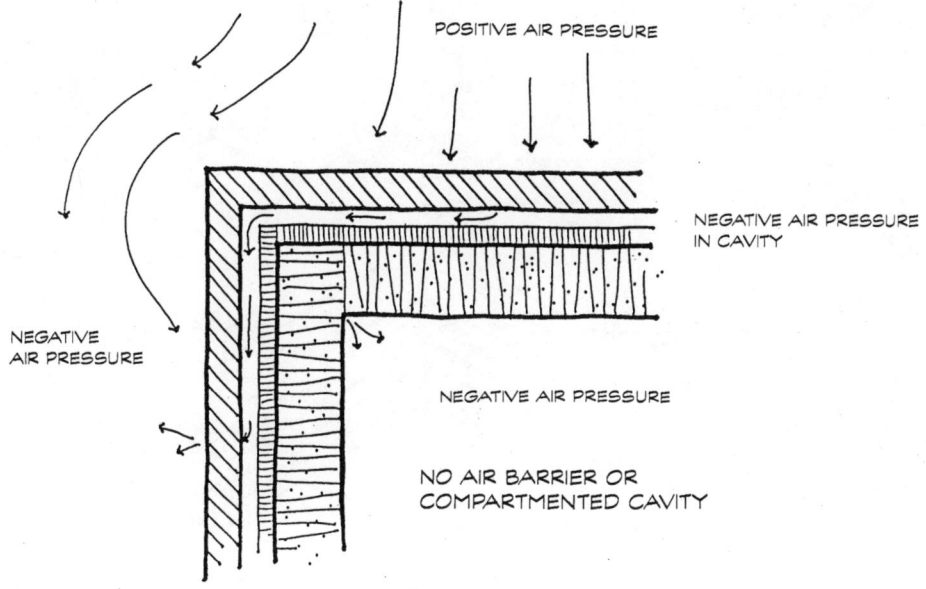

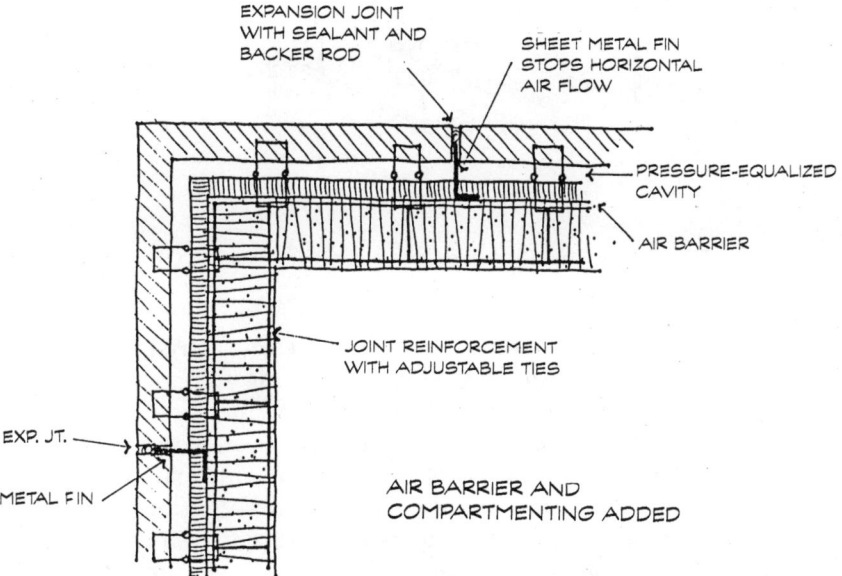

Figure 8.7 Compartmentalizing a masonry cavity wall at the corners. (*From Quirouette, Rain Penetration Control.*)

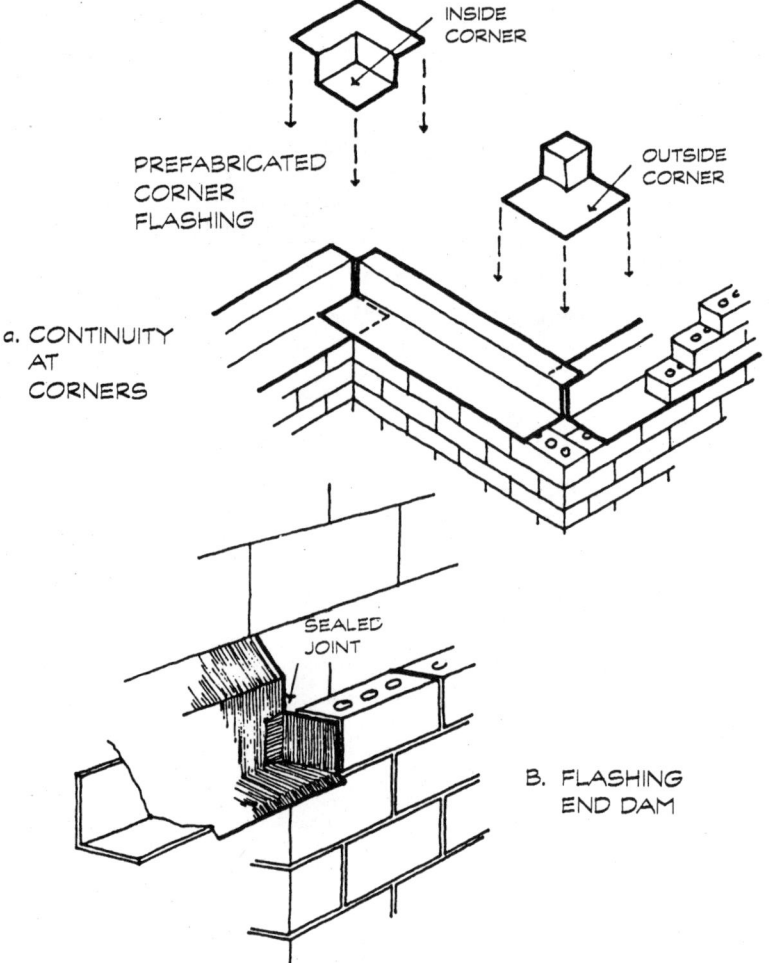

Figure 8.8 Corners and end dams in masonry flashing. (*From Beall, Masonry Design and Detailing, 4th ed.*)

between adjacent materials must be considered, and components selected and detailed accordingly to prevent cracking. Concrete products shrink irreversibly, clay products expand irreversibly, and metals alternately expand and contract with temperature changes. In conjunction with the flexible anchorage of backing and facing materials, control joints and expansion joints are used to alleviate the potential stresses caused by differential movement between dissimilar materials and by thermal and moisture movement within the masonry.

The terms control joint and expansion joint are not interchangeable. The two types of joints are different in both function and configuration (*Fig. 8.9*). *Control joints* are continuous, weakened joints designed to accommodate the permanent and irreversible moisture shrinkage of portland–cement based products such as concrete, concrete masonry, and stucco. When shrinkage stresses are high enough to cause cracks, the cracking will occur at these weakened joints rather than at random locations. Although moisture-controlled units and joint reinforcement can be used to limit shrinkage cracking in concrete masonry, strategically located control joints must also be used to prevent random cracking and the accompanying infiltration of water. Control joints must provide lateral stability between adjacent wall sections with a grout key or hard rubber shear key. Control joints must also be sealed at the outside face against moisture penetration. For standard units, the joints are first laid up in mortar just as any other vertical joint would be. After the mortar has stiffened slightly, the joints are raked out to a depth which will allow placement of a backer rod or bond-breaker tape, and a sealant joint of the proper depth. Concrete masonry shrinkage always exceeds its expansion because of the initial moisture loss experienced after manufacture. So even though control joints may contain hardened mortar, they can accommodate subsequent thermal expansion and contraction movements which are much smaller and occur after the joint opens from the initial curing shrinkage. *Expansion joints* in masonry are designed to accommodate the permanent and irreversible moisture expansion of brick and other clay units. As the brick on either side expands, an expansion joint closes to accommodate the increased volume, so it cannot contain mortar or other hard materials. Compressible fillers may be used to keep mortar out of expansion joints during construction, because even small mortar bridges can cause localized spalling of the unit faces when the joint tries to close. Filler materials should be at least as compressible as the joint sealant which will be used, and the compressibility of the sealant must be considered

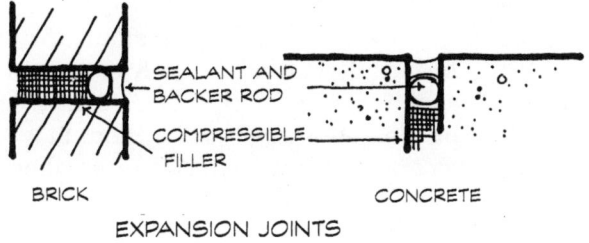

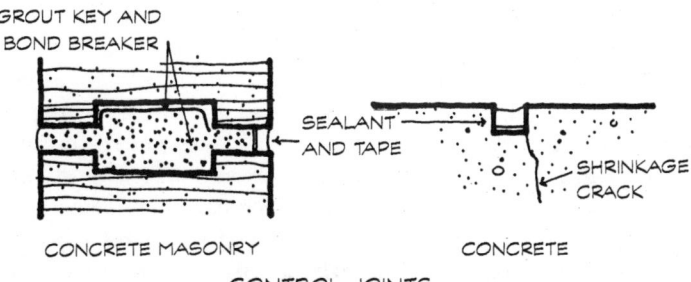

Figure 8.9 Expansion and control joints are different. (*From Beall, Masonry Design and Detailing, 4th ed.*)

in calculating joint width and spacing. To provide lateral stability, the brick must be anchored to its backing on either side of the joint.

Coefficients of thermal expansion and moisture expansion can be used to estimate the expected movement of various materials, and movement joints can be sized and located accordingly. If details do not sufficiently accommodate wall movement, excessive moisture can penetrate through the resulting cracks. Chapter 9 includes tables and formulas for calculating expected movement, recommended joint spacings, and details of masonry expansion and control joints. The exact locations of control and expansion joints will be affected by design features such as openings, offsets, and intersections. In brick walls, expansion joints should be located near corners because the opposing expansion of intersecting walls can cause cracking. For both brick and concrete masonry walls, joints should be located at points of weakness or high stress concentration such as abrupt changes in wall height; changes in wall thickness; columns and pilasters; and at one or both sides of windows and doors. Steel reinforcement can be used to restrain movement and reduce the spacing of control and expansion joints. Prefabricated wire joint reinforcement is routinely used in concrete masonry walls to reduce shrinkage, and is also sometimes used in brick walls to control expansion. For concrete block, prefabricated wire reinforcement is usually placed in every second or third bed joint and should be discontinuous at control joints. Wire joint reinforcement can reduce moisture shrinkage in concrete masonry but control joints are still required even though they may be spaced further apart. Masonry that is fully reinforced to resist structural or seismic loads, however, usually contains enough steel that cracking is not a problem and control joints are not required.

Cracking in masonry can result from restraining the natural expansion and contraction of the materials themselves, or from failure to allow for differential movement of adjoining or connected materials. When masonry walls are connected to steel or concrete frames, differential movement must be accommodated in the anchorage of one material to another. Even if the exterior masonry veneer carries its own weight to the foundation without shelf angles or ledges, the structural columns, floors and backing walls provide the lateral support which is required by code. Flexible connections should allow relative vertical movement without inducing excessive stress (i.e., they should resist the lateral tension and compression of wind loads, but allow a limited amount of in-plane shear movement).

Although brick and stone masonry do not typically require the application of decorative or protective coatings, some types of concrete masonry units are very porous and should be protected from excessive surface absorption. Applied *coatings* for masonry walls must be carefully selected on the basis of their permeability. A nonporous paint film will prevent some rainwater from entering a wall but, more significantly, it will impede the evaporation of moisture which might enter the wall by other means. Water may enter through pores in materials, partially filled mortar joints, improperly flashed copings, sills, or parapet walls, through capillary contact with the ground, failed sealant joints, or any number of other sources. Moisture escapes from a masonry wall in only two ways: (1) through continuous cavities outfitted with flashing and weepholes and (2) by evaporation at the wall face (breathing). Vapor permeance at the exterior face is important in all masonry construction, but in "barrier walls" where there are no cavities or weepholes, evaporation at the exterior face is the only method of moisture removal.

There are numerous types of paint suitable for masonry walls, including cement-based paints, water-thinned emulsions, fill coats, solvent-thinned paints, and high-build coatings. In selecting a paint finish system, there are several things to consider. Paint products that are based on drying oils may be attacked by free alkali from the units or mortar. Alkaline-resistant paints and primers are recommended to prevent this, or else the masonry must fully cure for at least 30 days before painting. Surface conditions must also be considered, and preparations suitable to the selected finish made. *Clear water repellents* are usually advertised as a cure-all for masonry moisture problems, and they are often incorrectly referred to as "sealers" or "waterproof" coatings—which they are not. Water repellents generally change the capillary angle of pores in the face of the masonry to repel rather than absorb water, but they will not bridge hairline cracks or separations at the mortar-to-unit interface. Clear water repellents can reduce absorption through the face of the masonry and prevent soiling on light-colored units and stone while still permitting the wall to breathe. There are three types of clear water repellents: stearates, acrylics, and silicones. No single type is equally suitable or effective on all masonry substrates, because the physical and chemical properties of clay brick, concrete masonry, and stone vary so widely. Compatibility of substrate and surface treatment should always be evaluated on an individual basis. Chapter 10 discusses a variety of paints and water repellents used to protect exterior building walls. Coatings on masonry, however, are not a substitute for good design and proper installation of the primary moisture protection system.

Parapet walls that extend above the roof line can be the most problematic element of a masonry cladding system. In fact, the single most important detail in a wall that is not protected by a roof overhang is the parapet coping. Copings constructed of masonry materials such as brick, precast concrete, or stone are problematic because the mortar in the joints can shrink and create water leakage paths. The mortar joints in masonry copings should be raked out and sealed with a backer rod and elastomeric sealant, and a membrane flashing should be provided below the coping. The best way to protect the top of a masonry parapet wall is with a metal cap flashing (*Fig. 8.10*). In high-risk exposures, redundant protection can be provided by placing a layer of membrane flashing below the metal cap. Joints in the metal cap should be constructed with a cover plate or gutter plate sealed with a non-hardening butyl sealant. The vertical leg of the cap flashing should overlap the top of the masonry by at least 2 in. so that wind-driven rain is not blown up underneath it. In high-wind areas, continuous rather than periodic cleats should be used to secure the metal. If sealant is used to seal the bottom edge of the cap flashing,

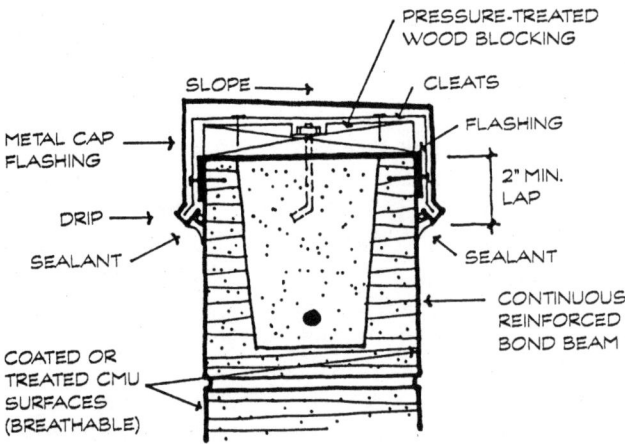

Figure 8.10 Metal cap flashing at masonry parapet. (*From Beall, Masonry Design and Detailing, 4th ed.*)

weep tubes should be incorporated to allow drainage of any water that does penetrate the surface. The back side of a masonry parapet should never be sealed with a roofing membrane or a vapor-impermeable coating. The wall must be able to breathe so that moisture can evaporate at the surface. Where the top edge of the roof base flashing must be protected, a two-piece metal counterflashing installed in a mortar bed joint provides for easy installation and maintenance of the roofing membrane (see *Fig. 6.38* in Chapter 6).

8.2.3 Curtain walls and windows

Curtain walls can be fabricated as "stick" systems that are assembled and glazed on site, or as unitized systems of prefabricated components that are erected on site. Curtain walls can also be designed as barrier systems, drainage systems, rain screen systems, or combination systems. Barrier systems use a wet sealant as the exterior glazing bead to prevent water from entering the framing (*Fig. 8.11A*). Drainage systems use dry glazing gaskets which are not watertight, so penetrated moisture is drained through integral weepholes in the horizontal framing members or through a separate sill flashing member (*Fig. 8.11B*). Rain screen systems are equipped with weeps and vented air chambers within the framing members for pressure equalization (*Fig. 8.11C*). Venting the air chamber to the outside and providing an air barrier between the chamber and the inside creates pressure equalization and reduces the suction of water through the exterior glazing seals in a wind-driven rain. Some curtain walls use a combination of protective strategies, particularly when different materials such as stone or metal panels are incorporated in the system.

Glass and metal curtain walls are typically specified by performance requirements and supplied as proprietary systems. The design of individual details is performed by the manufacturer or fabricator. Anchorage and joinery must

Cladding 369

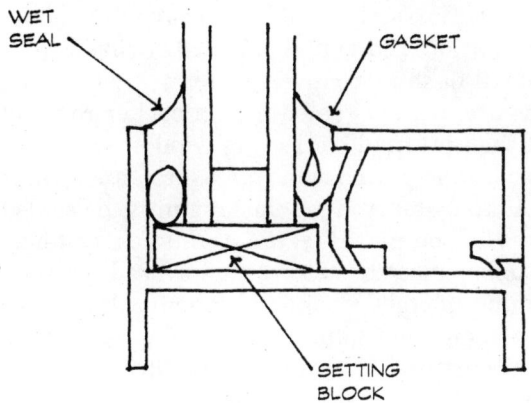

a. RAIN BARRIER SYSTEM

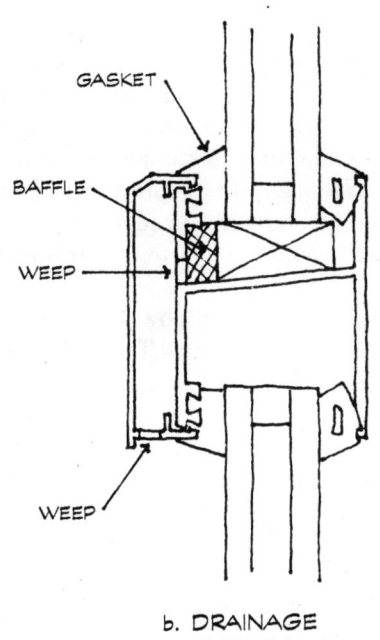

b. DRAINAGE SYSTEM

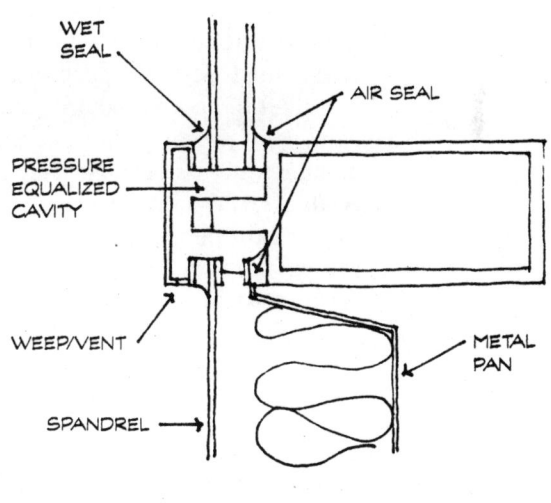

c. PRESSURE EQUALIZED RAIN SCREEN

Figure 8.11 Barrier, drainage, and rain screen curtain walls.

accommodate thermal expansion and contraction, and external fit-up joints must be protected against rain penetration. Individual punched or ribbon windows must be fitted with head and sill flashing to divert water away from the opening. Perimeter joints at jambs are usually sealed with a backer rod and elastomeric sealant. The height of the upturned back leg of window sill flashing must be adequate to accommodate expected wind pressures and water head (*Fig. 8.12*), and the corners of the frame and sill flashing must be sealed to complete the watertight integrity of the pan (*Fig. 8.13*). In stick systems, end dam blocks are composed of rubber inserts that are compressed between the vertical and horizontal framing members and silicone sealants are installed at the interface between the end dams and the glazing pocket members. These end dams are a critical part of the performance of drainage and rain screen walls.

The size and locations of weeps must be adequate to drain penetrated water effectively. Weeps that are too small can be temporarily blocked by the surface tension of the draining water itself. Round holes less than $1/4$ in. in diameter and slots less than $1/8$ in. wide can support a significant head of water before the surface tension breaks and the water drains out.

Expansion joints are required in curtain walls and large window systems to accommodate thermal expansion and contraction. In most parts of the United States, the seasonal range of metal surface temperatures is at least 150°F, and in some parts of the country it may be as much as 200°F. Since the aluminum used in window and curtain wall systems has a very high thermal expansion coefficient, these temperature ranges translate into movements of $1/4$ in. to $5/16$ in. over a 10-ft length of metal. Over the same temperature range, the glass movement will be only half as much. Slip joints and bellows action in the framing members accommodate much of this movement. Expansion joints and perimeter seals must also be adequately sized and spaced, and elastomeric sealants installed to accommodate expected movement and still provide protection from rain penetration. Silicone sealants are the most commonly used material in glazing systems because they have tenacious adhesion and excellent durability. Sealants and joint sizing are discussed in detail in Chap. 9.

The rain penetration and air infiltration resistance of window systems are critical parts of a building's overall moisture protection strategy. The American Architectural Manufacturers Association (AAMA) and the American National

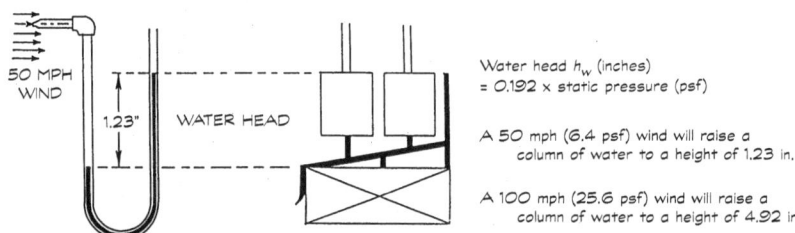

Figure 8.12 Wind can drive water uphill. (*From AAMA Aluminum Curtain Wall Design Guide Manual.*)

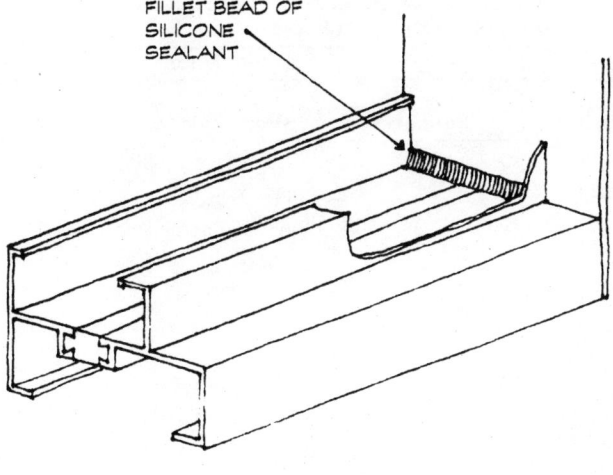

CORNER OF FRAME

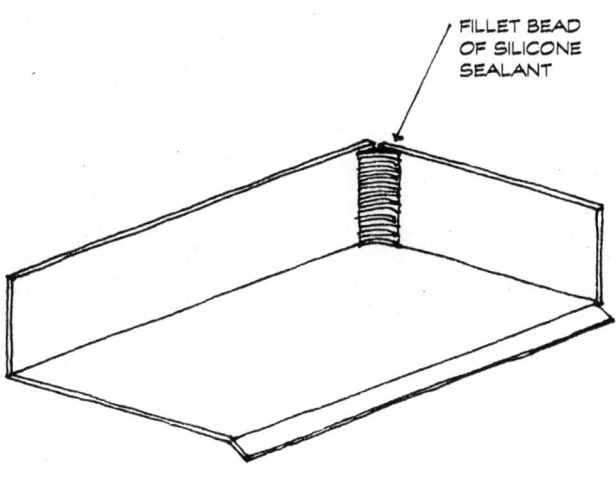

CORNER OF SILL PAN FLASHING

Figure 8.13 Seal corners of frame and sill flashing.

Standards Institute (ANSI) have developed tests to evaluate the performance of window systems and standards governing minimum performance requirements. Using the product symbols and letter designations shown in *Table 8.1*, ANSI/AAMA 101-88 *Voluntary Specifications for Aluminum Prime Windows and Sliding Glass Doors* lists the primary performance requirements for a variety of window types (*Table 8.2*). The minimum performance class to which a product must conform in order to be labeled residential, commercial, or

TABLE 8.1 Product Symbols and Letter Designations Used to Classify Performance of Aluminum Prime Windows and Sliding Glass Doors

Product type		Grade	Performance		class		
Product code	Product	Designation	Description	Class*	R	C	HC
A	Awning	R	Residential	15	•		
C	Casement	C	Commercial	20	•	•	
DH single-hung	Double- or	HC	Heavy commercial	25	•	•	
VS	Vertical sliding			30	•	•	
F	Fixed			35	•	•	
HS	Horizontal sliding			40	•	•	•
P	Projected			45	•	•	•
TH	Top-hinged			50	•	•	•
VP	Vertical pivoted						
GH	Greenhouse						
JA	Jal-awning						
J	Jalousie						
DA	Dual action						

*Class designation number corresponds to design wind pressure in pounds per square foot.

heavy commercial and the optional performance classes which may be specified for each of the three designations are shown in *Table 8.3*. Architectural windows, whether made from standard, modified, or custom extrusions, must be ruggedly built and have a long life expectancy in both structural stability and performance with little maintenance. Performance requirements for "architectural windows" for commercial- and monumental-type buildings are covered in AAMA *Guide Specification GS-001* and shown in *Table 8.4*. The performance requirements listed in the various tables specify maximum allowable air infiltration, but no water leakage, as defined in the test method, is permitted. Where an air barrier is required in the exterior envelope, choose a window type with low air infiltration rate, and seal the membrane or drywall air barrier to the window perimeter to assure the integrity of the system. Water penetration tests are performed in accordance with ASTM E331 *Standard Test Method for Water Penetration of Exterior Windows, Curtain Walls, and Doors by Uniform Static Air Pressure Difference.* Water leakage is defined in that test as "penetration of water into the plane of the innermost face of the test specimen under specified conditions of air pressure difference across the specimen during a 15-minute test period. In tests of windows and doors, it also occurs whenever water penetrates through the frame of the test specimen."

In addition to water and air infiltration resistance, the performance of windows and curtain walls is also judged on thermal resistance. A metal framing member that extends through from interior to exterior will form a thermal

Table 8-2 ANSI/AAMA 101-88 Voluntary Specifications for Aluminum Prime Windows and Sliding Glass Doors Primary Performance Requirements for Residential, Commercial, and Heavy Commercial Windows

Window/door designation*	Design pressure, lb/sq ft (Pa)	Structural test pressure,[†] lb/sq ft	Water resistance test pressure,[‡] lb/ft^2 (Pa)	Air infiltration test pressure lb/ft^2 (Pa)	Maximum rate[§]
Awning					
A-R15	15 (718)	22.5 (1077)	2.86 (137)	1.57 (75)	0.37
A-C20	20 (958)	30 (1437)	3.00 (143.6)	1.57 (75)	0.37
Casement					
C-R15	15 (718)	22.5 (1077)	2.86 (137)	1.57 (75)	0.37
C-C20	20 (958)	30 (1437)	3.00 (143.6)	1.57 (75)	0.37
C-HC40	40 (1915)	60 (2873)	6.00 (287.3)	6.24 (299)	0.37
Double-hung					
DH-R15	15 (718)	22.5 (1077)	2.86 (137)	1.57 (75)	0.37
DH-DW-R15	15 (718)	22.5 (1077)	2.86 (137)	1.57 (75)	0.37
DH-C20	20 (958)	30 (1437)	3.00 (143.6)	1.57 (75)	0.37
DH-DW-C20	20 (958)	30 (1437)	3.00 (143.6)	1.47 (75)	0.37
DH-HC40	40 (1915)	60 (2873)	6.00 (287.3)	1.57 (75)	0.37
Sliders					
HS-R15	15 (718)	22.5 (1077)	2.86 (137)	1.57 (75)	0.37
HS-DW-R15	15 (718)	22.5 (1077)	2.86 (137)	1.57 (75)	0.37
HS-C20	20 (958)	30 (1437)	3.00 (143.6)	1.57 (75)	0.37
HS-DW-C20	20 (958)	30 (1437)	3.00 (143.6)	1.57 (75)	0.37
HS-HC40	40 (1915)	60 (2873)	6.00 (287.3)	1.57 (75)	0.37
Projected					
P-R15	15 (718)	22.5 (1077)	2.86 (137)	1.57 (75)	0.37
P-C20	20 (958)	30 (1437)	3.00 (143.6)	1.57 (75)	0.37
P-HC40	40 (1915)	60 (2873)	6.00 (287.3)	6.24 (299)	0.37
Top-hinged					
TH-C20	20 (958)	30 (1437)	3.00 (143.6)	1.57 (75)	0.37
TH-HC40	40 (1915)	60 (2873)	6.00 (287.3)	6.24 (299)	0.37
Vertically pivoted					
VP-C20	20 (958)	30 (1437)	3.00 (143.6)	1.57 (75)	0.37
VP-HC40	40 (1915)	60 (2873)	6.00 (287.3)	6.24 (299)	0.37
Vertical slide					
VS-R15	15 (718)	22.5 (1077)	2.86 (137)	1.57 (75)	0.37
VS-DW-R15	15 (718)	22.5 (1077)	2.86 (137)	1.57 (75)	0.37
Greenhouse window					
GH-R15	15 (718)	22.5 (1077)	2.86 (137)	1.57 (75)	0.37
Jal-awning window					
JA-R15	15 (718)	22.5 (1077)	2.86 (137)	1.57 (75)	0.37
Jalousie window					
J-R15	15 (718)	22.5 (1077)	2.86 (137)	1.57 (75)	0.37
Fixed windows					
F-R15	15 (718)	22.5 (1077)	2.86 (137)	1.57 (75)	0.15
F-DW-R15	15 (718)	22.5 (1077)	2.86 (137)	1.57 (75)	0.15
F-C20	20 (958)	30 (1437)	3.00 (143.6)	1.57 (75)	0.15
F-DW-C20	20 (958)	30 (1437)	3.00 (143.6)	1.57 (75)	0.15
F-HC40	40 (1915)	60 (2873)	6.00 (287.3)	6.24 (299)	0.15
Dual action					
DA-R15	15 (718)	22.5 (1077)	2.86 (137)	1.57 (75)	0.37
DA-C20	20 (958)	30 (1437)	3.00 (143.6)	1.57 (75)	0.37
DA-HC40	40 (1915)	60 (2873)	6.00 (287.3)	6.24 (299)	0.37

*DW, where used, indicates dual window.

[†]Structural test pressures shown are for both positive and negative loads. After each specified loading, there shall be no glass breakage, permanent damage to fasteners, hardware parts, support arms, or actuating mechanisms or any other damage which could cause the window to be inoperable. There shall be no permanent deformation of any main frame, sash, or ventilator member in excess of 0.4%. Permanent deformation requirement applies to the primary window members only.

[‡]Where the manufacturer offers or specifies an insect screen, the water resistance test is to be performed both with and without the insect screen in place. In dual windows, where an exterior insect screen is offered by the manufacturer, the dual window unit must be tested with the screen in both the summer and winter modes.

[§]Air infiltration rate is cfm per foot of operating ventilator or sash crack length except for fixed windows, jalousie windows, and sliding glass doors which shall be cfm per square foot of area. In dual windows, only the operating sash of the designated primary window or the area of the designated fixed primary window is to be used in determining the sash crack or square footage.

SOURCE: AAMA Window Selection Guide.

TABLE 8.3 ANSI/AAMA 101-88 Performance Classes for Residential, Commercial, and Heavy Commercial Windows

Performance class	Product grade designation	Design pressure, lb/sq ft	Structural test pressure, lb/sq ft	Water resistance test pressure, lb/sq ft
Minimum Performance Classes				
15	R	15	22.5	2.86
20	C	20	30	3.00
40	HC	40	60	6.00
Optional Performance Classes				
20	R	20	30.0	3.00
25	R and C	25	37.5	3.75
30	R and C	30	45.0	4.50
35	R and C	35	52.5	5.25
40	R and C	40	60.0	6.00
45	R, C, and HC	45	67.5	6.75
50	R, C, and HC	50	75.0	7.50

TABLE 8.4 AAMA Guide Specification GS-001 Performance Requirements for Architectural Windows

Window type	Air pressure difference across test specimen, psf		Maximum allowed air, cfm/fcp[‡]
	Air[*]	Water[†]	
Double-hung, single-hung, triple-hung	6.24	8.00	0.30
Casement	6.24	8.00	0.10
Outswinging and inswinging			
Side-hinged inswing	6.24	8.00	0.10
Horizontal sliding	6.24	8.00	0.10
Vertical pivot/horizontal pivot (both 360° and 180° types)	6.24	8.00	0.10
Projected	6.24	8.00	0.10
Top-hung inswing	6.24	8.00	0.10
Dual action	6.24	8.00	0.10
Fixed	6.24	8.00	0.06[§]

[*]6.24 psf—approximately 50 mph wind velocity pressure.

[†]8.00 psf—approximately 56 mph wind velocity pressure; may be increased to 20% of positive design wind pressure up to 12 psf maximum.

[‡]For operating windows: Air infiltration in cubic feet per minute per linear foot of crack perimeter. 1.5 times this amount allowed in field test.

[§]For fixed windows: Air infiltration in cubic feet per minute per square foot of window area. 1.5 times this amount allowed in field test.

Uniform load deflection test—specify design wind pressure required by the governing building code, as calculated from ANSI A58.1, or as determined by boundary layer wind tunnel testing. Use minimum 30 psf. Under any circumstance, however, the design wind pressure must be at least equal that required by the governing code.

Uniform local structural test—Specify 1.5 the design wind pressure used in the uniform load deflection test. Use minimum 45 psf.

SOURCE: AAMA *Guide Specification* GS-001.

bridge and suffer conductive heat loss. In cold weather, if the temperature of the metal is below the dew point of the interior air, condensation will form on its surface. When a mullion section has minimum exterior exposure (*Fig. 8.14A*), its interior surface temperature will be warmer than if the ratio of exposed area is reversed (*Fig. 8.14B*). Thermal breaks of plastic or elastomeric materials inserted into the metal section divide it into exterior and interior components that are insulated from one another to prevent winter surface condensation (*Fig. 8.14C*). Some condensation can be expected on *all* windows under *some* circumstances, but thermally broken framing members and multi-pane glazing systems can perform well under even severe winter conditions. A relatively small amount of condensation under extreme conditions is generally considered tolerable, and will not significantly affect overall thermal performance. In a good window design, the U value of the framing will be equal to or less than that of the insulating glass.

Guidelines for maximum U values and minimum condensation resistance factors for various conditions of outdoor temperature and indoor humidity are published by AAMA. *Figure 8.15* shows surface temperatures at which condensation will form in an environment with a given humidity. For example, in a building with an indoor relative humidity of 40%, the indoor surface temperature of the glass and framing must remain above 42°F to prevent condensation. *Figure 8.16* shows the maximum allowable U value for any component of an exterior wall under a given set of indoor and outdoor conditions. For example, in an area with a winter design temperature of 0°F, a building with an expected indoor relative humidity of 25% should have an exterior wall with a maximum U value of 0.80 Btu/hr·ft^2·°F, or condensation will form on the interior surfaces. Since windows and curtain walls are made up of glazing and framing components, they are more difficult to evaluate for condensation resistance than homogeneous elements with a constant U value. AAMA has developed a standard test for evaluating the performance of window systems AAMA 1503.1 *Voluntary Test Method for Thermal Transmittance and Condensation Resistance of Windows, Doors and Glazed Wall Sections* that rates the overall performance of the system. The condensation resistance factor (CRF) is a rating number obtained under standard laboratory conditions with 68°F indoor

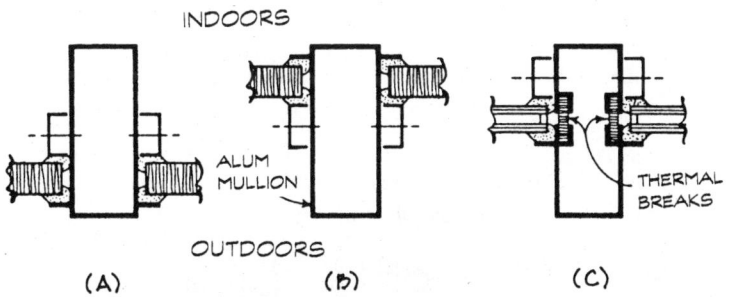

Figure 8.14 Amount of exterior exposure affects interior temperature of framing. (*From AAMA Aluminum Curtain Wall Design Guide Manual.*)

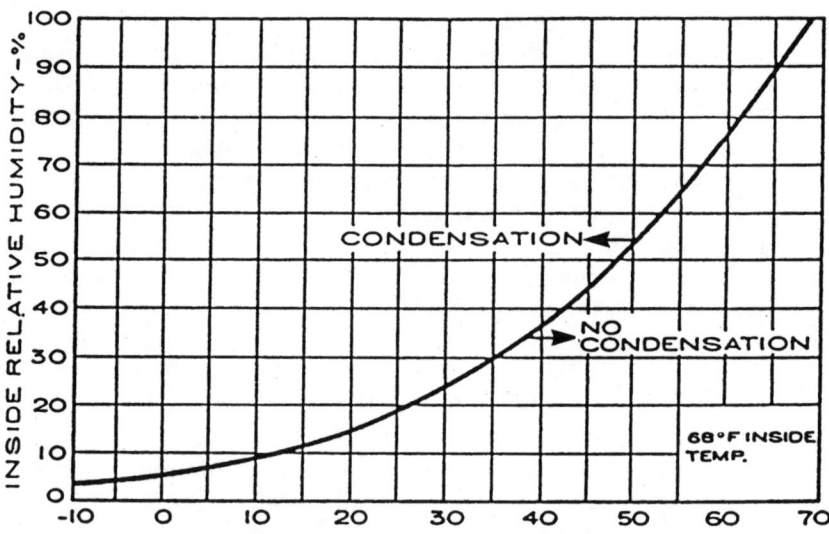

Figure 8.15 Relative humidity and surface temperature at which visible condensation will occur. (*From AAMA Aluminum Curtain Wall Design Guide Manual.*)

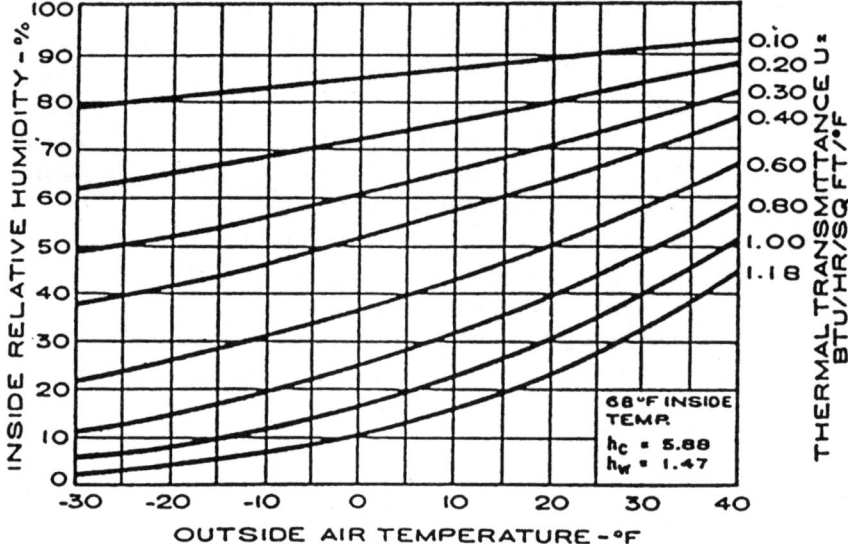

Figure 8.16 Outside air temperature at which condensation will occur on inside surface for different U values and indoor relative humidities. (*From AAMA Aluminum Curtain Wall Design Guide Manual.*)

temperature, 18°F outdoor temperature, 15 mph wind, and minimal air infiltration. The CRF is found by the following formula:

$$\text{CRF} = \frac{T_{is} - T_o}{T_i - T_o} \times 100 \tag{8.1}$$

where T_{is} = temperature of the inside surface
T_o = outside air temperature
T_i = inside air temperature

Using the map and graph in *Fig. 8.17,* it is easy to determine the minimum recommended CRF for any given climate in the contiguous 48 states. Converted to tabular form, *Table 8.5* also shows the maximum recommended indoor relative humidity for various climatic conditions. Occupancies with requirements for higher indoor humidities like computer facilities, museums, and hospitals represent extreme conditions, and condensation may occur on even the highest-rated windows unless there are additional protective measure such as supplemental air flow and perimeter heating along the glazed wall sections. Under such extreme conditions, the flow of warm air at the windows is critical to prevent condensation, and obstructions such as draperies and window blinds can cause problems.

8.2.4 Exterior insulation and finish systems (EIFS)

Introduced in the United States in the 1970s, exterior insulation and finish systems have enjoyed increasing popularity as a versatile and economical cladding for new construction and for renovation of existing buildings. EIFS consist of a polymer-modified or polymer-based synthetic coating applied as a reinforced base coat and finish coat over rigid thermal insulation board which mimics the appearance of traditional cement and lime stucco. The apparent simplicity and successful European performance of EIFS applied over masonry substrates belied the subtle complexities introduced by the predominant application in this country of EIFS over non-masonry substrates such as metal studs and sheathing. Our understanding of system behavior and the development of improved details and application procedures in fact, have been driven by field experience with system failures, maintenance, and repairs.

Conventional EIF systems are designed to act as moisture barriers, and do not incorporate a drainage cavity for penetrated moisture. Without such redundancy, sealant joints and perimeter flashings are critical elements in the cladding's success or failure. In areas where frequent or extended periods of rain or damp cold are common, the systems have little tolerance for moisture. Experience in some parts of the United States and Canada has been so negative that state and local building authorities have placed limitations on the use of EIFS, and some insurance underwriters refuse to bond contractors who install it.

The EIFS Industry Members Association (EIMA) defines two classes of systems, each of which consists of insulation board, glass fiber reinforcing mesh, weather barrier base coat and decorative finish coat (*Fig. 8.18*):

- Class PB systems
  - Molded expanded polystyrene (MEPS) insulation (ASTM C578, Type I), adhesively attached to substrate by using a notched trowel or ribbon and dab application
  - Polymer-modified cementitious base coat thickness varies depending on the thickness or the number of layers of reinforcing mesh, with a *minimum*

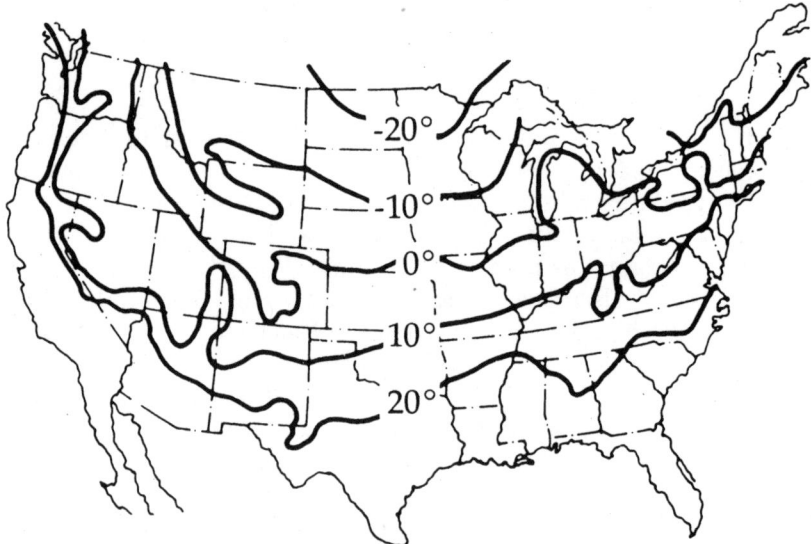

During the months of December, January, and February, the temperature will not be at or below the indicated design temperature 97.5% of the time. Temperature data for specific locations are given in Chapter 33, Weather Data and Design Conditions, *ASHRAE Handbook of Fundamentals,* published by American Society of Heating, Refrigerating and Air-Conditioning Engineers, Inc.

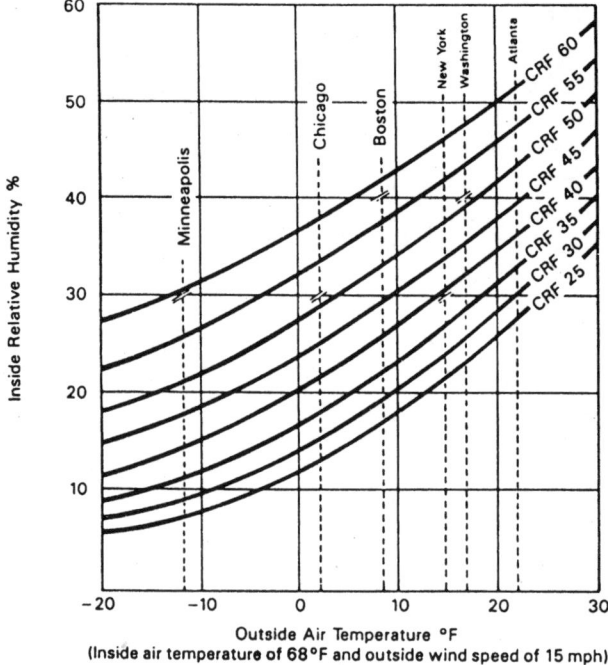

Figure 8.17 AAMA minimum recommended condensation resistance factor (CRF) for aluminum windows and curtain wall systems. (*From AAMA Aluminum Curtain Wall Design Guide Manual.*)

TABLE 8.5 AAMA Tabular Recommendations for Minimum Condensation Resistance Factor (CRF) for Aluminum Windows and Curtain Wall Systems

Interior air temperature=68°F, wind velocity=15 mph

Outside air temperature, °F	Maximum recommended inside relative humidity,* %	Inside relative humidity					
		15%	20%	25%	30%	35%	40%
−20	20	46	52	57	60		
−10	25	39	46	52	56	60	
0	30	30	39	45	52	57	61
+10	35	17	29	37	44	50	57
+20	40	0	16	25	34	40	48

*Relative humidities higher than those shown are not recommended for outside air temperatures equal to or lower than those shown unless specifically taken into account in the building design.
SOURCE: AAMA *Aluminum Curtain Wall Design Guide Manual.*

- Polymer-modified cementitious base coat thickness varies depending on the thickness or the number of layers of reinforcing mesh, with a *minimum* thickness of $1/16$ in
- Fiberglass reinforcing mesh in one or two layers fully embedded in the base coat at the time of installation
- Polymer-based, noncementitious finish coat
- Class PM systems
  - Extruded polystyrene (XEPS) insulation (ASTM C578, Type IV) mechanically attached to substrate
  - Metal lath or fiberglass reinforcing mesh installed over the surface of the insulation before the base coat is applied, and mechanically attached to substrate
  - Base coat applied over reinforcing mesh to a uniform thickness ranging from $1/4$ to $3/8$ in. (base coat thickness is not dependent on the thickness or on the number of layers of reinforcing mesh)
  - Polymer-modified cementitious finish coat
  - Include control joints to limit panel size and help prevent lamina cracking
  - Use metal or plastic accessories at terminations and intrasystem joints

PB systems are more flexible, lend themselves to the use of three-dimensional shaped detailing, require fewer control joints, and are much more widely used in the United States in both residential and commercial construction. This discussion will be limited primarily to the design, application and detailing of Class PB systems. EIMA publishes guide specifications for both Class PB and Class PM systems.

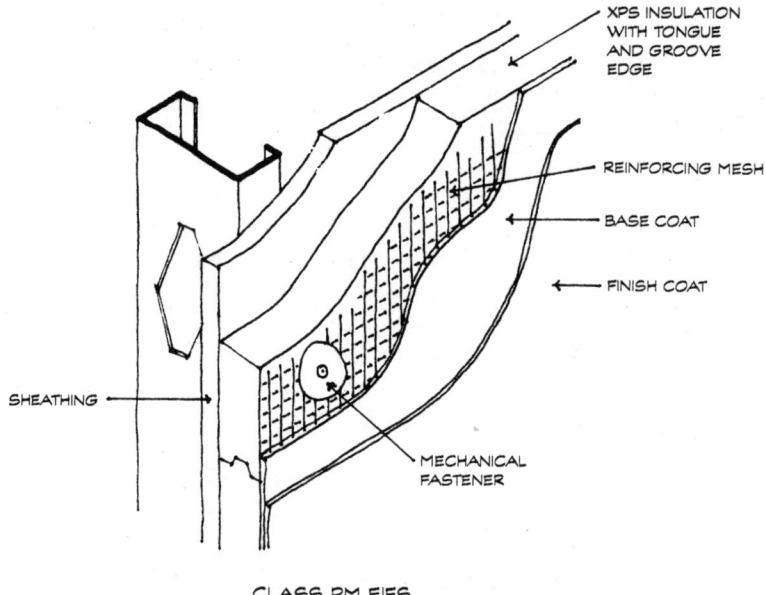

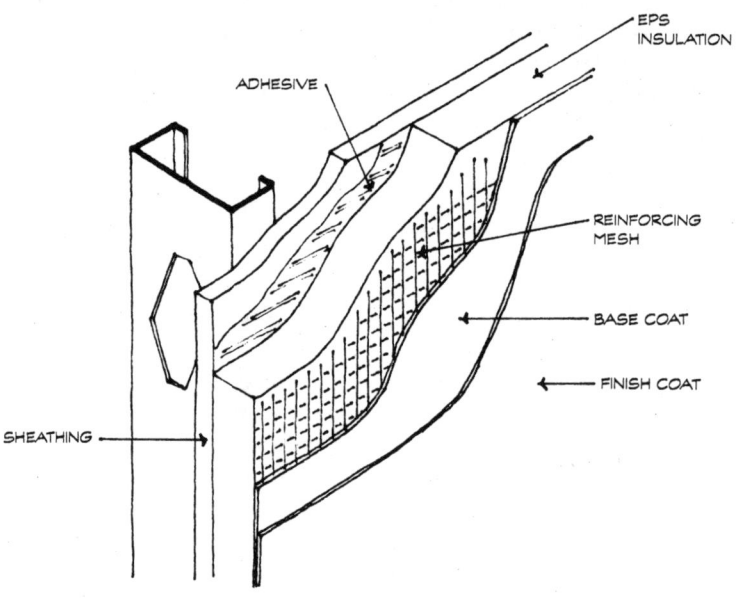

Figure 8.18 Two basic types of exterior insulation and finish systems. (*Adapted from Williams and Williams, Exterior Insulation and Finish Systems, Current Practices and Future Considerations, ASTM Manual 16.*)

Substrates. Most Class PB EIFS in the United States are installed over some type of sheathing board attached to metal or wood stud framing systems. Paper faced, water-resistant core gypsum sheathing (ASTM C79) has proved unreliable and is no longer recommended by industry experts. Gypsum sheathing which meets the requirements of ASTM C1177 *Standard Specification for Glass Mat Gypsum Substrate for Use as Sheathing* performs significantly better, but is still sensitive to trapped moisture, as is plywood and oriented-strand board (OSB). Prolonged exposure to moisture contents of 20% or more can lead to deterioration of gypsum substrates and decay of wood-based sheathing. The Army Corps of Engineers and the Department of Housing and Urban Development do not permit the use of gypsum-based sheathing on military construction projects or HUD program construction. Plywood sheathing should meet the requirements of Department of Commerce Standard PS1-83, Exterior or Exposure I rating. OSB should meet the requirements of the American Plywood Association (APA) standard PRP-108, Exterior or Exposure I rating. Because of moisture penetration, decay, and delamination problems that have been experienced in the past, many industry experts now recommend *fiber cement board* complying with the requirements of ASTM C1186 *Standard Specification for Flat Non-Asbestos Fiber Cement Board, Type A*. Backing walls of *masonry* or *concrete* construction of course also provide excellent substrates.

Joints. Expansion joints are installed in EIFS at the floor lines of wood frame buildings, at locations where dissimilar substrates are bridged, at building expansion joints, at changes in wall height, at penetrations, where the system abuts other materials, and wherever necessary to accommodate building movement. Joints are generally not required at window and door openings if the insulation board is properly cut and reinforced. Corners should be cut from a single piece of insulation board because the alignment of board joints with the corners of openings will cause cracking (*Fig. 8.19*). Insulation board joints also should not align with joints in the sheathing boards for the same reason.

Decorative grooves called "aesthetic" joints are sometimes used to delineate color changes, or to provide convenient stopping places for applicator crews. Cracking is less likely to occur at these locations when the grooves are rounded rather than V-shaped, and when they do not align with joints in the substrate sheathing. The insulation board thickness also should be maintained at not less than $3/4$ in. at the base of aesthetic joints. Some architects use sealant joints for the same purpose, combining the function of aesthetics and movement.

Since the insulation board itself does not provide a suitable substrate for sealant adhesion, the board edges have traditionally been covered with the EIFS lamina consisting of the base coat, reinforcing mesh, and finish coat. A common mode of failure in these joints, however, has been softening and delamination of the finish coat to which the sealant was adhered. Most Class PB EIFS finish coats are water-based materials that have a tendency to re-emulsify with prolonged exposure to moisture. Moisture can be absorbed into

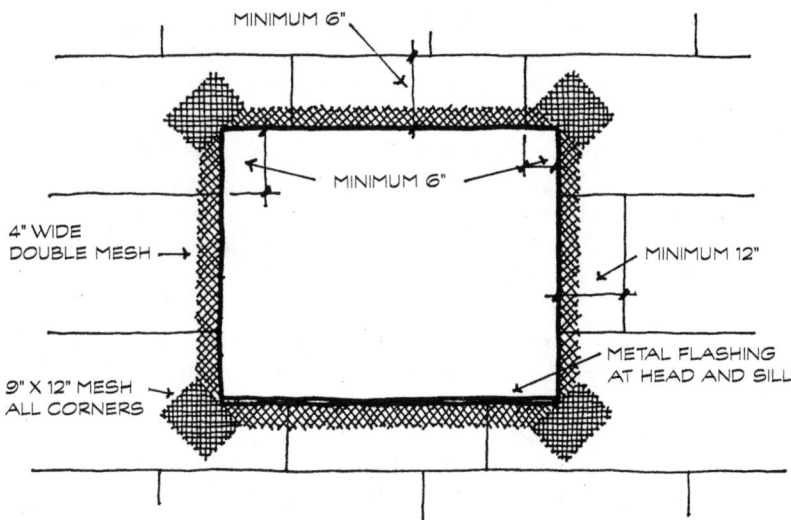

Figure 8.19 Corner reinforcing at window openings in EIFS.

the joint through the pores of the adjacent finish or may condense from water vapor passing through the system. Open-cell backer rods, which were once commonly used in EIFS sealant joints, also act as a sponge to hold trapped moisture. As the finish on the sides of the joint softens, the pull of the sealant against it causes separation from the base coat and moisture penetrates, leading eventually to delamination of the paper face of the gypsum sheathing and deterioration of its core, or rotting of wood sheathing.

Sealant joints should be constructed by one of two methods currently recommended. The first method is called *back-wrapping* (*Fig. 8.20A*). The base coat and a fine-gauge fiberglass reinforcing mesh are used to form the joint surfaces, providing a durable substrate for the sealant that is less sensitive to moisture than the finish coat. The mesh must be pulled tight around the insulation board edges and should extend onto the front and back faces a minimum of $2\frac{1}{2}$ in. After embedding in base coat material, this mesh is overlapped on the exposed face of the wall with a layer of standard EIFS reinforcing mesh and base coat. The finish coat is applied only up to the edge of the joint. A backer rod is then placed at the proper depth, sealant is applied, and the joint tooled to a watertight, concave configuration.

A second method of joint construction uses plastic J molds applied at the insulation board edges (*Fig. 8.20B*). The front flange of the molding is perforated to provide a mechanical interlock with the base coat, and the reinforcing mesh must overlap the flange. Metal edge moldings are unsatisfactory because of differences in the rate of thermal expansion and contraction between the metal and the EIFS coating. Even the plastic accessories can be somewhat problematic. They can cause cracking along the edge of the molding if the coating is thin, or a slight lump if the coating thickness is increased. This method of joint fabrication, however, does provide a durable surface for sealant adhesion, and will make future maintenance and sealant replacement much easier. Sealant manufacturers should be consulted as to the need for and proper type of primer required for maximum adhesion.

Most EIFS manufacturers now recommend that sealant joints be backwrapped with base coat and reinforcing mesh only, while the finish coat is held back at the wall surface. As an added safety factor, fiberglass-faced gypsum sheathing, cementitious backer board, or masonry substrates are recommended instead of standard paper-faced gypsum. Closed-cell backer rods or closed skin/open cell rods should be used in EIFS sealant joints, and low modulus silicone sealants are generally believed to place the least stress on the joint,

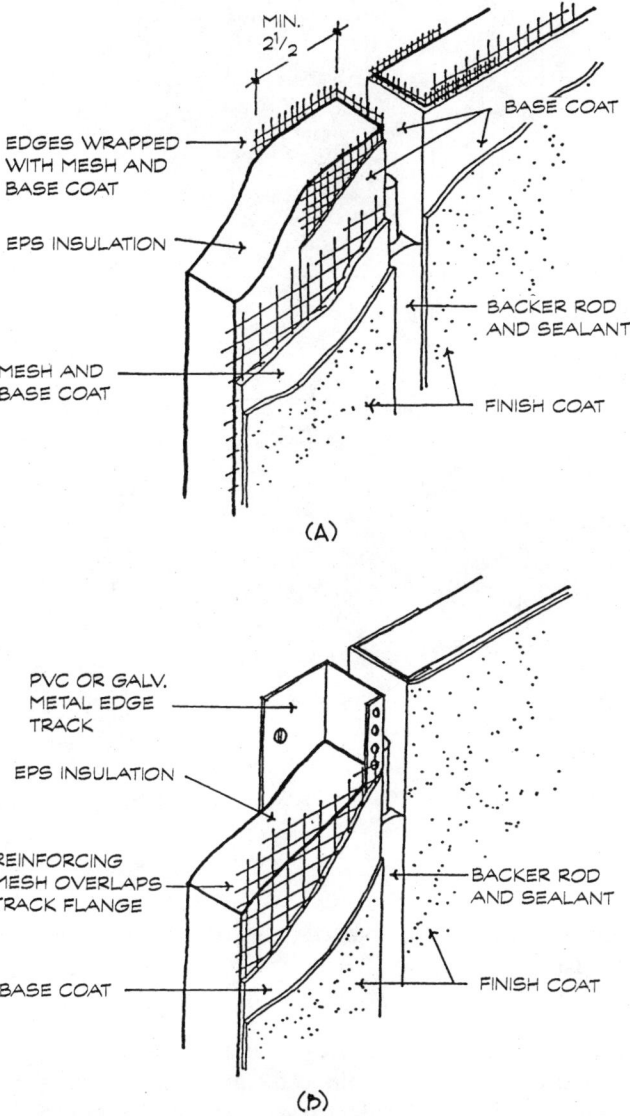

Figure 8.20 Sealant joints in EIFS.

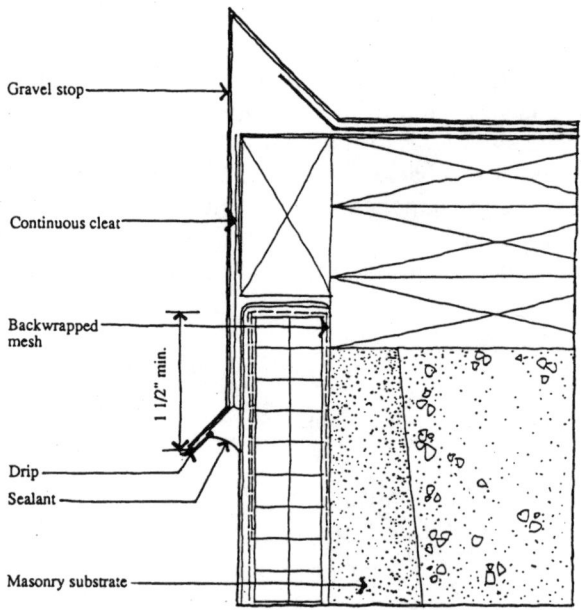

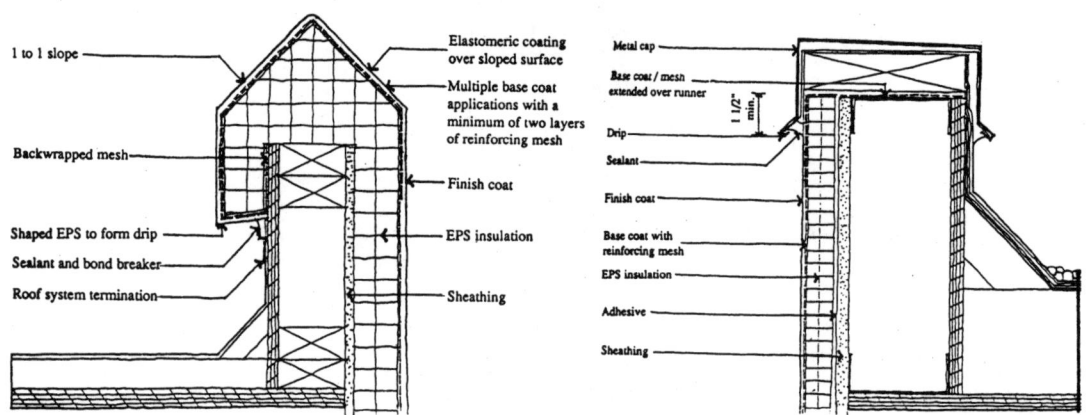

Figure 8.21 EIFS details at gravel stop and copings. (*From Williams and Williams, Exterior Insulation and Finish Systems, Current Practices and Future Considerations, ASTM Manual 16. Copyright ASTM. Reprinted with permission.*)

even at low temperatures. Joints should also be configured to provide the proper width-to-depth ratio.

Detailing. Since EIFS are proprietary systems, the manufacturer's specific recommended details should be followed, and all materials and components should be from the same manufacturer. For conditions not covered by the manufacturer's details, industry guide details can be used to develop specific custom designs. The details in *Figs. 8.21* through *8.24* are taken from ASTM Manual 16 *Exterior Insulation and Finish Systems.* They represent state-of-the-art technology and recommended practices based on extensive field experience with EIFS. Of particular concern is the protection of the system at edges and terminations where the vulnerability to water penetration is greatest. All customized details should be submitted to the system manufacturer for review and approval.

Quality assurance. Architects should secure the EIFS manufacturer's written approval of all details, require a manufacturer's technical representative to visit the job site before installation begins, specify that only manufacturer-approved EIFS applicators be allowed to perform the work, and that sealant applicators be trained professionals. Some architects also like to specify that a manufacturer's technical representative visit the job site before or during installation, and some EIFS manufacturers require job site visits for warranties to be issued. A pre-installation inspection will allow the manufacturer to approve the substrate,

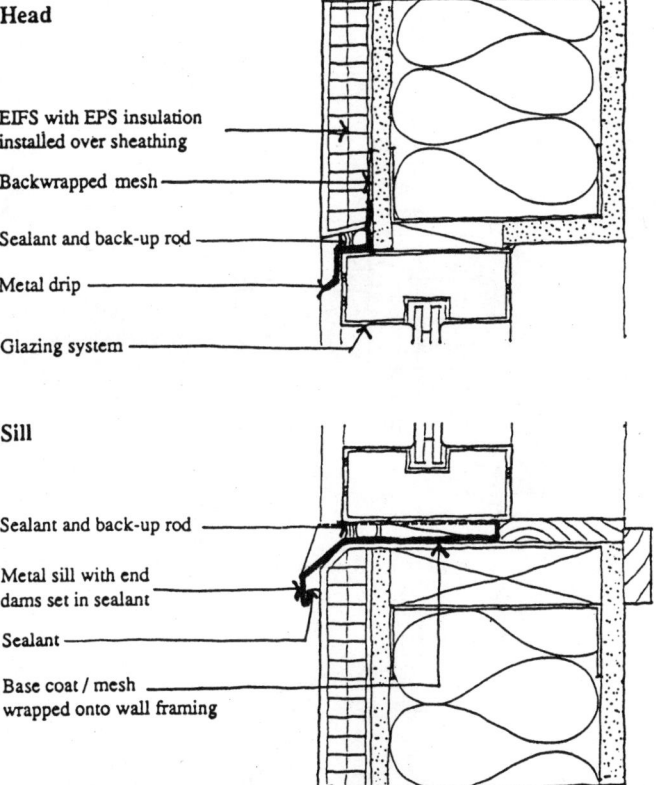

Figure 8.22 EIFS details at window head and sill. (*From Williams and Williams, Exterior Insulation and Finish Systems, Current Practices and Future Considerations, ASTM Manual 16. Copyright ASTM. Reprinted with permission.*)

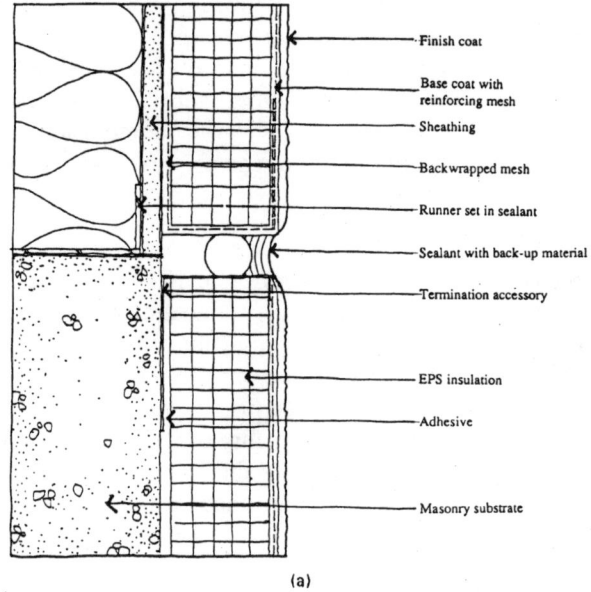

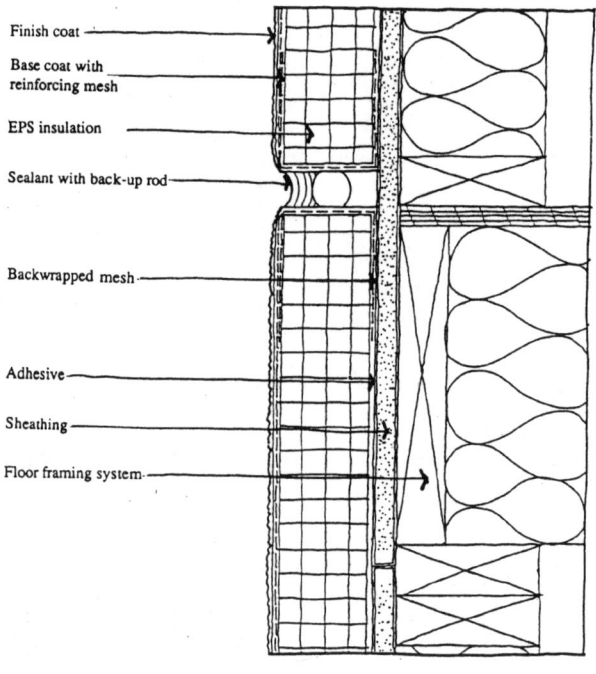

Figure 8.23 EIFS details at expansion joints. (*From Williams and Williams, Exterior Insulation and Finish Systems, Current Practices and Future Considerations, ASTM Manual 16. Copyright ASTM. Reprinted with permission.*)

Cladding 387

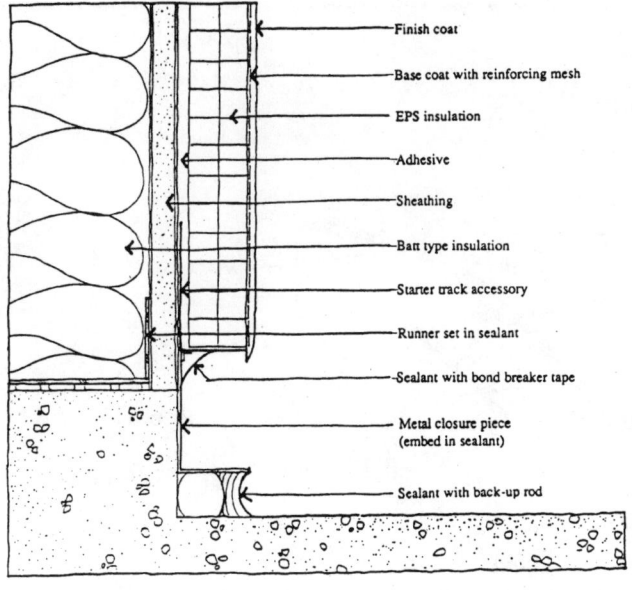

(a)

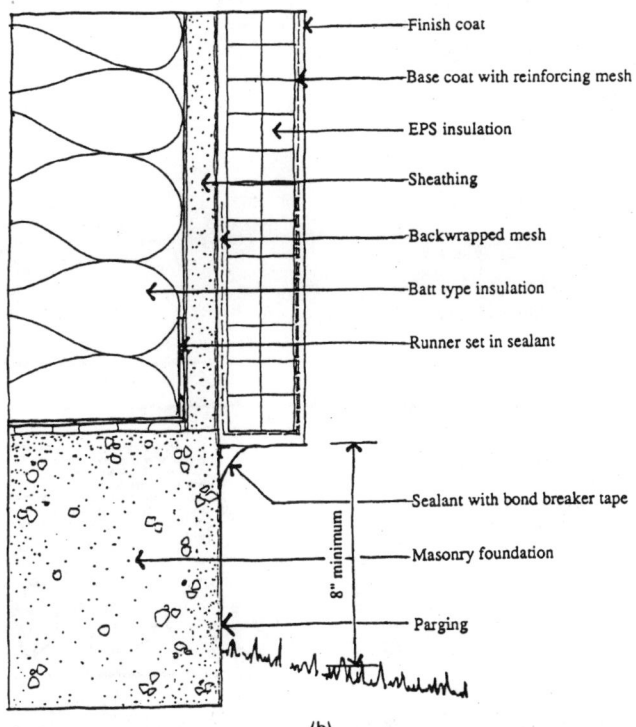

(b)

Figure 8.24 EIFS details at base of wall. (*From Williams and Williams, Exterior Insulation and Finish Systems, Current Practices and Future Considerations, ASTM Manual 16. Copyright ASTM. Reprinted with permission.*)

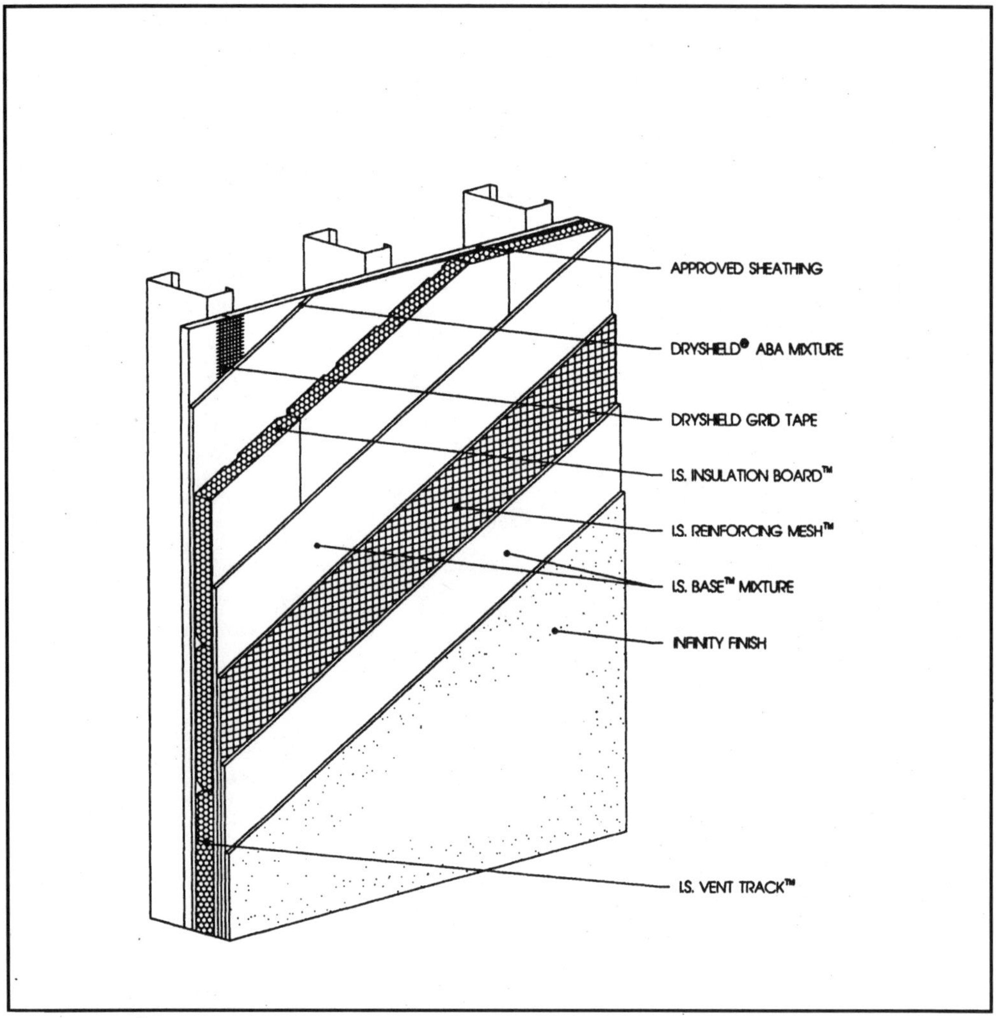

Figure 8.25 Pressure-equalized EIFS. (*From Dryvit Systems, Inc.*)

make recommendations, and review installation requirements with the applicator. For warranty work, a final inspection may also be required by the manufacturer. Critical performance issues include proper thickness of base coat and embedment of reinforcing mesh, filling of gaps between insulation boards, and elimination of offsets in adjacent board surfaces. Some code jurisdictions require, and many architects are specifying, third-party inspection of EIFS in a further effort to ensure quality.

Committees at ASTM are currently developing performance-based standards for the design and application of EIFS in an effort to improve the durability and extend the service life of these systems. To help reduce application errors, ASTM PS49 *Provisional Standard Practice for Application of Class PB Exterior Insulation and Finish Systems* can be used in the interim.

Some manufacturers have responded to the failure of the rain barrier strategy used in EIFS by developing "pressure-equalized EIFS" that adapt rain screen principles to these systems (*Fig. 8.25*). Some systems use a grooved MEPS board to provide drainage channels, a trowel-applied air barrier/adhesive layer and special vents located in the sealant joints, or a baffled rain screen joint. The systems must be engineered for each project, much like a curtain wall, and involve shop drawings as well as engineering calculations. Installation requires special training and skill, and the pressure-equalized systems are currently estimated to cost 20 to 40% more than the conventional systems. It remains to be seen whether these modified systems will solve the problems typically associated with EIFS, but such attempts at solving industry problems are needed if the system is to remain a viable one. Even with such modified systems, it is still critical that flashing at penetrations and proper termination details be an inherent part of all system designs.

Chapter 9

Joints and Sealants

Building walls are made up of many separate components and dissimilar materials isolated from one another by joints. We depend on sealants and other materials to fill these joints and to provide air and moisture barriers in the building envelope.

9.1 Building Movements

Contemporary structural frames are lighter and more flexible than their predecessors, and exterior envelope materials are often thinner, more rigid, and less forgiving. Wind loads, thermal movements, and structural deflections can wreak havoc on exterior joints, opening gaps and voids to the intrusion of rain. Buildings are in a state of constant motion, undergoing dimensional changes in response to fluctuations in temperature and moisture content, and moving under wind and structural loads. A failure to adequately provide for dimensional change or movement in sealant joints, or a failure of the sealant joint to maintain a continuous weather seal, invites the intrusion of moisture into the wall and subsequently into the building itself. Joint design, proper material selection, and good workmanship are all equally important in assuring the performance of a sealant joint system.

9.2 Joint Types

Various joint types such as butt joints, lap joints, and fillet joints are each appropriate to different types of materials and different field conditions. Butt joints are the most common type of site-constructed sealant joint. They are used between adjacent, parallel panels that are in the same plane, and may be hourglass-shaped, rectangular, or square (*Fig. 9.1*). Butt joints are typically used as expansion joints, control joints, and perimeter joints between similar or dissimilar components. A control joint is used to control shrinkage cracking in materials such as concrete, concrete masonry, and stucco. An expansion joint is a structural separation in a building, or a joint in or between components designed to absorb thermal and moisture movement and other effects. A perimeter joint separates one system, material, or component from another, and is also usually designed to accommodate thermal and moisture movement and other effects.

Triangular fillet joints may be used between adjacent panels in non-parallel planes such as the inside corners of walls, or at window and door setbacks (*Fig. 9.2*). Fillet joints may even be necessary where access is difficult for proper tooling of a butt joint.

Lap joints are frequently used in the framing of glazing systems, and with metal panels, and flashing (*Fig. 9.3*). Preformed tapes and liquid sealants are both used in lap joints, and are sometimes used together to provide primary and secondary seals. Lap joints are required with sheet metal and other very thin materials. The sealant must be able to resist shear stresses as the components move in adjacent parallel planes, and tensile and compressive stresses as they move perpendicular to one another.

9.3 Joint Sizing

Designing sealant joints requires more than just rules of thumb for spacing and width-to-depth ratios. It requires calculation of the different types of movement and tolerances that may affect the joint and therefore the sealant performance.

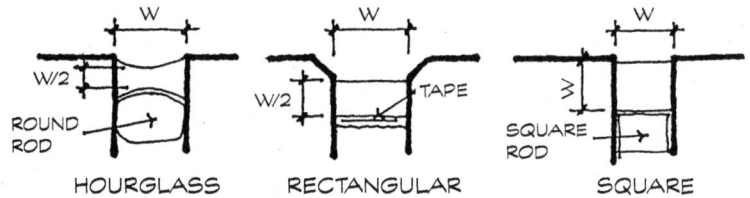

Figure 9.1 Butt joints. (*From CSI Monograph 07M900, Joint Sealers.*)

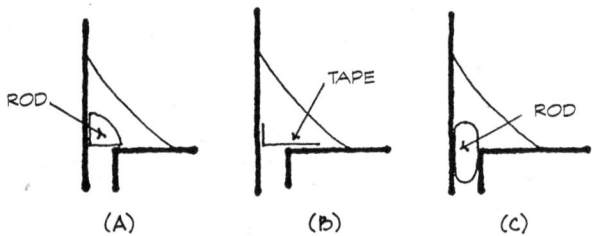

Figure 9.2 Fillet joints. (*Adapted from Myers, Behavior of Fillet Sealant Joints.*)

Figure 9.3 Lap joints. (*From CSI Monograph 07M900, Joint Sealers.*)

While rules of thumb may be adequate for simple joints between homogenous substrates with minimal movement, most construction applications are more complex. Joint design will often require consideration of thermal and moisture expansion and contraction, deflection, creep, building sway, vibration and construction tolerances. The basic formula for finding the required joint width J includes the width required for thermal movement J_t plus the width required for moisture movement J_m plus the width required for construction tolerances J_c.

9.3.1 Calculating thermal movement

The amount of thermal expansion and contraction varies for different building materials. The rate of thermal expansion and contraction of most building materials is known, and a standard coefficient can be used to calculate expected movement for a given set of conditions.

Table 9.1 lists linear coefficients of thermal expansion C_t for many common building materials. The amount of movement ranges from relatively small dimensional changes in masonry materials to very large dimensional changes in metals and plastics. Thermal expansion and contraction are calculated from the material's coefficient of thermal expansion and the difference in hot and cold service temperature conditions.

The service temperature used in calculating thermal movement must be the surface temperature of the material, and not the ambient temperature. Surface temperatures, which represent greater extremes, provide a more accurate prediction of movement.

Winter wall surface temperatures are usually established simply from ASHRAE winter design dry bulb air temperatures because the wall surface will be within a few degrees of ambient. The higher the R-value of the wall, the closer the surface temperature is to the exterior air temperature. (Refer to Chap. 3 for methods of calculating thermal gradients.) Summer wall surface temperatures are affected by both ambient temperature and solar radiation, so radiative heat gain must also be considered:

$$T_s = T_a + XS \qquad (9.1)$$

where T_s = extreme summer wall surface temperature
T_a = extreme summer air temperature (dry bulb)
X = constant for heat capacity of material (*Table 9.2*)
S = solar absorption coefficient (*Table 9.3*)

The total wall surface temperature differential ΔT is found by subtracting winter surface temperature from summer surface temperature. In Canada and the northern United States, the maximum extreme summer surface temperature is estimated to be 155°F and the maximum extreme winter surface temperature is estimated to be −45°F, for a total temperature differential of 200°F. In the southern United States, the maximum extreme summer surface temperature is estimated to be 180°F and the maximum extreme winter surface temperature

TABLE 9.1 Thermal Expansion Coefficients C_t

Material	in/in/°F (multiply by 10^{-6})
Brick	
Clay or shale	3.6
Fire clay	2.5
Concrete masonry	
Normal weight	
Sand and gravel aggregate	5.2
Crushed stone aggregate	5.2
Medium weight	
Air-cooled slag	4.6
Lightweight	
Coal cinders	3.1
Expanded slag	4.6
Expanded shale	4.3
Pumice	4.1
Stone	
Granite	2.8–6.1
Limestone	2.2–6.7
Marble	3.7–12.3
Sandstone	4.4–6.7
Slate	4.4–5.6
Travertine	3.3–5.6
Concrete	
Calcareous aggregate	5.0
Siliceous aggregate	6.0
Quartzite aggregate	7.0
EIFS lamina	
Light colored	7.5
Dark colored	10.0
Metals	
Aluminum	13.2
Steel, carbon	6.7
Steel, stainless	
301 alloy	9.4
302 alloy	9.6
304 alloy	9.6
316 alloy	8.9
410 alloy	6.1
430 alloy	5.8
Brass, 230 alloy	10.4
Bronze	10.0–11.6
Copper	9.4
Lead	15.9
Zinc	18.0
Plaster, gypsum	
Sand aggregate	6.5–6.75
Perlite aggregate	7.3–7.35
Vermiculite aggregate	8.4–8.6
Glass	4.9
Plastic	
Acrylic sheet	41.0
Polycarbonate sheet	38.0
Wood	
Parallel to fiber	
Fir	2.1
Maple	3.6
Oak	2.7
Pine	3.0
Perpendicular to fiber	
Fir	32.0
Maple	27.0
Oak	30.0
Pine	19.0

TABLE 9.2 Constant for Heat Capacity of Material X

$X = 100$	Low-heat-capacity materials*
$X = 130$	Solar radiation reflected on low-heat-capacity materials†
$X = 75$	High-heat-capacity materials*
$X = 100$	Solar radiation reflected on high-heat-capacity materials†

* Materials such as EIFS and well-insulated metal panel curtain walls have low thermal storage capacity and therefore low heat capacity. Materials such as precast panels and masonry walls, on the other hand, have high thermal storage capacity and therefore high heat capacity.
†If the wall surface receives reflected as well as direct radiation, use the larger constant. Reflected radiation can come from adjacent wall surfaces, roofs, and paving.
SOURCE: O'Connor, "Design of Sealant Joints."

$-10°F$, for a total temperature differential of $190°F$. Total wall surface temperature differential ΔT is used in the basic formula for thermal movement

$$M_t = (C_t)(\Delta T)(L) \tag{9.2}$$

where M_t = thermal movement
C_t = coefficient of thermal expansion (Table 9.1)
ΔT = total summer/winter surface temperature differential
L = panel length or joint spacing, inches

For example, thermal movement for a marble veneer panel with a thermal expansion coefficient of 0.0000086 (from *Table 9.1*), an estimated surface temperature differential of $125°F$, and a joint spacing of 10 ft would be calculated as

$$M_t = (0.0000086)(125°F)(120 \text{ in.}) = 0.129 \text{ in. (or about } \tfrac{1}{8} \text{ in.)}$$

The joint width required to accommodate this alternating thermal expansion and contraction must take into account the movement capability of the sealant itself. The sealant industry currently rates the movement capability of materials as ±12.5%, ±25%, or ±50%. A sealant rated ±25% can tolerate a maximum extension of +25% of the joint width and a maximum compression of −25% of the joint width. Since the sealant can compress only 25%, the joint width must be 4 times the expected movement. A more elastic sealant rated ±50% movement capability has a maximum compression of 50%, so the joint width must be twice the expected movement. A sealant advertised as +100%/−50% is still governed by its compressibility, so the joint width still must be twice the calculated movement to allow room for the compressed thickness of the sealant itself. ASTM C1193 *Standard Guide for Use of Joint Sealants* uses the formula

$$J_t = \left(\frac{100}{S_m}\right) M_t \tag{9.3}$$

where J_t = minimum joint width required for thermal movement alone
S_m = tested (ASTM C719) sealant movement capability
M_t = calculated thermal movement

TABLE 9-3 Solar Absorption Coefficients S

Material	Coefficient	Material	Coefficient
Optical flat black paint	0.98	Red oil paint	0.74
Glass, tinted (¼")	0.94–0.96	Brick, light buff (yellow)	0.50–0.70
Flat black paint	0.95	Surface color, light gray	0.65
Glass, clear (¼")	0.93–0.95	Concrete, natural	0.65
Glass, clear (low-e)	0.91–0.94	Mineral board, white	0.61
Glass, reflective	0.64–0.92	Aluminum, clear finish	0.60
Black lacquer	0.92	Medium light buff bricks	0.60
Dark gray paint	0.91	Medium dull green paint	0.59
Black concrete	0.91	Medium orange paint	0.58
Dark blue lacquer	0.91	Marble, white	0.58
Galvanized steel, unfinished	0.90	Medium yellow paint	0.57
Black oil paint	0.90	Glass, tinted (¼")	0.48–0.53
Stafford blue bricks	0.89	Medium blue paint	0.51
Dark olive drab paint	0.89	Medium kelly green paint	0.51
Dark brown paint	0.88	Brick, white	0.25–0.50
Dark blue-gray paint	0.88	Light green paint	0.47
Blue or dark green lacquer	0.88	Surface color, white	0.45
Brown concrete	0.85	Polished brass, copper	0.40
Brick, red	0.65–0.85	Aluminum paint	0.40
Medium brown paint	0.84	White semigloss paint	0.30
Glass, reflective (¼")	0.60–0.83	Galvanized steel, white	0.26
Surface color, dark gray	0.80	White gloss paint	0.25
Copper, tarnished	0.80	Silver paint	0.25
Medium light brown paint	0.80	White lacquer	0.21
Brown or green lacquer	0.79	Polished alum., chrome	0.20
Medium rust paint	0.78	Polished alum. reflector	0.12
Wood, smooth	0.78	Aluminized mylar film	0.10
Mineral board, natural color	0.75	Tinned surface	0.05
Light gray oil paint	0.75	Vapor-deposited coatings	0.02

SOURCE: O'Connor (1990).

Using the thermal movement calculated above for a marble veneer panel, and assuming a sealant with ±25% movement capacity, the minimum joint width required to accommodate thermal expansion and contraction would be

$$J_t = \left(\frac{100}{25}\right)(0.129) = 0.516 \text{ in. (or about } \tfrac{1}{2} \text{ in.)}$$

Some researchers propose that joints be sized using only a percentage of the rated sealant movement capacity. This adds a factor of safety to allow for imprecisions in establishing surface temperatures, imperfections in workmanship, and other limiting factors. The amount of reduction should depend on the particular circumstances of a joint design. If the sealant movement is calculated at only 80% of the rated capacity, the formula would then be

$$J_t = \left(\frac{100}{0.8 S_m}\right) M_t \qquad (9.4)$$

which increases the minimum required width of the joint to

$$J_t = \left[\frac{100}{(0.8)(25)}\right](0.129) = 0.645 \text{ in. (or about } 5/8 \text{ in.)}$$

ASTM C1193 provides a graph which can be used to find the recommended joint width for thermal movement for sealants with various movement capabilities (*Fig. 9.4*). This graph is based only on thermal movement, and does not consider moisture related movements or construction tolerances.

With some marble materials, thermal expansion is slightly greater than thermal contraction. In other words, the material never quite returns to its original size after a heating/cooling cycle. The cumulative effect of repeated heating/cooling cycles is a gradual and *permanent* expansion of the material. This phenomenon is called *hysteresis*. The effect of hysteresis is exaggerated when the marble panels are thin because there is no compensating mass behind the weather-exposed surface to provide stability. Hysteresis is believed to be the primary culprit in the multimillion dollar re-cladding of the Amoco Tower in Chicago. The thin ($1\frac{1}{4}$- to $1\frac{1}{2}$-in.) marble panels gradually expanded with repeated heating and cooling cycles and, because they were rigidly restrained, they bowed out of plane. Most of the bowed panels were located on the building's southeast corner where they were unshielded from the morning sun by adjacent buildings. Expansion at the outer surface of marble also increases the porosity of the stone and makes it more vulnerable to freeze-thaw damage. Thicker panels have not experienced the same failure. Residual thermal expansion of marble panels should be calculated at 20% of the original increase.

9.3.2 Calculating moisture movement

Porous materials such as concrete, concrete masonry, stucco, brick, stone, and wood expand and contract with fluctuations in moisture content as well as temperature changes. Some moisture movement is reversible, and some is permanent.

Reversible moisture expansion and contraction occurs when a material is alternately wetted and dried due to atmospheric or environmental conditions, and is usually offset by opposing thermal expansion and contraction which occur at about the same time. Moisture content generally decreases as the ambient temperature increases because the moisture in the material evapo-

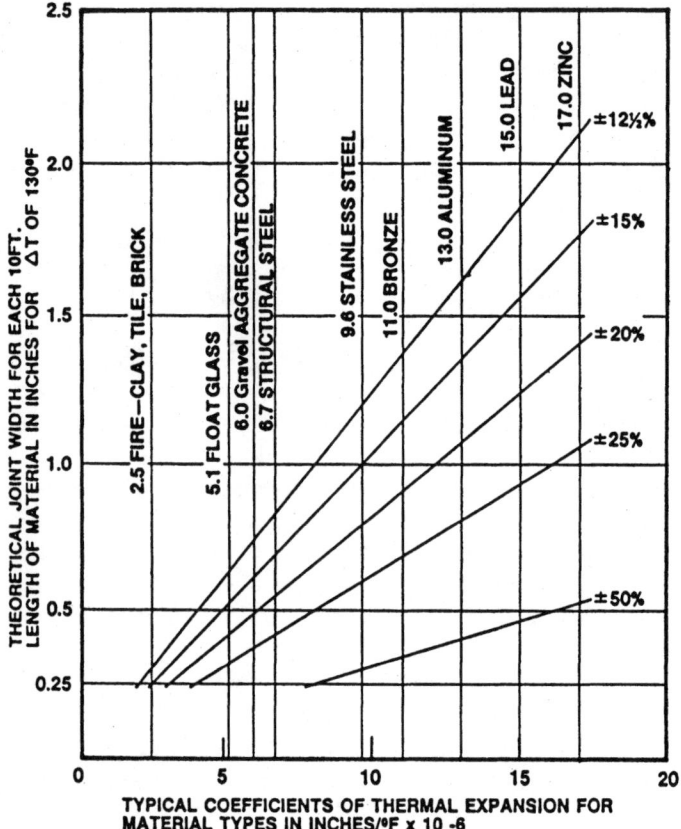

Figure 9.4 Recommended joint width for sealants with various movement capabilities *based only on thermal expansion.* (*From ASTM C1193, Standard Guide for Use of Joint Sealants. Copyright ASTM. Reprinted with permission.*)

rates into the warm air. The decrease in moisture content causes the material to shrink, but the accompanying increase in temperature causes the material to expand. Since the moisture shrinkage is small, the net effect is that the thermal expansion is somewhat less than it would have been without the moisture content changes. When the temperature falls and the material cools off again, water vapor condenses within the material, increasing its moisture content again. This increase in moisture content causes the material to expand, but the cooling process itself causes the material to shrink. The net effect again, is that the thermal contraction is somewhat less than it would have been without the moisture content changes. Even though the compensating thermal movements are not precisely simultaneous, the overall effect of reversible moisture expansion and contraction is generally considered negligible.

Irreversible moisture movements occur in concrete, concrete products, brick and other clay masonry materials. Irreversible moisture movements are not based on fluctuating moisture conditions in service, but on the manufacturing process and the age of the material. Cement-based products such as concrete, concrete masonry, and stucco experience initial moisture shrinkage as the cement hydrates and excess construction water evaporates. This is a *permanent shrinkage* and is in addition to subsequent reversible thermal and moisture expansion and contraction. Clay masonry products such as brick and terra cotta experience initial moisture expansion as the units re-absorb atmospheric moisture after firing. This is a *permanent expansion* and is in addition to subsequent reversible thermal and moisture expansion and contraction. Chapter 7 contains a discussion of shrinkage control joints in concrete walls and slabs, and Chap. 8 covers the use of expansion and shrinkage control joints in masonry.

Irreversible moisture expansion and contraction must be taken into account in sizing sealant joints. *Table 9.4* lists a range of moisture movement coefficients for several types of concrete and masonry. The minus signs indicate permanent shrinkage, and the plus signs indicate permanent expansion. To find the sealant joint width that is required to accommodate moisture movements, these coefficients are used in the formula

$$J_m = \left(\frac{C_m}{100}\right) L \qquad (9.5)$$

where J_m = minimum joint width required for moisture movement
C_m = coefficient of moisture movement
L = panel length or joint spacing, in inches

The amount of movement may range from $-1/4$ in. for a 25-ft panel section of lightweight aggregate concrete to $+3/16$ in. for a 25-ft panel section of clay masonry. This irreversible movement is added to the joint width calculated for thermal movement in Eq. (9.4).

9.3.3 Construction tolerances

Little is exact in the manufacturing, fabrication, and construction of buildings and building components. Tolerances allow for the realities of fit and misfit of

TABLE 9.4 Moisture Movement Coefficients C_m

Material	Coefficient	Type of movement
Concrete, gravel aggregate	−0.0003 to −0.0008	Shrinkage
Concrete, limestone aggregate	−0.0003 to −0.0004	Shrinkage
Concrete, lightweight aggregate	−0.0003 to −0.0009	Shrinkage
Concrete block, dense aggregate	−0.0002 to −0.0006	Shrinkage
Concrete block, lightweight aggregate	−0.0002 to −0.0006	Shrinkage
Face brick, clay	+0.0002 to +0.0007	Expansion

NOTE: If specific data for a particular unit is not available, use the maximum value given in the table. Minus sign denotes shrinkage, plus sign denotes expansion.
SOURCE: O'Connor, "Design of Sealant Joints."

the various parts as they come together in the field and ensure proper technical functions such as structural safety, joint performance, secure anchorage, moisture resistance, and acceptable appearance. Each construction trade or industry develops its own standards for acceptable tolerances based on economic considerations of what is reasonable and cost-effective. Few, if any, construction tolerances are based on hard data or engineering analysis. There has also never been any coordination among various industries to facilitate the fit of one system with another. Different materials and systems, because of the nature of their physical properties and manufacturing methods, have greater or lesser relative allowances for the manufacture or fabrication of components and the field assembly of parts.

Fabrication and erection tolerances must be considered when determining the required width of sealant joints. Construction tolerances can cause the actual width of a sealant joint to be wider or narrower than intended. Joints that are slightly wider than necessary do not usually present much problem except perhaps aesthetically. Joints that are too narrow in width, though, dramatically affect sealant stresses and can create problems of both constructability and performance.

If a $1/2$-in. joint is designed to accommodate $1/8$-in. movement in both compression and extension, this represents a 25% movement capacity or strain on the sealant. With a $\pm 1/4$-in. construction tolerance, however, the as-built joint could be as wide as $3/4$ in. or as narrow as $1/4$ in. At $3/4$ in., the anticipated movement will create only a 16% strain on the sealant. At $1/4$ in., though, a $\pm 1/8$-in. movement generates a 50% strain, and only the highest-performance sealants can tolerate that much movement. A sealant with only ±25% movement capacity will fail, and the joint will begin to leak within one seasonal cycle.

A lack of coordination among various system and material tolerances results in some inevitable problems of misfit that are taken up in the joints. Some allowance must be made in designing sealant joint width and/or determining sealant joint spacing to permit what the architect or engineer determines to be a "reasonable" combined or net construction tolerance for the materials and systems involved. Construction tolerance provisions in sealant joint calculations are normally added to the sum of the joint widths required for thermal and moisture movement. For some construction systems and materials, construction tolerances can be as much as $\pm 1/8$ in. For others, realistic expectations of the net construction tolerances affecting any given joint will often be closer to $\pm 1/4$ in. Actual tolerances should be determined for each joint design according to the specific materials and project conditions. An excellent book by David Kent Ballast *Handbook of Construction Tolerances* provides industry recommended tolerances for many different materials and systems.

9.3.4 Sizing vertical joints for walls

The width required for vertical sealant joints in walls is obtained by adding the joint widths calculated to accommodate thermal and moisture movements and the joint width estimated to accommodate construction tolerances in the formula

$$J = J_t + J_m + J_c \tag{9.6}$$

where J = total calculated joint width
J_t = minimum joint width required for thermal movement
J_m = minimum joint width required for moisture movement
J_c = minimum joint width required for construction tolerances

Figure 9.5 summarizes the steps explained in the previous sections for determining each of the components of the equation. If the calculated width is unacceptable from an aesthetic standpoint, decrease the panel length or joint spacing (L) and re-calculate J_t and J_m.

Selecting a sealant color that matches the adjacent materials will help camouflage the joint. If an exact match is not possible, a slightly darker sealant will be less noticeable than a slightly lighter sealant. On the other hand, if you want to accentuate a pattern of sealant joints in a wall, choose a sealant color that contrasts with the adjacent materials. Joints can also be articulated by shaping the adjacent substrate, or hidden in the shadow line of projecting architectural elements.

9.3.5 Sizing horizontal joints for walls

For horizontal sealant joints in building walls, some additional movements affect performance, and should be considered.

Slab or beam deflection at exterior perimeter walls is caused by both dead and live loads. Dead load deflection usually occurs before the joint is sealed, and most of the deflection will occur before the cladding is even installed. Live load deflection, though, occurs after the exterior envelope is complete and joints are sealed. This affects not only the sizing of the joints, but performance of the sealant as well because it may be placed in permanent compression. For example, if $L/360$ is used for live load deflection occurring after the cladding and sealant are in place, a joint below a 16-ft spandrel beam (or a shelf angle attached to the beam) would deflect about $\frac{1}{2}$ in. This much deflection is unacceptable from the sealant compression standpoint, and deflection limits will have to be reduced to an acceptable value for joint design and wall integrity.

Concrete column shrinkage will also have to be considered in the design of horizontal sealant joints. Slabs and beams can be cambered and formwork adjusted slightly in height to limit deflection and frame shortening, but a certain amount of vertical movement is unavoidable. Deflection from loads and long-term creep can be calculated by the structural engineer. The residual frame shortening that takes place after the cladding is installed depends on the time lapse after erection of the frame as well as many other, sometimes indeterminate, factors. In some instances, concrete frames may shorten as much as $\frac{1}{8}$ to $\frac{1}{4}$ in. in a 12-ft story height. Some researchers estimate that steel frames shorten about $\frac{1}{16}$ in. in 12 ft, and that concrete frames shorten about $\frac{1}{8}$ in. in 10 ft due to the combined effects of elastic deformation, shrinkage, and creep. Multi-story buildings should be evaluated for the effects of these conditions on horizontal sealant joints, and the width required to

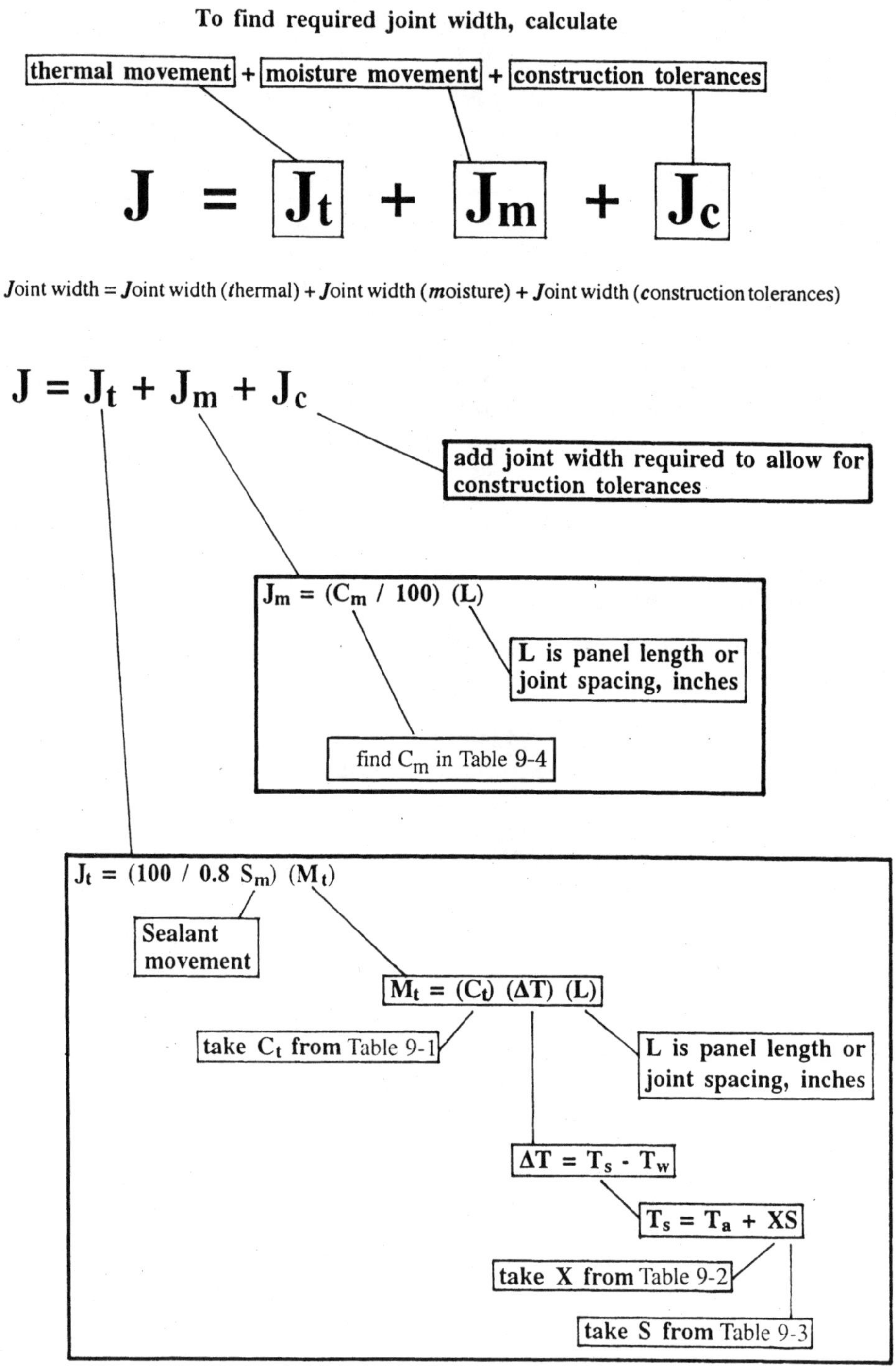

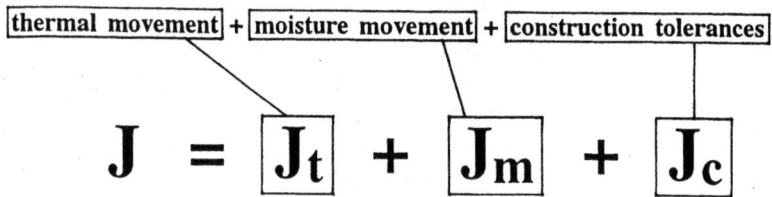

Figure 9.5 Steps in calculating required sealant joint width.

accommodate such movements added to the calculated joint width for thermal and moisture movements and construction tolerances.

9.3.6 Sizing joints for slabs and decks

Joints in horizontal slabs and decks are sized in the same way as joints in vertical wall elements. Horizontal and sloping surfaces, though, also experience radiational cooling with a clear night sky, and surface temperatures can be as much as 15 to 20°F cooler than the air temperature. This means that the total summer/winter temperature differential that the surface experiences is greater than for vertical surfaces. This additional service temperature range should be included in the formula for thermal movement.

9.4 Sealant Stresses

Sealant stresses may simply alternate between tension and compression, but can also include lateral shear stresses either in plane or perpendicular to the joint. Although thermal expansion and contraction are usually the primary contributors, all of the various joint movements produce corresponding movements in the sealant itself. There are four basic types of sealant movement: extension, compression, longitudinal shear, and transverse shear (*Fig. 9.6*).

9.4.1 Tension and compression

As temperatures rise in warm weather, adjacent materials expand and the joint closes, causing compressive strain in the sealant. With cooler fall and winter temperatures, the substrate materials contract, the joint opens, and the sealant experiences tensile strain.

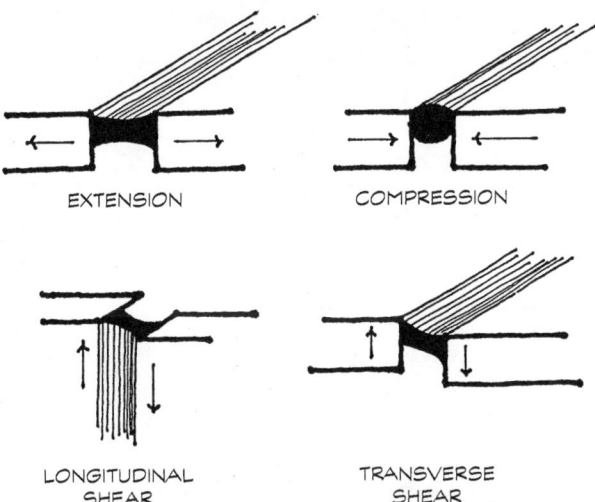

Figure 9.6 Sealant movement. (*From CSI Monograph 07M900, Joint Sealers.*)

The size of the joint in relation to expected thermal movement directly affects the amount of strain. If a ½-in. joint is designed to accommodate ±⅛ in. movement in both compression and extension, this represents a 25% strain on the sealant in either direction. With a joint depth of ¼ in., the relatively slender sealant bead experiences little strain. However, if construction tolerances are not included in the joint design and a ±¼ in. construction tolerance does occur, the as-built "half inch" joints may vary from ¼ in. to ¾ in. At ¾ in. wide, the anticipated ±⅛-in. thermal movement will create only a 16% strain on the sealant (*Fig. 9.7*). At ¼ in., however, a ±⅛-in. movement will generate a 50% strain, causing a 25% capacity sealant to fail.

Sealant in joints that have been undersized for expansion or frame shortening are often placed in permanent compression. If the stresses are great enough, the sealant (with compression limits of 10 to 50%) could eventually be squeezed out of the joint. In a properly sized and detailed joint, the sealant will be able to experience reversible thermal movement without failure even after permanent moisture expansion or contraction and elastic deformations have taken place.

9.4.2 Sealant shear stresses

Shear stresses may be longitudinal and/or transverse, and may also occur simultaneously with either extension or compression. In high-rise structures, wind loading may contribute to shearing stresses both in plane and perpendicular to the joint. More typically, though, differential thermal movement between adjacent components causes a diagonal extension of the sealant.

Longitudinal shear is a lengthwise displacement of one side of a joint relative to the other (*Fig. 9.6*). It can occur because of differential settlement at a

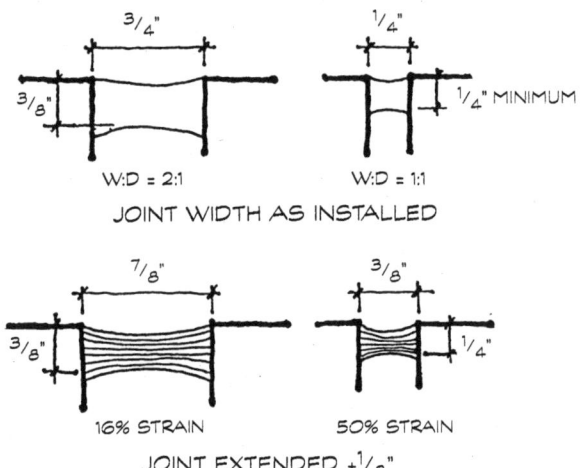

Figure 9.7 Strain is minimized if sealant width is equal to twice the sealant depth. (*From CSI Monograph 07M900, Joint Sealers.*)

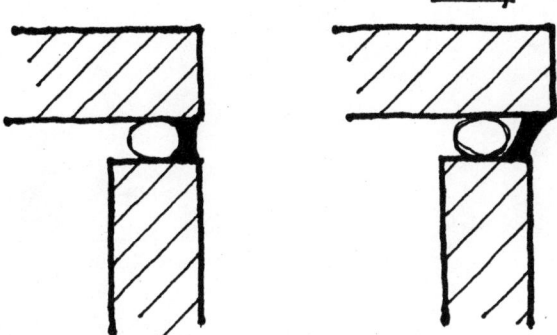

Figure 9.8 Transverse shear. (*From CSI Monograph 07M900, Joint Sealers.*)

structural expansion joint, or when two different materials with different coefficients of thermal movement form opposite sides of a joint (e.g., when aluminum curtain wall mullions with a coefficient of 0.0000132 abut a brick wall panel with a coefficient of 0.0000036). A diagonal lengthening of the joint occurs, the magnitude of which will depend on the type and spacing of anchorage or supports and the unrestrained length of the respective materials. Longitudinal shear is usually nonuniform, reaching a peak along only part of the length of the sealant joint.

Transverse shear is an out-of-plane movement crosswise to the joint face (*Fig. 9.6*). It typically occurs at joints between walls or elements that change plane, such as at corners and offsets. A west-facing surface, for instance, will expand as it absorbs solar radiation, and a shaded, north-facing surface will be relatively close to the ambient temperature (*Fig. 9.8*). Transverse shear is also typical in lap joints.

Allowable longitudinal and transverse shear movements can be calculated on the basis of the static, unstressed joint width and the movement capability of the sealant. The relationship is illustrated by a right triangle (*Fig. 9.9*) in which X is the normal joint width, Y is the allowable movement in shear, and Z is the allowable extension of the sealant. For a $1/4$-in. joint and a sealant with ±25% movement capability, then, $Z=1.25X$, or 0.3125 in. Using the pythagorean theorem ($X^2+Y^2=Z^2$), we solve for Y ($Y=\sqrt{Z^2-X^2}$). In this example, $Y=\sqrt{0.3125^2-.25^2}=0.1875$ in. ($3/16$ in.).

When the joint width is not known, the same equation can be used to calculate basic relationships for various classes of sealants. For a ±25% sealant, Z is still $1.25X$ even though X is not known: $Y=\sqrt{(1.25X)^2-X^2}=0.75X$, therefore the allowable shear movement is 75% of the installed joint width. If the sealant is rated ±50%, then $Y=\sqrt{(1.5X)^2-X^2}=1.12X$.

Standard calculations for four different classes of sealants are plotted on a graph in *Fig. 9.10,* and the results are tabulated in *Table 9.5*.

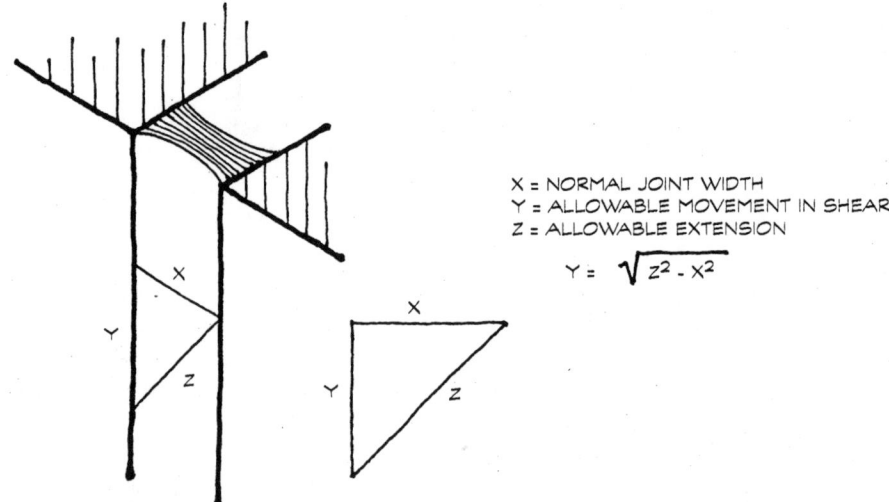

Figure 9.9 Allowable shear movement. (*From Klosowski, Sealant Movement in Shear.*)

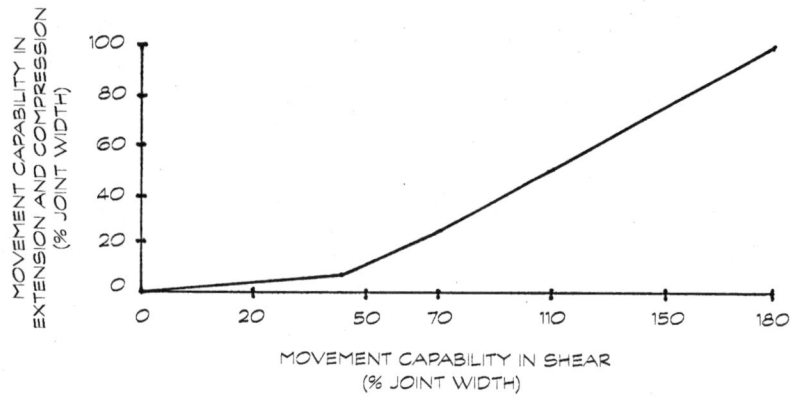

SOURCE: KLOSOWSKI (1987)

Figure 9.10 Movement capability in shear. (*From Klosowski, Sealant Movement in Shear.*)

9.5 Other Factors Affecting Sealant Performance

Other factors that affect sealant movement and performance are joint geometry (width to depth ratios), installation temperatures, movement of the joint during sealant curing, and three-sided joint adhesion.

9.5.1 Joint geometry

In order for sealants to function properly, they must have proper shape as well as proper size. The proper depth of the sealant depends on the width of the joint *at the time of application*. Since construction tolerances and ambient temperatures may affect the actual width of joints at the time of sealant application, the project specifications should include some guidelines on sealant depth for the applicator. Most industry sources recommend that for *butt joints* up to $1/2$ in. wide, joint depth should be less than or equal to the joint width, with 2:1 as a preferable ratio. Increasing the depth theoretically increases the strain in the sealant (*Fig. 9.11*). For some types of joints and sealants, manufacturers may recommend a 1:1 ratio. Although there is not complete agreement within the industry, ASTM C1193 provides the following guidelines.

TABLE 9.5 Shear Movement Ability

Sealant movement ability in extension/compression, % of joint width	Probable joint movement ability in shear, % of joint width
±10	45
±25	75
±50	112
+100/−50	175

SOURCE: Klosowski, *Sealant Movement in Shear.*

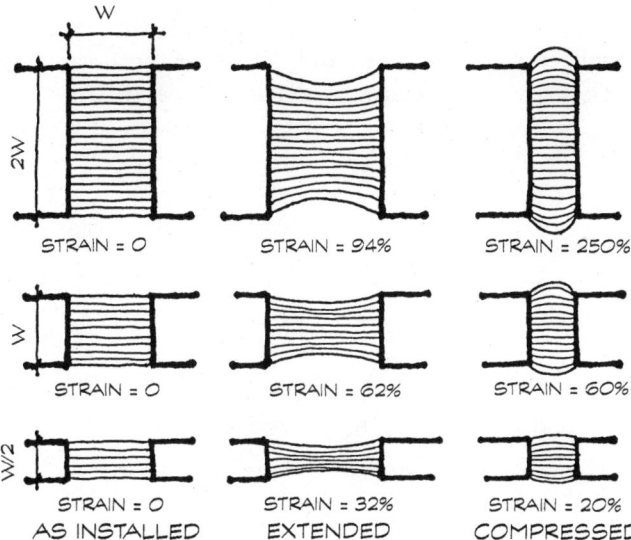

Figure 9.11 Sealant strain increases with sealant depth for a given sealant width. (*From ACI, Guide to Joint Sealants for Concrete Construction.*)

- For butt joints in concrete, masonry, and stone
  - Joints up to $1/2$ in. wide, sealant depth may be equal to sealant width
  - Joints $1/2$ to 1 in. wide, sealant depth should be one-half the width
  - Joints 1 to 2 in. wide, sealant depth should not exceed $1/2$ in.
  - Joints wider than 2 in., consult sealant manufacturer
- For butt joints in metal, glass, and other nonporous substrates
  - Joints $1/4$ to $1/2$ in. wide, sealant depth minimum $1/4$ in. or one-half the width up to a maximum of $1/2$ in.

Sealant depth should always be constant along the length of the joint and should never be less than a minimum of $1/4$ in. after the sealant has been tooled against the backing.

A triangular or quarter-round backer rod provides best sealant performance in *fillet joints* (*Fig. 9.2A*). Unfortunately, these materials are not readily available except in custom orders for large quantities. Conventional round backer rod can be inserted in fillet joints when there is sufficient room, with the rod left protruding 50% above the joint surface (*Fig. 9.2C*). It is generally best (and least expensive) to use conventional round rod whenever sufficient space permits its proper placement. When the size of the joint does not permit insertion of conventional backer rod, bond breaker tape can be used (*Fig. 9.2B*). The width of the bond breaker (*W*) can be calculated by dividing the combined thermal and moisture movement by the sealant movement capability. If the joint is expected to move $\pm 3/16$ in. and the sealant is classified as ±25%, the bond breaker or release tape must be $3/4$ in. wide (W=0.1875/0.25=0.75) (*Fig. 9.12*). The sealant thickness *T* should be a minimum of $1/4$ in., and the adhesion width *B* should be a minimum of $3/8$ in. Fillet joints may be tooled concave or convex. Some researchers believe the convex shape provides better performance because the feather edge is eliminated. This can be particularly important with sealants that are sensitive to ultraviolet degradation.

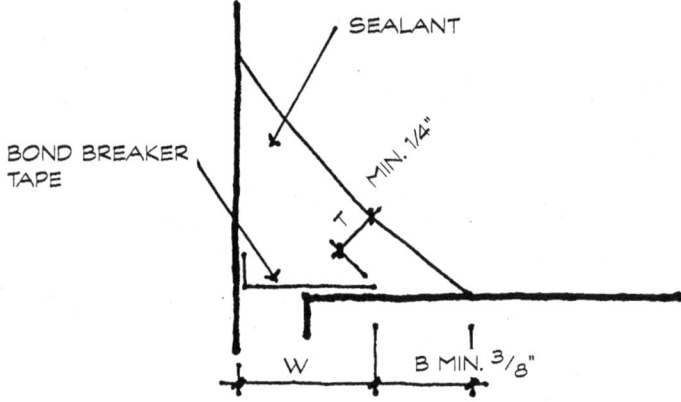

Figure 9.12 Detail of fillet joint (*From Myers, Behavior of Fillet Sealant Joints.*)

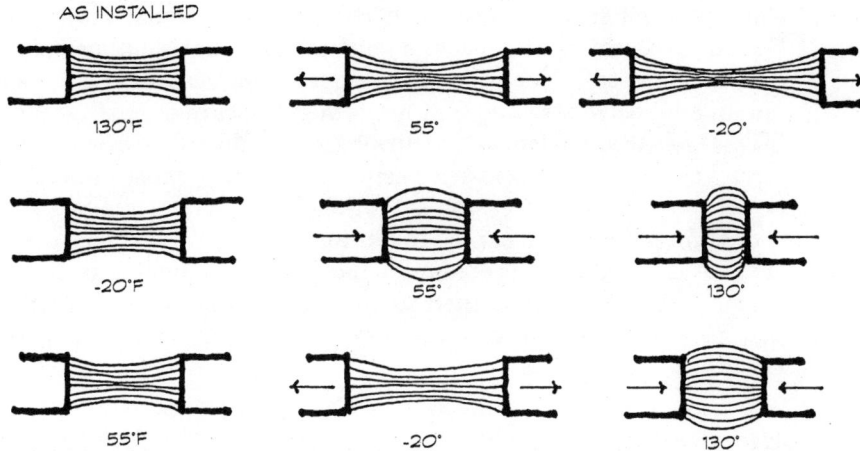

Figure 9.13 Moderate installation temperatures produce less sealant strain. (*From CSI Monograph 07M900, Joint Sealers.*)

The sealant in *lap joints* must be of sufficient thickness to withstand shear stresses. Generally, the thickness of the sealant should be twice the expected movement.

9.5.2 Installation temperatures

Surface temperatures at the time of sealant installation affect subsequent performance. It is not always possible to install sealants during moderate weather (i.e., in the spring or fall). Construction schedules often necessitate either high- or low-temperature installation which will affect the service demands made on the sealant. *Figure 9.13* illustrates the dynamics of joint movement and sealant performance at various installation temperatures. The more moderate the installation temperature, the less strain the sealant will undergo in service, and the better it will perform. Most sealant manufacturers recommend an installation temperature range of 40 to 90°F for most applications.

When the sealant is installed at moderate temperatures, cold weather causes sealant extension and warm weather causes compression. If installed at low temperatures, however, the sealant is in compression in all but extremely cold conditions, and may cause permanent deformation or extrusion from an undersized joint. If installed at high temperatures, the sealant is in extreme tension most of the time, and failure in adhesion or cohesion is possible if the sealant joint is not properly sized.

9.5.3 Movement during cure

Excessive movement of a joint during sealant cure can cause sealant failure. Movement during cure is most problematic in bridge deck joints, in aluminum cladding, and in climates which experience wide daily temperature cycles. Compression set results when a joint closes in hot weather before full sealant cure is completed. When the joint re-opens, the sealant experiences high internal and adhesive stresses that may cause rupture or spall the face of the joint (*Fig. 9.14*). Expansion set occurs when a joint opens in cold weather and stretches the sealant before it is cured. Reclosure causes the material to bulge out of the joint. Compression set failures are more common.

Certain precautions can be taken to successfully seal joints where compression set is a potential problem. Sealants may be applied in the late afternoon or early evening so that there is an extended period of time before the joint experiences significant compression. The sealant itself should be a rapid curing type.

9.5.4 Three-sided adhesion

Proper sealant performance requires backer rods or bond breaker tape (sometimes called release tape) to prevent sealant adhesion to the back of the joint. Three-sided adhesion imposes unwanted stress on the sealant, restricting proper elongation and causing joint failure (*Fig. 9.15*).

9.6 Sealant Properties

There are many different properties and physical characteristics of sealants which affect performance, including movement capability, modulus of elasticity, and hardness.

9.6.1 Movement capability

Movement capability is always stated as a ± percent value that indicates the amount of movement the sealant can take in either extension or compression from its original, installed joint width. ASTM C920 *Standard Specification for*

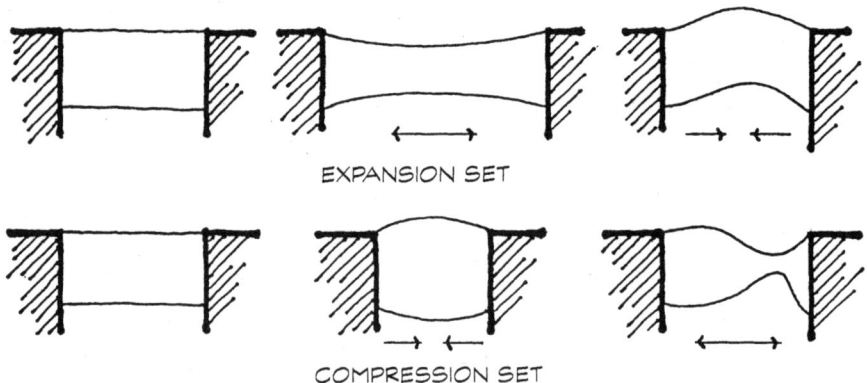

Figure 9.14 Excessive movement during cure can cause sealant failure. (*From CSI Monograph 07M900, Joint Sealers.*)

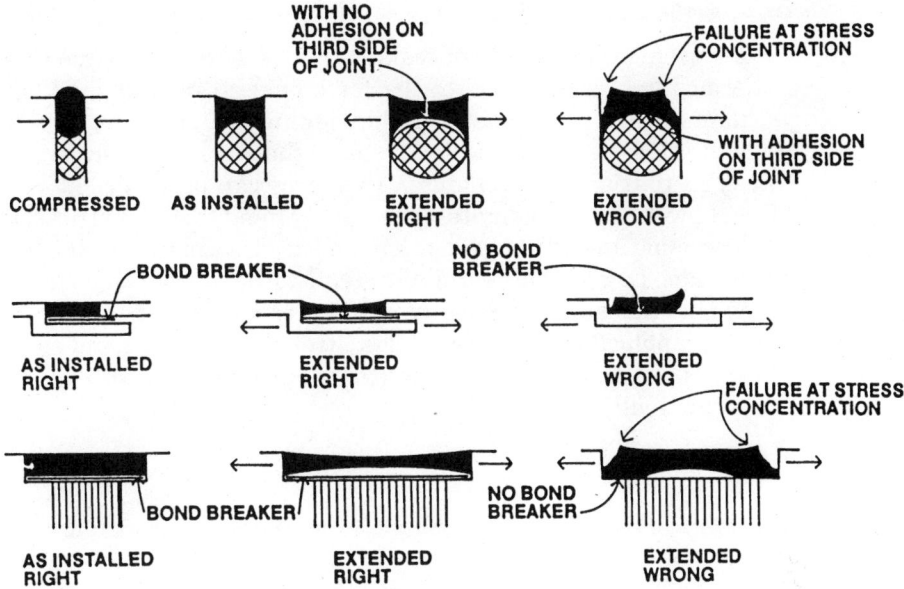

Figure 9.15 Three sided adhesion causes sealant failure. (*From ASTM C1193, Standard Guide for Use of Joint Sealants. Copyright ASTM. Reprinted with permission.*)

Elastomeric Joint Sealants includes only two movement classifications: those sealants with a minimum movement rating of ±12.5% (Class 12½), or a rating of ±25% or greater (Class 25). The ± percentage is not additive (e.g., ±25% rating does not equal 50% movement capacity). Non-elastomeric sealants (sometimes referred to as caulking compounds) are unsuitable for moving joints, so they do not have movement capability ratings. There are high-performance sealants on the market with reported ratings of ±50% and +100/−50% (extension/compression, respectively), but the ASTM Class is still only ±25 since the standard has not yet added higher classifications. Using sealants with a higher movement capability will reduce either the required joint width or the calculated spacing of joints.

9.6.2 Recovery and stress relaxation

To maintain movement capability over an extended service life, sealants must have good elasticity in cyclic movements. Elasticity is a measure of the sealant's ability to recover its original shape after elongation or compression. *Recovery* for good-quality elastomeric sealants may range from 75 to 95% or higher. *Stress relaxation* (the opposite of recovery) is a reduction in stress after prolonged conditions of strain. All sealants experience some stress relaxation. Materials with a high degree of stress relaxation recover their original shape slowly and incompletely. While a small amount of relaxation is good because it reduces the stress at the bond line, too much can cause sealant failure.

9.6.3 Modulus of elasticity

The movement capability of sealants is closely related to modulus of elasticity. The modulus is the ratio of the force (stress) needed to elongate (strain) a sealant to a certain point. High-modulus sealants are so strong, they put a high stress on the adhesive bond at the substrate. On substrates that are weak in tension (like concrete), this force can be sufficient to spall the joint face. Low-modulus sealants, on the other hand, put less stress on the bond line for the same amount of elongation, and are therefore less likely to spall a weak substrate. Low-modulus sealants tend to have high movement ratings, and high-modulus sealants usually have lower movement capability. In structural glazing applications, where sealant strength, adhesion and elongation characteristics are important to the integrity of the system, only medium- and high-modulus sealants are used.

9.6.4 Hardness

Sealants are tested for hardness on the Shore A durometer, which is for softer materials (harder materials are tested on the Shore D durometer). On a scale of 0 to 100, bubble gum has a Shore A hardness of about 5, while a pencil eraser has a Shore A of about 70. Sealants can range anywhere from 10 or 15 to 50 or 60, with hardness generally increasing with age (heat aging). As the hardness increases, the modulus of elasticity also increases, causing higher adhesive strain in the joint. Some adhesive failures that are blamed on poor installation actually result from this increased strain, as indicated by a change in hardness. Hardness can also be a measure of abrasion resistance, which is particularly important in horizontal traffic joints. ASTM C920 describes sealants for Use T (traffic areas) and Use NT (non-traffic areas) based in part on hardness and abrasion resistance.

9.6.5 Ultraviolet and ozone resistance

Many of the polymers used in organic building materials are composed of long-chained molecules which can be degraded by radiation within the ultraviolet wavelength range. Only ultraviolet (UV) radiation possesses sufficient energy to break the primary bonds. The only chemical effect of visible and infrared radiation is to speed up the rate of reactions that may be occurring from other causes. The amount of heat in solar radiation is not sufficient to raise the temperature to the point where chemical bonds can be thermally broken. Chemical deterioration caused by ultraviolet radiation can take one of two forms. In the first form, the polymer may completely revert to a soft, gummy consistency. This *reversion* usually occurs very slowly when UV radiation is the only factor. When moisture and UV radiation act together, reversion occurs more quickly. In the second form, UV deterioration can cause sealant hardening and embrittlement. Both hardening and reversion of a sealant prevent proper extension and compression in the joint in response to temperature and moisture movements.

Ozone in concentrations as small as 50 parts per hundred thousand (which is not uncommon in urban areas) can cause surface degradation such as crazing and cracking in many sealants. Ultraviolet radiation also causes similar surface degradation in some sealants. When a sealant is installed in properly designed joints between opaque materials such as concrete or masonry, there is little likelihood of failure of high-performance sealants. In glazing applications, however, ultraviolet rays travel through the glass and are reflected back to the sides of the sealant joint where surface deterioration can cause adhesive failure. Reflective films and coatings on glass may also increase the intensity of this effect.

9.6.6 Chalking and dirt pickup

Weathering can cause surface oxidation or chalking on the surface of some sealants. Discoloration often results from dirt pickup on the sealant surface. Chalking and the subsequent sloughing off of surface particles helps keep dirt from sticking to urethanes and polysulfides, but the dirt may wash down on other building surfaces causing stains. Slow-curing urethane and polysulfide formulations can pick up blowing dust during construction that becomes permanently embedded in the surface before it is cured. Good surface skinning characteristics, short tack-free time, and rapid cure are important in controlling dirt pickup.

Silicones experience neither oxidation nor slow cure, but do experience a degree of discoloration and soiling that is dependent on the modulus of the sealant. Static charges on the surface attract dirt which is not rinsed away by rain, but only by scrubbing, after which soiling recurs since the static charge remains. Low-modulus silicones pick up more dirt than high-modulus formulations. With sufficient dirt buildup, rain may cause surface staining below the sealant joint as contaminants wash down across adjacent building surfaces. This dirt pickup phenomenon is not restricted to silicones, however. Other types of sealants can also experience this type of soiling. The problem is an aesthetic one, and does not affect sealant performance.

9.6.7 Substrate compatibility

Some materials are very poor substrates for almost any sealant. Wet, frozen, contaminated, deteriorated, weak, unsound, and unstable surfaces will prevent adhesion, as will concrete-form release compounds, rust, and polyethylene coatings on glass spandrel panels. Some sealants may also stain porous substrates by leaching of solvents, others may cause localized edge discoloration or delamination of laminated glass, or react chemically with component materials in the joint assembly. Manufacturers should be consulted on compatibility issues and recommended substrates for various sealants.

9.6.8 Paintability

Although many sealants are listed by their manufacturers as "paintable," the subject merits some discussion. The manufacturer's recommendation is generally based only on whether paint will stick to the surface of the material. In addition to this consideration, however, several other factors affect the application of paints and coatings to sealant joints. Sealant compounds which contain oils or resins may leach residue to the surface and cause staining if paint is applied before the sealant is thoroughly and completely cured. Compounds which cure by release of solvents may be paintable, but only after the solvent is evaporated. Most important to consider, however, is movement. Inelastic paint applied to elastomeric sealants in moving joints will tear as the sealant stretches. Elastomeric coatings have some tolerance for movement, but may not be capable of the amount of extension required at joints. Since sealant joints are often painted to disguise their appearance, paint failure defeats that purpose. Sealants are available in many colors, and a better solution to aesthetic considerations is the selection of a compatible color or the articulation of the joints as part of the design. Routine maintenance painting of adjacent surfaces should require masking of the sealant joints to prevent inadvertent coverage.

9.6.9 VOC compliance

Regulations governing volatile organic compounds (VOCs) have had an effect on the sealant industry. Of the sealant materials themselves, only solvent acrylics and butyl rubbers are affected, but the impact on cleaners and primers is more universal. Manufacturers are working on new VOC-compliant primer formulations, and alternative cleaning materials are being tested. In states with strict VOC regulations, proper substrate preparation may be more difficult or less effective, and therefore have a significant effect on sealant adhesion.

9.7 Sealant Materials

The most commonly used sealants are oil-base, butyl rubber, acrylic latex, solvent acrylic, urethane, polysulfide, and silicone compounds. *Table 9.6* is excerpted from ASTM C1299 *Standard Guide for Use in Selection of Liquid-Applied Sealants.* Each individual sealant formulation, however, will vary from one manufacturer and one product to the next. Polysulfides, urethanes, and silicones are the three high-performance elastomeric sealants. Each has advantages and limitations which make them more or less suitable for any given application. *Table 9.7,* from the Sealant, Waterproofing and Restoration Institute (SWRI) manual *Sealants: The Professionals' Guide,* summarizes some of the primary performance characteristics of each generic sealant type.

9.7.1 Forms

Joint sealers used in building construction come in several forms. Tape seals are mastic materials furnished in roll form with a paper backing that is removed just before installation. Preformed gaskets may be either dense rubber or foam, and are supplied in many different extruded shapes for applications such as glazing installations. Expanding foam seals are usually furnished in self-adhesive strips, and are precompressed so that they expand in place to as much as 4 times their installed thickness. Gaskets and expanding foam seals can be used only in joints where they will be under continuous compression.

Liquid joint sealants (or bulk sealants) are chemical mixtures that are either pourable or gunnable. Pourable liquids are self-leveling, and are used for horizontal joints such as those in decks and pavements. Gunnable liquids are used both horizontally and vertically. They must be "non-sag" when used in vertical or overhead joints so that they will hold their shape until curing is complete. Sealant curing may be by solvent release or chemical reaction.

There are two categories of gunnable liquid sealants: elastomeric and non-elastomeric. Non-elastomeric sealants (sometimes called caulking compounds) are limited to use in non-moving joints—that is, joints that have less than 10% expected movement. Elastomeric sealants, on the other hand, have elastomeric properties and may be used in moving joints. Elastomeric sealants are covered by ASTM C920, *Standard Specification for Elastomeric Joint Sealants,* and must have a minimum movement capability of $12\frac{1}{2}$%. Sealants are available in single-component and multi-component types. Single-component sealants are furnished ready for application, and require no mixing. Multi-component sealants are mixed at the job site just prior to application. Multi-component sealants of some generic types cure more rapidly and perform better than their single-component counterparts, but also have a shorter working time. Incomplete or improper mixing of the ingredients reduces performance capability and usually results in sealant failure. Multi-component sealants are generally less expensive than their single-component counterparts.

9.7.2 Oil-based compounds

Oil-based (or oleoresinous) compounds, like all joint sealants, come in many different proprietary formulations. Performance depends on the types and amounts of additives or modifiers in the mix. Basic formulations consist primarily of oils, fillers, and colorants. Drying oils such as linseed and soy will quickly oxidize and harden. To counteract this tendency, polybutene, polyisobutylene, or butyl rubber can be added as a plasticizer.

Oil-based compounds come in gun grade for general caulking and knife grade for back-bedding and face glazing of wood and steel sash. Since primers are required to prevent oil from migrating into porous substrates, oil-based sealants should not be used on light-colored stone that is susceptible to staining. Oil-based compounds are fast-skinning, and can generally be painted in 24 hours. The less expensive products have only a ±2% movement capability at best and, in exterior applications, will last only 1 or 2 years. Some of the plasticized products can tolerate ±5% movement and, in a protected environment, may last up to 10 years. Oil-based sealants are covered by ASTM C570, *Standard Specification for Oil- and Resin-Base Caulking Compound for Building Construction.*

TABLE 9.6 Properties and characteristics of Sealant Materials

Sealant type	Volume shrinkage	Curing characteristics	Service temperature range	Adhesion, ASTM C794	Tack-free time, ASTM C679	Substrate staining, ASTM C510
Oleoresinous ASTM C669, ASTM C570	<10%	Skins rapidly, slow cure	−20 to 158°F	Fair	24 to 72 hr.	Slight to some
Butyl (mastic)	<30%	Non-curing	−20 to 180°F	Fair	Remains tacky	None
Butyl (solvent) ASTM C1085	<30%	Skins rapidly, slow cure	−20 to 158°F	Good	24 to 72 hr.	None to slight
Acrylic (solvent)	15–30%	Slow cure	−20 to 180°F	Good	24 to 72 hr.	None
Acrylic (emulsion) ASTM C834	Clear 40–50% Pigmented 20–30%	Slow cure	−20 to 180°F	Good to excellent	30 min to 4 hr.	None
Polysulfide (1 part) ASTM C920	1–2%	Dependent on relative humidity and temperature, slow cure	−40 to 194°F	Excellent on most substrates	<72 hr.	None
Polysulfide (2 part) ASTM C920	1–2%	Variable from <1 hr. to >24 hrs.	−40 to 194°F	Excellent on most substrates	Variable <24 hr.	None
Polyurethane (1 part) ASTM C920	1–2%	Humidity dependent, slow cure	−40 to 300°F	Excellent on most substrates	<72 hr.	None
Polyurethane (2 part) ASTM C920	3–5%	72-Hr minimum cure	−40 to 300°F	Excellent on most substrates	<24 hr.	None
Silicone (low modulus) ASTM C920	5–10%	Slow cure, generally neutral chemical release	−60 to 300°F	Excellent on most substrates	<2 hr.	None to some
Silicone (medium modulus) ASTM C920, ASTM C1184	5–10%	Slow to fast cure, neutral cure or acid release	−60 to 266°F	Excellent on most substrates	<2 hr.	None to some
Silicone (1 part high modulus) ASTM C920	5–10%	Medium to fast cure, neutral cure, or acid release	−60 to 300°F	Excellent on most substrates	<1 hr.	None to some
Silicone (2 part high modulus) ASTM C920, ASTM C1184	10–20%	Fast cure, neutral, alcohol, or no chemical release	−60 to 300°F	Excellent on most substrates	<30 min.	None to some

Sealant type	Dirt pickup resistance	Low temperature flexibility, ASTM C711, C718, C734	Application temperature range, surface	Ultraviolet resistance	Movement capability, ASTM C719	Life expectancy, years
Oleoresinous ASTM C669, ASTM C570	Poor to fair	Poor	50 to 104°F	Poor	±2%	1–5
Butyl (mastic)	Good	Good	50 to 122°F	Fair to good	±7.5%	5–10
Butyl (solvent) ASTM C1085	Fair to good	Fair to good	40 to 122°F	Fair to good	±7.5%	5–10
Acrylic (solvent)	Good	Fair	40 to 122°F	Excellent	±10%	5–10
Acrylic (emulsion), ASTM C834	Good to excellent	Good to excellent	40 to 122°F	Excellent	±7.5 to ±25%	5–20, minimum
Polysulfide (1 part) ASTM C920	Good	Excellent	40 to 122°F	Fair to good	±25%	10–20
Polysulfide (2 part) ASTM C920	Good	Excellent	40 to 122°F	Fair to good	±25%	10–20
Polyurethane (1 part) ASTM C920	Good	Excellent	40 to 122°F	Good	±25% to ±50%	10–20
Polyurethane (2 part) ASTM C920	Good	Excellent	40 to 122°F	Good	±25% to ±50%	10–20
Silicone (low modulus) ASTM C920	Fair	Excellent	−20 to 122°F on frost free surfaces	Excellent	±25% to +100%–50%	20 minimum
Silicone (medium modulus) ASTM C920, ASTM C1184	Fair	Excellent	−20 to 122°F on frost free surfaces	Excellent	±25% to ±50%	20 minimum
Silicon (1 part high modulus) ASTM C920	Fair	Excellent	−20 to 122°F on frost free surfaces	Excellent	±25%	20 minimum
Silicone (2 part high modulus) ASTM C920, ASTM C1184	Poor to fair	Excellent	Varies with manufacturer	Excellent	±12.5% to ±25%	20 minimum

SOURCE: Excerpted from ASTM C1299 Standard Guide for Use in Selection of liquid-Applied Sealant. Copyright ASTM. Reprinted with permission.

TABLE 9.7 **Advantages and Limitations of Various Sealant Types**

Sealant type	Advantages	Limitations
Polysulfide	• Two-part formulations have fast cure • Long-term performance • Capable of stress relaxation • Good adhesion • Good weathering and UV resistance • Good low-temperature properties • Excellent chemical resistance • Paintable, nontoxic, nonpermeable	• High viscosity • May require primer • Poor abrasion resistance for horizontal joints • Limited high-temperature performance • Poor tear strength • One-part formulations have slow cure • Subject to compress set because of slow cure
Polyurethane (urethane)	• Excellent adhesion to and compatibility with a wide variety of substrates • Capable of handling a broad range of joint sizes • Wide range of movement capabilities • Excellent elongation and recovery properties • Nonstaining • Fair chemical and weather resistance • Paintable	• Not recommended for structural glazing • Primer may be required on some architectural metal finishes and natural stones • Limited high-temperature performance • Ultraviolet reversion • Variety of formulations can cause wide differences in performance among brands
Silicone	• Many chemistries from which to choose • Widest range of modulus and movement-capability formulations • Longest life expectancy compared to other elastomeric sealants • Performance range from -50 to $300°F$, with minimum modulus change • Color stability • Adhesion to all construction surfaces except bitumen, Teflon, and polyethylene • No surface cracking or checking due to UV	• Plasticizer in some formulations can stain porous substrates such as masonry • Surface dirt retention • Higher price compared to other elastomeric sealants • Discoloration of cured sealant when it comes in contact with asphaltic membranes, bitumen waterproofing materials, and dense neoprenes • Not paintable

SOURCE: Sealant, Waterproofing, and Restoration Institute, *Sealants–The Professional's Guide*.

9.7.3 Butyl sealants

Butyl rubber forms the basis for these compounds. Butyl sealants may be solvent-release, or may be modified with polybutene oils, polyisobutylene mastic, or isoprene or butylene polymers to soften, plasticize, improve cure, or decrease shrinkage.

Regardless of additives or modifiers, butyl sealants are limited to a maximum of ±7.5% movement, and are widely used where a non-hardening cure is desired, such as in lap joints. They are relatively soft, slow-curing, products that will stain almost any porous substrate. Lesser-quality compounds with low solids content experience significant shrinkage. Butyls are also thermoplastic, thus having low recovery from extension and relatively high resistance to movement at low temperatures. They are non-priming, show good adhesion to most material, and have good water resistance. Butyls are relatively inexpensive, and can last 5 to 10 years depending on specific product quality and severity of exposure. Butyl sealants are covered by ASTM C1085, *Standard Specification for Butyl Rubber-Based Solvent-Release Sealants*.

9.7.4 Acrylic latex sealants

Latex sealants are formulated from acrylic polymers, fillers, and surfactants dispersed in water. Additives include silanes for adhesion, ethylene glycol for freeze/thaw stability, and other agents to control flow and slump.

Acrylic latex sealants should be limited to indoor use. They are thermoplastic, and in fact are so stiff in cold weather that they are not even extrudable at temperatures below 40°F. Maximum joint movement is ±7.5% to ±12.5% of the joint width. Shrinkage is high (20 to 50%), tear resistance is poor, water resistance is low, and there is little recovery from extension. Before cure is complete, latex sealants can freeze at temperatures below zero, and are also susceptible to rain washout in the first 12 to 18 hours after application.

For indoor use, these compounds are fast-skinning, and can be painted over in 30 minutes to an hour. They are non-staining, do not harden with age, and will adhere to glass, ceramics, drywall, gypsum plaster, and most metals without priming. Concrete, stucco, masonry, plastics, and most woods, however, must be primed before caulking. Latex sealants can also be applied to damp substrates. They are inexpensive and have, for the most part, replaced oil-based sealants for interior residential and commercial applications. Acrylic latex sealants are covered by ASTM C834, *Standard Specification for Latex Sealing Compounds*.

9.7.5 Solvent-release acrylics

Solvent-release acrylics are based on acrylic acid combined with fillers, catalysts, plasticizers, and solvent. Unlike acrylic latex sealants, solvent acrylics are better suited to exterior use. Their odor during cure is so obnoxious that they cannot be used in closed areas, occupied buildings, or near food or foodstuffs (which would absorb the odor).

Movement capability is ±10% to ±12.5%. Adhesion to most substrates is excellent, with only minimal cleaning and no priming required. Solvent acrylics have good weathering characteristics, resist UV and ozone deterioration, are non-staining and color stable, but water resistance is poor and shrinkage is relatively high.

Solvent acrylics also have low recovery and are thermoplastic. Some, in fact, must be heated before application in cold temperatures. When the manufacturer recommends heating before application, it generally indicates a product with low solvent content, which means less shrinkage and hardening. Surface skinning will take 36 hours or more, during which time the sealant picks up dirt, noticeably discoloring the lighter-colored compounds.

9.7.6 Polysulfides

Polysulfides were the first of the high-performance sealants on the U.S. market. In general, they are reported as having movement capability of ±25%, and are available in one- and two-part formulations, in a non-sag consistency for vertical building joints, and in a pourable consistency for horizontal joints. Both types are based on polysulfide polymers mixed with fillers, curing agents, plasticizers, and stabilizers. Silane is added to improve adhesion, and manganese oxide to resist heat and ultraviolet degradation.

Adhesion to steel, aluminum, and glass is very good, and generally does not require priming. Porous substrates such as marble, limestone, granite, brick, concrete, concrete block, wood and plastics, however, must be primed to achieve serviceable adhesion and to avoid staining. Primers for polysulfides are made and furnished by the sealant manufacturer. All manufacturers have their own primers, developed for use with their own specific sealant formulation, and these are not interchangeable from one proprietary product to another.

Two-part polysulfides have good extensibility, and good resistance to weathering and aging. At 75%, recovery is less than that of urethanes and silicones. Special formulations can provide excellent solvent resistance to many chemicals and petroleum products such as gasoline and jet fuel. Tack-free time is less than 24 hours, which will allow some initial dirt pickup. Full cure is achieved within a week. In temperatures below 40°F, though, the cure rate is slower.

One-part polysulfides have similar properties except that recovery is poorer and cure rate is much slower. Curing depends on ambient moisture and oxygen. With moderate temperatures and high humidity, tack-free time is reported by manufacturers to be as long as three days. Full cure may take 30 to 45 days under optimum conditions, and much longer in cold, dry weather. After the sealant has been properly tooled into the joint, a light water spray can speed initial skinning and help prevent excessive dirt pickup. Because of the time factor, though, the sealant may deform because of joint movement before curing is complete. Newer formulations of one-part polysulfides are being developed and tested in hopes of accelerating the cure rate.

9.7.7 Urethanes

Urethane sealants are made from urethane polymers combined with fillers, plasticizers, silane adhesion additives, colorants, and solvent. Movement range is reported generally as ±25% of joint width. Some manufacturers advertise high performance mixes at ±50% movement capability and low-modulus formulations as high as +100/−50%. Both one- and two-part systems are marketed.

Urethanes have good extensibility, excellent resistance to abrasion, long service life, good adhesion to most substrates (especially concrete, masonry, and metals), and paintability. Resistance to ultraviolet and ozone deterioration is fairly good, but most urethanes have a tendency to craze. Primers are required for concrete contaminated with form oils or other substances, and urethanes generally are not recommended for glazing applications. At 80 to 90%, recovery is much better than that of polysulfides, but not as good as that of silicones. Two-component urethanes are well suited for horizontal traffic joints. Some urethanes also have good chemical and solvent resistance. Shrinkage is negligible, and tear resistance is excellent if sealant depth is sufficient.

Some light-colored urethanes will discolor with time, but dirt pickup is a problem only with the one-part systems. Tack-free time is shorter for single-component urethanes, but curing is much slower (12 to 36 hours in good weather, several days in cold, dry weather). Hot, humid conditions will produce a faster cure. If urethanes do not pick up dirt during cure, they will generally stay clean, since surface oxidation continuously sloughs off any accumulated dirt.

Some urethanes may begin to deteriorate at service temperatures of 170 to 180°F, which are not uncommon on south-facing walls in southern climates. To prevent such deterioration, designers sometimes recess the sealant deeper in the joint than usual to minimize sun exposure and heat buildup. All urethanes are extremely sensitive to moisture before and during cure. A damp substrate will cause bubbles at the bond line, which can destroy adhesion.

Research and field investigations during the past several years have pinpointed a relatively frequent problem of hardening or reversion in some urethane sealants. The combination of moisture and ultraviolet radiation can cause either hardening or softening of the sealants, particularly on the southern exposure of buildings in warm and temperate climates. A field report of long-term sealant performance published by ASTM (see App. C, Bibliography, Gorman, 1995) also found urethane reversion problems in a high altitude desert climate where ultraviolet exposure alone was thought to be the cause of reversion, without the presence of moisture. Reversion and hardening in urethane sealants had previously been attributed to improper field mixing of multi-component materials. Accelerated weathering tests, though, have now duplicated the failures even with strict laboratory controls on mixing. Although not all urethanes have exhibited this failure mechanism, some architects are switching to other types of sealants unless manufacturers can satisfactorily demonstrate that their material is not subject to premature hardening or reversion. Some urethane manufacturers have changed their formulations to improve the ultraviolet resistance of their sealants.

9.7.8 Silicones

Silicone sealants come in a variety of formulations that include single- and multi-component mixes. The multi-component materials generally have a faster cure rate than one-part systems, and therefore a shorter working time for joint tooling (15 minutes to 3 hours).

Made from silicone polymers, fillers, cross-linkers, and any number of additives used to produce specific properties, silicones are gunnable even in very cold weather, and have excellent extension and movement in the cold as well. They also retain their shape in hot weather, and the opaque sealants are highly resistant to ultraviolet and ozone deterioration. Once cured, silicones experience very little hardening or shrinkage over time.

The reported movement capability of silicone sealants will vary with the proprietary chemical formulation. Some high-modulus products fall in the low range of ±12.5%, others in the higher range of ±25%. High-performance, low-modulus silicones can tolerate ±50% movement, and sometimes as much as +100%/−50%. Recovery with high-modulus silicones is 85 to 99%. Low-modulus mixes range from 70 to 90% recovery. Silicones are chemically stable and extremely resistant to weathering and aging, and the low-modulus sealants have relatively good tear resistance when punctured. Silicones have unprimed adhesion to most substrates but, as with any sealant, surface preparation is important. Silicones develop a tenacious bond to glass, and are the only sealants permitted for structural glazing applications. Anodized or coated aluminum and Kynar-coated metals should be tested for sealant adhesion before a particular silicone sealant is selected for use.

Substrate compatibility is also a potential problem. Acetoxy silicones (those that release acetic acid) should not be used on marble, galvanized steel, copper, concrete or other cementitious materials, or on any surface prone to attack or corrosion by weak acids. Silicones that release amines should not be used on copper. Neutral cure silicones are compatible with most substrates.

Silicones abrade more easily than urethanes, so traffic-bearing joints should be recessed sufficiently to avoid abrasion wear. Most silicones will not accept paint. Those that do are usually limited to joint movements of ±12.5% to ±25%. Plasticizers in some silicones will migrate and stain porous substrates, so preconstruction testing is recommended. Dirt pickup can also be a problem. Fast skinning characteristics prevent silicones from imbedding blowing dust at the job site into an uncured surface but, unlike urethanes and polysulfides, they do attract airborne dirt after cure because of static charges on the surface.

9.7.9 Tapes and gaskets

Preformed rubber gaskets and wedges are used extensively in the glazing industry, but other types of preformed sealants are available as well. Compression seal glazing gaskets are made from EPDM, neoprene, silicone and

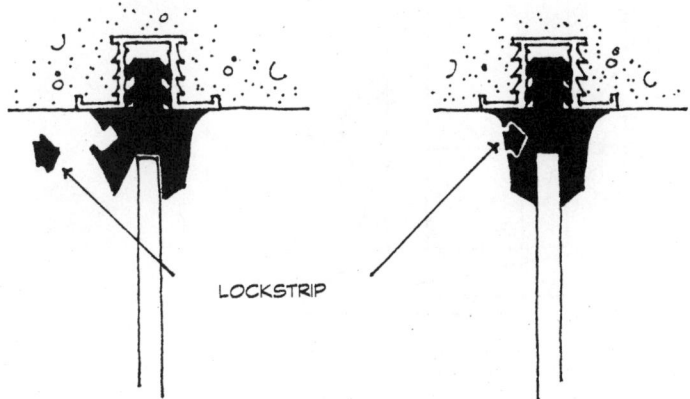

Figure 9.16 Lockstrip glazing gaskets. (*From Allen, Architectural Detailing: Function, Constructability, Aesthetics.*)

vinyl. Lock-strip or zipper gaskets are similar in function to compression seal gaskets, except that they also form an integral part of the mechanical attachment of the glass to the framing system (*Fig. 9.16*). Lock-strip gaskets are made from neoprene and EPDM rubber. Hardening of the rubber over time and loss of resiliency can cause water infiltration in glazing systems that rely on gaskets alone as the primary weather seal.

Glazing tapes of expanded cellular material with pressure-sensitive adhesives are commonly made from PVC, TPE (thermoplastic elastomer), polyethylene, or polyurethane. Primary tape seals should be affixed to the inside surface of the outside rabbet so that they expand outward, and slightly over the lip of the rabbet (*Fig. 9.17A*), while secondary tape seals should be recessed and capped with a bead of liquid-applied sealant (*Fig. 9.17B*). In high-rise buildings, where pressure differentials make the control of water penetration more critical, a double-seal system is often used. Double seals consist of a primary seal at the exterior, with a secondary "heel bead" behind the glass on the interior side (*Fig. 9.17C*). The heel bead acts as an air barrier, and as a dam to prevent any water that may have entered the channel from leaking to the inside, and forcing it to drain through the weeps.

Cellular glazing tapes are intended to be maintained under a 25 to 50% compression. This means that for a system with a face clearance of $^{3}/_{16}$ in., a $^{1}/_{4}$-in.-thick tape is necessary. *Table 9.8* shows tape thickness recommendations for some common opening sizes.

Butyl tape sealants are used extensively in the glazing industry, but are also widely used in the roofing industry to form lap joints in sheet metal flashing. Butyl tapes are non-hardening and provide an immediate seal upon installation, because no curing is required. The highest-quality compounds have 100% solids content so that shrinkage is minimal.

AAMA Publication 850.1 *Fenestration Sealants Guide Manual* classifies two types of expanded cellular glazing tapes. Type I tapes are intended to function as the primary weather seal, while Type II tapes are intended for use with a

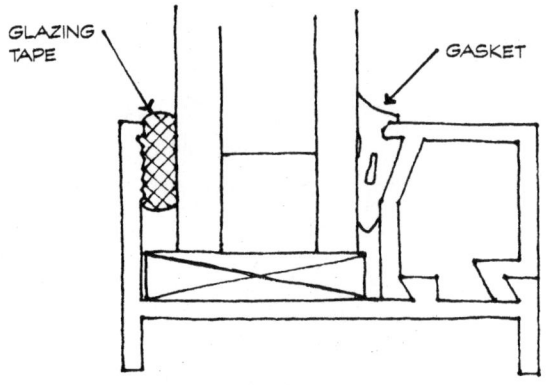

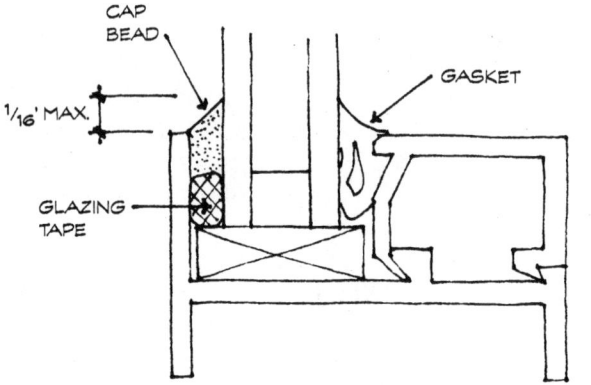

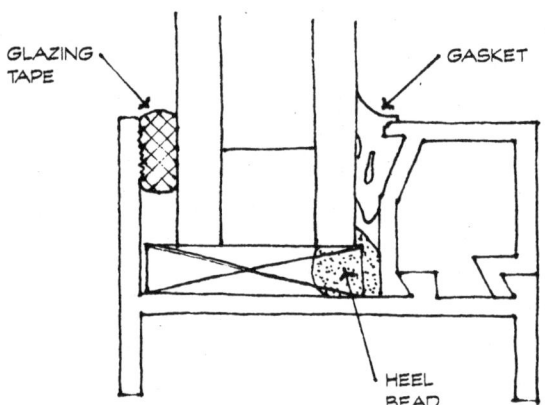

Figure 9.17 Glazing tapes and gaskets. (*From AAMA 850-91, Fenestration Sealants Guide Manual.*)

TABLE 9.8 Recommended Cellular Glazing Tape Thickness

Joint size, in.	Tape thickness, in.
0.025	$1/32$
0.046 ($3/64$)	$1/16$
0.094 ($3/32$)	$1/8$
0.125 ($1/8$)	$3/16$
0.187 ($3/16$)	$1/4$
0.250 ($1/4$)	$3/8$

SOURCE: AAMA 850-91 *Fenestration Sealants Guide Manual.*

full bead of wet sealant as the primary seal. Butyl mastic glazing tapes are classified for moderate movement, limited movement, or maximum movement.

Preformed expanding foam sealants are often used as secondary seals in two-stage joints, particularly when the back side of the joint is inaccessible for wet sealing. These expanding foam sealants are made from impregnated open-cell polyurethane foam. Under compression to a certain density (usually 25%), the foam sealant becomes watertight. The foam is precompressed to a thin profile for insertion in the joint, and has an adhesive strip for positioning the foam on one side of the joint. The foam then expands to full thickness and creates the compression necessary for weathertightness. When used as a backing for a wet sealant, a bond breaker tape is required to prevent adhesion. Some expanding foam sealants are made with a factory-installed low-modulus silicone facing so that they may be used as primary weather seals.

9.7.10 Accessories

Accessory materials used in sealant joints must be compatible with the sealant, and unaffected by any solvents contained in it. Backup materials containing asphalt, coal tar, or polyisobutylene should never be used, as they can cause loss of adhesion and may stain porous substrates. Butyl shims and spacers used in glazing applications may also bleed through some sealants and cause residue staining. Neoprene setting blocks can stain porous substrates through a silicone sealant and turn the sealant itself yellow.

Open-cell backer rods, made primarily of polyurethane foam, are porous and quite flexible. If not adequately compressed in the joint, open-cell rods are easily displaced during sealant application and tooling. Air can also penetrate open-cell rods. This can be an advantage, especially in arid and/or cold climates, in that it provides for faster curing of one-part sealants, but open-cell rods also absorb and hold water. They can wick moisture from the open cavity behind the joint which is then driven through the sealant or along the bond line by vapor pressure, often causing joint failure.

Closed-cell backer rods of polyethylene foam are popular with applicators because they are more rigid and do not displace during tooling. (Polyethylene is also used to make bond-breaker tape because it will not adhere to any sealant.) Closed-cell foams do not permit water to penetrate the back of the joint, but may present problems of outgassing. If the surface skin of the rod is punctured or abraded during installation, subsequent compression in the joint during movement can cause bubbles to form behind the sealant bead. The gas may break a hole through the sealant and cause immediate failure, or the thin skin of sealant stretched across the bubble may suffer accelerated ultraviolet degradation, or may fail cohesively with joint movement.

A proprietary backer rod of extruded polyolefin foam provides a non-absorbent outer skin with an interior network of both open and closed cells. It provides the moisture resistance of traditional closed cell rods, but is non-gassing when punctured. Cut ends, however may wick moisture into the internal cell structure where it is trapped by the vapor-resistant skin. The rod also has the high compression recovery associated with traditional open-cell rods, but must be placed carefully to avoid folding or creasing in the joint.

9.8 Sealant Installation

Successful sealant applications rely heavily on workmanship. While many joint failures are caused by poor design or inadequate allowance for movement, many more are caused by field installation problems.

9.8.1 Climatic conditions

Weather conditions at the job site affect both the substrate materials and the sealants themselves. Cold, damp surfaces inhibit adhesion, and moisture that is trapped in a joint can cause sealant failure. Rising temperatures dry substrates fairly rapidly, and for most sealants, dry to the touch is sufficient. Urethanes are more sensitive to moist surfaces than other sealants, though, so it is best to allow at least a full day's drying time after a rain. Concrete should be dry to the touch for at least 24 hours before any urethane sealant is applied. After a rain, allow one full day for drying before applying multi-component urethanes, and 2 to 3 days' drying time before applying single-component urethanes. Frost, surface condensation, and shaded north elevations can also be a problem in cold or humid conditions. The generally recommended minimum installation temperature of 40°F is intended to minimize the potential for these conditions. Substrate materials that get very hot may hamper the installation of some sealant types during warm weather, and especially on southern and southwestern exposures where surface temperatures are highest.

Because of changes in joint shape due to expansion and contraction of the substrate materials, sealants are best applied when temperatures are moderate. Temperature and humidity also affect curing time. High temperatures and humidity generally cause more rapid curing than cooler temperatures and lower humidity. Windy conditions are conducive to blowing dust, which can be permanently embedded in the surface of uncured sealants.

9.8.2 Substrate contaminants

The same dust that can contaminate the surface of wet sealants is also deposited on open joint surfaces. Cleaning with solvents, brushes or oil-free compressed air (depending on the substrate) is required to obtain good adhesion to the sides of the joint. The Sealant, Waterproofing and Restoration Institute estimates that most sealant failures result from poor joint preparation. Failure, indeed, is inevitable if the sealant cannot make adequate wetting contact with the substrate because of dust, dirt, mortar particles, surface laitance, rust, form-release oils, coatings, or other contaminants.

9.8.3 Joint cleaning

Solvent cleaning removes many contaminants from non-porous substrates such as metals and glass. Most sealant manufacturers provide detailed instructions for joint preparation, but untrained applicators too often ignore these guidelines. The following are general recommendations for solvent cleaning:

- Use clean white rags so that the dirt removed is clearly visible on the rag.
- Use clean, fresh solvent of the type recommended for the surface being cleaned.
- Provide clean containers for solvent use and storage.
- Use a "two-rag wipe" technique: one rag to apply the solvent, and a second to wipe it off before the solvent evaporates.
- Pour the solvent on the rag—never dip it in the container, as this will contaminate the solvent.
- Discard rags when they become soiled so that dirt is not redeposited on the surface.
- Be careful not to spread dirt or contaminants over exposed surfaces because residue can stain many materials.

Different non-porous substrates require different kinds of solvents, and the manufacturer of the material being cleaned should always be consulted for specific recommendations. *Table 9.9* lists several common substrate materials and the solvents most often effective in cleaning. Oily dirt must be removed with a cleaner containing trisodium phosphate. If detergent cleaners are used on any substrate, the detergent residue itself must be removed with a solvent wipe. Not every kind of contaminant is effectively removed by every solvent, and some substrates can be damaged by using the wrong solvent. If the manufacturer does not have specific recommendations, always test for compatibility and effectiveness before use.

For porous substrates, brushing, oil-free compressed air, mechanical scraping, grinding, sandblasting, or saw cutting is necessary. Grinding, however, can drive oily contaminants farther into the substrate, and sandblasting can aggressively erode the surface of porous masonry materials.

TABLE 9.9 Typical Solvent Cleaners

Surface	Commonly used solvent
Granite and marble	MEK (methyl ethyl ketone)
Glass and ceramics	MEK or alcohol
Aluminum:	
Anodized	MEK, xylene, or 1:1:1 TCE (trichloroethylene)
Mill finish	Xylene or TCE (solvents are often not effective, though; may require cleaners containing phosphoric acid and detergents)
Fluoropolymer coatings	Isopropyl alcohol
Lead	Xylene or MEK
Steel:	
Unpainted	Xylene or naptha
Stainless	MEK
Galvanized	Xylene or toluene
Paints and plastics	Naptha or alcohol (test for compatibility)

NOTE: Isopropyl alcohol may be substituted for the above solvents when there are concerns or regulations about volatile organic compounds (VOC) and safety issues.

9.8.4 Priming

In addition to joint cleaning, some sealants require surface priming on certain substrates to assure good adhesion. Sealant manufacturers usually recommend specific primers for each sealant and type of substrate, and materials cannot be substituted because of chemical incompatibility.

The primer must adequately coat the sides of the joint, and then be allowed to dry. Primer that inadvertently covers surfaces which will remain exposed can stain and are often difficult to remove. Only as much joint should be primed as can be sealed within the time period recommended by the sealant manufacturer. Primed joints that have been left overnight or that become dirty must be re-primed before sealant can be applied. Some epoxy primers, if left to dry overnight, may not adhere to themselves and must be completely removed by sandblasting or grinding before the surface can be re-primed.

9.8.5 Sealant backing

Shallow, three-sided joints require a bond-breaker or release tape to prevent adhesion to the back of the joint. Deep joints require placement of a backer rod to control sealant depth and to give a firm surface against which the sealant can be applied and tooled. Installing backer rod to the correct depth in the joint is critical. If the rod is positioned too deep, high sealant stresses will cause adhesive failure. If the rod is installed too shallow, cohesive failure will result. Many installers judge placement depth visually, but tools designed or adapted to control placement can assure proper width-to-depth ratios and better sealant performance. Tools should be smooth-surfaced to avoid puncturing or tearing closed-cell rods (*Fig. 9.18*).

Closed-cell backer rods should be sized one-quarter to one-third wider than the joint so that they have enough resistance to displacement during sealant application and tooling. Open-cell backer rods should be sized one-third to one-half wider than the joint so they will have sufficient compression in the opening to remain in place during sealant application and tooling.

9.8.6 Sealant application

Masking tape is sometimes applied along the sides of joints to assure neat application and to prevent staining of porous materials. If tape is used, it should be applied after cleaning the joint and before using primers which might cause staining of adjacent surfaces. Tape should be removed immediately after sealant placement and tooling. Leaving it in place too long, especially in warm weather, can make removal difficult or cause surface staining. Do not tape joints that cannot be sealed in the same day, and remove all tape at the end of each day.

Multi-component sealants require job-site mixing. To ensure proper curing and development of physical properties, the manufacturer's instructions for mixing should be strictly followed since both undermixing and overmixing can lead to joint failure.

The nozzle of the caulking gun should be a plastic, or standard metal tip with an extrusion width equal to that of the joint. The plastic tips of small sealant cartridges can be cut to the appropriate width. Metal tips for bulk guns come in different sizes. The angle of the nozzle tip should be kept parallel to the plane of the joint surface to avoid over-or underfilling the joint.

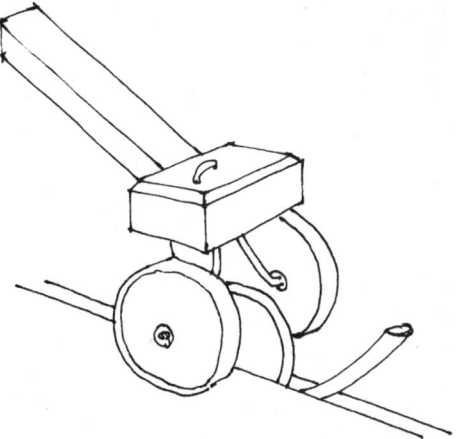

Figure 9.18 Tool for controlling placement depth of sealant backer rod.

Holding the nozzle tip tightly to the surface also prevents the sealant from squeezing out. Sealant must fill the joint completely.

All sealant joints must be tooled. Tooling seals the joint by pressing the sealant firmly against the joint sides, eliminating voids and air pockets, and creating a neat, clean, surface. Most joints are tooled to a concave shape, but flush and recessed shapes are also acceptable (*Fig. 9.19*). Wetting agents such as solvents, soap solutions, and water are not recommended as aids to tooling, but they are too often used. Dry tooling is the safest technique. Wet tooling can inhibit the curing process if solvent wetting agents are used, or cause excessive dirt pickup if solvents or soaps are used. Because of the frequent misuse of wet tooling products and misapplication of wet tooling techniques, most manufacturers (and SWRI) recommend against wet tooling.

9.8.7 Repair and remedial caulking

Remedial caulking operations must always begin with an analysis of the cause for sealant failure so that it is not repeated. Cohesive failure, or splitting, could indicate improper mixing, joint movement greater than the elastic capacity of the sealant, adhesion to the back of the joint, a poor-quality sealant, or improper joint geometry among others. Adhesive failure could indi-

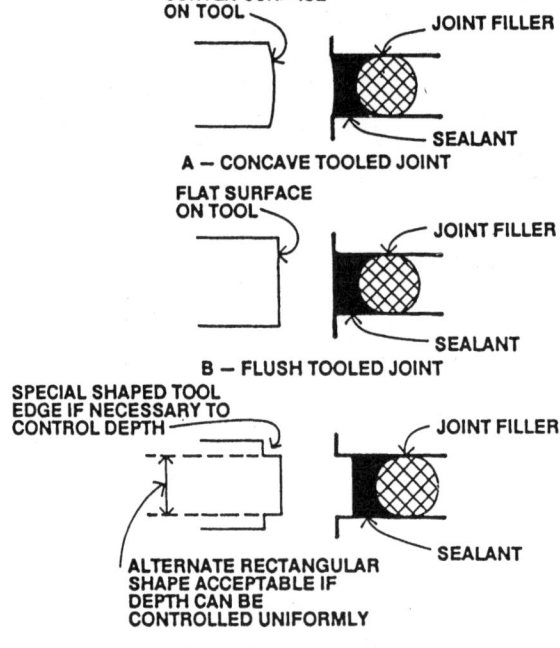

Figure 9.19 Tooled joint configurations. (*From ASTM C1193, Standard Guide for Use of Joint Sealants. Copyright ASTM. Reprinted with permission.*)

cate poor joint cleaning, moisture or other contaminants on the substrate at the time of installation, failure to prime the joint surfaces when recommended, absence or improper installation of backer rod, excessive sealant depth, excessive joint movement, moisture pressure from the substrate, or use of an improper sealant.

Joint movement should be measured in service. If measurement is not possible, movement should be recalculated to make sure the joint is of adequate size. Service and installation conditions should be analyzed, and a repair sealant selected. If necessary, enlarge the joint to provide the required movement ability and to permit proper sealant configuration. Identify the existing sealant by brand name if possible from job records, or by generic type from laboratory analysis or the following simple field tests:

- *Silicones* wipe clean when dirt is rubbed off their surface.
- *Polysulfides* emit a sulfurous odor when burned.
- *Urethanes* have an acetic odor when burned.
- *Butyls* smell like old tires when burned.
- *Silicones* give off white smoke when burned.

The old sealant must be removed *completely* by solvent cleaning, cutting, scraping, or grinding (depending on the substrate type) to assure proper adhesion of the new material. This is especially true if the new sealant is a different type from the old. All traces of primer, water repellents, coatings, and contaminants must also be removed, and the joint cleaned and primed (if required) as for new applications. The extent of repairs needed will depend on the extent of failure, aesthetic considerations, practicality, and cost. Partial repairs or spot caulking may be unattractive, may complicate record keeping of completed and uncompleted work, and usually do not work.

When it is impossible or impractical to properly clean or prepare the receiving surfaces, or when the joint is too narrow, a "bridge" or "bandage" joint may be necessary (*Fig. 9.20*). The existing joint must be isolated with bond-breaker tape, and a new surface fillet or preformed strip of sealant applied that will bond on either side of the joint. An alternative solution for difficult-to-clean joints is the use of a preformed, expanding foam sealant. Expanding foam sealants are not dependent on clean joint surfaces since they are held in place by continuous compression. They conform to irregularities in the joint sides, and some are designed to take a silicone sealant on the exterior surface to provide a color match.

9.9 Quality Assurance and Quality Control

Quality assurance and quality control are critical in attaining good performance of sealant joints. The project specifications should fully develop (1) a quality assurance program that includes submittal of complete manufacturer's data and (2) a program of quality control testing to assure compliance with the project requirements.

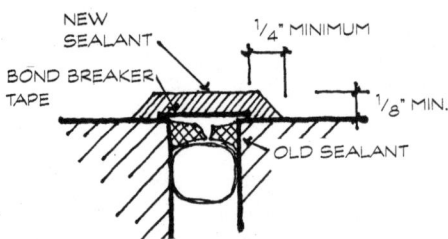

Figure 9.20 Bridge or bandage joint. (*From CSI Monograph 07M900, Joint Sealers.*)

9.9.1 Design considerations

Quality assurance for sealant joints must begin with the design concept. System performance dramatically increases when the sealant is not the first and only line of defense in keeping water out of the building. A secondary means of resisting, or collecting and expelling, infiltrated water should be included in the design of wall systems. Joint design should also include consideration of adequate accessibility for component installation and sealant tooling for both initial construction and maintenance.

The rain screen principle works on the theory that it is impossible to form a perfect seal at all of the openings on the outside of a building. Two-stage joints form a "rain screen" in which the outer seal repels or diverts most of the water hitting the outside surface, but is not required to act as a total barrier. Weep tubes through the outer sealant bead allow penetrated moisture to drain back to the outside. Additional vents, in combination with the weeps, help equalize pressure in the cavity or air space between the inner and outer seals so that moisture is not sucked into the building by pressure differentials (*Fig. 9.21*). The system requires an air barrier on the inside to maintain pressure equalization and to prevent warm moist indoor air from passing into the outer extremities of the wall where it can condense and then freeze. However, it is often difficult to maintain such an air barrier in a two-stage sealant joint. Field installation can also be difficult due to joint size variations. Remedial caulking operations have revealed that, after an extended period of service, some backer rods lose their ability to recover from joint compression and fall into the cavity, obstructing the air chamber. Many two-stage joint designs require accessibility from both the interior and the exterior of the building, making them impractical to maintain.

Joint designs that have natural water shedding rather than water collecting capabilities are inevitably more successful than those in which the sealant is the first and only rain barrier. Lapped or protected joints in building panels can not only protect sealant joints from moisture exposure, but minimize ultraviolet and ozone deterioration as well (*Fig. 9.22*). Joints on horizontal surfaces such as copings and sills have much greater exposure than vertical joints. One proprietary product used on historic restoration projects provides

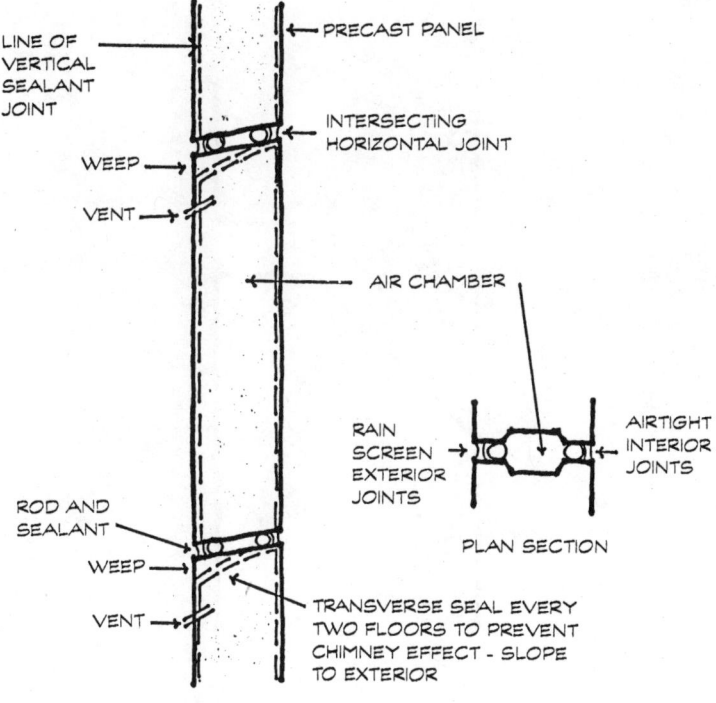

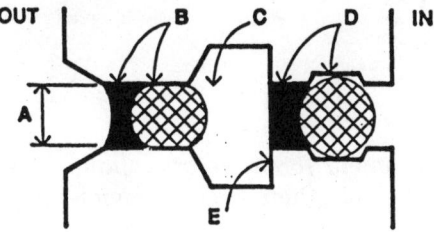

A — 1 in. (25 mm) minimum for access to interior air seal
B — sealant and joint-filler preferred for rain screen; preformed compression seal also used
C — pressure equalization chamber; vent to outside, and chamber baffles at every second floor vertically and same distance horizontally
D — sealant and joint-filler installed from outside to facilitate continuity of air seal; building framework hinders application of continuous air seal from interior
E — concrete shoulders required for tooling screed

Figure 9.21 Two-stage joints used in rain screen walls. (*From ASTM C1193, Standard Guide for Use of Joint Sealants. Copyright ASTM. Reprinted with permission.*)

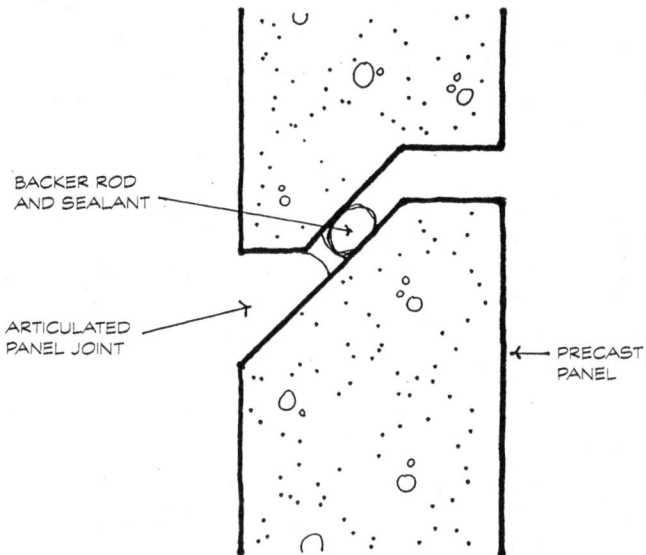

Figure 9.22 Ship-lap joint.

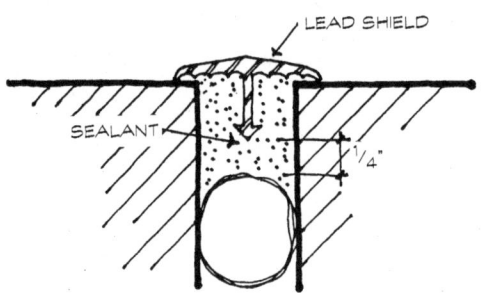

Figure 9.23 Lead shield at coping joint. (*From Weathercap, Inc.*)

a decorative lead shield for masonry copings to shed water away from the sealant joints and shield them from ultraviolet exposure (*Fig. 9.23*).

9.9.2 Testing

ASTM standards should be used to specify sealants because they establish minimum acceptable criteria for different types of sealants. ASTM tests for sealants are, for the most part, laboratory procedures that verify minimum acceptable properties and are most useful in comparing products with one another. These tests, however, usually do not include extended, long-term durability data and have little relationship to actual field performance. Although many manufacturers still cite conformance with Federal TT-S specifications, the government stopped reviewing and updating these standards in

1979, opting for ASTM and other industry standards instead. Federal sealant specifications are outdated and should not be used.

Comparing manufacturers' data sheets for product evaluation can sometimes be misleading if non-standard tests or "modified" versions of ASTM tests are used. To help facilitate product comparisons on an apples-to-apples basis, SWRI has instituted a validation program in which independent laboratory verification of product compliance with three key performance criteria are submitted to SWRI for certification. The standardized testing methods, standardized format for presenting test data, and the SWRI copyrighted "Seal of Validation" assures users that they are comparing products on a clear and equal basis.

On large projects with high liability exposure, full scale mock-up panels should be tested by an independent laboratory, in a facility that can simulate anticipated wind loading and rain conditions. On smaller projects with smaller budgets, a simple "field" adhesion test can help reduce the chances of sealant failure and rejection of work. This test is similar to ASTM C794 *Test Method for Adhesion-In-Peel of Elastomeric Joint Sealants,* except that the substrates used are actual job-site materials prepared as they will be in actual application. If a water repellent will be applied to the materials before the sealant is installed, it should also be applied to the test sample. If the joint will be saw-cut, so must the surface of the sample, and so on. The test procedure is as follows:

1. Clean and prepare the surface as specified by the sealant manufacturer for actual application.
2. If performing the test on a formed joint, install backer rod or release tape as specified. Form the joint to the proper depth for a length of 4 or 5 in., then install and tool the sealant.
3. If performing the test on a flat surface, apply sealant and tool so that a bead $1/4$ in. thick, 1 in. wide, and 4 or 5 in. long is formed.
4. Allow the sealant to cure for the time recommended by the manufacturer (about 2 weeks for silicones, 3 to 4 weeks for urethanes and polysulfides).
5. Undercut or side cut the sealant so that a 1 in. tab is formed.
6. Pull the tab slowly.
7. For sealants used as structural glazing adhesives, the tab should break as it is slowly pulled. If the tab pulls away from the surface before it breaks, the sealant fails.
8. For sealants used as weather seals, mark the 1-in. tab by holding a ruler perpendicular to the surface.
9. Pull the sealant tab straight up along the edge of the ruler, and hold it for 1 minute at twice the required extension (12.5%, 25%, etc.), and note if there is any loss of adhesion.
10. If the sealant fails in adhesion, the test is over.

11. If the sealant does not fail in adhesion, continue the test by immersing the test pieces in water for 1 day. Then re-pull the tab to the specified extension and note any loss of adhesion. If the sealant passes, it is considered acceptable in dry climates and for applications other than structural glazing.
12. If the project site is in an area where prolonged rain or high humidity is common, immerse the sample in water again for 6 more days, and re-pull the tab. If there is no loss of adhesion, the sealant passes the test.

Have the field adhesion test performed by an independent testing laboratory sufficiently in advance of sealant application so that adjustments can be made if necessary. Adhesion tests without the water immersion can be performed on installed sealants in the field by randomly selecting a test location, cutting a tab, and measuring the sealant pull as described above.

9.9.3 Applicator qualifications

Through various testing programs, sealants and their substrates regularly undergo close scrutiny resulting in a wide choice of quality products. Applicator training, however, is much less prevalent. Each linear foot of joint requires informed decision making and skilled techniques. Because of the risk factors involved and the cost of scaffolding a building facade to perform repairs, project specifications should require that each applicator be trained and certified in programs such as those sponsored by the SWRI.

9.9.4 Submittals

In the case of joint failure, the first question the manufacturer will ask is "What is the batch number of the sealant and primer used?" It is therefore prudent, particularly on critical jobs, to require that the applicator supply batch numbers of the products as an attachment to the warranty.

9.10 Avoiding Common Problems

There are some common design and installation problems which are found repeatedly in remedial sealant work. The first is the difficulty presented in achieving good sealant bond between precast concrete panels with a rough surface profile such as exposed aggregate (*Fig. 9.24*). Because of the rough surface, the sealant cannot be adequately tooled to make contact with the substrate surfaces on both sides of the joint. To achieve weather resistance with exposed aggregate finishes, precast panel edges should be fabricated smooth, or the joint surfaces must be ground smooth in the field before sealant application. Alternatively, the system can be designed as a rain screen where face sealing is not required to be 100% perfect.

Another frequent problem area is at the corners of window installations. The cut-and-piece method of assembling some framing systems results in the absence

Joints and Sealants **437**

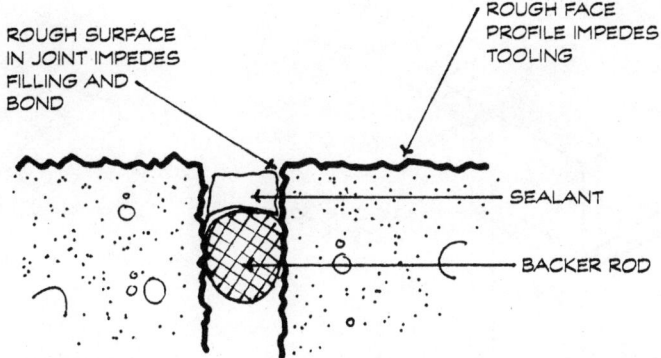

Figure 9.24 Defective sealant joint in rough-surfaced precast panel. (*From Nicastro, Difficult Sealant Joints.*)

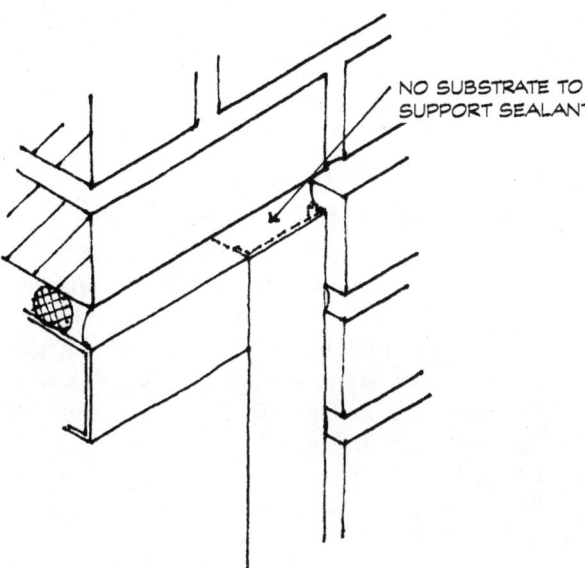

Figure 9.25 Sealant cannot be installed against thin metal edge. (*From Nicastro, Difficult Sealant Joints.*)

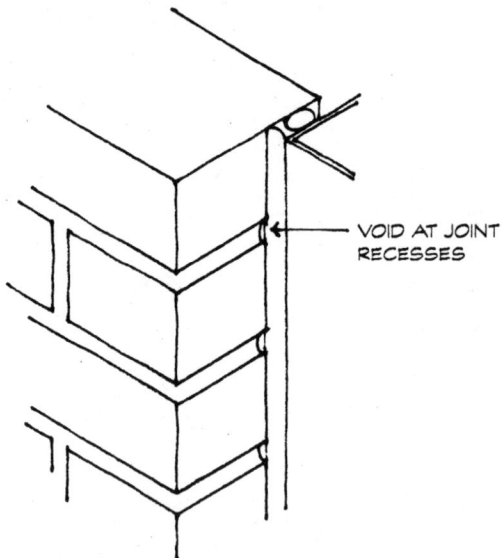

Figure 9.26 Mortar joints prevent adequate tooling to eliminate voids at bond surface. (*From Nicastro, Difficult Sealant Joints.*)

of a surface to hold the backer rod in place and against which to apply the sealant (*Fig. 9.25*). A mullion cap or integral return is required at the perimeter of all glazing installations to ensure adequate support for the sealant.

Sealant installation can also be a problem where raked mortar joints form one side of a vertical sealant joint. This occurs usually at window jambs (*Fig. 9.26*). Because of the profile of the mortar joint, the sealant cannot be adequately tooled to eliminate voids at the interface. Even concave mortar joints can be a problem for the same reason if they are tooled very deep. Where masonry must be bounded by a vertical sealant joint, the mortar joint profile should be kept as flat as possible until it is outside the area of intersection with the window jamb sealant joint.

Chapter 10

Coatings

True barrier membranes of the type used in roofing and waterproofing are not normally applied to exterior wall surfaces because of both aesthetic and economic considerations. Decorative coatings, however, can be used as part of an overall moisture protection system. In some environments, coating of wood, concrete, and masonry is unnecessary. In other exposures, paints and coatings are needed to provide a protective layer between the substrate and the environment. To be effective, the coating must be continuous, pinhole free, and thick enough to separate the substrate from the environment. Coatings are not a substitute for good design, but can serve as an adjunct in preventing damage from metal corrosion, wood rot, and excessive moisture absorption in porous materials.

This chapter covers exterior coatings including *paints and primers* and *clear water-repellent coatings*.

10.1 Paints and Primers

Paints and primers can be both decorative and protective because they have the ability to hide the substrate or alter its color through pigmentation. Different types of paints and primers are suitable for different substrates and different purposes.

10.1.1 Paint and varnish

Paint is basically a mixture of a hiding pigment, a material to bind the pigments together (binder), and a thinner (vehicle). The binder may be a vegetable oil or a synthetic resin, and, to aid in application, it may be thinned in an organic solvent or dispersed in water. *Varnish* is essentially a resin binder without pigment. Only spar varnish is durable in exterior environments. *Stains* are a hybrid between paint and varnish, containing a small amount of pigment. Stains provide better protection for wood than clear varnishes, but allow some of the wood grain to remain visible. Although properly formulated stains are more durable than even the best clear finishes, they will not last as long as fully pigmented paint coatings. Primers are used as a base coat to prepare a substrate for the application of paint top coats. Paints contain the components that provide, in varying degrees, resistance to weather, chemicals, dirt, scrubbing, and staining, and are available in several different sheens—flat, eggshell, semigloss, and gloss. An *enamel* is a paint capable of forming a very smooth, hard film, often using varnish as the vehicle.

Most paints can be classified according to the type of binder used in the formulation. Solvent-thinned paints include oil paints, alkyds, epoxies, silicone alkyds, and urethanes. Water-based emulsions include acrylics, epoxies, and urethane latexes as well as polyvinyl acetates.

Oil paints are not widely used anymore because they dry so slowly. Several different kinds of oils, including soy oil, tung oil, and castor oil are used in making alkyd resins, but are not usually used as binders by themselves. Oil paints have generally been replaced by alkyds.

Alkyds are coatings produced by reacting a drying oil acid with an alcohol. Drying of the surface occurs by evaporation of the solvent. Curing of the resin occurs by oxidation. The more oil there is in the formula, the longer it takes to dry, the better the wetting properties, and the better the elasticity. Alkyds can be used as interior or exterior trim paints, machinery enamels, or durable wall finishes. To avoid saponification, they should not be used on masonry or other alkaline surfaces except over an alkali-resisting primer or sealer, or on galvanized metal surfaces except over a recommended primer.

Catalyzed epoxies are produced by combining an epoxy resin with a curing agent. Solvent evaporation causes the surface to dry, while a chemical cross-linking is the curing mechanism. The mixture has a limited pot life or time of workability which may vary from a few minutes to several hours depending on the formulation. When properly cured, catalyzed epoxies have excellent solvent and chemical resistance. They are excellent wall coatings, producing a surface that is highly resistant to abrasion, chemicals, and cleaning. Many epoxies can be used on floors in high traffic areas. Most epoxies develop a nonprogressive chalking on exterior exposures, but otherwise have good durability. The name of a catalyzed epoxy should include the type of catalyst, i.e., polyamide epoxy, polyamine epoxy, and so on.

Epoxy esters are produced by reacting a drying oil with an epoxy resin. Epoxy esters dry by solvent evaporation and cure by oxidation. Neither as hard nor as chemically resistant as catalyzed epoxies, epoxy esters have a good intermediate degree of toughness and chemical resistance and can be used in areas subject to occasional spills of aggressive liquids. They are easy to apply and are available as conventional coatings in a single package and do not require the use of a hardener or catalyst.

Silicone alkyds are alkyds modified with silicone resins in amounts up to 30%. Silicone alkyd paints dry, cure, and perform as alkyds, but have greatly improved color and gloss retention. They apply easily and have found widespread use in coastal areas and environments subjected to intense sunlight.

Urethanes form tough, hard, flexible, chemical-resistant films by moisture curing or by the addition of a catalyst. Urethanes are light-stable, gloss-retentive, and non-yellowing. For maximum performance, they are often used over epoxy primers or zinc-rich primers with epoxy intermediate coats to protect chemical plants, bridges, water and wastewater facilities, and other industrial sites.

Latex paints combine synthetic resins and pigments dispersed in water. The two types of latex binders used in exterior coatings are *acrylic* and polyvinyl acetate (PVA or *vinyl*). As the water evaporates, coalescing solvents allow the particles of resin to fuse together forming a continuous coating. Latex paints have excellent adhesion, color and gloss retention, long-term flexibility, and toughness. Other advantages include ease of application and cleanup, safety, and VOC compliance. Latex paints must be protected from freezing and applied at a minimum temperature of 50°F.

Water-based acrylic epoxies approach the durability and performance of their solvent-based counterparts. They offer the added advantage of low odor and can be used over conventional paints on interior applications.

Table 10.1 shows a comparison of the performance and applications of various paint products formulated with different types of binders.

10.1.2 Primers

Primers may serve one of several purposes. Primers are usually designed to promote adhesion between the substrate and the finish. Sometimes primers are part of the protective system, particularly in corrosive environments where rust-inhibitive or zinc-rich primers are commonly used. Primers can also serve as sealers for porous substrates.

Primers for wood substrates include oil, oil-and-resin (oleoresinous), alkyd, and latex formulations. Oil primers are slow drying and are no longer commonly used. Oil-and-resin primers (sometimes called fortified primers) have added phenolic or alkyd resins for faster drying, shorter recoating time, resistance to bleeding, resistance to mold and moisture, and controlled penetration. Alkyd and latex primers are most commonly used today.

Primers for metal substrates include alkyd-based, latex, zinc-chromate, zinc-rich, and wash primers. *Alkyd-based* and *latex primers* are used for most architectural applications except where extra corrosion protection is required. *Zinc-chromate primers* usually contain iron oxide and phenolic resin binders. *Zinc-rich primers* contain at least 80% zinc particles by volume. The zinc provides cathodic protection for steel substrates in the same way that galvanizing serves as a sacrificial coating. To be effective, the zinc pigment particles must be in direct contact with the steel. Sandblasting cleans the surface and provides texture to promote adhesion. *Wash primers* usually contain phosphoric acid and clean as well as prime the surface. They are frequently used to protect freshly sandblasted surfaces, and are available in both one- and two-part formulations.

10.1.3 Cementitious coatings

Portland cement paints are permeable, pigmented, cement wash coatings that have properties similar to the concrete they are used to protect. For this reason, they are used for filling in and leveling minor imperfections in concrete surfaces. Portland cement paints are supplied as dry powders that must

TABLE 10.1 Comparison of Coatings with Various Binders

	Oil	Alkyd	Latex	Epoxy	Urethane	Vinyl
Film Properties						
Gloss retention	P	E	E	P	F	VG
Yellowing	Cons	S	N	Cons	Cons	S
Adhesion	E	G	G	E	G	F
Hardness	P	G	F	E	E	G
Flexibility	F	G	E	G	E	E
Moisture permeability	Mod	Mod	High	Low	Low	Low
Resistance to:						
Abrasion	P	F	F	E	E	E
Water	F	G	G	E	E	E
Detergents	F	FG	FG	E	E	E
Acid	P	F	G	E	E	E
Alkali	P	F	G	E	E	E
Heat	F	G	F	G	G	P
Strong solvents	F	F	P	E	E	P
Use and Service						
Use on:						
Wood (pigmented)	G	E	E	VG	E	NR
Fresh concrete	NR	NR	E	E	G	E
Metal (pigmented)	G	E	VG	E (primer)	F	E
Metal (clear)	NR	E	NR	NR	E	NR
Surface prep. for metal	1	2	2	3	4	4
Service:						
Interior	NR	E	E	E	E	NR
Normal exterior	G	E	E	E (primer)	F	E
Marine	P	G	F	E (primer)	E	E
Corrosive environment	NR	F	NR	E	E	E

E = excellent
VG = very good
G = good
FG = fairly good
F = fair
P = poor
NR = not recommended
N = none
S = slight
Mod = moderate
Cons = considerable

Surface preparation:
1 = easiest
4 = most demanding

be mixed with water before application. After the addition of water, thorough mixing is required to obtain a creamy consistency that facilitates uniform application. The coating must be used within 4 hours after adding the water. Cement-based coatings have not been used as extensively since the introduction of latex paints.

Surface preparation includes removal of form-release agents, curing compounds, surface laitance, and efflorescence. Cleaning can be accomplished by chemical or mechanical means, or by acid etching.

Cementitious coatings adhere reasonably well to concrete, but not to other surfaces, and should not be applied over old paint. Colors are limited to earth colors which darken when they are wet, and only a flat finish can be obtained. The surface of cement paints is soft and granular, tends to show stains, and is difficult to clean. Portland cement paints must be applied to a damp surface, but there should be no free surface moisture. After application, the paint should be moist-cured for 48 to 72 hours. For two-coat applications, the second coat should be applied within 24 hours of the first.

10.1.4 Elastomeric coatings

Acrylic elastomeric coatings are water-based, flexible, membrane-type materials which can expand and contract with movements in the substrate to which they are applied. The elasticity of the coating must be able to accommodate the expected expansion and contraction of the substrate, or cracking may permit water to penetrate freely through the resulting voids. Elastomeric coatings can span small cracks in the substrate (up to about $1/32$ in.), but realistic expectations of performance should take into account the coating's ability to expand and contract at *low temperatures* when substrate cracks are widening and the coating itself is brittle. Elastomeric coatings are often tested for elongation at 75°F which is not the service temperature at which this property is important. It would be much more realistic to test elastomeric coatings at or below freezing.

To provide good performance, elastomeric coatings should be at least 45 to 50% solids by volume. Formulations based on 100% acrylic polymers have generally shown better performance in elasticity, resistance to hydrolysis, and ultraviolet stability than those made with styrenated acrylic or vinyl acetate. To achieve a pinhole-free, monolithic membrane requires a two-coat application.

Newer formulations of silicone elastomeric coatings have recently been introduced into the market, and offer some different performance characteristics. A silicone elastomeric coating incorporates an emulsion of small silicone rubber globules suspended in water. When the water evaporates, the rubber globules coalesce and react together to form a continuous silicone rubber film. These coatings can bridge cracks up to $1/16$ in. wide, and have excellent permeability to allow walls to breathe.

An important difference between silicone and acrylic coatings is their resistance to alkaline substrates. Acrylic coatings cannot be applied to alkaline substrates such as concrete, stucco, and masonry until the substrate has cured for at least 30 days. Silicone coatings, on the other hand, are naturally alkali-resistant, and can be applied to new concrete, stucco, and masonry surfaces within 48 hours without the damaging effects of alkali hydrolysis or saponification which could occur with acrylics. Silicone coatings will not sustain fungal growth like acrylics, so they do not need an added fungicide, but silicones may attract and hold more surface dirt. The handling and application characteristics of silicone and acrylic elastomeric coatings are similar. Formulations based on 100% silicone will perform better than silicone-modified acrylics in which only a small amount of silicone emulsion is present.

10.1.5 Coating selection

Paint selection will depend on the type of substrate and the environment from which it must be protected. The coatings industry generally defines four environmental classifications:

- Type A—aggressively corrosive (chemical process plants)
- Type C—corrosive (industrial plants, paper mills, refineries)
- Type M—moderate (light manufacturing)
- Type P—protected (architectural)

Most architectural applications on buildings will fall into either the Type M or Type P exposure classifications. *Table 10.2* provides a simplified guide to coatings that are effective in these environments. *Tables 10.3* through *10.8* provide more detailed recommendations for coatings on exterior masonry and concrete, wood, and metal.

10.2 Clear Water-Repellent Coatings

Clear coatings can be applied as *water repellents* which inhibit surface absorption, but allow vapor transmission. Their purpose is not to "seal" the surface, but to keep moisture in its liquid form from penetrating the surface. Clear water repellents are often used on porous concrete and concrete masonry substrates.

TABLE 10.2 General Coating Recommendations

Type of application	Environment	Recommended coatings
Architectural	Protected (P)	Alkyd, latex
Light manufacturing	Moderate (M)	Water-based acrylic, acrylic epoxy, modified urethane
Industrial plants, paper mills, refineries	Corrosive (C)	Epoxy, zinc-rich coatings, two-part urethane
Chemical process plants	Aggressively corrosive (A)	Coal-tar epoxy, vinyl ester, polyester, polyamine epoxy

SOURCE: Glidden Paint Company.

TABLE 10.3 **Coating Recommendations for Exterior Masonry, Concrete, and Stucco**

Coating type	Application
Water-based:	
Latex block filler	To fill holes and joints in concrete and block surfaces prior to painting. May be used under latex or solvent-based top coats.
Latex paint	Self-priming on masonry, concrete, and stucco. Normal film thickness.
High-build latex coating	Self-priming, contains aggregate that provides texture to obscure minor surface imperfections, transmits moderate amounts of water vapor.
Cement paint	Inexpensive, but chalks excessively if not properly applied. Difficult to repaint. Not to be used on brick or other clay masonry substrates.
Solvent-based:	
High-build alkyd coating	Self-priming, contains aggregate that provides texture to obscure minor surface imperfections, transmits moderate amounts of water vapor. Not for use on fresh concrete or on damp surfaces.

SOURCE: Adapted from H. E. Ashton, *General Recommendations for Painting Buildings,* Canadian Building Digest No. 172, Division of Building Research, National Research Council of Canada, Ottawa, 1975.

TABLE 10.4 **Coating Recommendations for Exterior Wood**

Coating type	Application
Primer:	
Alkyd	For use under alkyd or latex top coats. Dries overnight. Should not be applied over old oil paint.
Latex	May be used under alkyd or latex top coats. Redwood and cedar substrates can stain through primer, but will transmit color to latex top coat. Should not be applied over chalking paint.
Top coat:	
Alkyd	Gloss and semigloss finish with moderate gloss retention. Good leveling but more difficult to apply than latex. Good flexibility and color retention. Should not be applied over old oil paint.
Latex	Eggshell or semigloss finish. Not self-priming on wood (use latex or alkyd primer). Quick drying and can be applied on damp surfaces, but not at low temperatures. Excellent flexibility. Recommended where previous coatings have tended to blister.
Stain:	
Fortified oil	Light colors do not perform as well as dark colors. Two coats are much better than one.
Alkyd	Two coats required on exterior.
Clear finish:	
Phenolic varnish	Not recommended because low durability requires frequent reapplication.
Knot sealer	For sealing knots before painting to reduce staining or resin exuding.

SOURCE: Adapted from H. E. Ashton, *General Recommendations for Painting Buildings,* Canadian Building Digest No. 172, Division of Building Research, National Research Council of Canada, Ottawa, 1975.

TABLE 10.5 Coating Recommendations for Exterior Metals

Coating type	Application
Primer:	
Ferrous metals	
Oil-alkyd:	
Zinc chromate–iron oxide	50% oil, overnight dry. Requires at least very thorough hand preparation, but preferably power cleaning.
Alkyd	
Zinc chromate–iron oxide	Fast dry (6 hours) automotive type. Requires very good surface preparation, usually chemical pretreatment. Normally applied by spray.
Zinc-rich	So-called cold galvanizing. Very good performance in rural areas. Grit blasting required for most surfaces.
Zinc (including galvanized metals):	
Vinyl wash	Two-component for application to new surfaces to ensure good adhesion to zinc. No other primer required in normal exposures, but should be top-coated.
Zinc dust-zinc oxide	Two-component for application to new zinc. Can also serve as finish coat where flat gray is acceptable
Top coat:	
Alkyd enamel	General-purpose enamel for brush application. Durable, with fairly good gloss retention in normal exposures.
Acrylic enamel	Excellent color and gloss retention, but must be factory-applied because baking is required to cure.
Urethane enamel	Non-chalking, non-yellowing, two-component type normally applied by spray. Excellent color and gloss retention but expensive. Applied over epoxy primer.
Latex paint, flat	Suitable where flat finish desired on primed metal.
Latex enamel, gloss	Initial gloss not as high as that of alkyd enamel, but good gloss retention.
Clear finish:	
Brass and bronze	Acrylic lacquers give best performance.

SOURCE: Adapted from H. E. Ashton, *General Recommendations for Painting Buildings,* Canadian Building Digest No. 172, Division of Building Research, National Research Council of Canada, Ottawa, 1975.

Some types form a surface film while others inhibit absorption by penetrating surface pores to change the capillary angle from absorption to repellency. Neither type will "seal" the surface, or bridge cracks. Water will still penetrate freely through small cracks and voids.

Water repellents are not generally required on new brick masonry, but are often used in remedial and restoration work. Exposed concrete masonry walls, however, are much more porous than brick and usually do need either integral water repellent admixtures or surface treatments to reduce absorption. In remedial applications to masonry walls, it is imperative that other sources of water penetration be located and repaired before a water repellent is applied. Roof leaks, parapet leaks, window leaks, failed sealant joints, and bond line separations at the mortar joints must be corrected so that the coating does not exacerbate the moisture problems by slowing down the drying process.

Clear water repellents are often recommended as a solution to efflorescence problems. However, if the water repellent is applied to a wall that still contains both moisture and salts, the resulting problems may be even more damaging than the efflorescence which is merely an aesthetic problem. The water in the wall will still take the salts into solution, and as it migrates toward the outer face, most of it will stop at the inner depth of the water repellent. The water will then evaporate through the surface as a vapor and deposit the salts inside the masonry unit. This interior crystalline buildup (sometimes called *subflorescence*) can exert tremendous pressure capable of spalling the unit face (*Fig. 10.1*). Clear water repellent applications are not recommended as a treatment for efflorescence unless the chain of contributory conditions (moisture, salts, and migration paths) is also broken.

The active ingredients in clear water repellents can be broadly characterized as either film formers or penetrants.

10.2.1 Film formers

Film formers deposit their primary water-repellent components on the substrate surface. The most common types of film-forming water repellents are acrylics, stearates, and silicone resins.

Acrylics are the most widely used of this type of water repellent. All acrylics will darken the surface of the substrate and change the finish from matte to glossy. The higher the solids content, the higher the gloss. Acrylics are sometimes used to accentuate the colored aggregates in burnished concrete block.

Stearates used in making water repellents are aluminum or calcium salts of

TABLE 10.6 Coating Systems for Exterior Masonry, Concrete, and Stucco

Surface	System type	Sheen	Primer	Top coat
Stucco	Acrylic	Flat	1 coat concrete and masonry primer	1 or 2 coats acrylic latex
		Semigloss	1 coat concrete and masonry primer	1 or 2 coats acrylic semigloss latex
EIFS	Acrylic	Flat or semigloss	Acrylic primer to help hide repair patches	2 coats acrylic elastomeric coating
Poured concrete, precast concrete	Acrylic	Flat	1 coat concrete and masonry primer	1 or 2 coats acrylic latex
	Acrylic textured coating	Flat	1 coat concrete and masonry primer-sealer	1 coat rough- or coarse-textured acrylic coating
Concrete block	Acrylic	Flat	1 coat PVA block filler	1 or 2 coats acrylic latex or 1 or 2 coats acrylic elastomeric coating

SOURCE: Weismantel, *Paint Handbook*.

TABLE 10.7 Coating Systems for Exterior Wood

Surface	System type	Sheen	Primer	Top coat
Wood siding	Alkyd	Gloss	1 coat alkyd exterior primer	1 or 2 coats alkyd gloss house and trim paint
	Acrylic latex	Satin	1 coat exterior latex primer	1 or 2 coats satin latex house paint
	Acrylic latex	Semigloss	1 coat exterior latex primer	1 or 2 coats semigloss latex house and trim paint
	Acrylic latex	Flat	1 coat exterior latex primer	1 or 2 coats flat latex house paint
	Alkyd or latex	Flat stain		1 or 2 coats semitransparent alkyd or latex stain
Door and trim	Alkyd	Semigloss	1 coat alkyd exterior primer	2 coats alkyd semigloss enamel
	Acrylic latex	Semigloss	1 coat exterior latex primer	2 coats latex semigloss enamel
Hardboard siding	Alkyd	Semigloss	1 coat alkyd exterior primer	1 or 2 coats alkyd semigloss house and trim paint
	Acrylic latex	Satin	1 coat alkyd exterior primer	1 or 2 coats satin latex house and trim paint
Rough wood and shakes	Alkyd or latex	Flat stain		1 or 2 coats semitransparent alkyd or latex stain
Flooring and decking	Phenolic alkyd	Gloss	1 coat porch and deck enamel, thinned 10%	1 coat porch and deck enamel without thinning
	Acrylic latex	Satin flat	1 coat exterior latex primer	1 or 2 coats latex floor enamel

SOURCE: Weismantel, *Paint Handbook.*

fatty acids, and are sometimes referred to as metallic soaps. Stearates impart water repellency by reacting with free salts in mineral building materials, and form a film by filling the substrate pores. Silicone resins react with moisture to form an elastomeric film.

Film formers create a physical barrier on the substrate surface. If there are no discontinuities in the coating, water repellency is determined by the permeability of the coating. When applied correctly, most acrylics, stearates, and silicone resins are water repellent.

10.2.2 Penetrants

Penetrants are the most popular type of water repellents because they do not change the appearance of the substrate surface in any way. They deposit their primary water repellent component on the walls of the substrate pores, in the pores themselves, or both. Penetration is effective to a depth of $1/16$ to $1/4$ in., depending on the formulation. The most common penetrants are the reactive types such as siloxanes, silanes, silicates, and siliconates.

Siloxanes are polymerized compounds that are best suited for dense surfaces such as brick. Silanes are not pre-polymerized, so they rely on a chemical reaction with atmospheric moisture and substrate alkalis to form the polymer chain after application. Because concrete and concrete masonry are alkaline, they speed up the reaction time of silanes regardless of the ambient humidity or the amount of moisture in the substrate. Silanes are therefore most suitable for concrete and concrete masonry substrates.

Silicates may be either water-based or solvent-based. Ethyl silicates are used in historic preservation work as consolidants for deteriorated brick and natural stone. Sodium silicates are used as set accelerators for concrete, as curing compounds, and in numerous other applications. Sodium silicates are promoted as water repellents, but their effectiveness has not been proven.

TABLE 10.8 Coating Systems for Exterior Metals

Surface	System type	Sheen	Primer	Top coat
Structural steel	Alkyd	Gloss or semigloss	1 coat red oxide–zinc chromate	1 or 2 coats alkyd gloss or semigloss enamel
	Zinc	Gloss	1 coat inorganic zinc-rich	1 coat polyamide epoxy
General purpose steel and accessories	Alkyd	Semigloss	1 coat red oxide–zinc chromate	2 coats alkyd semigloss enamel

SOURCE: Weismantel, *Paint Handbook.*

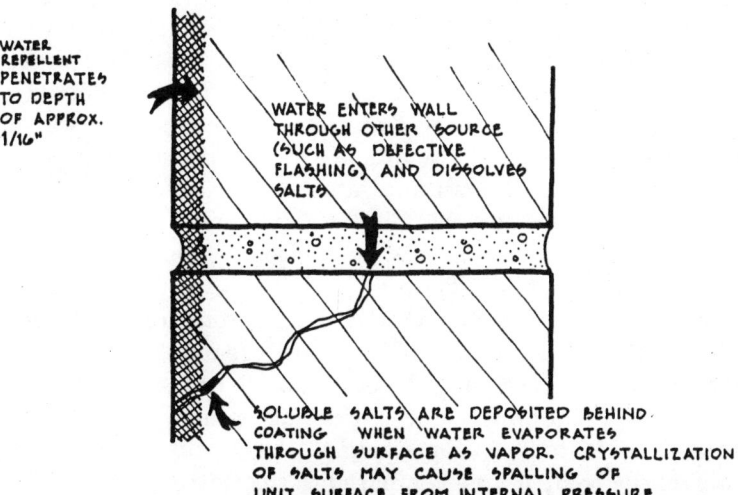

Figure 10.1 Water repellents will not cure efflorescence if source of moisture and salts are still present. (*From Beall, Masonry Design and Detailing, 4th ed.*)

Methyl siliconates are a soluble form of methyl silane which bonds with silica-containing materials through a reaction with carbon dioxide. Bonding with the substrate is slow compared to other reactive penetrants because of the low amounts of carbon dioxide in the atmosphere. Washout of the material is possible if it rains shortly after application. Reaction is even slower in alkaline environments, so siliconates should not be used on concrete or concrete masonry. Silicates and siliconates are not as effective as silanes and siloxanes, but they are water-soluble and low in volatile organic compounds.

Reactive penetrants impart water repellency by changing the substrate's surface tension and eliminating capillary suction. This "anti-capillarity" can be overcome by applying a force such as wind or hydrostatic pressure. Water vapor passes freely through a substrate treated with silanes, siloxanes, or siliconates under the action of temperature or pressure differentials.

10.2.3 Water repellent selection and field tests

The chemistry of silicone-based water repellents is confusing enough as it is, but matters are made even more complicated by the fact that many manufacturers make products that are blends of more than one type of penetrant. The best way to choose a clear water repellent is to consult the manufacturer for recommendations on the particular substrate in question.

Following the selection of one or more recommended products, field testing of sample areas is the best way to determine the effectiveness of the coating. Some of the variables that affect the performance of water repellents on concrete wearing surfaces are contaminants, curing compounds, surface preparation, weather conditions, and the finish and quality of the substrate. On masonry walls, performance is affected by contaminants, surface preparation, weather conditions, and the porosity and integrity of the surface.

To conduct a field-test sample for a concrete wearing surface, apply the water repellent in accordance with the coating manufacturer's recommendations to an area at least 5 ft square. After 7 days of curing, remove two 3-×4-in. core samples and perform a modified ASTM C642 *Test for Specific Gravity, Absorption, and Voids in Hardened Concrete*. The test should be modified by sealing the sides and bottom of the cores with a paraffin wax or other impervious material. Acceptable results are a reduction in absorption of 85% of the treated core compared to an untreated core.

To conduct a field-test sample for a masonry or concrete wall, apply the water repellent in accordance with the coating manufacturer's recommendations to an area at least 5 ft square. After 7 days of curing, use a RILEM tube to test surface absorption. (RILEM is an acronym for the French expression for International Union of Testing and Research Laboratories for Materials and Structures.) Place one tube on the brick or block and one tube on a mortar joint. After 30 minutes of testing, acceptable results are a reduction in absorption of 80 to 85% (depending on the manufacturer's warranty requirements) of the treated area compared to the untreated area. The RILEM tube is a calibrated plastic tube (*Fig. 10.2*). When attached to the wall and filled with water, the hydrostatic pressure simulates an 85 mph wind-driven rain.

10.2.4 Application procedures

The substrate surface must be clean and dry before water repellents are applied. Most manufacturers recommend a low-pressure power wash to clean dirt and atmospheric contaminants from the surface. Allow 48 hours' drying before proceeding with the application. Silanes and siloxanes can be applied to a damp surface, but they cannot penetrate pores that are already full of water. Wait at least 48 hours after a rain before beginning application. On new construction, allow at least 30 days' curing time at 75°F. This lowers the alkalinity of the surface and allows excess construction moisture to evaporate. Allow longer curing during winter months when evaporation is slower. If the surface feels cool or damp, it is too wet, especially for acrylics or silicone resins. If in doubt, tape a 12-in.-square piece of clear plastic over the surface, being careful to seal all the edges. If vapor condenses under the plastic overnight, there is still too much moisture in the wall or slab.

Joint sealant installation should be completed at least 72 hours before application because some water repellents can inhibit adhesion of the sealants or affect their cure. With solvent-based coatings, protect adjacent surfaces such as bituminous or rubber products, glass, metal door and window frames, and vegetation. Recommended application temperatures generally fall between 50 and 90°F.

Figure 10.2 RILEM tube used to test surface absorption.

Water repellents are usually applied by low-pressure airless sprayers, but small jobs can be done with a brush or roller. On two-coat applications, follow the manufacturer's instructions for the time lapse between coats. Avoid spraying on windy days to prevent drift. On occupied buildings where solvent-based materials are used, temporarily shut down air intakes to avoid contaminating the indoor air. Better yet, spray after hours or on weekends.

Some manufacturers provide a 3-, 5-, 7-, or even 10-year warranty on various types of clear water repellents. Building owners should be made aware that such coatings require periodic re-application, just as paint coatings require periodic maintenance.

10.3 Surface Preparation

Many paint failures are caused by inadequate or improper surface preparation. No coating will adhere to a surface that is soiled, contaminated, or in poor condition. Each type of substrate and each type of contaminant has cleaning methods that are effective in preparing the surface to receive a protective coating.

10.3.1 Concrete

Concrete must cure for a minimum of 30 days at 75°F, longer at lower temperatures. The pH of the surface should be between 6 and 9 (*Fig. 10.3*). Moisture content should not exceed 15%. Either check the surface with a moisture meter, or tape an 18-in. square of plastic to the surface and allow it to sit for 16 hours. If moisture collects on the underside of the plastic or on the concrete surface, the moisture content is too high. Concrete surface temperature should be at least 55°F before application of any coating and until the coating is cured. Patch all hollows, bug holes, honeycombs, and voids with a cement patching compound, and grind smooth fins, protrusions, and rough edges. Remove all grease, dirt, oil, loose paint, tar, glaze, laitance, efflorescence, loose mortar, cement, hardeners, sealers, form-release agents, curing compounds, and other treatments.

Method A: Blast cleaning (ASTM D4259). Brush blasting or sweep blasting includes dry blasting, high-pressure water blasting, water blasting with abrasives, and vacuum blasting with abrasives. Water blasting at 3500 to 4500 psi is acceptable on sound concrete where abrasive blasting is not permitted. Water blasting will remove loose concrete, mortar, andcontaminants faster and more economically than hand or power tool cleaning, but will not provide a profile or open voids as effectively as abrasive blasting. Follow this procedure:

```
1 ←———— 7 ————→ 14
Acid    Neutral   Alkaline
```
Figure 10.3 pH scale.

1. Use 16- to 30-mesh silica sand and oil-free air.
2. Remove all surface contamination by Method D below.
3. Stand approximately 2 ft from the surface to be blasted, and move nozzle at a uniform rate to remove laitance and open bug holes.
4. Vacuum- or air-blast to remove dust and loose particles.
5. Surface should have a texture similar to medium grit sandpaper.
6. Allow to dry and check moisture content before applying coating.

Method B: Acid etching (ASTM D4260). Acid etching is used for floors and horizontal decks but is impractical and difficult for vertical surfaces. Acid etching is often required on floors to roughen a trowel finish. Follow this procedure:

1. Remove all surface contamination by Method D below.
2. Saturate surface with clean water.
3. Apply a 10 to 15% muriatic acid or 50% phosphoric acid solution at the rate of 1 gal per 75 sq ft and scrub with a stiff brush.
4. Allow sufficient time for scrubbing until bubbling stops. If no bubbling occurs, surface is contaminated with grease, oil, or other substance. Remove the contamination in accordance with Method D below. After contamination is removed, wet the surface and re-apply the acid treatment.
5. Rinse the surface two or three times, removing the acid/water mixture after each rinse.
6. Surface should exhibit a texture similar to that of medium-grit sandpaper.
7. If necessary, repeat the etching process several times until a suitable texture is achieved. Bring the pH of the surface to neutral with a 3% solution of trisodium phosphate or similar alkaline cleaner, and flush with clean water to achieve a clean, sound surface.
8. Allow to dry and check moisture content before applying coating.

Method C: Power tool cleaning or hand tool cleaning (ASTM C4259)

1. Use needle guns or power grinders equipped with a suitable grinding stone of appropriate size and hardness which will remove concrete, loose mortar, fins, projections, and surface contaminants. Hand tools may also be used for small or hard-to-reach areas.
2. Vacuum- or air-blast to remove dust and loose particles.
3. Allow to dry and check moisture content before applying coating.

Method D: Surface cleaning (ASTM D4258)

1. Broom clean, vacuum clean, air-blast clean, water-blast clean, or steam clean as described in ASTM D4258.

2. Concrete curing compounds, form-release agents, and concrete hardeners may not be compatible with recommended coatings. Check for compatibility by applying a 2 to 3 sq ft test patch of the coating. Allow the coating to dry for 1 week and then test for adhesion in accordance with ASTM D3359. If the coating system is incompatible, prepare the concrete surface as outlined in ASTM D4259 for power tool or blast cleaning.

10.3.2 Stucco

Stucco must be clean and free of any loose material. Allow portland cement plaster (stucco) to cure for at least 30 days at 75°F (longer at cooler temperatures). The pH of the surface should be between 6 and 9.

10.3.3 Concrete block

Remove loose mortar and foreign material. Surfaces must be free of dust, dirt, laitance, form-release agents, moisture-curing membranes, loose cement, and other contaminants. Block and mortar must be cured for at least 30 days at 75°F (longer at cooler temperatures). Surface pH should be between 6 and 9.

10.3.4 Brick

Remove dirt, loose mortar, foreign material, and loose paint. Allow to weather for at least 1 year, followed by wire brushing to remove any efflorescence.

10.3.5 Exterior wood

Must be clean, well aged or kiln-dried, and sanded smooth prior to painting. Pressure-treated lumber should be allowed to age a minimum of 6 months. Prime and paint other exterior woods as soon as possible after installation. No painting should be done immediately after a rain or during foggy weather, or when the temperature is below 50°F. Knots and pitch streaks must be scraped, sanded, and spot-primed or sealed before a full priming coat is applied. Patch all nail holes and imperfections with a wood filler or putty and sand smooth.

If a coating with low permeability is specified, all wood should be back-primed on the unexposed side to avoid blistering the coating. Cut ends of boards, plywood edges, and the tops and bottoms of wood doors exposed to moisture should be primed and painted to prevent excessive absorption.

10.3.6 Exterior hardboard

Whether factory-primed or unprimed, hardboard siding must be cleaned thoroughly and primed with an alkyd primer.

10.3.7 Steel

Steel must be cleaned by one of the Steel Structures Painting Council (SSPC) surface preparation methods described below. Most steel cleaning and priming is done in the shop, but the painting contractor should check and approve the primer coatings before beginning application of field coatings. A simple test of adhesion requires placing a piece of tape on the surface and then removing it to see if the primer peels off with it. Dirt from the construction site (and salt deposits in coastal areas) should be washed off the surface with clean water before painting. Where shop primers are damaged or abraded during steel erection at the site, clean and touch up with a compatible primer before painting.

Solvent cleaning: SSPC-SP1. Solvents such as water, mineral spirits, xylol, and toluol are used to remove oil, grease, dirt, soil, drawing compounds, and other contaminants. But this does not remove rust or mill scale. Change rags and cleaning solution frequently so that deposits of oil and grease are not spread over additional areas in the cleaning process. Do not use hydrocarbon solvents to clean grease, oil, or waxes because they will spread rather than remove these substances.

Hand tool cleaning: SSPC-SP2. This is a mechanical method of surface preparation involving wire brushing, scraping, chipping, and sanding used to remove loose rust, mill scale, and other contaminants from the surface *after* solvent cleaning in accordance with method SSPC-SP1. This does not require removal of intact rust or mill scale.

Power tool cleaning: SSPC-SP3. This mechanical method of surface preparation involves power sanders or wire brushes, power chipping hammers, abrasive grinding wheels, and needle guns to remove loose rust, mill scale, and other contaminants from the surface *after* solvent cleaning in accordance with method SSPC-SP1. Does not require removal of intact rust or mill scale.

White metal blast cleaning: SSPC-SP5. This method removes all visible rust, mill scale, paint, and contaminants, leaving the metal uniformly white or gray in appearance. Used where maximum performance of protective coatings is necessary due to exceptionally severe conditions.

Commercial blast cleaning: SSPC-SP6. Oil, grease, dirt, rust scale, and foreign matter are completely removed from the surface and all rust, mill scale, and old paint are completely removed by abrasive blasting except that slight shadows, streaks, or discolorations caused by rust stain, mill scale oxides, or slight, tight residues of paint or coating may remain. If the surface is pitted, slight residues of rust or paint may be found in the bottom of pits. At least two-thirds of each square inch of surface area must be free of all visible residues and the remainder must be limited to the light discoloration, slight staining, or light residues described.

Brush-off blast cleaning: SSPC-SP7. Oil, grease, dirt, rust scale, loose mill scale, loose rust, and loose paint or coatings are removed completely. Tight mill scale and tightly adhered rust, paint, and coatings are permitted to remain, but all mill scale and rust must have been exposed to the abrasive blast pattern sufficiently to expose numerous flecks of the underlying metal fairly uniformly distributed over the entire surface.

Near white blast cleaning: SSPC-SP10. Oil, grease, dirt, mill scale, rust, corrosion products, oxides, paint, and other foreign matter are completely removed from the surface by abrasive blasting, except for very light shadows, very slight streaks, or slight discolorations caused by rust stain, mill scale oxides, or slight, tight residues of paint or coatings. At least 95% of each square inch of surface area must be free of all visible residues and the remainder is limited to the light discoloration described.

10.3.8 Galvanized steel

Allow to weather 6 months prior to coating. When weathering is not possible, prepare the surface in accordance with SSPC-SP1 solvent cleaning as described above.

10.3.9 Aluminum

Remove oil, grease, dirt, oxides and other foreign material by SSPC-SP1 solvent cleaning, or scrubbing with brushes. Adhesion to aluminum can be difficult, so it is wise to prepare a small test area before general coating application. If adhesion is not good, mechanically abrade the surface of the aluminum by sanding or wire brushing.

10.4 Common Problems

The primary cause (60 to 80%) of coating failures is improper surface preparation. The second most common cause is improper application. A distant third is improper material specification, and fourth is product failure. Coating performance and service life is generally proportional to the degree of surface preparation. With more specialized coatings, application is even more critical and less forgiving.

10.4.1 Alligatoring

Alligatoring is a pattern of surface cracking caused by incompatibility of top coat and primer, or coating over an incompletely cured primer.

10.4.2 Blistering

Blisters are caused by heat and moisture. They can occur when paint is applied at high temperatures, allowing the vehicle to evaporate before the binder can adhere. Blisters can also occur when moisture trapped in a wall follows a heat source (the sun) trying to escape by evaporation. Blistering is closely associated with the permeability of a coating and its application to surfaces that are too wet. Coatings with poor wet adhesion and low permeability (oil paints) blister badly when ambient conditions cause condensation at the coating-substrate interface. Those with high permeability but poor wet adhesion (latex paints) tend to resist blistering in all but severe conditions because they allow moisture to escape through the surface film. The opposite combination of good wet adhesion and low permeability (alkyd paints) usually provides good blister resistance also (*Table 10.9*). Blistering and peeling can also occur when green or wet wood is coated with a primer or sealer that has a low permeance to water vapor.

10.4.3 Checking, cracking, and flaking

These problems occur when paint is applied to an existing coating that has lost its elasticity and is unable to expand and contract with climatic changes. Cracking and checking can also be caused by shrinkage or embrittlement of the coating, application of a hard film over a flexible film, ultraviolet light sensitivity, excessive dry film thickness of zinc-rich primers, surface freezing of fresh latex coatings, or application of an excessive number of coats. Although time-consuming, removal of the existing paint is the best corrective measure.

10.4.4 Delamination or peeling

Delamination of coatings can be caused by poor surface preparation, failure to remove chalk, contamination between coats, incompatible coatings, failure to comply with recoat time limits, thermal shock, cathodic currents, moisture transmission, or improper application.

10.4.5 Flatting

Flatting is a loss of gloss caused by rain, fog, high humidity, or damp surfaces. Recoating the surface will correct the problem. Overthinning or the wrong solvent thinner can also cause flatting. When this is the cause, the coating may require removal before recoating if film properties or adhesion are affected.

TABLE 10.9 Permeability and Resistance to Blistering

Coating type	Wet adhesion	Permeability	Blister resistance
Oil paints	Poor	Low	Poor
Alkyds	Good	Low	Good
Latex paints	Poor	High	Good

SOURCE: Adapted from H. E. Ashton, *Exterior Coatings for Wood*, Canadian Building Digest No. 91, Division of Building Research, National Research Council of Canada, Ottawa, 1967.

10.4.6 Oxidation, chalking, and fading

Oxidation of solvent-based coatings is caused by insufficient film thickness, improper priming, and pigment breakdown under UV exposure. Organic colors such as red, blue, and yellow are generally not as lightfast as earth colors.

10.4.7 Pinholing

Pinholes are small (even microscopic) discontinuities in the coating film. Pinholing can be caused by solvent migration through the film after it has begun to set. If pinholes are only occasional, touch up or recoat areas as necessary. If pinholing is extensive, it indicates that the coating was applied and cured under adverse environmental conditions. Apply a thin mist coat to fill surface voids, followed by a full wet coat when conditions are suitable for recoating.

10.4.8 Premature rusting

Premature rusting of steel may be caused by insufficient dry film thickness; contamination of the substrate with weld splatter, dirt, or other construction site materials; or poor application with skips and holidays.

10.4.9 Wrinkling

Wrinkling can be caused by insufficient cure of the previous coat, slow drying conditions, or by excessive film thickness. The surface can usually be sanded smooth and then recoated.

10.5 Maintenance Painting

Maintenance painting is required to correct defects and failures in existing coatings and surfaces, and also as a matter of renewal when the expected service life of a coating is ended. With each successive coating application, vapor permeability decreases, so precautions should be taken in selecting maintenance coatings for maximum breathability. Depending on the condition of the existing surface, different cleaning techniques and surface preparations are required.

10.5.1 Concrete, stucco, and masonry surfaces

Fill hairline cracks with a cement patching compound flush with the surface. Clean out and fill larger structural cracks with a patching compound or sealant. If large cracks exhibit movement, grind them out and install a backer rod and joint sealant.

Remove loose, blistering, and peeling paint with hand or power tools. If 25% or more of the surface area is affected, completely remove the coating by water blasting. If moisture problems are the cause of blistering or peeling, determine the source of the moisture and correct the problem before recoating the surface.

The same is true for efflorescence. If there is a continuing source of moisture which causes recurring efflorescence, the problem must be corrected before proceeding. Remove powdery efflorescence by scrubbing with a brush and water. Remove crusty efflorescence with specially formulated commercial cleaners. Remove mildew with a solution of one-third household bleach and two-thirds water. Do not add detergents or ammonia to the bleach water solution. Apply the solution and scrub the mildewed area. Allow the solution to remain on the surface for 10 minutes. Rinse thoroughly with clean water and allow the surface to dry for 48 hours before recoating. A fungicide additive can be mixed with the new coating, but mildew may recur if environmental conditions are favorable to its growth.

10.5.2 Painted wood and metal surfaces

Remove dirt and dust by washing and/or dusting with a stiff bristle brush. Remove oil and grease with mineral spirits, followed by washing with trisodium phosphate and water and a clean water rinse. Remove rust by mechanical abrasion with hand or power tools. Remove chalking paint by hand or power tools to a sound surface. Use abrasive or water blasting if necessary. Remove loose paint by scraping, sanding, wire brushing, or other abrasion methods. Spot prime bare wood or metal areas with an appropriate primer. Dull glossy surfaces by scuff sanding. Fill nail holes and cracks in wood with wood putty. Fill cracks between adjoining surfaces with wood putty or a paintable acrylic sealant.

Remove mildew with a solution of one-third household bleach and two-thirds water. Do not add detergents or ammonia to the bleach-water solution. Apply the solution and scrub the mildewed area. Allow the solution to remain on the surface for 10 minutes. Rinse thoroughly with clean water and allow the surface to dry before recoating. A fungicide additive can be mixed with the new coating, but mildew may recur if environmental conditions are favorable to its growth.

10.5.3 Stained wood surfaces

Brown stains on redwood and cedar surfaces are caused by moisture in the wood which leaches water-soluble tannic acid or resins. The source of moisture as well as the stains must be removed before recoating. Clean the surface with a solution of equal parts denatured alcohol and water. Results will vary with the severity of the condition. Repeat the cleaning process if necessary.

Appendix A

Glossary

A

absorptance The ratio of the radiant flux absorbed by a body to that incident upon it (ASTM C168*).

absorption, thermal Transformation of radiant energy to a different form of energy by interaction with matter (ASTM C168).

absorption As applied to building materials and insulations, the accumulation of water in a material or in its cells or fibers, accompanied by a physical or chemical change such as softening of the fibers, dissolving of a binding agent, or the swelling of wood. (Also see **adsorption**)

Absorptivity The ability of a material, independent of its geometry or surface condition, to absorb radiant energy.

acid rain Rain having a pH of less than 5.65 (ASTM E631).

acrylic A group of thermoplastic resins or polymers formed by polymerizing the esters of acrylic acid.

acrylic emulsion Clear, water-based repellents which form a film.

acrylic latex Latex-modified acrylic sealant.

acrylic solutions Clear, solvent-based repellents which form a film.

adhesion The clinging or sticking together of two surfaces; the state in which two surfaces are held together by forces at the interface; the tendency of a material to bond to another substance or material when under a separating stress.

adhesion in peel Force required to peel a sealant, caulk, adhesive, or coating from the surface; usually expressed in pounds per inch.

---

*ASTM designations indicate definitions taken from ASTM standard terminology documents. See Appendix B for document titles.

adhesive failure Failure of the bond between the sealant, adhesive, or coating and the substrate surface (ASTM C717).

adsorption As applied to building materials and insulations, the accumulation of water in or on the surface of a material or on its fibers or cell walls, without any chemical or physical change.

aged insulation value Thermal resistance (R value) of a thermal insulation material as determined after standard conditioning to simulate service exposure (ASTM E631).

aggregate (1) Any granular mineral material. (2) Crushed stone, crushed slag, or water-worn gravel used for surfacing a built-up roof. (ASTM D1079.)

air barrier A system or network of materials that prevents air movement through a building enclosure. (See also **vapor retarder.**)

air leakage The passage of uncontrolled air through cracks or openings in the building envelope or its components, such as ducts, because of air pressure or temperature difference (ASTM E631).

albedo Deprecated term (see **reflectance**).

alligatoring The cracking of the surfacing bitumen on a built-up roof, producing a pattern of cracks similar to an alligator's hide; the cracks may not extend through the surfacing bitumen (ASTM D1079).

aluminized Armco Steel trade name for a commercially pure aluminum coating for steel sheet that is applied by continuous hot dipping.

ambient temperature Temperature of the surrounding air.

anode The positive terminal of an electrolytic cell.

anhydrous Being without water, especially water of hydration.

application life See **POT life.**

application rate The quantity (mass, volume, or thickness) of material applied per unit area (ASTM D1079).

architectural metal roofing Metal roofing panels and flashings, whether site-formed, shop-formed, or preformed, with the purpose of aesthetic enhancement of a building.

asphalt A dark brown to black cementitious material in which the predominating constituents are bitumens which occur in nature or are obtained in petroleum processing (ASTM D1079).

asphalt felt An asphalt-saturated felt (ASTM D1079).

asphalt mastic A mixture of asphaltic material and graded mineral aggregate that can be poured when heated, but requires mechanical manipulation to apply (ASTM D1079).

B

backnailing The practice of blind nailing roofing felts to a substrate in addition to hot mopping to prevent slippage (ASTM D1079).

backer rod See **sealant backing.**

bandage joint   See **joint, bridge.**

base ply   The bottom or first ply in a built-up roofing membrane when additional plies are to be subsequently installed (ASTM D1079).

base sheet   A product intended to be used as a base ply in a built-up roofing system (ASTM D1079).

batten seam   Longitudinal metal roof panel seam consisting of upstands that are spaced some distance apart and supported by a structural batten of wood or some other material, and then covered by a cap seamed to the upstands. Modified designs for use with harder metals like steel and aluminum may be referred to as "applied cap" or "integrated battens."

bitumen   (1) A class of amorphous, black or dark-colored (solid, semisolid, or viscous), cementitious substances, natural or manufactured, composed principally of high-molecular-weight hydrocarbons, soluble in carbon disulfide, and found in asphalts, tars, pitches, and asphaltites. (2) A generic term used to denote any material composed principally of bitumen. (ASTM D1079.)

bituminous   Containing or treated with bitumen (ASTM D1079).

bituminous emulsion   (1) A suspension of minute globules of bituminous material in water or in an aqueous solution. (2) A suspension of minute globules of water or of an aqueous solution in a liquid bituminous material (invert emulsion). (ASTM D1079)

blackbody   The ideal perfect emitter and absorber of thermal radiation. It emits radiant energy at each wavelength at the maximum rate possible as a consequence of its temperature, and absorbs all incident radiance (ASTM C168).

blind nailing   The use of nails that are not exposed to the weather in the finished roofing (ASTM D1079).

blister   (1) A raised portion of a roofing membrane resulting from local internal pressure. (2) The similarly formed protuberances in coated prepared roofing. (ASTM D1079)

blocking   (1) Wood built into a roofing system above the deck and below the membrane and flashing to (*a*) stiffen the deck around an opening, (*b*) act as a stop for insulation, or (*c*) serve as a nailer for attachment of the membrane or flashing. (2) Wood cross-members installed between rafters or joists to provide support at cross-joints between [roof] deck panels. (3) Cohesion or adhesion between similar or dissimilar [roofing] materials in roll or sheet form that may interfere with the satisfactory and efficient use of the material. (ASTM D1079)

bond   The adhesive and cohesive forces holding two components in intimate contact.

bond breaker   Material to prevent adhesion at designated interface (ASTM C717).

breather finish   Coating system allowing the passage of water vapor. *Note:* A breather finish has water-vapor permeance greater than that acceptable for a water-vapor retarder (ASTM E631).

brooming   Embedding a ply [of roofing felt] by using a broom to smooth it out and ensure contact with the adhesive under the ply (ASTM D1079).

Btu   British thermal unit. The quantity of heat required to raise the temperature of one pound of water one degree Fahrenheit.

building envelope The outer elements of a building, both above and below ground, that divide the external from the internal environments (ASTM E631).

building fabric Elements, components, parts, materials, or systems of a building separately or in combination (ASTM E631).

built-up roofing (BUR) A continuous, semiflexible membrane consisting of plies of saturated felts, coated felts, fabrics, or mats assembled in place with alternate layers of bitumen, and surfaced with mineral aggregate, bituminous materials, or a granule-surfaced sheet (ASTM D1079).

butt joint See **joint, butt.**

butyl rubber A copolymer of isobutylene with a small amount of isoprene, variously manufactured as sheet goods, blended with other rubbers, and used to make sealants.

C

cant strip A beveled strip used under flashings to modify the angle at the point where the roofing or waterproofing membrane meets any vertical element (ASTM D1079).

cap sheet A granule-surfaced coated felt used as the top ply of a built-up roofing membrane (ASTM D1079).

capillarity A wick-like action whereby a liquid will migrate through a porous material because of surface tension.

capillary migration (of water) Movement of water induced by the force of molecular attraction (surface tension) between the water and the material it contacts (ASTM E631). See also **rising damp.**

carbonation A process of chemical weathering whereby minerals that contain sodium oxide, calcium oxide, potassium oxide, or other basic oxides are changed to carbonates by the action of carbonic acid derived from atmospheric carbon dioxide and water (ASTM E631).

caulk (noun) See **sealant** (ASTM C717). *Note:* The term *caulk* has traditionally referred to non-elastomeric sealant compounds used where little or no movement capability is required (usually less than 10%); caulking compounds are a type of sealant.

caulk (verb) To install or apply a sealant across or into a joint, crack, or crevice (ASTM C717). To fill joints, cracks, or crevices in order to prevent the passage of air or water (ASTM E631).

chalking Formation of a powdery surface condition caused by disintegration of a surface binder or elastomer due to weathering or other such destructive process.

change of state The process whereby liquid is heated to the point of evaporation, changing the liquid into a gas; the condensation of a gas on a cooler surface returning it from gaseous to liquid or solid form; the cooling of a liquid or gas below its freezing point, changing it to a solid.

chimney effect See **stack effect.**

cladding system Material assembly applied to a building as a non-loadbearing [exterior] wall, or attached to [an exterior] wall surface as a protective and ornamental covering (ASTM E631).

class ASTM C920 sealant classification of dynamic movement capability; a Class 25 sealant must have the ability to absorb at least ±25% joint movement.

climate The characteristic weather conditions of a specific place, locale, or region, averaged over an extended period of time.

closed cell A cell totally enclosed by its walls and hence not interconnecting with other cells (ASTM C717).

closed-cell material A cellular material in which substantially all cells in the mass are closed cells (ASTM C717).

coal tar A dark brown to black, solid cementitious material obtained as residue produced by the destructive distillation of coal (ASTM D1079).

coal-tar felt A felt that has been saturated with refined coal tar (ASTM D1079).

coal-tar pitch A dark brown to black, solid cementitious material obtained as residue in the partial evaporation or distillation of coal tar (ASTM D1079).

coated sheet (or felt) (1) An asphalt felt that has been coated on both sides with harder, more viscous asphalt. (2) A glass fiber felt that has been simultaneously impregnated and coated with asphalt on both sides. (ASTM D1079.)

coating A liquid, liquefiable, or mastic composition that, after application in a thin layer, is converted to a solid protective, or decorative, or functional adherent film (ASTM E631).

cohesion The molecular attraction that holds the body of a sealant or adhesive together; the tendency of a material to maintain its integrity without separating or rupturing within itself when subjected to external forces.

cohesive failure Failure characterized by rupture within the sealant, adhesive, or coating (ASTM C717).

coil coating Automated process of applying films and laminates to steel and aluminum sheet in which a long coil of metal is fed through a line that cleans, degreases, preheats, primes, cures, finish-coats, cures, and rewinds the coil, all in a non-stop operation. Texture effects are often added during this process.

cold-applied Capable of being applied without heating as contrasted to hot-applied. *Note:* Cold-applied products are furnished in a liquid state, whereas hot-applied products are furnished as solids that must be heated to liquefy them. (ASTM C1127).

cold joint See **joint, cold.**

cold-process roofing A continuous, semiflexible membrane consisting of plies of felts, mats, or fabrics laminated on a roof with alternate layers of roof cement and surfaced with a cold-applied coating (ASTM D1079).

compatibility (1) The capability of two or more materials to be placed in contact or close proximity with one another, and each material maintaining its usual physical or chemical properties, or both (ASTM C717). The ability of two or more substances to exist in harmony when mixed together or when brought into intimate contact without any adverse physical or chemical reaction (ASTM C981).

compatible materials Compounds or substances that can exist in close proximity to one another without detrimental effects on either (ASTM C717).

component An individually distinguishable product that forms part of a more complex product (i.e., subsystem of a system).

compound An intimate mixture of all the ingredients necessary for a finished material or product (ASTM C717).

compression gasket A gasket designed to be used under compression (ASTM C717).

compression seal A seal which is attained by a compressive force on the sealing material (ASTM C717).

compression set The amount of permanent set that remains in a specimen after removal of a compressive load; mode of failure in sealant that is stressed in compression before it cures; the change occurring in a sealant when deformed that prevents full recovery.

condensation The process of changing water vapor to liquid water or ice by taking away heat; the opposite of evaporation (which requires the addition of heat).

conductance, surface The time rate at which heat is exchanged between the surface of a material and the surrounding air.

conductance, thermal See **thermal conductance.**

conduction, thermal See **thermal conduction.**

conductivity, thermal See **thermal conductivity.**

construction joint See **joint, construction.**

contaminants Foreign material such as dust, dirt, oils, or rust.

continuous hot dipping Automated hot-dip process for coating steel coil in which the steel is cleaned, degreased, pickled, coated, cooled, and rewound in a non-stop operation. Zinc, aluminum, and aluminum-zinc alloy coatings are all applied by this method.

control joint See **joint, control.**

convection The transfer of heat by the circulation of fluid (liquid or gas).

convection, natural Heat transfer of a fluid such as air or water that results from the natural rising of the lighter, warmer fluid and the sinking of the heavier, cooler fluid.

coping A covering on top of a wall exposed to the weather, usually sloped to carry off water (ASTM D1079).

copolymer A combination of two monomers.

corrosion The electrochemical degradation of metals due to reaction with their environment.

counterflashing Formed metal or elastomeric sheeting secured on or into a wall, curb, pipe, rooftop unit, or other surface, to cover and protect the upper edge of a base flashing and its associated fasteners (ASTM D1079).

crack A flaw (building defect) consisting of complete or incomplete separation within a single element or between contiguous elements of constructions. *Note:* Occasionally the basic design, or the material characteristics, of a building element will be such that minor cracking may occur. Such cracks are not flaws or defects. (ASTM E631)

creep (1) The dimensional change with time of a material under load, following the initial instantaneous elastic or rapid deformation (ASTM C981). (2) The time-dependent part of a strain resulting from stress (ASTM C717).

cricket A relatively small, elevated area of a roof constructed to divert water from a horizontal intersection of the roof with a chimney, wall, expansion joint, or other projection (ASTM D1079).

crushed stone The product resulting from the artificial crushing of rocks, boulders, or large cobblestones, substantially all faces of which have resulted from the crushing operation (ASTM D1079).

cure (noun) The process by which a compound attains its intended properties through evaporation, chemical reaction, heat, radiation, or combinations thereof (ASTM C717).

cure (verb) To attain the intended performance properties of a compound by means of evaporation, chemical reaction, heat, radiation, or combinations thereof (ASTM C717).

cured Pertaining to the state of a compound that has attained its intended performance properties by means of evaporation, chemical reaction, heat, radiation, or combinations thereof (ASTM C717).

curing Chemical process of developing ultimate properties of a finish or other material over a specified period of time (ASTM E631). (See also **drying**.)

curing method Method by which a compound cures, i.e., solvent evaporation, chemical reaction, heat, or combinations thereof.

curing time Time in which a compound attains its intended properties; time required to produce vulcanization at a given temperature.

cutback Solvent-thinned bitumen used in cold-process roofing adhesives, flashing cements, and roof coatings (ASTM D1079).

cutoff A detail designed to prevent lateral water movement into the insulation where the [roofing] membrane terminates at the end of a day's work, or used to isolate sections of the roofing system. It is usually removed before the continuation of the work. (ASTM D1079.)

D

dampproofing Treatment of a surface or structure to resist the passage of water in the absence of hydrostatic pressure (ASTM D1079).

dead level Absolutely horizontal, or zero slope (ASTM D1079).

dead-level asphalt A roofing asphalt conforming to the requirements of ASTM D312, Type I (ASTM D1079).

dead-level roofing   A roofing system applied on a surface with a 0 to 2% incline (ASTM D1079).

deck   The structural surface to which the roofing or waterproofing system (including insulation) is applied (ASTM D1079).

degradation   Deterioration, usually in the sense of a physical or chemical process rather than a mechanical one.

degree days (DD)   A temperature-time unit used in estimating building heating requirements. For any given day, the number of DD equals the difference between the reference temperature (usually 65°F) and the mean temperature of the outdoor air for that day. DD per month or per year are the sum of the daily DD for that period.

delta T (ΔT)   A difference in temperature (T).

density   The mass of a substance, expressed in pounds per cubic foot (lb/cu ft).

design temperature   A designated temperature close to the most severe winter or summer temperature extreme of an area, used in estimating heating and cooling demand.

desorption   The separation of an adsorbate as such from a sorbent. Opposite of adsorption, absorption, or both.

dew point   (1) The temperature corresponding to 100% relative humidity for an air-vapor mixture at constant pressure. (2) The temperature at which condensation of water vapor begins in a given space for a given humidity as the temperature of the vapor is reduced.

diffuse radiation   Sunlight that is scattered by air molecules, dust, water vapor, and translucent materials.

direct radiation   Light that has traveled a straight path from the sun, as opposed to diffuse radiation.

drainage, positive   The drainage condition in which consideration has been made for all loading deflections of the deck, and additional roof slope has been provided to ensure complete drainage of the roof area within 24 hours of rainfall precipitation (NRCA).

drainage course   See **percolation layer.**

drainage hole   An opening in a construction provided for the escape of unwanted liquid, as in a retaining wall (ASTM E631). See also **weephole.**

dry bulb temperature   A measure of the sensible temperature of the air.

dry-film thickness   Thickness of cured film, coating, or membrane (ASTM C717).

drying   Process of developing, solely by evaporation of volatile ingredients, ultimate properties of a finish or other material over a specified period of time (ASTM E631). (See also **curing.**)

drying time   The time required for solvent dissipation.

ductility   The ability of a material to be plastically deformed by elongation without fracture.

durability The capacity of a material, product, component, assembly, or construction to remain serviceable as intended with prudent maintenance during the designed service life under anticipated internal and external environments (ASTM E241).

durometer Instrument used to measure hardness on the Shore A or Shore D scale; may also refer to the relative hardness rather than the instrument itself.

E

edge stripping Application of felt strips cut to narrower widths than the normal felt-roll width to cover a joint between flashing and built-up roofing (ASTM D1079).

edge venting The practice of providing regularly spaced protected openings at a roof perimeter to relieve water vapor pressure in the insulation (ASTM D1079).

EIFS See **exterior insulation and finish system.**

elasticity The ability of a material to return to its original shape after deformation such as stretching, compression, or torsion.

elastomer A macromolecular material that returns rapidly to approximately the initial dimensions and shape after substantial deformation by a weak stress and release of the stress (ASTM C717).

elastomeric Having the characteristics of an elastomer (ASTM C717).

elongation Extension produced by a tensile stress (ASTM C717).

electrolyte A substance that is capable of forming a conducting liquid medium.

embedment (1) The process of pressing a felt, aggregate, fabric, mat, or panel uniformly and completely into hot bitumen or adhesive to ensure intimate contact at all points. (2) The process of pressing granules into coating in the manufacture of factory-prepared roofing, such as shingles. (ASTM D1079.)

emittance, effective (E) The combined effect of emittances (ϵ) from two parallel boundary surfaces of a dimension much larger than the distance between them.

emittance (ϵ) The ratio of the radiant flux emitted by a specimen to that emitted by a blackbody at the same temperature and under the same conditions (ASTM C168). Emittance values range from 0.05 for brightly polished metals to 0.96 for flat black paint.

emulsion An intimate mixture of bitumen and water, with uniform dispersion of the bitumen or water globules, usually stabilized by an emulsifying agent or system (ASTM D1079).

envelope A continuous membrane edge seal formed at the [roof] perimeter and at penetrations by folding the base sheet or ply over the plies above and securing it to the top of the membrane. The envelope prevents bitumen seepage from the edge of the membrane. (ASTM D1079.)

EPS Expanded polystyrene (ASTM E631). See also **cellular polystyrene insulation** and **rigid cellular polystyrene (RCP) insulation** under **insulation, thermal.**

equinox Either of the two times during the year when the sun crosses the celestial equator and when the length of day and night are approximately equal. These are the autumnal equinox on or about September 22 and the vernal equinox on or about March 22.

equilibrium moisture content (1) The moisture content of a material stabilized at a given temperature and relative humidity, expressed as a percent of moisture by weight. (2) The typical moisture content of a material in any given geographical area. (ASTM D1079.)

equiviscous temperature (EVT) The temperature at which the viscosity of an asphalt is 125 cSt; the recommended asphalt temperature ±25°F at the time of application to the substrate (ASTM D1079).

evaporation The process of changing liquid water to water vapor by adding heat; the opposite of condensation (which requires taking away heat).

exfiltration The uncontrolled outward air leakage through cracks and interstices in any building element and around windows and doors of a building, caused by the pressure effects of differences in the indoor and outdoor air density from temperature, stack effect, or ventilation pressures.

exfoliation Peeling, swelling, or scaling of stone or mineral surfaces in thin layers caused by chemicals, physical weathering, or heat.

expansion joint See **joint, expansion.**

exterior insulation and finish system (EIFS) Non-loadbearing outdoor wall finish system consisting of a thermal insulation board, an attachment system, a reinforced base coat, exterior joint sealant, and a compatible finish (ASTM E631).

extreme climate For evaluating the desired properties of vapor retarders for a specific application, extreme climate is considered to require heating above 5000 Fahrenheit degree-days, and extreme moisture exposure is high temperature with high relative humidity for which continual air conditioning is recommended (ASTM E241).

F

fabric A woven cloth of organic or inorganic filaments, threads or yarns, used in roofing and waterproofing.

face fastening Method of fastening metal roofing or flashing using gasketed screws that pierce the panels to attach them to each other or to the building structure or deck. Other terms used to describe this fastening method include through-fastening, screw-down, nested lap, and exposed fastening.

felt A flexible sheet manufactured by the interlocking of fibers with a binder or through a combination of mechanical work, moisture, and heat. Felts are manufactured principally from vegetable fibers (organic felts), asbestos fibers (asbestos felts), or glass fibers (glass fiber felts); other fibers may be present in each type (ASTM D1079).

fiberglass felt See **glass felt.**

fillet joint   See **joint, fillet.**

film forming   Treatment that fills pores, forming a continuous film on the surface.

film laminate   Dry, precured film (usually colored) heat-applied to metal panels during the coil coating process.

finger blisters   Finger-shaped blisters or wrinkles in the plies of a built-up roofing or waterproofing membrane (ASTM D1079).

finish   (1) The final treatment or coating of a surface; (2) the fine or decorative work required to make a building or its parts complete (ASTM E631).

fishmouth   (1) A half-cylindrical or half-conical opening formed by an edge wrinkle or failure to embed a roofing felt. (2) In shingles, a half-conical opening formed at a cut edge. (ASTM D1079.)

flashing   (1) A generic term describing the transitional area between the waterproofing membrane and surfaces above the wearing surface. (2) A terminal closure or barrier to prevent ingress of water into the system (ASTM C898, C981, and C1127). (3) The system used to seal membrane edges at walls, expansion joints, drains, gravel stops, and other places where the membrane is interrupted or terminated. Base flashing covers the edges of the membrane. Cap flashing or counterflashing shields the upper edges of the base flashing. (ASTM D1079.)

flashing cement   A trowelable mixture of cutback bitumen and mineral stabilizers including asbestos or other inorganic fibers (ASTM D1079).

flat asphalt   A roofing asphalt conforming to the requirements of ASTM D312, Type II (ASTM D1079).

flat seam   Method of joinery normally associated with handcrafted metals in which adjacent metal sections are folded so that they hook together and are oriented in a horizontal direction flush with the surface of the metal sheet. Concealed attachment cleats are used within the seams.

flexible   Indicates pliability as a permanent condition.

floated finish   A concrete finish provided by consolidating and leveling the concrete with only a power-driven or hand float, or both. *Note:* A floated finish is more coarse than a troweled finish (ASTM C1127.).

flood coat   The top layer of bitumen used to hold the aggregate on an aggregate-surfaced, built-up roofing membrane (ASTM D1079).

flow-through principle   An expression used to describe the performance of a construction such as a wall, in which water vapor that enters one side meets little impedance so that it may flow completely through the construction and out of the opposite side. No seriously deteriorating accumulation of moisture is expected if there are no conditions that cause condensation within the construction (ASTM E241).

fluid-applied elastomer   An elastomeric material, fluid at ambient temperature, that dries or cures after application to form a continuous membrane. Such systems normally do not incorporate reinforcement. (ASTM D1079.)

fluorocarbon   Broad chemical group, including fluoropolymers, that is sometimes used to specify premium paint resins.

fluoropolymer Generic chemical term referring to a polymer containing fluorinated atoms.

freeze-thaw cycle The freezing and subsequent thawing of a material.

functional metal roofing Term that usually denotes a metal roofing system that is both hydrostatic and structural, and normally installed at very low slopes.

G

Galvan Trade name for a 5% aluminum–zinc alloy coating for steel sheet applied by continuous hot dipping.

Galvalume Trade name for a 55% aluminum–zinc alloy coating for steel sheet applied by continuous hot dipping.

galvanizing Applying a coating of zinc to steel by either hot-dipping or electrodeposition.

gasket Any preformed, deformable device designed to be placed between two adjoining parts to provide a seal (ASTM C717).

gauge A number designating a specific thickness of metal sheet, or diameter of wire, cable, or fastener shank tabulated in a standardized series, each of which represents a decimal fraction of an inch (or millimeter) (ASTM E631).

glass felt Glass fibers bonded into a sheet with resin and suitable for impregnation in the manufacture of bituminous waterproofing, roofing membranes, and shingles (ASTM D1079).

glass mat A thin mat of glass fibers with or without a binder (ASTM D1079).

glaze coat (1) The top layer of asphalt in a smooth-surfaced built-up roof assembly. (2) A thin protective coating of bitumen applied to the lower plies or top ply of a built-up membrane, when application of additional felts, or the flood coat and aggregate surfacing, is delayed. (ASTM D1079.)

glazing Installation of glass or other materials in prepared openings (ASTM C717).

gloss The luster, shininess, or image reflection of a surface, directly related to the smoothness of that surface.

grade (1) Level or elevation of a land or water surface (ASTM E631). (2) ASTM C920 classification for consistency of sealant, i.e., pourable (Grade P), or non-sag (Grade NS). (3) ASTM C216 classification for clay face brick, e.g., moderate weathering (Grade MW) or severe weathering (Grade SW).

gravel Coarse, granular aggregate, with pieces larger than sand grains, resulting from the natural erosion of rock (ASTM D1079).

gravel stop A flanged device, frequently metallic, designed to prevent loose aggregate from washing off the roof and to provide a continuous finished edge for the roofing (ASTM D1079).

grout, concrete Concrete containing no coarse aggregates; a thin mortar (ASTM C981).

grout, masonry A mixture of cementitious materials, aggregates, and water, with or without admixtures, used to fill voids in masonry; initially mixed to a consistency suitable for pouring or pumping without segregation of constituents.

gun consistency Compound formulated in a degree of softness suitable for application through the nozzle of a caulking gun; also referred to as *gunnable* or *gun grade*.

H

hardness The resistance of a material to indentation as measured under specified conditions (ASTM C717).

headlap The minimum distance, measured at 90° to the eave along the face of a shingle or felt as applied to a roof, from the upper edge of the shingle or felt, to the nearest exposed surface (ASTM D1079).

heat The form of energy that is transferred by virtue of a temperature difference.

heat capacity The property of a material defined as the quantity of heat needed to raise one cubic foot of the material one degree Fahrenheit. Numerically, the density multiplied by the specific heat.

heat sink A substance that is capable of accepting and storing heat, and therefore can also act as a heat source.

heat transmission The quantity of heat flowing through unit area due to all modes of heat transfer induced by the prevailing conditions (ASTM C168).

holiday An area where a liquid-applied material is missing (ASTM D1079).

hot-applied sealant A compound that is applied in a molten state and cures primarily by cooling to ambient temperatures. *Note:* A hot-applied sealant is sometimes called a *hot-melt sealant*. (ASTM C717.)

hot-dip In steel mill practice, a process whereby ferrous alloy base metals are dipped into molten metal (usually zinc, aluminum, tin, or terne) for the purpose of fixing a rust-resistant coating.

humidity, absolute The weight of water vapor in a given volume of air in pounds per cubic foot.

humidity ratio The ratio of the mass of water vapor to the mass of dry air in a given air-vapor mixture.

humidity, relative (RH) (1) The ratio of the partial pressure of the water vapor in a given air-vapor mixture to the saturation pressure of water vapor at the same temperature. (2) The amount of water in vapor form in a given volume of air at a fixed temperature, as a percentage of the amount of water vapor the same air could hold if saturated. Since warm air can hold more water vapor at saturation than cooler air, temperature is a critical factor.

hydrogenesis Another term for condensation. The term is especially applied to base and soil substrates under highway pavements, where the barometric pump causes the inhalation of humid air, which then condenses in those structures, causing an ever-increasing moisture content and sometimes instability.

hydrokinetic Joinery of metal roofing panels or flashings in a way that enables them to perform only if natural, gravitational water shedding is possible on steep slope roofs—not submersion in water. (See also **hydrostatic**.)

hydrologic cycle The hydrologic cycle consists of the evaporation of water from oceans and other bodies of open water; condensation to produce cloud formations; precipitation of rain, snow, sleet or hail upon land surfaces; and dissipation of rain or melted solids by direct runoff into lakes and by seepage into the soil, thereby producing a continuing endless source of water in the subgrade.

hydrolysis Decomposition or alteration of a chemical substance by water.

hydrophilic Having an affinity for, attracting, adsorbing, or absorbing water.

hydrophobic Lacking affinity for, repelling, or failing to adsorb or absorb water.

hydrostatic Joinery of metal roofing panels or flashing in a way that renders them able to withstand the hydrostatic pressure of submersion under ponding water on low-slope roofs.

hydrostatic pressure A state of stress in which all the principal stresses are equal (and there is no shear stress), as in a liquid at rest; the product of the unit weight of the liquid and the difference in elevation between the given point and the free liquid elevation (ASTM C717).

hygroscopic (1) Attracting, absorbing, and retaining atmospheric moisture (ASTM D1079). (2) Pertaining to water absorbed by hydrophilic porous materials.

hysteresis Irreversible loss of strength, usually manifested as visible deformation of thin (less than 2-in) marble cladding panels due to anisotropic thermal expansion.

I

ice dam A mass of ice formed at the transition from a warm to a cold roof surface. Frequently formed by refreezing meltwater at the overhang of a steep roof, an ice dam may cause ice and water to back up under shingles or other roofing materials (ASTM D1079).

impermeable Having a permeance of zero.

incident angle The angle between the incident ray from the sun and a line drawn perpendicular to the receiving surface.

incline The slope of a roof expressed in percent or in the number of vertical units of rise per horizontal unit of run (ASTM D1079).

infiltration The uncontrolled inward air leakage through cracks and interstices in any building element and around windows and doors of a building, caused by the pressure effects of differences in the indoor and outdoor air density from temperature, stack effect, or ventilation pressures.

inorganic Being or composed of matter other than hydrocarbons and their derivatives, or matter that is not of plant or animal origin (ASTM D1079).

insolation The total amount of solar radiation incident upon an exposed surface measured in Btu per hour per square foot (Btu/h·ft^2) (langleys).

installation temperature Temperature at time of installation.

insulating concrete A lightweight concrete made with lightweight coarse aggregate and having relatively low insulating characteristics (ASTM C981).

insulation, thermal A material or assembly of materials used to provide resistance to heat flow (ASTM C168).

  beadboard Molded expanded polystyrene thermal insulation board; also called MEPS (ASTM E631).

  blanket insulation A relatively flat and flexible insulation in coherent sheet form furnished in units of substantial area (ASTM C168).

  block insulation Rigid insulation preformed into rectangular units (ASTM C168).

  board insulation Semirigid insulation preformed into rectangular units having degree of suppleness particularly related to their geometrical dimensions (ASTM C168).

  cellular glass insulation Insulation composed of glass processed to form a rigid foam having a predominantly closed-cell structure (ASTM C168).

  cellular polystyrene insulation Insulation composed principally of polymerized styrene resin processed to form a rigid foam having a predominantly closed-cell structure (ASTM C168). See also **rigid cellular polystyrene insulation** under **insulation, thermal.**

  cellular polyurethane insulation Insulation composed principally of the catalyzed reaction product of polyisocyanurate compounds, processed usually with a gas blowing agent to form a rigid foam having a predominantly closed-cell structure.

  cellulosic fiber insulation Insulation composed principally of cellulose fibers usually derived from paper, paperboard stock, or wood, with or without binders (ASTM C168).

  loose-fill insulation Insulation in granular, nodular, fibrous, powdery, or similar form designed to be installed by pouring, blowing, or hand placement (ASTM C168).

  mineral fiber insulation Insulation composed principally of fibers manufactured from rock, slag, or glass, with or without binders (ASTM C168).

  perlite insulation Insulation composed of natural perlite ore expanded to form a cellular structure (ASTM C168).

  reflective insulation Insulation depending for its performance upon reduction of radiant heat transfer across air spaces by use of one or more surfaces of high reflectance and low emittance (ASTM C168).

  rigid cellular polystyrene (RCPS) insulation Rigid thermal insulation board formed by expansion of polystyrene resin beads or granules in a closed mold (EPS), or by the expansion of polystyrene resin in an extrusion process (XPS). *Note:* Ad hoc abbreviations such as MEPS and XEPS are deprecated. The term **beadboard** (see under **insulation, thermal**) should not be used for commercial EPS. (ASTM E631)

vermiculite insulation Insulation composed principally of natural vermiculite ore expanded to form an exfoliated structure (ASTM C168).

wood fiber insulation Insulation composed of wood fibers, with or without binders (ASTM C168).

isolation joint See **joint, isolation.**

isotropic Having the same value for a property in all directions.

J

joint The space or opening between two or more adjoining surfaces (ASTM C717).

bridge joint Sealant joint composed of bond-breaker tape over the joint movement area with an overlay of sealant lapping either side of the tape sufficiently to bond well to the surfaces (often used where extreme joint movement occurs and conventional joint design is not possible). Sometimes called *bandage joint*.

butt joint (1) A joint having the edge or end of one member matching the edge, end, or face of another member without overlap. *Note:* An edge-to-face butt joint may also be called a *tee joint* or an *ell joint*. (ASTM E631). (2) A joint having the edge or face of one member spaced from and sealed to the edge or end of another member without overlap (ASTM C717).

cold joint (1) Boundary between later-applied and previously-applied coatings, plaster, mortar, or concrete. *Note:* At the boundary there can be less than the desired union of materials (ASTM E631). (2) A plane of weakness in concrete caused by an interruption or delay in the placing operation, which permits the first batch to start setting before the next batch is placed, resulting in little or no bond between the two batches (ASTM C717).

construction joint (1) In the construction of members intended to be continuous, a predetermined, intentionally created discontinuity between or within constructions and having the ends of the discontinuous members fastened to each other to provide structural continuity (ASTM E631). (2) A formed or assembled joint at a predetermined location where two successive placements ("lifts") of concrete meet. *Note:* Frequently a keyway or reinforcement is placed across the joint. With proper design, this joint may also function as a control or an isolation joint (ASTM C717).

control joint (1) In concrete, concrete masonry, stucco, or coating systems, a formed, sawed, or assembled joint acting to regulate the location of cracking, separation, and distress resulting from dimensional or positional change (ASTM E631). (2) A formed, sawed, tooled, or assembled joint acting to regulate the location and degree of cracking and separation resulting from the dimensional change of different elements of a structure. *Note:* This joint is usually installed in concrete and concrete masonry construction to induce controlled cracking at preselected locations, or where a concentration of stresses is expected (ASTM C717).

expansion joint (1) A discontinuity between two constructed elements or components, allowing for differential movement (such as expansion) between them without damage (ASTM E631). (2) A formed or assembled joint at a predetermined location, which prevents the transfer of forces across the joint as a result of movement or dimensional change of different elements of a structure or building (ASTM C717).

field joint A connection between adjoining members or parts, made at the time of installation (ASTM E631). See also **construction joint.**

fillet joint A triangular sealant bead at the internal corner of two intersecting planes; a rounded bead of sealant over the edges of two adjacent or overlapping surfaces.

isolation joint A formed or assembled joint specifically intended to separate and prevent the bonding of one element of a structure to another and having little or no transference of movement or vibration across the joint (ASTM C717).

lap joint A joint in which the units being joined override one another so that with movement, the sealant is in shear between the joint faces.

perimeter joint A joint formed by the outer edge of one panel or material and the leading edge of another.

reinforced joint A concrete joint bridged by reinforcing steel embedded in both joining parts (ASTM C717).

slip joint A joint allowing axial sliding movement of joined parts (ASTM E631).

joint sealing system A combination of joint cleaners, primers, fillers, backer rods, bond breakers, caulking compounds, sealants, gaskets, or tapes used to close joints between building components, sections, panels, or dissimilar materials.

joint backing See **sealant backing.**

joint filler A compressible material used in a partially or totally filled expansion, control, or isolation joint by its permanent placement in or between building materials such as concrete or masonry during construction; sometimes used as a sealant backing in a partially filled joint (ASTM C717).

L

laitance A weak layer of cement and aggregate fines on a concrete surface that is usually caused by an overwet mixture, overworking the mixture, improper or excessive finishing, or combinations thereof (ASTM C717).

lap joint See **joint, lap.**

latex sealant A compound that cures primarily through water evaporation. *Note:* The terms *latex sealant* and *emulsion sealant* are sometimes used interchangeably (ASTM C717).

liquid bituminous material One having a definite volume but no definite form, except as provided by its container (ASTM D1079).

lock-strip gasket A gasket in which sealing pressure is attained by inserting a keyed locking strip into a mating keyed groove in one face of the gasket (ASTM C717).

loose-laid membrane A ballasted roofing [or waterproofing] membrane that is attached to the substrate only at the edges and penetrations through the roof [or deck] (ASTM D1079).

M

mastic Usually refers to the consistency of a given material or substance; also commonly accepted to mean a type of material that may or may not skin, but will remain elastic and pliable with age; mastics are non-elastomeric. (See **flashing cement** and **asphalt mastic**.)

membrane A flexible or semiflexible roof covering or waterproofing, whose primary function is the exclusion of water (ASTM D1079).

metal flashing Frequently used as through-wall, cap-, or counterflashing (ASTM D1079). See **flashing**.

microclimate (1) The climate in the immediate vicinity of an object or organism. (2) The climate of a defined local area, such as a house or building site, formed by a unique combination of factors such as wind, topography, solar exposure, soil, and vegetation.

mineral fiber felt A felt with rock wool as the principal component (ASTM D1079).

mineral-surfaced roofing Built-up roofing whose top ply consists of a granule-surfaced sheet (ASTM D1079).

mock-up A section of a structure or assembly, built full size or to scale, for the purpose of studying construction details, testing performance, judging appearance, or any combination thereof (ASTM E631).

modified bitumen Roofing material consisting of bitumen that has been modified through the inclusion of one or more polymers and may contain stabilizers and other additives.

modulus As related to sealants used in building construction, the stress (force per unit area) at a corresponding strain (elongation) expressed as a percent of the original dimension (ASTM C717).

modulus of elasticity Stress/strain ratio; the ratio of the force (stress) needed to elongate (strain) a material to a certain point.

moisture See **water**.

monomer Material composed of single molecules; building block in the manufacture of polymers.

mopping The application of hot bitumen with a mop or mechanical applicator to the substrate or the plies of a built-up roof. There are four types of mopping: (1) *solid*—a continuous coating; (2) *spot*—bitumen is applied in roughly circular areas, generally about 18 in. in diameter, leaving a grid of unmopped, perpendicular areas; (3) *strip*—bitumen is applied in parallel bands, generally 8 in. wide and 12 in. apart; (4) *sprinkle*—bitumen is shaken onto the substrate from a broom or mop in a random pattern. (ASTM D1079.)

mud cracking Surface cracking resembling a dried mud flat (ASTM D1079).

muntin A secondary intermediate member subdividing a glazed area (ASTM C717).

N

nailing (1) Exposed nailing of roofing wherein nail heads are concealed from the weather. (2) Concealed nailing of roofing wherein nail heads are concealed from the weather. (ASTM D1079.) (See **blind nailing.**)

necking The localized reduction in cross section that may occur in a material under stress (ASTM C717).

neoprene A synthetic rubber (polychloroprene) used in liquid- or sheet-applied elastomeric roofing [and waterproofing] membranes or flashing (ASTM D1079).

nocturnal cooling Cooling by nighttime radiation, convection, and evaporation.

non-sag sealant A compound that exhibits little or no flow when applied in vertical or inverted joints (ASTM C717).

non-skinning Descriptive of a product that does not form a surface skin after application, and usually remains tacky or sticky.

O

oil based Combination or mixture of natural or synthetic resins mixed with drying oils, as in caulking compounds, paints and varnishes.

oil canning Rippled, uneven effect in the plane of a metal coil or finished sheet, especially noticeable in panels that have wide, flat areas.

oleoresinous See **oil based.**

one-on-one See **ply-on-ply.**

opaque Impenetrable by light.

open cell A cell not totally enclosed by its walls and hence interconnecting with other cells (ASTM C717).

open-cell material A cellular material in which substantially all cells in the mass are open cells (ASTM C717).

organic Being or composed of hydrocarbons or their derivatives, or matter of plant or animal origin (ASTM D1079).

outgassing The emission of occluded gases from a material by vacuum, heat, or pressure. *Note:* As applied to sealant backing materials contained within a filled joint, outgassing may occur when the material is compressed or punctured causing gas bubbles to form in the overlaying sealant (ASTM C717).

overall heat transfer coefficient (U value) See **thermal transmittance.**

oxidation Formation of an oxide; deterioration of rubbery materials due to the action of oxygen or ozone.

P

parging The process of troweling a coat of cement mortar onto a masonry wall surface.

patina Discoloration resulting from metal's natural weathering and oxidation.

penetrant Treatment that lines pores without forming a film on the surface.

percent elongation/compression Amount of tensile and compressive deformation a sealant reportedly will withstand without adhesive or cohesive failure; usually expressed as a plus-or-minus percentage such as ±25%, meaning that the sealant will elongate 25% and compress 25% of its original length.

percolation layer (drainage course) A layer of washed gravel or of a manufactured drainage media that allows water to filter through to the drain (ASTM C898 and C981).

perlite An aggregate used in lightweight insulating concrete and in preformed perlite insulating board; formed by heating and expanding siliceous volcanic glass (ASTM D1079).

perm (1) Time rate of water vapor migration through a material or a construction of one grain per hour per square foot, per inch of mercury pressure difference (ASTM E241). (2) Empirical unit of water-vapor permeance (mass flow rate), equal to one grain (avoirdupois) of water vapor per hour flowing through one square foot of a material or construction induced by a vapor-pressure difference of one inch of mercury between the two surfaces. *Note:* This mass flow rate can be stated in other desired or convenient units. A maximum value of one perm is the moisture vapor migration rate below which there is low probability of induced moisture problems in conventional buildings in climates not exceeding 5000 heating degree days (65°F base), and not so hot and humid that continual air conditioning would be required. (ASTM E631.)

perms Grains per square foot per hour per inch of mercury difference in vapor pressure at standard test conditions (ASTM E96).

permeability The ability of a material to transmit water vapor through a thickness of one inch. It is measured in perm-inches.

permeance The rate of water-vapor transmission per unit area at a steady state through a membrane or assembly, expressed in grains per square foot per hour per inch of mercury ($gr/ft^2 \cdot h \cdot in\ Hg$) (ASTM D1079).

permanent set The amount by which an elastic material fails to return to its original form after a deformation.

permeability, vapor See **vapor permeability.**

permeance, vapor See **vapor permeance.**

phased application The installation of a roofing or waterproofing system during two or more separate time intervals; a roofing system not installed in a continuous operation (ASTM D1079). (See **ply-on-ply.**)

pickling The process of chemically removing oxides and scale from the surface of a metal by the action of water solutions of acids.

picture framing A rectangular pattern of ridges in a membrane over insulation or deck joints (ASTM D1079).

pinhole A tiny hole in a film, foil, or laminate comparable in size to one made by a pin (ASTM D1079).

pitch An inclination or slope measured in degrees or percent, or by the ratio of rise and run (ASTM E631). (See **coal-tar pitch**.)

pitch pocket A flanged, open-bottomed metal container placed around a column or other roof penetration, and filled with hot bitumen or flashing cement to seal the joint (ASTM D1079).

ply (1) A single layer of membrane reinforcement in a bituminous membrane waterproofing system (ASTM C981). (2) A layer of felt in a built-up roofing membrane; a four-ply membrane has at least four plies of felt at any vertical cross section cut through the membrane (ASTM D1079).

ply-on-ply The application of a single ply of roofing over the substrate, followed by the application of a second single ply over the first, a third single ply over the second, and so on (phased application), as opposed to application of multiple plies in a single progressive operation by the shingle method.

polymer A very long chemical chain of monomers prepared by means of an addition and/or a condensation polymerization; the units may be the same or different; there are copolymers, dipolymers, terpolymers, quadripolymers, and high polymers.

polymerization Chemical reaction in which two or more monomers are linked to form a more complex compound with a higher molecular weight.

polymerized Treated by heating or cooking so that molecules of different substances unite into larger molecules of a different substance with individual characteristics.

polysulfide A synthetic, rubber-like elastomer that is practically insoluble in oils and solvents.

polyurethane See **cellular polyurethane** (under **insulation, thermal**) and **urethane**.

pond A surface which is incompletely drained (ASTM D1079).

porosity Ratio of pore space to the total volume of a material, expressed as a percent.

POT life POT stands for potential open time. (See **working life**.)

preformed metal roofing Metal panels (usually steel or aluminum) and/or flashings that are factory-formed, and typically furnished cut to length and with clips and other hold-down devices, closure pieces, and sealants.

primer (1) A compatible coating designed to enhance adhesion (ASTM C717). (2) A thin liquid bitumen applied to a surface to improve the adhesion of heavier applications of bitumen and to absorb dust (ASTM D1079).

protection board See **protection course**.

protection course Semirigid sheet material placed on top of a waterproofing membrane to protect it against damage during subsequent construction and to provide a protective barrier against compressive and shearing forces induced by materials placed above [or against] it (ASTM C898).

psychrometrics A branch of physics dealing with the measurement or determination of atmospheric conditions, and particularly the study of the behavior of moist air under various temperature and humidity conditions.

R

R **value** See **thermal resistance.**

radiant flux Power emitted, transferred, or received in the form of electromagnetic waves or photons (ASTM E772).

radiation The process by which energy is emitted or transferred in the form of photons or electromagnetic waves (ASTM E772).

rake The sloped edge of a roof at the first or last rafter (ASTM D1079).

receiving surface See **substrate.**

recovery Percent of original sealant shape regained after elongation or compression.

re-entrant corner An inside corner of a surface, producing stress concentrations in the roofing or waterproofing membrane (ASTM D1079).

reflectance The fraction of the incident radiation upon a surface that is reflected from the surface (ASTM C168).

refraction The change in direction of light rays as they enter a transparent medium such as water, air, or glass.

reglet (1) A continuous groove, slot, or recess within a building component surface which receives other components such as flashing, gaskets, or anchors; a continuous prefabricated metal or plastic device containing a groove, slot, or recess which can be cast into (as a form) or mounted onto a building component surface (ASTM C717). (2) A groove in a wall or other surface adjoining a roof surface for the attachment of counterflashing (ASTM D1079).

reinforced membrane A roofing or waterproofing membrane reinforced with felts, mats, fabrics, or chopped fibers (ASTM D1079).

relative humidity See **humidity, relative.**

release tape See **bond breaker.**

resistance See **thermal resistance.**

resilience The tendency of a material to return to its original shape after the removal of a stress that has produced elastic strain.

re-temper To add more water to a hydraulic-setting compound after the initial mixing, but before partial set has occurred (ASTM E631).

ridging An upward, tenting displacement of a membrane, frequently over an insulation joint (ASTM D1079).

rigid Not flexible.

rising damp Upward-moving moisture in a wall or other structure standing in water or in wet soil (ASTM E631). See also **capillary migration.**

roll roofing Coated felts, either smooth or mineral-surfaced (ASTM D1079).

roof assembly An assembly of interacting roofing components (*including* the roof deck) designed to weatherproof and, normally, to insulate a building's top surface (NRCA).

roof system A system of interacting roof components (*not including* the roof deck) designed to weatherproof and, normally, to insulate a building's top surface (NRCA).

roofing system (1) Assembly of interacting components designed to weatherproof, and sometimes to insulate, the roof surface of a building (ASTM E631). (2) An assembly of interacting components designed to weatherproof, and normally to insulate, a building's top surface (ASTM D1079).

rubber A material that is capable of recovering from large deformations quickly and forcibly, and can be, or already is, modified to a state in which it is essentially insoluble (but can swell) in boiling solvent such as benzene, methyl ethyl ketone, and ethanol-toluene azeotrope. *Note:* A rubber in its modified state, free of diluents, retracts within 1 minute to less than 1.5 times its original length after being stretched at room temperature to twice its length and held for 1 minute before release. (ASTM D1079.)

S

saddle In roofing, a small structure that helps to channel surface water to drains. Frequently located in a valley, a saddle is often constructed like a small hip roof, or like a pyramid with a diamond-shaped base (ASTM D1079). See **cricket.**

saponification The deposit of a gray scum or gray dust on the inside surface of a subgrade wall or floor, as the result of moisture moving through the concrete and washing certain chemicals from the concrete mass.

saturated felt A felt that has been immersed in hot bitumen; the felt adsorbs and absorbs as much bitumen as it can retain under the processing conditions, but remains porous and contains voids (ASTM D1079).

scaling See **spalling.**

seal A barrier against the passage of liquids, solids, or gases (ASTM C717).

sealant A material that has the adhesive and cohesive properties to form a seal (ASTM C717).

sealant backing A compressible material placed in a joint before application of a sealant. *Note:* The purpose of sealant backing is to assist in providing the proper sealant configuration, to limit the depth of the sealant, and in some cases, to act as a bond breaker (ASTM C717).

sealer Liquid coating applied to surfaces to fill pores, voids, or hairline cracks.

self-leveling sealant A compound that exhibits flow sufficient to seek gravitational leveling (ASTM C717).

sensible heat Heat that results in a temperature change.

serviceability The capacity of a material, product, component, assembly, or construction to perform the functions for which it was designed and constructed (ASTM E241).

setting Process by which, after application, a liquid (wet-state) material changes to a serviceable condition by **curing** or **drying**. *Note:* Generally, curing implies a chemical reaction, while drying implies evaporation of volatile constituents (ASTM E631).

setting time Term used loosely to describe that period when a material has either dried sufficiently through solvent release, or cured sufficiently through chemical reaction, to reach a specified condition.

service life Life expectancy in service.

service temperature Expected ambient temperature range in which product or material must perform.

shark fin An upward-curled felt sidelap or endlap (ASTM D1079).

shelf life The maximum time packaged materials can be stored under specified conditions and still meet the performance requirements specified (ASTM C717).

shingle (noun) A small unit of prepared roofing designed to be installed with similar units in overlapping rows on inclines normally exceeding 25% (NRCA).

shingle (verb) (1) To cover with shingles. (2) To apply any sheet material in overlapping rows like shingles. (ASTM D1079.)

shingling (1) The procedure of laying parallel felts so that one longitudinal edge of each felt overlaps and the other longitudinal edge underlaps an adjacent felt. Normally, felts are shingled on a slope so that the water flows over rather than against each lap. (2) The application of shingles to a sloped roof. (NRCA.)

shop drawing A drawing prepared by the fabricator based on a working drawing and used in a shop or on a site for assembly (ASTM E631).

Shore A hardness Measure of firmness (0 to 100) of a substance or material by means of a Shore A hardness gauge (see **durometer**), used on elastomers and natural rubbers; a Shore A reading of 80 equals a Shore D reading of 30.

Shore D hardness Measure of hardness (0 to 100) of a substance or material by means of a Shore D hardness gauge, used on rigid and semirigid materials.

shrinkage A decrease in length, area, or volume (ASTM C717).

silane Generally refers to alkyltrialkoxysilanes. A monomeric organosilicon compound with one unhydrolyzable silicon-carbon bond, which forms a chemical bond with siliceous minerals providing water-repellent protection to the substrate. Silanes are penetrants.

siliconate Organic modified alkali silicates. Siliconates are generally applied in aqueous solution to harden and/or protect masonry substrates. Siliconates are penetrants.

silicone resin Any of the organopolysiloxanes applied to materials for water repellency. Silicone water repellents are generally highly polymerized resins applied in any of several organic solvents. Application is accompanied by chemical bonding to the substrate if silicate minerals are present. The size and shape of the polymer of which the resin is composed determines whether the silicone is classified as a film former or a penetrant.

silicone rubber One of the family of polymeric materials in which the receiving chemical group contains silicone and oxygen atoms in links in the main chain.

silicone sealant A liquid-applied curing compound based on polymers of polysiloxane structures (ASTM C717).

siloxane Generally refers to alkylalkoxysiloxanes that are oligomerous (i.e., siloxane of low molecular weight with the polymer consisting of two, three, or four monomers). As with other silicones, application is accompanied by chemical bonding to the substrate if silicate minerals are present. Oligomerous siloxanes are properly classified as penetrants.

skinning Formation of a dry film on the surface of a compound.

slip sheet Sheet material such as reinforced kraft paper, polyester scrim, or polyethylene sheeting placed between two components of a roofing system (such as membrane and insulation) to ensure that no adhesion occurs between them, and to prevent possible damage from chemical incompatibility.

slippage Relative lateral movement of adjacent components of a built-up membrane. It occurs mainly in roofing membranes on a slope, sometimes exposing the lower plies or even the base sheet to the weather. (ASTM D1079.)

slope The tangent of the angle between the roof surface and the horizontal plane, expressed as a percentage or in inches of rise per foot of horizontal distance (ASTM D1079). (See **incline**.) The Asphalt Roofing Manufacturers Association (ARMA) ranks slope as follows:

Level slope   Up to $1/2$ in./ft

Low slope $1/2$ to $1\,1/2$ in./ft

Steep slope   Over $1\,1/2$ in./ft

smooth-surfaced roof A built-up roof without mineral aggregate surfacing (ASTM D1079).

softening point The temperature at which a bitumen becomes soft enough to flow as determined by an arbitrary, closely defined method (ASTM D1079).

solar absorptance The fraction of incident solar energy absorbed by a surface.

solar altitude The angle of the sun above the horizon, measured in a vertical plane.

solar radiation Electromagnetic radiation emitted by the sun.

solar absorptance See **absorptance**.

solar reflectance See **reflectance**.

solar transmittance The fraction of incident solar energy transmitted by a surface.

solid bituminous material One having a viscosity of over 1×10^5 cSt at 40°C or an equivalent viscosity at an agreed-upon temperature. This includes powders and granular materials. (ASTM D1079.)

solvent Having the power of dissolving or forming a solution with something.

solvent-release sealant A compound that cures primarily through solvent evaporation (ASTM C717).

spall (noun) A fragment or chip as from concrete, brick, stone, or other similar materials (ASTM C717).

spall (verb) To break off fragments or chips, as from concrete, brick, stone or other similar materials by water freezing within the material, corrosion expansion of embedded metal, movement pressures, or other physical or chemical processes (ASTM C717).

spalling (1) The development of spalls (ASTM C717). (2) Crumbling or chipping of a masonry or concrete surface due to freezing of absorbed water, corrosion of embedded steel, cement-aggregate reaction, restraint against movement, or other causes (ASTM C981).

spangle Metal flake appearance in zinc-rich metallic coatings.

specific heat The number of Btu required to raise the temperature of one pound of a material one degree Fahrenheit in temperature.

specification A precise statement of a set of requirements to be satisfied by a material, product, system or service. *Note:* It is desirable that the requirements, together with their limits, should be expressed numerically in appropriate units. (ASTM E631.)

split A membrane tear resulting from tensile stress (ASTM D1079).

spud To remove the roofing aggregate and most of the bituminous top coating by scraping and chipping (ASTM D1079).

square A roof area of 100 sq ft, or enough material to cover 100 sq ft of deck (ASTM D1079).

stack effect Phenomenon of warm air rising in buildings to create pressure differentials at the top and bottom of the building envelope.

stack vent A vertical outlet in a built-up roofing system to relieve the pressure exerted by water vapor between the roofing membrane and the vapor retarder or deck (ASTM D1079).

standing seam Generic term used to describe any metal roofing joinery that uses an upturned portion of the metal to marry adjacent metal sections and is held in place with a concealed clip or cleat within the seam.

stearate Salt or ester of stearic acid that functions as a water repellent by forming a "soap" within the pores of the material. Stearates are generally classified as film formers, but can be considered penetrants in modified form.

steep asphalt A roofing asphalt conforming to the requirements of ASTM D312, Type III (ASTM D1079).

strawberry A small bubble or blister in the flood coating of a gravel-surfaced membrane (ASTM D1079).

stripping Strip flashing: (1) the technique of sealing a joint between metal and built-up membrane with one or two plies of felt or fabric and hot- or cold-applied bitumen;

(2) the technique of taping joints between insulation boards or deck panels (ASTM D1079).

structural panel roofing   Metal panel roofing that can span over open supports at significant spacings (usually 3 ft), while carrying both negative and positive loads to those supports.

structural sealant   A sealant capable of transferring dynamic or static ("live" or "dead") loads, or both, across joint members exposed to service environments typical for the structure involved (ASTM C717).

structural sealant glazing   A glazing system wherein a structural sealant is used to transfer loads between a lite or panel and a supporting framework, without mechanical fasteners or other methods of attachment (ASTM C717).

sublimation   The volatilization or evaporation of a solid directly to the vapor state, without passing through the liquid state.

substrate   (1) A material upon which films, treatments, adhesives, sealants, membranes, and coatings are applied. (2) Materials that are bonded or sealed together by adhesives or sealants (ASTM C717). (3) The surface upon which the roofing or waterproofing membrane is placed (structural deck or insulation) (ASTM D1079).

substrate failure   Joint failure caused by breaking, tearing, or spalling of the substrate.

sump   A depression around a drain (ASTM D1079).

surface heat transfer coefficient (h)   The rate of heat transfer from a unit area of a surface to the adjacent air and environment caused by a temperature difference of one degree Fahrenheit between the surface and the air.

super-steep asphalt   A roofing asphalt conforming to the requirements of ASTM D312, Type IV (ASTM D1079).

T

tack   Sticky or adhesive quality of a surface.

tack-free   Not sticky.

tack-free time   Time required for surface dryness.

tape sealant   A sealant having a preformed shape, and intended to be used in a joint initially under compression (ASTM C717).

tapered edge strip   A tapered insulation strip used to elevate the roofing at the perimeter and at penetrations of the roof (ASTM D1079).

tar   A brown or black bituminous material, liquid or semisolid in consistency, in which the predominating constituents are bitumens obtained as condensates in the processing of coal, petroleum, oil-shale, wood, or other organic materials (ASTM D1079).

temper   In hydraulic-setting compounds, to bring to a usable state by mixing in or adding water (ASTM E631).

tension leveling Mill process that pulls and stretches a metal coil to produce better consistency in flatness and reduce oil canning.

terne Coating of 20% tin–lead alloy usually applied to 28- or 30-gauge copper-bearing steel.

therm A unit of thermal energy equal to 100,000 Btu.

thermal break An element of low thermal conductivity placed in such a way as to reduce or prevent the flow of heat in a larger element.

thermal capacity See **heat capacity.**

thermal conductance (C) The time rate of steady-state heat flow through a unit area of a material or construction induced by a unit temperature difference between the body surfaces; $C=q/\Delta T$ (ASTM C168).

thermal conduction (1) The movement of heat through a material, as through a metal rod with one end in a fire. (2) The direct transfer of heat from one substance to another, including from surfaces such as walls to the air.

thermal conductivity (k) The time rate of steady-state heat flow through a unit area of a homogeneous material induced by a unit temperature gradient in a direction perpendicular to that unit area, expressed as Btu/h·ft^2·°F·in (ASTM C168). The reciprocal of thermal resistivity.

thermal resistance (R) The quantity determined by the temperature difference, at steady state, between two defined surfaces of a material or construction that induces a unit heat flow rate through a unit area (ASTM C168).

thermal resistivity (r) The quantity determined by the temperature difference, at steady state, between two defined parallel surfaces of a homogeneous material of unit thickness, that induces a unit heat flow rate through a unit area (ASTM C168). The reciprocal of thermal conductivity.

thermal shock The stress-producing phenomenon resulting from sudden temperature drops in a roof membrane when, for example, a rain shower follows brilliant sunshine (ASTM D1079).

thermal storage capacity The ability of a material, per square foot of exposed area, to absorb and store heat. Numerically, the density times the specific heat times the thickness.

thermal transmittance (U) The heat transmission in unit time through unit area of a material or construction and the boundary air films, induced by unit temperature difference between the environments on each side (ASTM C168). The U-value is the reciprocal of the total R-value.

thermocirculation The convective circulation of a fluid such as water or air that occurs when warm fluid rises and is displaced by denser, cooler fluid in the same system.

thermoplastic A characteristic of materials which soften readily with the application of heat long before reaching their point of decomposition, and regain their original properties after cooling. Capable of being repeatedly softened by heat and hardened by cooling.

thermosetting A property of materials whose point of softening with the application of heat is very near their point of decomposition, and whose properties will not return to their starting point after cooling.

through-wall flashing A water-resistant membrane or material assembly extending totally through a wall and its cavities, positioned to direct any water within the wall to the exterior (ASTM D1079).

tolerance The allowable deviation from a value or standard; especially the total range of variation permitted in maintaining a specified dimension in machining, fabricating, or constructing a member or assembly (ASTM E631).

tooling (1) The act of compacting and contouring a sealant in a joint (ASTM C717). (2) The act of compacting and shaping a mortar joint.

tooling time In a sealant, the time interval after application of a one-component sealant or after mixing and application of a multi-component sealant during which tooling is possible (ASTM C717).

traffic surface A surface exposed to traffic, either pedestrian or vehicular (ASTM C1127).

translucent Having the characteristic of transmitting light but causing sufficient diffusion to eliminate the perception of distinct images.

transmittance See **solar transmittance** and **thermal transmittance.**

transparent Having the characteristic of transmitting light so that objects or images can be seen as if there were no intervening material.

trapezoidal rib Metal roofing panel profile having a seam or intermediate longitudinal formation whose cross section has the geometric shape of a trapezoid.

troweled finish A concrete finish provided by smoothing the surface with power-driven or hand trowels, or both, after the float finishing operation. *Note:* A troweled finish is smoother than the floated finish. (ASTM C1127.)

U

U **value** See **thermal transmittance.**

ultraviolet resistance Resistance to degradation by ultraviolet radiation.

urethane Generic term for ethylene carbonate; used extensively as a base polymer in elastic sealants.

V

vapor barrier See **vapor retarder.**

vapor migration The movement of water vapor from a region of high vapor pressure to a region of lower vapor pressure (ASTM D1079).

vapor permeability The property of a material that permits migration of water vapor under the influence of a difference in vapor pressure across the material.

vapor permeance Time rate of water-vapor transmission through a unit area of a flat material or construction induced by unit vapor-pressure difference between the two specified surfaces, under specified temperature and humidity conditions (ASTM E631). (See **perm**.)

vapor pressure (Pv) The partial pressure of the water vapor in an air-vapor mixture. It is determined by the dew point temperature or by the dry bulb temperature and the relative humidity of the mixture. The units are pounds per square inch (psi) or inches of mercury.

vapor resistance The reciprocal of vapor permeance. A rating of the resistance of a material to the passage of water vapor. The unit of measure is a rep.

vapor resistivity The reciprocal of vapor permeability. A measure of the resistance of a one inch thickness of material to the passage of water vapor. The unit of measure is rep per inch.

vapor retarder (1) Material or system that impedes the transmission of water vapor under specified conditions (ASTM E631). (2) A layer of material or a laminate used to appreciably reduce the flow of water vapor into the roofing system (ASTM D1079). (3) Material or construction that retards water vapor migration, generally not exceeding one perm for ordinary houses in non-extreme climates (ASTM E241). (4) A layer of material which appreciably reduces the diffusion of water vapor. A material, film, or coating made for this use is generally expected to have a vapor permeance of 1 perm or less. (See also **air barrier**.)

vent An opening designed to convey water vapor or other gas from inside a building or a building component to the atmosphere (ASTM D1079).

venting Providing circulation of air or ventilation of an enclosed space by use of tubes, breather vents or openings.

vermiculite An aggregate used in lightweight insulating concrete and loose fill insulation, formed by heating and expanding a micaceous mineral.

viscosity The internal resistance to flow exhibited by a fluid.

volatile loss Weight loss by vaporization of solvent.

volatile organic compound Compounds of carbon which are emitted into the atmosphere during drying.

W

water Water as liquid, vapor or solid (ice) in any combination or in transition (ASTM E241).

water absorption Process in which a material takes in water through its pores and interstices and retains it wholly, without transmission.

water infiltration Process in which water passes through a material or system and reaches a space which is not directly or intentionally exposed to the water source.

water leakage Water infiltration which is uncontrolled, exceeds the resistance, retention, or discharge capacity of the system, or causes subsequent damage or premature deterioration.

water penetration Process in which water gains access into a material or system by passing through the surface exposed to the water source.

water permeation Process in which water enters, flows, and spreads within and discharges from a material.

water repellent A material or treatment for surfaces to provide resistance to penetration by water (ASTM E631).

water saturation The maximum amount of water a material or system can retain without discharge or transmission.

water vapor Water in the state of an invisible gas, diffused in the air.

waterproofing Treatment of a surface or structure to prevent the passage of liquid water under hydrostatic, dynamic, or static pressure (ASTM C717).

wearing surface See **traffic surface.**

weather The general atmospheric conditions at a given place and a given time with respect to temperature, humidity, precipitation, wind, radiation, and other meteorological events.

weather sealer Form of coating applied to the outer surface of a construction to augment its weather resistance (ASTM E631).

weathertight Impermeable to the passage of water or air or both under certain conditions as determined by test (ASTM C717).

weephole Small hole allowing drainage of fluid (ASTM E631).

weight loss See **shrinkage.**

wet bulb temperature The lowest temperature attainable by evaporating water into the air.

wet-film thickness The thickness of a liquid coating as it is applied (ASTM C898).

working life In a sealant, the time interval after opening a container of a single component sealant, or after mixing the components of a multi-component sealant, during which application and tooling is possible (ASTM C717).

Appendix B

ASTM Standards

ASTM A167	Specification for Stainless and Heat-Resisting Chromium Nickel Steel Plate, Sheet, and Strip
ASTM A308	Specification for Steel Sheet, Terne (Lead-Tin Alloy) Coated by the Hot-Dip Process
ASTM A361	Specification for Steel Sheet, Zinc-Coated (Galvanized) by the Hot-Dip Process for Roofing and Siding
ASTM A463	Specification for Sheet Steel, Cold-Rolled, Aluminum-Coated, Type 1 and Type 2
ASTM A525	Specification for General Requirements for Steel Sheet, Zinc-Coated (Galvanized) by the Hot-Dip Process
ASTM A653	Specification for Steel Sheet, Zinc-Coated (Galvanized) or Zinc-Iron Alloy-Coated (Galvannealed) by the Hot-Dip Process
ASTM B209	Specification for Aluminum and Aluminum Alloy Sheet and Plate
ASTM B370	Specification for Copper Sheet and Strip for Building Construction
ASTM C168	Definitions of Terms Relating to Thermal Insulating Materials
ASTM C208	Specification for Insulating Board (Cellulosic Fiber), Structural and Decorative
ASTM C406	Specification for Roofing Slate
ASTM C552	Specification for Cellular Glass Thermal Insulation
ASTM C570	Specification for Oil- and Resin-Base Caulking Compound for Building Construction
ASTM C578	Specification for Preformed Cellular Polystyrene Thermal Insulation
ASTM C591	Specification for Unfaced Rigid Polyisocyanurate Thermal Insulation Board
ASTM C612	Specification for Mineral Fiber Block or Board Thermal Insulation
ASTM C665	Specification for Mineral Fiber Blanket Thermal Insulation
ASTM C717	Terminology of Building Seals and Sealants

ASTM C719	Test Method for Adhesion and Cohesion of Elastomeric Joint Sealants Under Cyclic Movement (Hockman Cycle)
ASTM C726	Specification for Mineral Fiber Roof Insulation Board
ASTM C728	Specification for Perlite Thermal Insulation Board
ASTM C755	Practice for Selection of Vapor Retarders for Thermal Insulations
ASTM C793	Test Method for Effects of Accelerated Weathering on Elastomeric Joint Sealants
ASTM C794	Test Method for Adhesion-in-Peel of Elastomeric Joint Sealants
ASTM C834	Specification for Latex Sealing Compounds
ASTM C836	Specification for High Solids Content, Cold Liquid-Applied Elastomeric Waterproofing Membrane for Use with Separate Wearing Course
ASTM C898	Guide for Use of High Solids Content, Cold Liquid-Applied Elastomeric Waterproofing Membrane with Separate Wearing Course
ASTM C920	Specification for Elastomeric Joint Sealants
ASTM C957	Specification for High-Solids Content, Cold Liquid-Applied Elastomeric Waterproofing Membrane with Integral Wearing Surface
ASTM C981	Guide for Design of Built-up Bituminous Membrane Waterproofing Systems for Building Decks
ASTM C984	Specification for Perlite Board and Rigid Cellular Polyurethane Composite Roof Insulation
ASTM C1013	Specification for Membrane-Faced Rigid Cellular Polyurethane Roof Insulation
ASTM C1050	Specification for Cellular Polystyrene-Cellulosic Fiber Composite Board Insulation
ASTM C1079	Specification for Spray-Applied Rigid Cellular Polyurethane Thermal Insulation
ASTM C1085	Specification for Butyl Rubber-Based Solvent-Release Sealants
ASTM C1127	Guide for Use of High Solids Content, Cold Liquid-Applied Elastomeric Waterproofing Membrane with an Integral Wearing Surface
ASTM C1167	Specification for Clay Roof Tiles
ASTM C1177	Specification for Glass Mat Gypsum Substrate for Use as Sheathing
ASTM C1184	Specification for Structural Silicone Sealants
ASTM C1186	Standard Specification for Flat Non-Asbestos Fiber Cement Board
ASTM C1193	Guide for Use of Joint Sealants
ASTM C1281	Specification for Preformed Tape Sealants for Glazing Applications
ASTM C1289	Specification for Faced Rigid Cellular Polyisocyanurate Thermal Insulation Board
ASTM C1299	Guide for Use in Selection of Liquid-Applied Sealants
ASTM C1311	Specification for Solvent Release Sealants
ASTM D41	Specification for Asphalt Primer Used in Roofing, Dampproofing and Waterproofing

ASTM D43	Specification for Creosote Primer Used in Roofing, Dampproofing and Waterproofing
ASTM D173	Specification for Bitumen-Saturated Cotton Fabrics Used in Roofing and Waterproofing
ASTM D225	Specification for Shingles, Asphalt, Surfaced with Mineral Granules (Organic Felt)
ASTM D226	Specification for Asphalt-Saturated Organic Felt Used in Roofing and Waterproofing
ASTM D227	Specification for Coal-Tar-Saturated Organic Felt Used in Roofing and Waterproofing
ASTM D312	Specification for Asphalt Used in Roofing
ASTM D449	Specification for Asphalt Used in Dampproofing and Waterproofing
ASTM D450	Specification for Coal-Tar Pitch Used in Roofing, Dampproofing, and Waterproofing
ASTM D491	Specification for Asphalt Mastic Used in Waterproofing
ASTM D1079	Definitions of Terms Relating to Roofing, Waterproofing, and Bituminous Materials
ASTM D1227	Specification for Emulsified Asphalt Used as a Protective Coating for Roofing
ASTM D1668	Specification for Glass Fabrics (Woven and Treated) for Roofing and Waterproofing
ASTM D2178	Specification for Asphalt Glass Felt Used in Roofing and Waterproofing
ASTM D2626	Specification for Asphalt-Saturated and Coated Organic Felt Base Sheet Used in Roofing
ASTM D2822	Specification for Asphalt Roof Cement
ASTM D3462	Specification for Shingles, Asphalt, Made from Glass Felt and Surfaced with Mineral Granules
ASTM D3468	Specification for Liquid-Applied Neoprene and Chlorosulfonated Polyethylene Used in Roofing and Waterproofing
ASTM D3747	Specification for Emulsified Asphalt Adhesive for Adhering Roof Insulation
ASTM D4022	Specification for Coal Tar Roof Cement
ASTM D4068	Specification for Chlorinated Polyethylene (CPE) Sheeting for Concealed Water Containment Membrane
ASTM D4263	Test Method for Indicating Moisture in Concrete by the Plastic Sheet Method
ASTM D4434	Specification for Polyvinyl Chloride Sheet Roofing
ASTM D4586	Specification for Asphalt Roof Cement, Asbestos Free
ASTM D4601	Specification for Asphalt Coated Glass Fiber Base Sheet Used in Roofing
ASTM D4637	Specification for Vulcanized Rubber Sheet Used in Single-Ply Roof Membranes

ASTM D4811	Specification for Nonvulcanized Rubber Sheet Used as Roof Flashing
ASTM D4897	Specification for Asphalt-Coated Glass Fiber Venting Base Sheet Used in Roofing
ASTM D4990	Specification for Coal Tar Glass Felt Used in Roofing and Waterproofing
ASTM D5019	Specification for Reinforced Nonvulcanized Polymeric Sheet Used in Roofing Membranes
ASTM D5469	Guide for Application of New Spray Applied Polyurethane Foam and Coated Roofing Systems
ASTM D5643	Specification for Coal Tar Roof Cement—Asbestos Free
ASTM E96	Test Method for Water Vapor Transmission of Materials
ASTM E154	Test Method for Water Vapor Retarders Used in Contact with Earth Under Concrete Slabs, on Walls, or as Ground Cover
ASTM E241	Recommended Practices for Increasing Durability of Building Constructions Against Water-Induced Damage
ASTM E283	Method of Test for Rate of Air Leakage Through Exterior Windows, Curtain Walls, and Doors
ASTM E331	Test Method for Water Penetration of Exterior Windows, Curtain Walls, and Doors by Uniform Static Air Pressure Difference
ASTM E514	Test for Water Permeance of Masonry
ASTM E772	Terminology of Solar Energy Conversion
ASTM E936	Practice for Roof System Assemblies Employing Steel Deck, Preformed Roof Insulation, and Bituminous Built-up Roofing
ASTM E1105	Test Method for Field Determination of Water Penetration of Installed Exterior Windows, Curtain Walls, and Doors by Uniform or Cyclic Static Air Pressure Difference
ASTM E1514	Specification for Structural Standing Seam Steel Roof Panel Systems
ASTM E1643	Practice for Installation of Water Vapor Retarders Used in Contact with Earth or Granular Fill Under Concrete Slabs

Appendix C

Bibliography

Allen, Edward, *Architectural Detailing: Function, Constructability, Aesthetics,* New York, John Wiley & Sons, 1993.
American Architectural Manufacturers Association, *Aluminum Curtain Wall Design Guide Manual,* Palatine, Ill., AAMA, 1979.
American Concrete Institute, *Guide to Joint Sealants for Concrete Construction,* ACI 504R, ACI, Detroit, 1986.
American Concrete Institute, *A Guide to the Use of Waterproofing, Dampproofing, Protective and Decorative Barrier Systems for Concrete,* ACI 515.1R, ACI, Detroit, 1985.
American Society of Heating, Refrigeration and Air Conditioning Engineers, *Handbook of Fundamentals,* ASHRAE, Atlanta, 1989.
Ballast, David Kent, *Handbook of Construction Tolerances,* McGraw-Hill, New York, 1994.
Balliet, K. E., and Julian Panek, "Back-Up Materials," *Building Seals and Sealants,* ASTM STP 606, Julian Panek, ed., American Society for Testing and Materials, Philadelphia, 1976.
Banov, Abel, *Paints and Coatings Handbook,* 2d ed., Structures Publishing Co., Farmington, Mich., 1978.
Beall, Christine, "Selecting and Specifying Caulking and Sealants," *The Construction Specifier,* March 1990, pp. 54–61.
Beall, Christine, "Building Joint Movement," *Proceedings of the American Society of Civil Engineers 1990 Materials Engineering Congress,* Denver, August 1990.
Beall, Christine, *Masonry Design and Detailing,* 3rd ed., McGraw-Hill, New York, 1997.
Beall, Christine, "Sealant Joint Design," *Water in Exterior Building Walls: Problems and Solutions,* ASTM STP 1107, Thomas A. Schwartz, ed., American Society for Testing and Materials, Philadelphia, 1991.
Borzillo, Angelo R., "Standing Seam Roofs," *The Construction Specifier,* November 1993, pp. 94–106.
Brand, Ronald, *Architectural Details for Insulated Buildings,* Van Nostrand Reinhold, New York, 1990.
Brantley, L. Reed and Ruth T., *Building Materials Technology,* McGraw-Hill, New York, 1996.
British Standards Institution, *Code of Practice for Design of Joints and Jointing in Building Construction,* Building Research Station, Garston, Watford, U.K, 1981.
Brower, James R., "Surface Preparation—The Key to Proper Sealant Adhesion," *The Construction Specifier,* July 1976, pp. 46–49.
Brower, James R., "Silicone Sealants," *The Applicator,* first quarter 1987, pp. 1–5.
Brower, James R., "Compatibility," *The Applicator,* second issue 1988, pp. 6–7.
Callendar, John H., *Timesaver Standards for Architectural Design Data,* 6th ed., McGraw-Hill, New York, 1982.
Canadian Home Builders Association, *Builders' Manual,* CHBA, Ottawa, 1994.
Concrete Masonry Association of California and Nevada and Masonry Institute of America, *Clear Water Repellent Treatments for Concrete Masonry,* MIA/CMACN, 1993.

Construction Specifications Institute, monograph 07M900, *Joint Sealers,* CSI, Alexandria, Va., 1991.

Construction Specifications Institute, monograph MS-1, *Understanding Air and Vapor Control in Cold Climates,* CSI, Alexandria, Va., 1992.

Copper Development Association, *Sheet Copper Applications,* CDA, New York, 1985.

Downey, Patrick L., "When You're Hot, You're Hot. When You're Not, You're...Cool Construction Materials," *Interface: Journal of the Roof Consultants Institute,* September 1996, pp. 10–12.

Egan, M. David, *Concepts in Thermal Comfort,* Prentice-Hall, Englewood Cliffs, N.J., 1975.

Gates, David M., *Man and His Environment: Climate,* Harper and Row, New York, 1972.

Gibb, J. F., "Hidden, but Essential: a Technical Review of Backer Rods," *The Construction Specifier,* March 1980, pp. 40–45.

Glidden Paint Company, *Principles of Corrosion and Materials Selection.*

Gorman, Patrick, "Weathering of Various Sealants in the Field—A Comparison," in *Science and Technology of Building Seals, Sealants, Glazing and Waterproofing,* vol. 4, STP 1243, American Society for Testing and Materials, Philadelphia, 1995.

Griffin, C. W., and Richard Fricklas, *The Manual of Low-Slope Roof Systems,* 3d ed., McGraw-Hill, New York, 1996.

Grimm, Clayford T., "Probablistic Design of Expansion Joints in Brick Cladding," *Proceedings of the Fourth Canadian Masonry Symposium,* Fredericton, New Brunswick, Canada, June 1986.

Grimm, Clayford T., "Movement and Volume Change in Masonry," manuscript, 1990.

Grimm, C. T., "A Driving Rain Index for Masonry Walls," *Masonry: Materials, Properties and Performance,* ASTM STP 778, J. G. Borchelt, ed., American Society for Testing and Materials, Philadelphia, 1982, pp. 171–177.

Haddock, Rob, "Metal Roofing Tips," *The Construction Specifier,* April 1994, pp. 50–60.

Henshell, Justin, "Watertight Details: A Checklist," *The Construction Specifier,* May 1995, pp. 125–136.

Hogan, Lyle, "Designing Roofs to Avoid Air Invasion and Positive Pressure from Within," *Interface,* the Journal of the Roof Consultants Institute, May 1996, pp. 10–15.

Holland, Jim, "Attaching SSMRs," *The Construction Specifier,* April 1994, pp. 63–67.

Kelly Moore Paint Company, *Maintenance Painting Guide.*

Kerr, Dale D., "Controlling Rain and Wind," *Architecture,* October 1994, pp. 117–119.

Klosowski, Jerome M., "Sealant Movement in Shear," *The Applicator,* second quarter, 1987, pp. 8–13.

Klosowski, Jerome M., *Sealants in Construction,* Marcel Dekker, New York, 1989.

Klosowski, Jerome M., ed., *Science and Technology of Building Seals, Sealants, Glazing and Waterproofing Second Volume,* ASTM STP 1200, American Society for Testing and Materials, Philadelphia, 1992.

Lally, Heshmat O., "Flashings and Counterflashings," *The Construction Specifier,* November 1993, pp. 57–65.

Lacasse, Michael A., ed., *Science and Technology of Building Seals, Sealants, Glazing and Waterproofing, vol. 5,* ASTM STP 1271, American Society for Testing and Materials, Philadelphia, 1996.

Landsberg, R., *Weather, Climate and Human Settlements,* Special Environmental Report 7, World Meteorological Organization, Geneva.

Latta, J. K., *Walls, Windows and Roofs for the Canadian Climate,* National Research Council of Canada, Ottawa, 1973.

Lstiburek, Joseph, and John Carmody, *Moisture Control Handbook,* Van Nostrand Reinhold, New York, 1993.

Maloney, Jim, *Advanced Air Sealing: Simple Techniques for Air Leakage Control in Residential Buildings,* Iris Communications, Eugene, Ore., 1993.

Maslow, Philip H., "Sealants, Joints, and Membranes for Concrete," *The Construction Specifier,* December 1985, pp. 60–69.

McGettigan, Edward, "Selecting Clear Water Repellents," *The Construction Specifier,* June 1994, pp. 121–132.

Meadows, W. R., Inc., *The Hydrologic Cycle and Moisture Migration,* Elgin, Ill., 1989.

Merritt, Frederick S., ed., *Standard Handbook for Civil Engineers,* 3d ed., McGraw-Hill, New York, 1983.

Merritt, Frederick S., ed., *Building Design and Construction Handbook,* 5th ed., McGraw-Hill, New York, 1994.

Miller, Michael, "Selecting the Right Sealant," *The Construction Specifier,* January 1984, pp. 72–84.
Myers, James C., "Sealant Configurations and Performance," *Architectural Record,* January 1990, pp. 150–151.
Myers, James C., "Behavior of Fillet Sealant Joints," *Building Sealants: Materials, Properties and Performance,* ASTM STP 1069, Thomas F. O'Connor, ed., American Society for Testing and Materials, Philadelphia, 1990.
Myers, James C., ed., *Science and Technology of Building Seals, Sealants, Glazing and Waterproofing,* vol. 3, ASTM STP 1254, American Society for Testing and Materials, Philadelphia, 1994.
Nashed, Fred, *Time-Saver Details for Exterior Wall Design,* McGraw-Hill, New York, 1996.
National Institute of Standards and Technology, *Wind Loads on Buildings and Structures,* Building Science Series 30, U.S. Department of Commerce, NIST, Gaithersburg, MD, 1970.
National Oceanic and Atmospheric Administration, *Climatic Averages and Extremes for U.S. Cities,* Robert Quayle and Wayne Presnell, eds., Historical Climatology Series 6-3, U.S. Department of Commerce, NOAA, National Climatic Data Center, Asheville, N.C., 1991.
National Research Council of Canada, *Roofs That Work,* Building Science Insight '89, NRCC, Ottawa, 1989.
National Research Council of Canada, *An Air Barrier for the Building Envelope,* Building Science Insight '86, NRCC, Ottawa, 1986.
National Roofing Contractors Association, *Roofing and Waterproofing Manual,* 4th ed., NRCA, Rosemont, Ill., 1996.
Nicastro, David, "Difficult Sealant Joints," *Building Seals, Sealants, Glazing and Waterproofing,* ASTM STP 1200, Jerome Klosowski, ed., American Society for Testing and Materials, Philadelphia, 1992.
Nicastro, David, ed., *Science and Technology of Building Seals, Sealants, Glazing and Waterproofing, Fourth Volume,* ASTM STP 1243, American Society for Testing and Materials, Philadelphia, 1995.
Nicastro, David, ed., *Failure Mechanisms in Building Construction,* ASCE Press, New York, 1997.
O'Connor, Thomas F., "Design of Sealant Joints," *Building Sealants: Materials, Properties and Performance,* ASTM STP 1069, Thomas F. O'Connor, ed., American Society for Testing and Materials, Philadelphia, 1990.
Olin, Harold B., *Construction Principles, Materials and Methods,* 5th ed., Van Nostrand Reinhold, New York, 1990.
Olsen, Robert H., "A Brief Study of Bentonite," *The Applicator,* Sealant, Waterproofing and Restoration Institute, Kansas City, Mo., vol. 11, no. 2, Winter issue, 1989.
Panek, Julian R., ed., *Building Seals and Sealants,* ASTM STP 606, American Society for Testing and Materials, Philadelphia, 1976.
Panek, Julian R., and John P. Cook, *Construction Sealants and Adhesives,* 2d ed., John Wiley & Sons, New York, 1984.
Persily, Andrew, "Envelope Design Guidelines for Federal Office Buildings." *Progressive Architecture,* March 1992.
Quirouette, R. L., *The Difference Between a Vapour Barrier and an Air Barrier,* National Research Council of Canada, Building Practice Note 54, Ottawa, 1985.
Quirouette, R. L., "Rain Penetration Control," *The Construction Specifier,* November 1994, pp. 48–56.
Revere Copper Products, Inc., *Copper and Common Sense,* 7th ed., New York, 1982.
Rogers, Tyler Stewart, *Thermal Design of Buildings,* John Wiley & Sons, New York, 1964.
Rosen, Harold J., and Tom Heineman, *Architectural Materials for Construction,* McGraw-Hill, New York, 1996.
Rousseau, Madeleine, "Facts and Fictions of Rain Screen Walls," *Construction Canada,* March/April 1990, pp. 40–47.
Russell, James S., "Sorting Out Building Sealants," *Architectural Record,* January 1990, pp. 152–153.
Schwartz, Thomas A., ed., *Water in Exterior Building Walls, Problems and Solutions,* ASTM STP 1107, American Society for Testing and Materials, Philadelphia, 1991.
Scharff, Robert, *Roofing Handbook,* McGraw-Hill, New York, 1997.
Schroeder, Edward K., "Regarding Roof Edges," *The Construction Specifier,* November 1993, pp. 40–54.

Schroeder, Michael J., and Hovis, E. E., "Sealant Back-up Material," *Building Sealants: Materials, Properties and Performance,* ASTM STP 1069, Thomas F. O'Connor, ed., American Society for Testing and Materials, Philadelphia, 1990.

Shisler III, Franklin W., and Klosowski, J. M., "Sealant Stress in Tension and Shear," *Building Sealants: Materials, Properties and Performance,* ASTM STP 1069, Thomas F. O'Connor, ed., American Society for Testing and Materials, Philadelphia, 1990.

Stein, J. Stewart, *Construction Glossary,* John Wiley & Sons, New York, 1980.

SWRI, *Applicator Training Manual,* Sealant, Waterproofing and Restoration Institute, Kansas City, Mo., 1986.

SWRI, *Sealants—The Professional's Guide,* Sealant, Waterproofing and Restoration Institute, Kansas City, Mo., 1995.

Thomas, Robert G., Jr., *Exterior Insulation and Finish System Design Handbook,* CMD Associates, Vashon Island, Wash., 1992.

Tobiasson, Wayne, "Vapor Retarders to Control Summer Condensation," *Proceedings of the ASHRAE/DOE/BTECC/CIBSE Conference on Thermal Performace of the Exterior Envelope of Buildings IV,* Orlando, Fla., 1989.

Tobiasson, Wayne, and James Buska, "Snow, Ice and Standing Seam Roofs," *Architectural Specifier,* Fall 1993, pp. 50–54.

Tobiasson, Wayne, and Marcus Harrington, "Vapor Drive Maps of the U.S.," *Proceedings of the ASHRAE/DOE/BTECC Conference on Thermal Performace of the Exterior Envelope of Buildings III,* Clearwater Beach, Fla., 1985.

Tobiasson, Wayne, "Retarders Keep Moisture, the Enemy Within, At Bay." *RSI,* Aug. 1990, pp. 34–40.

Trechsel, Heinz, *Moisture Control in Buildings,* ASTM Manual 18, American Society for Testing and Materials, Philadelphia, 1994.

Tuluca, Adrian, et al., *Energy Efficient Design and Construction for Commercial Buildings,* McGraw-Hill, New York, 1997.

Warseck, Karen, "Why Sealant Joints Fail," *The Applicator* second quarter 1987, pp. 4–5.

Watson, Donald, and Kenneth Labs, *Climatic Building Design,* McGraw-Hill, New York, 1983.

Watts, Mike, "Thermal Conductivity in Mechanically Fastened Roof Systems," *Interface,* Journal of the Roof Consultants Institute, May 1996, pp. 6–9.

Weismantel, Guy E., *Paint Handbook,* McGraw-Hill, New York 1981.

Williams, Mark F., and Barbara Lamp Williams, *Exterior Insulation and Finish Systems, Current Practices and Future Considerations,* ASTM Manual 16, American Society for Testing and Materials, Philadelphia, 1994.

Wright, Gordon. "Wall Systems Strive to Foil Moisture Intrusion." *Building Design and Construction,* May 1991, pp. 62–66.

Yarosh, Kenneth, "New Developments in Elastomeric Coatings," *The Applicator,* Sealant, Waterproofing and Restoration Institute, December 1995, pp. 3–4, 16–17.

Index

absorptance, solar, 63, 66, 67
absorption, moisture, 126, 321–322
absorption, thermal, 81
adhesion-in-peel, 435
adhesive failure, 14–15, 413, 428, 430
aged insulation value, 16, 91
air barriers, 139–141, 186–187, 190–211, 218,
    222–224, 262, 264, 353, 355, 357, 362, 368, 372,
    389, 423, 432
air leakage, 117, 156–162, 176, 186–187, 191, 192,
    195–196, 216–217, 218, 234, 353, 370–371
albedo, 63
    (*See also* solar reflectance)
attic ventilation, 185, 235–236, 262–263

backer rod, 408, 425–426, 428–429
bandage joint (*see* joint, bridge)
barrier walls (*see* rain barrier systems)
basement:
    insulation, 112–118
    waterproofing, 320–323
beadboard (*see* insulation, polystyrene)
blackbody, 65
bond-breaker tape, 392, 408, 410–411
bridge joint, 431–432
building orientation, 33, 47, 49
built-up roofing (*see* roofing, built-up)
butt joint, 391–392, 408–409
butyl sealant, 248, 367, 416–417, 418

capillary suction, 123, 131, 134, 135, 249, 285–286,
    288, 289, 450
carbonation, 9
caulk (*see* sealant)
cementitious coatings, 314, 441, 443
chalking, 243, 413, 440, 458
chimney effect (*see* stack effect)
cladding systems, 351–389

climate, 21–29, 51–58, 149, 180
coatings, 314–315, 366, 443–463
cohesive, failure 428, 430
compatible metals, 12, 257
compression set, 409–410
condensation, 143–212, 262–267
conduction, thermal, 59, 72, 79–80
construction tolerances, 399–400
control joint, 364–365
convection, thermal 60, 72, 79–80, 152
corrosion, 9–13, 257
counterflashing, 278
creep, 214, 401
curtain walls, 205, 368–379

dampproofing, 283, 297, 321
design temperatures, 166–174
dew, 41
dew point, 143–145, 147, 148, 176
drainage walls, 139, 352–353, 359–362, 368–369
dry bulb temperature, 146, 147
durability, 5–21
durometer, 412

efflorescence, 16–18, 451, 453, 458
EIFS, 377–389
elastomeric coatings, 443–444
emittance, effective, 80–81
emittance, 65–67, 80–81
equilibrium moisture content, 93, 96
evaporation, 60–61
expansion and contraction, 6–8, 397–403
expansion joint, 223, 271, 279, 364–366, 370
exterior insulation and finish system, 377–389

fillet joint, 391–392, 408
flashing, 128–129, 226–230, 270–271, 315–317, 329,
    353–354, 364, 371

foundation insulation, 112–120
freeze-thaw, 14, 115
frost depth, 115, 118–119

galvanic corrosion, 9–13, 257
galvanizing, 10, 13
groundwater, 121, 284–287, 293–296

humidity, absolute, 144
humidity ratio, 144
humidity, relative, 144, 149, 152
hydrologic cycle, 40–41
hydrostatic pressure, 122, 126, 135, 138, 142, 285
hysteresis, 397

ice dam, 231–234, 249
insects, 19
insulation, thermal, 84–85, 86–98, 103, 108–112, 224–225
    batt, 90
    beadboard, 91
    blanket, 90
    board, 90–92
    cellular glass, 91, 224, 340
    cellulosic fiber, 90
    composite, 92
    fiberglass, 90
    foamed, 92
    foundation, 112–120
    loose fill, 87
    mineral fiber, 90
    perlite, 89, 91
    phenolic, 92
    polystyrene, 91–92, 112, 119, 220, 224, 261–262
    polyisocyanurate, 91–93, 224
    polyurethane, 91–93, 224
    reflective, 67
    slab insulation, 112–120
    sprayed, 92
    vermiculite, 89, 220
    wet, 15–16, 93–95, 97–98
    wood fiber, 90, 92

joint:
    bridge joint, 431–432
    butt joint, 391–392, 408–409
    control joint, 364–365
    expansion joint, 271, 279, 364–366, 370
    fillet joint, 391–392, 408
    lap joint, 249, 391–392
    two-stage joint, 355–356, 432–433

lap joint, 249, 391–392
latex sealant, 419
lock-strip gasket, 422–423

masonry, 357–368
microclimate, 46–51
modulus of elasticity, 412
moisture (*see* water)
mold and mildew, 9, 180, 235

oil based sealant, 415–417
oilcanning, 246–247
oleoresinous (*see* oil based)

paint, 439–441
parapets, 123, 216, 270, 367–368
parging, 320
perlite insulation, 89, 91, 220, 224
permeance, vapor, 154, 156–158, 174, 187, 264, 267
polystyrene insulation, 91–92, 112, 119, 220, 224, 261–262
polysulfide sealant, 416–417, 420
polyurethane:
    insulation, 91–93
    sealant, 416–418, 421
precast concrete, 355–357, 358, 359
protection course, 334
psychrometrics, 143–146

R-value (*see* thermal resistance)
radiant barrier, 66–68, 84
radiant heat loss, nocturnal, 47, 65–66
rain, 40–41, 124–126, 237–238
rain barrier systems, 138, 283, 352, 355, 359, 368–369, 377
rain screens, 140–141, 188, 190, 296, 353, 355–357, 359, 362, 363, 368, 369, 370, 389
reflectance, solar, 63, 66, 67, 68–71
relative humidity, 144, 149, 152
release tape (*see* bond-breaker tape)
rising damp (*see* capillary suction)
roof assembly, 218
roof system, 219
roof venting, 263
roofing, 213–282
    built-up, 267–280
    low-slope, 256–282
    metal, 239–256
    modified bitumen, 271–280
    protected membrane, 259–262
    shingle, 236–242
    single-ply, 280–283
    steep, 230–256

saturation vapor pressure, 144–145
sealant backing, 408, 425–426, 428–429
sealants, 391–438
    acrylic, 416–420
    butyl, 248, 416–417, 418

sealants, (Cont.):
  oil-based, 415–417
  polysulfide, 416–418, 420
  silicone, 416–418, 422
  tape, 414, 422–423, 424, 425
  urethane, 416–418, 421
shore "A" hardness, 412
shore "D" hardness, 412
silicone sealant, 416–418, 422
slab insulation, 112–120
slope, roof, 213–214
snow, 41–46
solar absorptance, 63, 66, 67
solar absorption coefficient, 101, 103
solar radiation, 61–71, 95, 101, 164, 166
solar reflectance, 63, 66, 67, 68–71
solar transmittance, 65
stack effect, 60, 159, 160, 162–164, 165
structural sealant glazing, 424
sublimation, 40, 60, 234
subsurface drainage, 293–296

tape sealant, 414, 422–423, 424, 425
termites, 19, 319
thermal break, 375
thermal bridge, 85–86, 89, 101–102, 104–108, 250, 255, 372, 375
thermal gradient, 95–96, 99–112, 174–176
thermal conduction, 59, 72, 79–80
thermal convection, 60, 72, 79–80, 152
thermal radiation, 61–71
thermal resistance, 71–86
thermal resistance, loss of, 93–95
thermal resistance ratio, 93–95, 97–98
thermal storage capacity, 81, 101, 102
thermal transmittance, 72, 375–376
tolerance, construction, 399–400

transmittance:
  solar, 65
  thermal, 72, 375–376

U-value (see thermal transmittance)
ultraviolet resistance, 269, 412, 422
urethane:
  insulation, 91–93
  sealant, 416–418, 421

vapor barrier (see vapor retarder)
vapor migration:
  airborne, 156–165, 218
  diffusion, 154–156, 164, 166, 218, 288, 290, 322
vapor permeance, 154, 156, 157–158, 267
vapor pressure, 144–145, 176, 177
vapor resistance, 154, 156, 166, 174
vapor retarders, 186–190, 191, 222–224, 301–305, 318
vermiculite insulation, 89, 220

water:
  absorption, 126, 321–322
  infiltration, 126, 127–137
  leakage, 126
  penetration, 126, 127–137, 370, 372
  permeation, 126
  saturation, 126
water repellents, 357, 367, 444–452
water vapor (see vapor)
waterproofing, 284–349
weather, 21
weepholes, 128–129, 354–355, 370
wet bulb temperature, 146, 147
wind, 24–35, 124–126
wind-driven rain, 51, 124–126
windows, 368–379

ABOUT THE AUTHOR

Christine Beall is a consulting architect based in McQueeney, Texas. She has 25 years experience in the design, specification, and construction of residential, commercial, institutional, and industrial buildings. Ms. Beall is also the author of McGraw-Hill's *Masonry Design and Detailing,* 4th edition, as well as a contributing author to *Architectural Graphic Standards,* 9th edition, the *Masonry Designer's Guide to the ACI/ASCE 530 Masonry Code and Specifications,* and the *McGraw-Hill Encyclopedia of Science and Technology,* 8th edition.